ANCIENT NEAR EASTERN STUDIES

SUPPLEMENT 34

ANCIENT NEAR EASTERN STUDIES

SUPPLEMENT 34

A PROMETHEAN LEGACY: LATE QUATERNARY VEGETATION HISTORY OF SOUTHERN GEORGIA, THE CAUCASUS

Simon Edward CONNOR

PEETERS
LEUVEN – PARIS – WALPOLE, MA.
2011

Series Editors: Antonio Sagona and Claudia Sagona

A catalogue record for this book is available from the Library of Congress.

ISBN 978-90-429-2350-8
D/2011/0602/2

PRINTED IN BELGIUM

Peeters N.V., Warotstraat 50, B-3020 Herent

CONTENTS

CHAPTER FOUR

MODERN POLLEN-VEGETATION RELATIONSHIPS

CHAPTER FIVE

THE TBILISI AREA

CHAPTER SIX

THE TSALKA PLATEAU

CHAPTER SEVEN

HOLOCENE PALAEOVEGETATION MAPS OF GEORGIA

PREFACE AND ACKNOWLEDGEMENTS

This project on the environmental history of the Caucasus region originally arose from The University of Melbourne's archaeological investigations at Sos Höyük in the Pasinler Valley, Eastern Anatolia, headed by Professor Antonio Sagona and Dr Claudia Sagona. Interest in the palaeoenvironments of the Pasinler Valley led Sarah Howell-Meurs and Jennifer Piro (New York University) to investigate prehistoric faunal exploitation, and both Jennifer Newton and Catherine Longford to examine Bronze and Iron Age plant use. Dr Ian Thomas made the first attempts to find out about the broader vegetation history of the Pasinler Valley, but it became apparent at an early stage that palynological fieldwork in Eastern Anatolia was subject to considerable constraints.

For this reason, the late Dr Tamaz Kiguradze suggested coming to Georgia to help provide an environmental context to his important excavations of early agricultural sites. My first visit to Georgia took place in 2001, when I was fortunate to meet Dr Zaal Kikodze†, Dr Eliso Kvavadze, Dr Henrikh Avakov, Dr Oleg Bendukidze, Dr Giorgi Arabuli and Dr Vakhtang Todria, all of whom also had an enormous input into my fieldwork and shaped many of my conceptions about Georgia's environment.

In particular, I would like to acknowledge the tireless efforts of Dr Eliso Kvavadze. She has been an unfailing source of dialogue, information, assistance and extraordinary hospitality. I express my deep gratitude to Dr Ian Thomas and Prof. Antonio Sagona from the University of Melbourne for their expert guidance through every stage of the project. The Todria family, Dr Maia Jejelava and Nana Saganelidze made my frequent visits to Georgia a source of great joy and lasting friendship. I thank Prof. Brigitta Ammann, Dr Pim van der Knaap and Jacqueline van Leeuwen from the Institute of Plant Sciences, Bern, for their insightful comments and constant encouragement. The project was made possible through the cooperation of the L. Davitashvili Institute of Paleobiology in Tbilisi and an Australian Institute of Nuclear Science and Engineering postgraduate research award.

Finally, my sincerest thanks go to my family for their constant love. I dedicate this work to them.

CHAPTER ONE

INTRODUCTION

The history of vegetation is shaped by the myriad forces of environmental change. Perhaps the most important challenge for the study of vegetation dynamics is unearthing their causes, whether related to climate change, soil formation, ecological succession, pathogen outbreaks or human impacts. Some of the most energetic and long standing debates in palaeoecology have revolved around the question of climatic or anthropogenic agency in the changes observed. Climate changes, rather than being a bland backdrop to cultural events, have impacted on the vegetation and hence shaped the way humans interacted with their environment in the past. Humans have also impacted on the environment directly; such that the task of disentangling the causes of past vegetation changes becomes an important one for understanding human history.

The Caucasus region, located at the crossroads of Europe, the Middle East, the Mediterranean and Central Asia, is home to a great diversity of plants including some that are numbered amongst history's most important agricultural resources — the vine, wheat, barley, flax and several kinds of fruit tree. Humans have lived in the Caucasus for more than 30,000 years and region's mountainous environment has allowed the evolution of many distinct cultural groups within a relatively small area, making this one of the richest ethno-linguistic areas in the world. In the Caucasus we find a diversity of cultures born of and contextualised by a complex and unique geography.

Archaeology and palaeoenvironmental research have a long history in the Caucasus region, but rarely have these two strands of knowledge been drawn together to provide a complete picture of prehistoric humans in their environment. Caucasian botanists have often argued that what are today treeless steppes and grasslands were once covered thickly in forest, which was destroyed by recent human activity. But evidence for if, when and how these forests were destroyed remains much more elusive. Part of the problem is the difficulty in interpreting palaeoenvironmental data from semi-arid areas where the dry climate can often distort or destroy the evidence. This constraint is often seen as an insurmountable barrier to the study of vegetation history in dry environments, and yet it was in these same environments that agriculture and cities first emerged in Western Asia; hence their environmental history is of considerable interest.

General aims

This study aims to shed light on the causes of vegetation change in Southern Georgia during the last 14,000 years.

Drawing on evidence for past climatic changes, the long-term impacts of grazing and burning, and the archaeological record, I will attempt to determine which of these forces made the greatest impression on Southern Georgia's vegetation history.

In doing so, I hope to reveal the character of the vegetation during various archaeological periods; examine ecological consequences associated with the adoption of agriculture and subsequent agro-pastoral activities; determine the past occurrence and impact of fire in the vegetation; trace the changing extent of forests and grasslands; gauge the effects of changes in precipitation and temperature; and question current thinking on how Southern Georgia's vegetation history has contributed to its rich biodiversity and varied landscapes.

Approach

To achieve these aims, Chapter Two introduces the study area and some of the current debates and uncertainties regarding its human and environmental history; Chapter Three describes the method of pollen analysis adopted in satisfying the project aims and the numerical techniques to overcome problems associated with palaeoenvironmental data from semi-arid environments; Chapter Four analyses the relationships between the present-day vegetation and the pollen that it produces, as a means of reconstructing past environmental changes; Chapters Five and Six present new data on the vegetation history of two contrasting areas of Southern Georgia (the Tbilisi area and the Tsalka Plateau) and, using the techniques developed in previous chapters, these chapters discuss the results in terms of past climate, forest cover and land-use; Chapter Seven places these results in a regional context by attempting to reconstruct the past vegetation for the whole of Georgia using a large dataset of published work, and then uses the same data to estimate the past climate of different parts of Georgia through time; Chapter Eight compares the vegetation history of Southern Georgia to the palaeoclimatic estimates from the previous chapter and to the area's archaeological record to make an informed judgement of the causes for past vegetation change; and finally, Chapter Nine outlines the significance of the project's findings and proposes directions for future research.

CHAPTER TWO

BACKGROUND

> And at day-dawn they looked eastward, and midway between the sea and the sky they saw white snow-peaks hanging, glittering sharp and bright above the clouds. And they knew they were come to Caucasus, at the end of all the earth.[1]

Language and Myth, Agriculture and Environment

The Caucasus has long been a source of fear and fascination, myth and legend in Western consciousness, a rugged and diversified landscape where Europe and Asia collide in an explosion of language and culture.

It was to the distant Caucasus that Jason and his Argonauts sailed on their mythical quest for the Golden Fleece. Here Prometheus, the titan of Greek legend, was punished for stealing fire from the gods. Zeus, furious that his planned destruction of the human race had been thwarted thus, shackled Prometheus to the Caucasus where an eagle came daily to feed on his liver (**Fig. 2.1**). By giving fire to humankind, Prometheus became revered as the great benefactor of farmers, artisans and metallurgists alike.

The Prometheus myth may indeed be an ancient Caucasian tale transmitted to the Greeks through early contacts.[2] Variations on the tale are spread through the region, from Abkhazia to Armenia. In some local variants, Abrskil, a Promethean figure, helps humankind rid the soil of choking vines, brambles and bracken so that crops can be planted.[3] The parallels with fire as an agricultural tool are obvious.

The question of when farming originated in the Caucasus presents a challenge to theories of agricultural evolution and language dispersal in Western Asia. Renfrew has argued that that Indo-European languages disseminated throughout Europe from the centres of agricultural development in Western Asia,[4] a claim subsequently reinforced by genetic evidence for a 'wave of advance' (**Fig. 2.2**) supplanting the pre-existing Pleistocene-age languages.[5] The Caucasian languages, which are unrelated to the Indo-European group, seem to have survived or countered the Neolithic Indo-European

[1] Charles Kingsley, *The Argonauts*, quoted in Allen 1942, p. 228
[2] Braund 1994; Lang 1966
[3] Lang 1966, p. 71.
[4] Renfrew 1992; Renfrew 1996.
[5] Cavalli-Sforza 1996; Cavalli-Sforza *et al.* 1994.

Figure 2.1. Dawn at Mt Kazbegi (5034 m), the Caucasus Mountains. Greek legend tells us that Zeus shackled Prometheus to this, Georgia's highest peak, for stealing fire from the gods. Prometheus' torment was to have an eagle peck out his liver each day. One may feel some empathy for the Greek titan after being treated to the wine-soaked hospitality that welcomes every visitor to this land of extraordinary beauty and cultural riches.

agricultural diffusion. Renfrew provides two possible explanations for this linguistic resilience:[6] (1) that a local agricultural system evolved independently in the Caucasus prior to or coeval with events in the Fertile Crescent; or (2) that the Caucasian people adopted the farming technology of the Indo-Europeans without the usual linguistic change.

In support of the first hypothesis, we might look to the great diversity and degree of endemism in the Caucasian flora,[7] which has undoubtedly been the source of many cultigens, particularly rosaceous fruits like apples, pears, plums and cherries.[8] The Caucasus lies within the natural range of many endemic cereals and the wild grapevine.[9] Wine-making, in particular, has an ancient history in the Caucasus, and especially so in Georgia, where grape cultivation may have originated.[10] Did the Caucasians stave off Indo-European colonisation by having a specialised proto-agricultural

[6] Renfrew 1992.
[7] Dolukhanov 1966; Gagnidze 2005.
[8] Menabde 1961; Vavilov 1992.
[9] Lisitsyna and Prishchepenko 1977; Zohary and Spiegel-Roy 1975.
[10] Burney and Lang 1971.

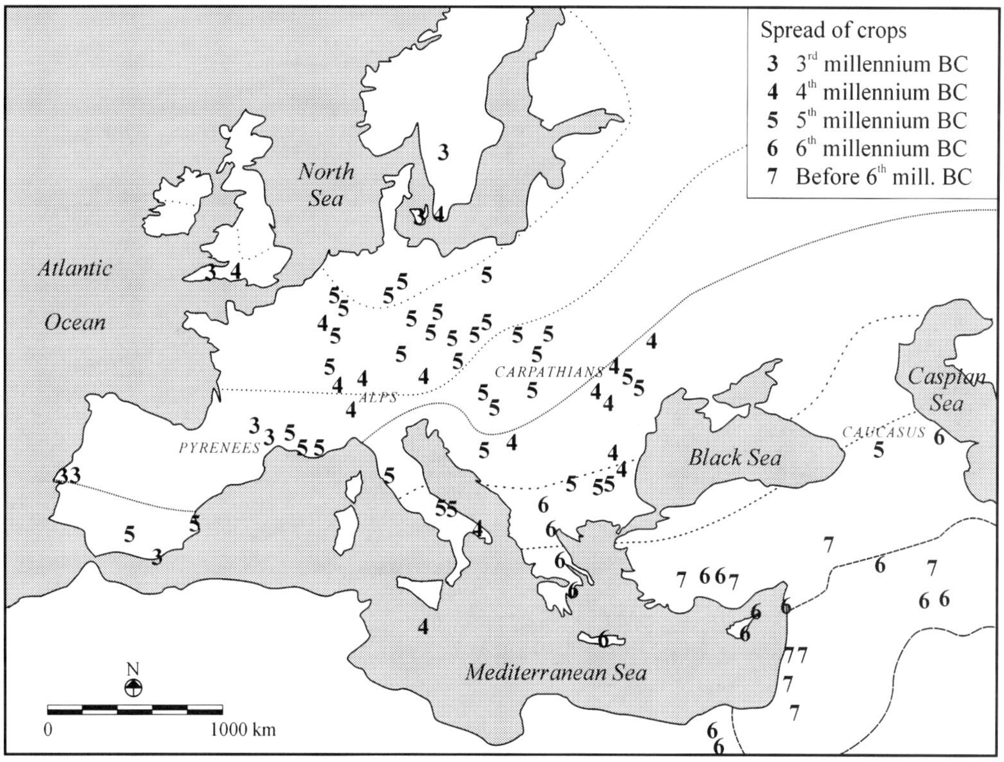

Figure 2.2. Map of NW Eurasia with the spread of the Near Eastern Neolithic crop assemblage from Western Asia indicated by numbers and the genetic 'wave of advance' indicated by lines emanating from the southeast. Simplified from Zohary and Hopf (1993) and Cavalli-Sforza (1996). The Caucasus region is located between the Black and Caspian Seas.

system emplaced? Or is linguistic stability here merely the result of rugged topography, which necessarily inhibits inter-cultural exchange?

Toward the second hypothesis, that the Caucasians adopted the Indo-European agricultural model without inheriting its language, the weight of archaeological evidence can be advanced. By and large, macrofossil remains of crops cultivated by the Neolithic people of the Caucasus were the same as those 'founder' crops of the Fertile Crescent and Levant.[11] These included einkorn, emmer, bread wheat, barley, oats, rye, millet, grapes and various legumes.[12] Purported finds of endemic wheat species (*Triticum carthlicum* Nev. and *T. macha* Dec. et Men.)[13] at early agricultural sites in Georgia might suggest that Neolithic settlements in this region were sited to take advantage of wild cereal stands.[14]

[11] Chataigner 1995; Lisitsyna and Prishchepenko 1977; Zohary and Hopf 1993.
[12] Chataigner 1995.
[13] Yanushevich and Rusishvili 1984.
[14] Kiguradze and Sagona 2003.

One of the most problematic aspects of archaeobotanical evidence is, however, the differential preservation of cereal grains and legumes compared to fruits, tubers and vegetables. The impact of this problem on archaeological interpretation and the question of agricultural origins cannot be underestimated, explaining as it does the 'Levantine Primacy'[15] in such discussions. It may be that an endemic Caucasian farming system, if it ever existed, was of an entirely different nature to that of the Fertile Crescent, inheriting its characteristic crop assemblage only through later cultural exchange. And, because cereal-based agriculture is often subsidiary to pastoralism in mountainous terrain,[16] the earliest farming in the Caucasus may have taken place in the rangelands.

Palynological Research in Western Asia — a Synopsis

These broad questions of agricultural origins were the impetus for early pollen-based research efforts in Western Asia. The multidisciplinary archaeological excavations at Jarmo, directed by R. J. Braidwood from the University of Chicago, challenged the idea that the traditional hearth of agriculture was situated in the Levant. Their work gave rise to the 'hilly flanks' theory: that cereal cultivation began in the rain-fed foothills of the Zagros and Taurus Mountains, better known as the Fertile Crescent.[17] The complimentary pollen diagram[18] from Lake Zeribar in Iran (**Fig. 2.3**) has remained pivotal in discussions of the vegetation history of Western Asia. Covering the last 22,000 years, it shows that the landscape of the Zagros prior to the terminal Pleistocene was one of treeless *Artemisia*-Chenopodiaceae steppe. Climate changes at the beginning of the Holocene triggered the slow expansion of oak woodland, reaching its fullest extent during the mid Holocene (**Fig. 2.4**).

The same general pattern is observed in pollen diagrams from Lake Mirabad in southwestern Iran and Lake Urmia in northwestern Iran.[19] That these climatic changes were broadly coincident with the emergence of cereal cultivation has raised questions about the element of 'climatic determinism' involved in this cultural event.[20]

In the Levant, the expansion of oak (*Quercus*) and terebinth (*Pistacia*) woodlands commenced earlier (**Fig. 2.4**), during the late Pleistocene in the Ghab Valley and Lake Huleh sediments.[21] In addition, this expansion took place at a much greater rate than at the Zagros sites, suggesting a wetter early Holocene climate in the coastal regions of

[15] Watson 1995.
[16] Newton 2004.
[17] Braidwood and Howe 1960.
[18] van Zeist and Wright 1963.
[19] Bottema 1986; van Zeist 1967.
[20] Wright 1970; Wright 1993.
[21] Niklewski and van Zeist 1970; van Zeist and Bottema 1991; Yasuda *et al.* 2000.

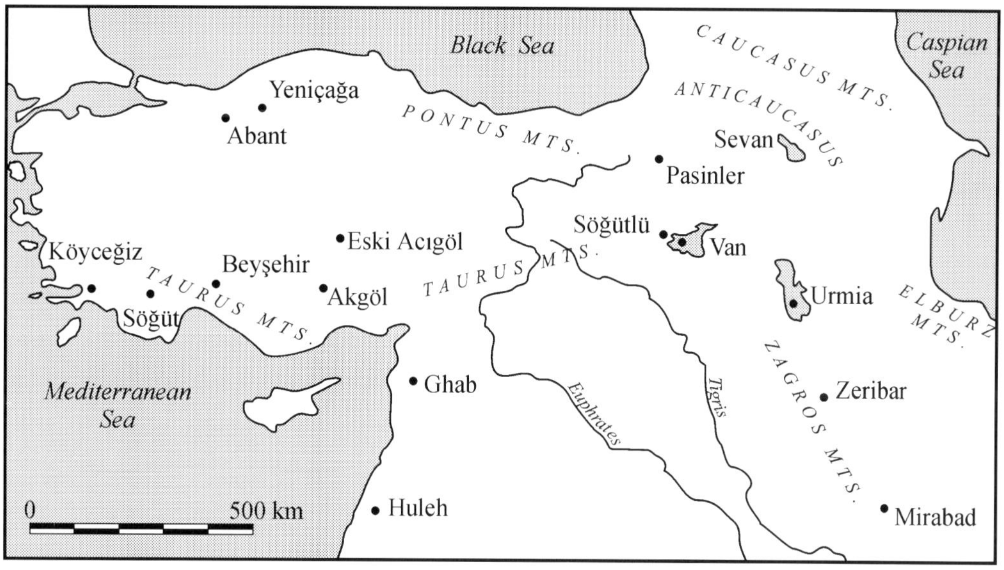

Figure 2.3. Map of Western Asia showing the pollen sites and mountain ranges mentioned in the text. Adapted from van Zeist and Bottema (1991).

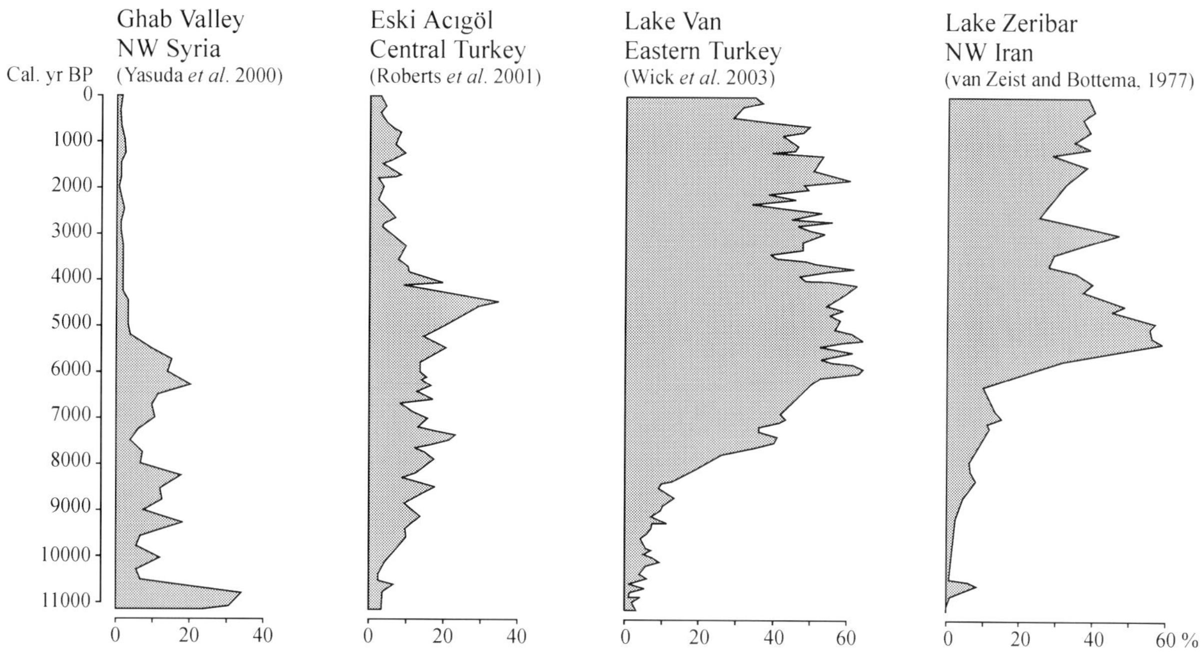

Figure 2.4. *Quercus* (oak) pollen curves from some key sites in Western Asia plotted on similar timescales. The diagrams indicate the existence of oak woodland in Syria during the early Holocene, spreading to the interiors of Anatolia and Iran during the mid Holocene. Adapted from Roberts (2002) and Yasuda *et al.* (2000).

the Eastern Mediterranean and Arabian Peninsula.[22] This can be related to the intensification of the African monsoon in these areas, compared to the seasonally droughty conditions prevalent in the Zagros and interiors of Anatolia during the early-mid Holocene.[23]

Anatolia, of all of Western Asia, has been the most intensively studied palynologically (**Fig. 2.3**). From eastern Turkey, whence palynological data are very scanty, the first record[24] from Lake Van has long been plagued by dating problems arising from the use of varve counting.[25] The original dating model indicated that oak woodland progressively replaced steppe vegetation between 6400 and 3400 years ago, leading to the conclusion that the early-to-mid Holocene climate of eastern Anatolia was much drier than in Iran.[26]

Since then the chronology of the Lake Van pollen record has been revised considerably, firstly by the study of a neighbouring peat bog and more recently by a new pollen diagram and varve count from Lake Van itself.[27] The new diagram shows that oak expansion took place between 10,460 and 6250 years ago,[28] corresponding much better to the Zagros records (**Fig. 2.4**). Even so, questions have been raised recently on whether the revised Lake Van varve ages are still somewhat young, and, more importantly, if the slow expansion of oak woodland during the early-mid Holocene was, in part, the result of Neolithic human impacts rather than climatic aridity alone.[29]

Pollen records of southern and southwestern Anatolia are of considerable interest from an archaeological perspective. Like the sequences discussed above, they exhibit an early Holocene *Quercus* expansion at the expense of *Artemisia*-Chenopodiaceae steppe.[30] However, the subsequent appearance of distinct phases of human impact, indicated by *Plantago, Olea, Fraxinus, Juglans, Vitis, Cerealia*-type and other anthropogenic indicator pollen,[31] provides a startling legacy of cyclical land-use and population change.[32] The cause of these vegetation changes was almost certainly political instability rather than climate change or volcanic activity.[33] Following these so-called 'Beyşehir Occupation Phases',[34] landscapes were often left deforested and eroded to

[22] Roberts and Wright 1993.
[23] Roberts *et al.* 2001; Stevens *et al.* 2001.
[24] van Zeist and Woldring 1978.
[25] Bottema 1986; Wick *et al.* 2003.
[26] van Zeist and Woldring 1978.
[27] Bottema 1995; Wick *et al.* 2003.
[28] Wick *et al.* 2003.
[29] Roberts 2002.
[30] Bottema and Woldring 1984; Bottema and Woldring 1990; Eastwood *et al.* 1999; Roberts *et al.* 2001; van Zeist and Bottema 1991; van Zeist *et al.* 1975; Vermoere *et al.* 2002.
[31] Behre 1990.
[32] Roberts 1990.
[33] Roberts *et al.* 1997.
[34] Bottema and Woldring 1990.

the point that oak woodland was unable to return, especially under the rapacious grazing habits of goats. Instead, pine forests became widespread around some sites.[35]

This trend is reflected in pollen diagrams from the northwestern Anatolian sites of Abant and Yeniçağa.[36] A late Holocene pollen diagram from Pasinler in northeastern Anatolia likewise exhibits a strong anthropogenic signal that varies through time as populations waxed between sedentism and transhumance under the influence of political, population and environmental pressures.[37] The small scale of the basin from which the Pasinler pollen record was obtained contrasts starkly with Lake Van, which, although in the same geographic region, has a surface area of over 3500 km^2 and therefore represents change on a much more regional scale, masking localised changes in land-use.[38]

Far from the results of Western Asian pollen analyses seeming "disappointing",[39] they have great potential to reveal the variety of human experience through time. Whilst the seminal records from Lakes Zeribar, Urmia, Van and Mirabad have set a general climatic backdrop for understanding agricultural innovation, they can tell us little about the environmental impacts of prehistoric cultures or those that dwelt in Western Asia more recently. The success of the southwestern Anatolian pollen analyses in this regard is to a large degree attributable to the comparatively small scale of the lake and wetland basins studied. The sheer number of sites analysed in southwestern Anatolia allows for differentiation of changes affecting one lake catchment from those exerting influence across a larger area.

If we accept that the emergence of an agricultural economy during the Neolithic was probably small in scale, with highly localised impacts on the soil, vegetation and fauna,[40] then it is clear that these changes are unlikely to manifest themselves in pollen records from large lakes. It could be more enlightening to use a suite of locally representative records from small lakes and wetlands and correlate this with nearby archaeological sites where the source area or catchment scales of both are comparable.[41]

With the exception of southwestern Anatolia, knowledge of prehistoric human interactions with their local environment in this region is largely based on macrofossil evidence, which, as noted, is only representative of the surrounding vegetation in the most general sense.[42] This leaves enormous scope for establishing a 'footprint' of prehistoric impact in vast areas of Anatolia, the Levant, Mesopotamia, Iran and the

[35] Roberts 1990.
[36] Beug 1967; Bottema and Woldring 1990.
[37] Thomas 1999.
[38] *Sensu* Jacobson and Bradshaw 1981; Prentice 1985; Sugita 1993.
[39] Behre 1990, p. 227.
[40] Buitenhuis 1990; Roberts 1998, p. 153.
[41] See, for instance, Flannery 1976; Roberts *et al.* 1997.
[42] Wasylikowa 1986.

Caucasus. The latter is of interest because of the ethno-linguistic and agro-botanical aspects discussed earlier, and also because of its turbulent political history, resulting in boom and bust cycles that doubtless left a considerable palaeoenvironmental legacy. In the following section, I introduce Georgia's present environment as a precursor to understanding its vegetation history.

Georgia's Geography and Vegetation

For a country with an area comparable to Ireland, Georgia is remarkably diverse. Situated on the eastern shore of the Black Sea, the country forms a great dissected valley running from west to east between the anticlinal Main Caucasus Range (maximum elevation 5642 m at Mt Elbrus) to the north and the volcanic Armenian highlands to the south. Georgia is, and has historically been, subdivided into two major regions: Colchis (Western Georgia) and Iberia (Eastern Georgia). These two are separated by the Likhi Range, which also forms the watershed between Colchian rivers flowing westward to the Black Sea, and catchments draining eastward to the Caspian Sea via the Kura (or Mtkvari) Valley (**Fig. 2.5**).

Botanically, the Caucasus sits at the junction of three major phyto-geographical provinces: the Euro-Siberian, Irano-Turanian and Euxinian territories.[43] During the Oligocene, the Caucasus existed as a chain of islands in the Tethys Sea, a temporary isolation that had far-reaching effects on the subsequent development of the flora,[44] and goes some way toward explaining the Caucasus region's great botanical diversity, which with 6350 species, is equivalent to the plant species diversity of the Iberian Peninsula or the Balkans.[45] Georgia is home to 4130 known species of vascular plants, of which 20 percent are endemic to the Caucasus region.[46] The country is particularly rich in Asteraceae (566 species, 44 endemic to Georgia), Poaceae (399 spp., no endemics), Fabaceae (317 spp., 34 endemics), Rosaceae (237 spp., 63 endemics), Brassicaceae (186 spp., 11 endemics) and Scrophulariaceae (181 spp., 14 endemics).[47]

Georgia is a mountainous country and this determines, in concert with climate and soils, the general character of the vegetation. Of Georgia's total area of 6,548,600 hectares, less than 25 percent is classed as lowlands and plains, and only 13 percent of the land area is between sea level and 200 metres.[48] The Main Caucasus Range shields Georgia from the continental climates of the southern Russian steppes, and intercepts rainfall from the dominant westerlies originating over the Black Sea.

[43] Zohary 1973.
[44] Tumajanov 1971.
[45] Dolukhanov 1966.
[46] Gagnidze 2005; Gagnidze *et al.* 2002; Sakhokia 1961.
[47] Gagnidze 2005.
[48] Ketskhoveli 1959.

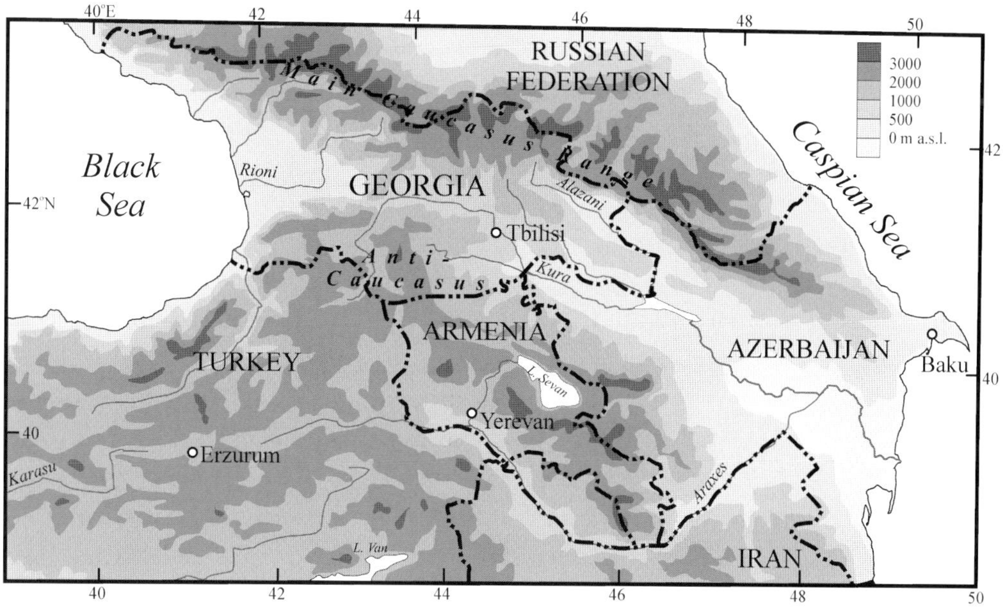

Figure 2.5. Topographic map of the South Caucasus Region. Source: *Атлас Грузинской ССР* (AN GSSR, 1964).

An example of the interplay between topography and climate, and its effect on plant life, is the Alazani Valley in Eastern Georgia (see **Fig. 2.5**). This valley carves its way south from the Main Caucasus Range and flows out to the southeast through the semideserts and steppes of Azerbaijan. The valley is warmer than much of Eastern Georgia, as it occupies the zone between mountains and semidesert, and the sheer wall of southwest-facing slopes promotes orographic rainfall. Climatic conditions are ideal for viticulture, and the Kakheti region is famous for its wines. The region is also known for its unusual forests, which contain thermophiles that are otherwise found only in proximity to the Black and Caspian Seas. Tertiary-relict trees, such as *Zelkova carpinifolia* and *Pterocarya fraxinifolia*, grow in forests that are more typical of humid western Georgia or the moist mountain slopes of northern Iran than of a region with less than 800 mm annual rainfall. The Alazani forests are also the only ones in Eastern Georgia to contain forests of yew (*Taxus baccata*) and chestnut (*Castanea sativa*).

Aspect is an important influence on vegetation composition in a mountainous terrain. Vegetation composition can vary dramatically from north- to south-facing slopes. On the mountains around Tbilisi, south-facing slopes are often treeless, whereas shaded northern slopes may be clothed in dense forest. Ketskhoveli[49] notes that,

[49] Ketskhoveli 1959.

amongst herbaceous plants, north-facing slopes generally have a prevalence of mesophytes with palmate leaves (*e.g. Anemone, Trollius, Geranium*), while graminoids prevail on south-facing slopes. Amongst the conifers, pine (*Pinus*) has a preference for southern aspects; fir (*Abies nordmanniana*) and spruce (*Picea orientalis*) prefer more sheltered northern slopes. Beech (*Fagus orientalis*) likewise prefers the moister conditions on north-facing slopes, leaving sunnier slopes for oak-hornbeam associations.[50] **Figure 2.6** shows some examples of the altitudinal distribution of important taxa in relation to aspect in Georgia.

The major climatic and topographic division in Georgia, however, is the Likhi Range – the 950 m-high granitic massif that links the Greater Caucasus Range to the Anticaucasus Mountains of Southern Georgia. Together, these mountains create an orographic rain shadow over Eastern Georgia. Western Georgia is a land of subtropical rains, dark coniferous forests, alder swamps, liana-festooned trees, giant sphagnum bogs and rivers flowing into the Black Sea. In Eastern Georgia the rivers drain into the Caspian Sea, and salt lakes, semideserts, xerophytic shrublands and hot, dry summers are more characteristic.

The humid, warm-temperate climate of Colchis led Soviet scholars to term this a 'subtropical' region.[51] Areas close to the Black Sea coast, such as the coastal resort of Batumi, receive approximately 2500 mm of annual precipitation (**Fig. 2.7**).[52] Average temperatures in lowland Colchis vary between 5 and 7 °C in winter, with rare frosts, and around 22–23 °C in summer.[53] The Colchis lowland is a large, often swampy plain formed by fluvio-glacial sediments eroded from the surrounding mountains and the effects of sea level rise during the Holocene.[54] Small mountain glaciers are distributed on the western parts of the Main Caucasus Range, in Svaneti and Abkhazia, but their past extent is as yet poorly understood due to a lack of systematic, well-dated studies.[55]

Colchis is regarded as a glacial refugium for thermophilous plants,[56] hence the presence of *Vaccinium myrtillus, Viburnum orientale, Laurocerasus officinalis, Rhododendron ponticum, Ilex* spp., *Hedera colchica, Quercus pontica, Betula medwedewii, Corylus colchica* and other species in lowland Colchic forests.[57] Many other species were present during the early Pleistocene; thermophiles such as *Aesculus, Juglans, Tsuga, Sequoia, Carya* and *Liquidambar* were forced into extinction during the

[50] Ketskhoveli 1959.
[51] Berg 1950.
[52] AN GSSR 1964.
[53] Tatashidze 2000.
[54] Dzhanelidze 1980.
[55] Serebryanny 1984; Serebryanny and Orlov 1993.
[56] Denk *et al.* 2001; Röhrig 1991.
[57] Dolukhanov 1966; Nakhutsrishvili 1999.

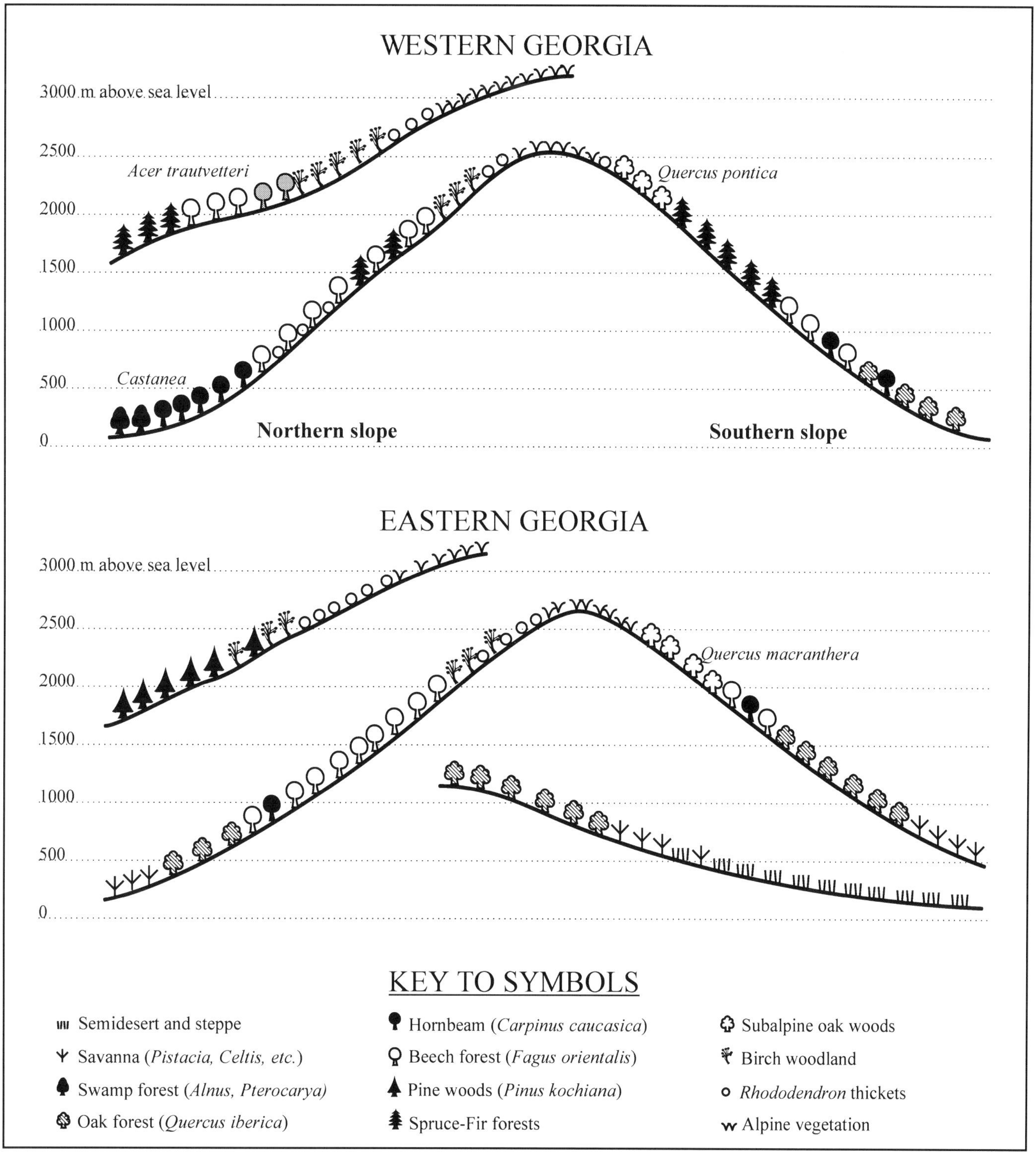

Figure 2.6. Schematic representation of vegetation zones in Western and Eastern Georgia in relation to altitude and aspect. Redrawn and simplified from Nakhutsrishvili (1999).

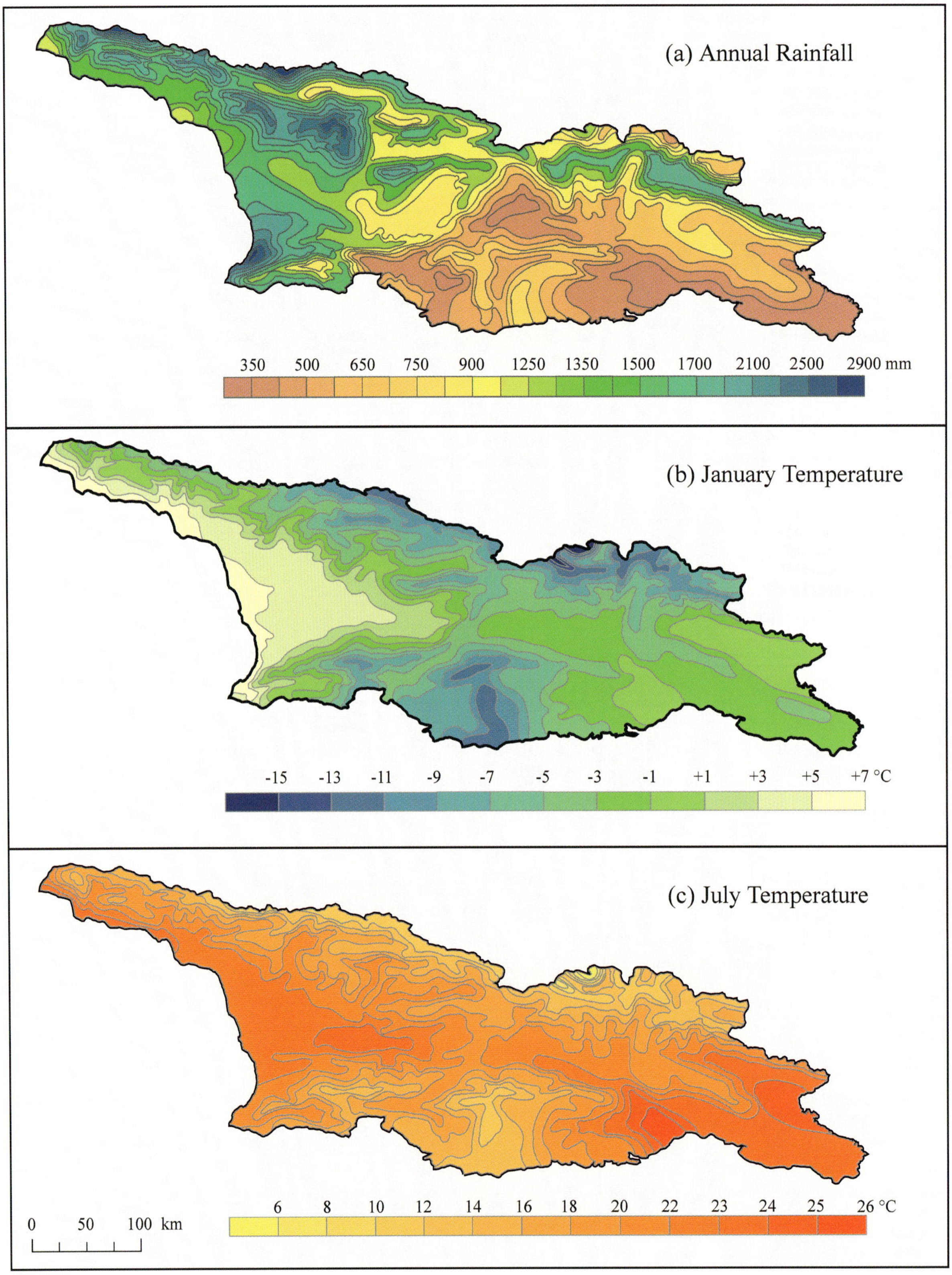

Figure 2.7. Maps of average annual precipitation (a), and average temperatures for January (b) and July (c) in Georgia. Adapted from *Атлас Грузинской ССР* (AN GSSR, 1964). Projection: World Geodetic Survey 1984.

succession of Pleistocene glacial periods.[58] A handful of Mediterranean plants are also found in Colchis forests, such as the strawberry tree (*Arbutus andrachne*) and box (*Buxus colchica*).

Vegetation in Western Georgia is arranged in altitudinal belts. In the lowlands, *Alnus barbata* and the Tertiary-relict *Pterocarya fraxinifolia* form dense swamp forests tangled with lianas.[59] In better-drained locations and with increasing elevation, *Quercus-Carpinus, Quercus-Castanea* and *Fagus-Carpinus* associations occur. *Fagus orientalis* becomes the dominant tree in forests above elevations of 600–700 metres (**Fig. 2.8**). Higher still, at 1200–1400 metres, beech gives way to coniferous forests of *Picea orientalis* and *Abies nordmanniana* and these prevail until the upper tree line around 2000 m, where subalpine herbaceous meadows or krummholz thickets of *Betula, Sorbus, Acer* and other species appear.[60] Human impact in the subalpine zone is intense, under the combined effects of livestock grazing, hay-making and wood cutting.[61]

The landscape of Eastern Georgia (Iberia) is entirely different, with large areas of steppe and semidesert vegetation maintained by a rain shadow cast by the Anticaucasus (or Lesser Caucasus) Mountains. Annual precipitation in the lowland steppe zone amounts to less than 400 mm, average winter temperatures are just below freezing and around 25 °C in summer (**Fig. 2.7**).[62] The Georgian capital, Tbilisi, is located at the northwestern corner of the lowland steppe zone, at the foot of the Trialeti Range. The soils of Eastern Georgia are generally chernozems and saline kastanozems in the semidesert and steppe zone, brown earths in the forest belt and peaty soils in the subalpine and alpine zones.[63] Chernozems also occur on the volcanic plateaux of the South Georgian Uplands where the vegetation is predominantly steppic grassland.

The vegetation of Eastern Georgia falls broadly into the Irano-Turanian group,[64] widespread through the steppes and arid woodlands of Iran, Eastern Anatolia and Central Asia. Oak-hornbeam forests of *Quercus iberica, Carpinus betulus* and *C. orientalis* are common in the foothill zone, replaced by *Fagus orientalis* or mixed forests at higher altitudes. On sunny slopes, *Quercus macranthera* grows from 1400 m elevation up into the subalpine zone. *Picea orientalis* and *Abies nordmanniana* forests are common only in the westernmost part of Southern Georgia (Akhaltsikhe Basin), and disjunct pockets of pine forest exist in rocky gorges and on steep, eroded slopes. At the lower tree line, xeric communities known as *shibliak* and *phrygana* are common.[65]

[58] Chochieva and Mamatsashvili 1996; Shatilova 1984; Shatilova and Mchedlishvili 1980; Shatilova *et al.* 2003; Shatilova *et al.* 2002; Shatilova and Stuchlik 1996; Tumajanov 1971.

[59] Denk *et al.* 2001; Ketskhoveli 1959.

[60] Dolukhanov 1966; Ketskhoveli 1959; Nakhutsrishvili 1999; Walter 1974.

[61] Badenkov *et al.* 1990; Bock *et al.* 1995; Dolukhanov 1978.

[62] AN GSSR 1964; Tatashidze 2000.

[63] Urushadze 1999.

[64] Zohary 1973.

[65] Dolukhanov 1966.

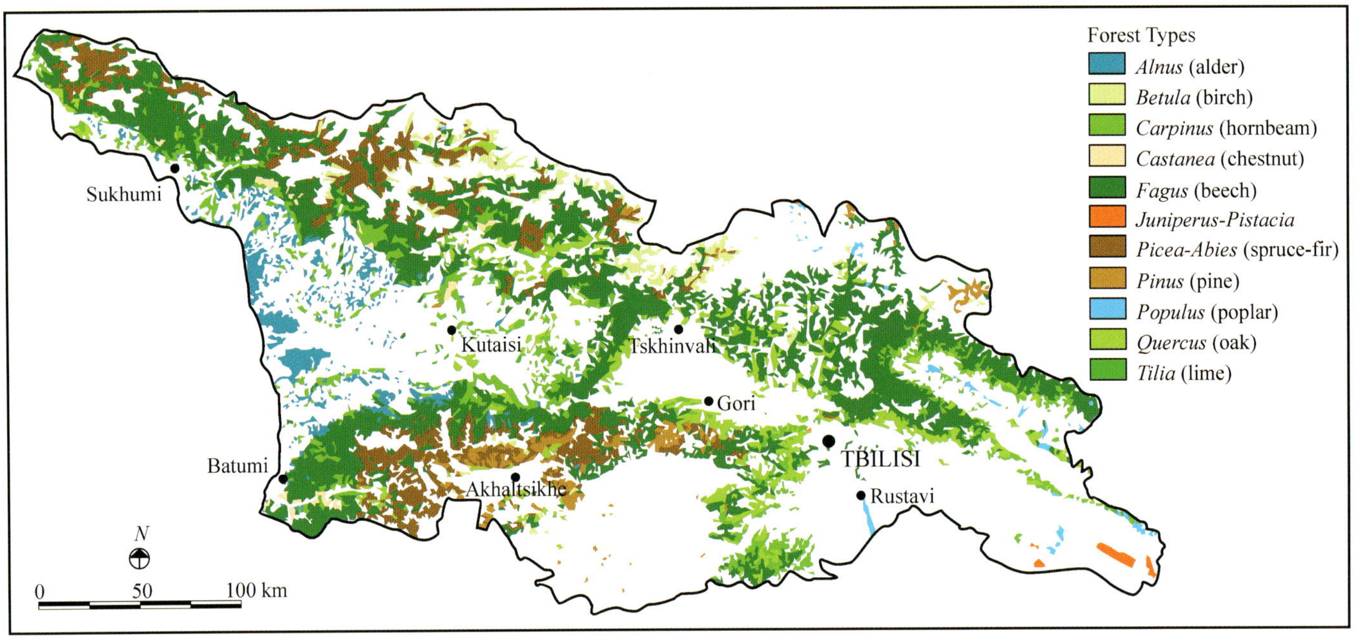

Figure 2.8. Forest map of Georgia, with areas classified according to the dominant trees present. Source: *Атлас Лесов СССР* (Kruchinin, 1973). Projection: Pulkovo.

These correspond to the Mediterranean vegetation groups called *maquis* or *garrigue*, and are regarded by many Caucasian botanists as the result of forest degradation.[66]

Large areas of the lowland steppe and semidesert zone have been appropriated for irrigated agriculture, but vegetation associations dominated by various chenopods, *Artemisia* species and *Bothriochloa ischaemum* are widespread. More restricted are areas of arid woodland, composed of *Pistacia mutica, Pyrus salicifolia, Juniperus foetidissima, J. polycarpos, Celtis caucasica, Quercus iberica* and *Carpinus orientalis.*[67] Gallery forests dominated by *Quercus pedunculiflora, Populus canescens* and *P. hybrida* occur alongside some rivers in Eastern Georgia, accompanied by large numbers of lianas.[68]

It must be remembered that although the Caucasus today tends to be synonymous with Russia and the former Soviet Union, its environment has always existed within the regional sphere of Western Asia, so great a barrier did the Main Caucasus Range present to human and biotic migration.[69] Thus we can regard Eastern Georgia, Armenia, Azerbaijan and Eastern Anatolia as being a varied yet distinct mountainous geographical unit.

[66] Badenkov *et al.* 1990; Ketskhoveli 1959; Sakhokia 1961.
[67] Arabuli 2003; Dolukhanov 1966.
[68] Ketskhoveli 1959; NACRES 2005.
[69] *E.g.* Seddon *et al.* 2002.

LATE QUATERNARY VEGETATION HISTORY OF GEORGIA

To understand the present-day complexity of Georgia's vegetation, it is necessary to look to the past to determine the causes of vegetation change. The final section of this chapter aims to introduce some of the debates and uncertainties that surround Georgia's vegetation history during the Late-glacial and Holocene periods. It is not an attempt to describe the results of previous studies in detail, but a discussion of the theoretical issues associated with interpretation of these results, with particular emphasis on causation.

Botanical Debate on Georgia's Vegetation History

For many years, botanists have argued that highland grassland, lowland steppe, shibliak scrub and phrygana vegetation have appeared in the Georgian landscape as a result of recent deforestation.[70] As one prominent writer recently stated, "during the Holocene, when climatic conditions were similar to nowadays, the entire territory of Georgia was occupied by forests".[71]

Around 40 percent of Georgia's territory is currently classified as forest (**Fig. 2.8**),[72] making it by far the most forested nation in the South Caucasus. Almost everything that has been written on the forest vegetation of this region is prefaced by a statement such as "these forests have also suffered from the agricultural activities of man",[73] 'woodland was degraded to shrubby vegetation or completely destroyed as a result of excessive wood-cutting and grazing',[74] 'profoundly modified as a result of human impact',[75] or 'strongly disturbed by felling'.[76] Some have suggested that this widespread deforestation took place during the last few centuries in Eastern Georgia,[77] and even more recently in Western Georgia.[78]

Evidence for these past forests comes in several forms. One is the presence of isolated forest trees in steppe landscapes, thought to represent remnants of a formerly continuous forest cover.[79] Another is the climatic and edaphic similarity between forested and non-forested areas, which suggest that steppe or scrub prevails where forest

[70] Badenkov *et al.* 1990; Ketskhoveli 1959; Nakhutsrishvili 1999; Sakhokia 1961; Sosnovskii 1915; Tumajanov 1971.
[71] Nakhutsrishvili 1999, section 6.
[72] Papava 2000.
[73] Nakhutsrishvili 1999, section 6.6.
[74] Walter 1974, p. 381.
[75] Gulisashvili 1964, p. 159.
[76] Sakhokia 1961, p. 17.
[77] Ketskhoveli 1959; Matcharashvili *et al.* 2004; Tumajanov 1971.
[78] Badenkov *et al.* 1990; Dzhanelidze 1980.
[79] Grossgeim 1948; Ketskhoveli 1959; Matcharashvili *et al.* 2004; Nakhutsrishvili 1999.

is capable of growing.[80] Others have looked to the historical record to find reference to the nature of the vegetation in the past.[81]

The weight of opinion therefore supports the contention that forests were formerly much more widespread than at present. But Avakov has questioned whether Eastern Georgia's highly diverse xeric scrub and steppe communities could have arisen as recently as other botanists contend.[82] He argues that steppe and scrub have a longer history that is intertwined with disturbances such as fire and grazing. Avakov supports this view with reference to the antiquity of farming in Georgia, as well as early drawings and photographs of Tbilisi that show a city surrounded by bare hills.[83] Steppe fauna were present in parts of Eastern Georgia during the early-mid Holocene,[84] and this indirect evidence lends support to Avakov's belief that the region's xerophilous vegetation has a very considerable ancestry indeed.

The question of if and when Georgia's steppe and scrub landscapes were forested is a critical one for understanding the present-day complexity of Georgia's vegetation, for determining how the biodiversity of xeric plant communities came about, for setting in context the events of history and prehistory, and for deciding how best to direct vegetation conservation and management efforts today. Pollen analysis, as a means of reconstructing past vegetation, is perhaps the most appropriate means of resolving this problem. As we shall see, however, the palynological literature is not without certain problems of its own.

General Comments on the Palynological Literature

The majority of Georgia's pollen diagrams were prepared during the Soviet period. Perhaps the most conspicuous aspect of Soviet palynology was the almost universal use of the Blytt-Sernander scheme to define chronological subdivisions for the Holocene (**Fig. 2.9**). The timing of these subdivisions was based on peat stratigraphical changes observed in Northern European raised bogs,[85] and the scheme's application relies on the assumption that vegetation shifts were driven exclusively by climate changes, synchronous across Europe. The picture that has emerged with the widespread adoption of radiocarbon dating is far more complex than the Blytt-Sernander scheme would suggest.[86] Indeed, temperature variations in southeastern Europe during the Holocene may follow an entirely different trend to the northern regions where

[80] Gulisashvili 1964; Ketskhoveli 1959.
[81] Dzhanelidze 1980; Ketskhoveli 1959; Maruashvili 1956; Matcharashvili *et al.* 2004; Sosnovskii 1915.
[82] Avakov 1982.
[83] Tournefort 1718.
[84] Bendukidze 1979.
[85] Roberts 1998, p. 118.
[86] Mangerud *et al.* 1982.

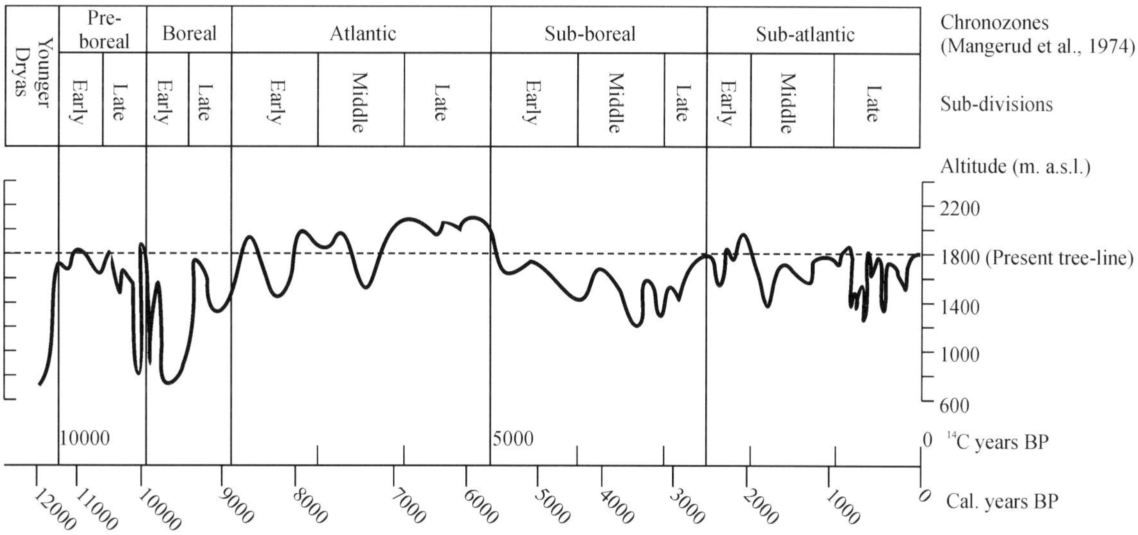

Figure 2.9. The Blytt-Sernander chronostratigraphic scheme (top) shown in relation to radiocarbon and calendar timescales (bottom) and reconstructed tree line fluctuations in Abkhazia, Western Georgia (curve, middle). Compiled from Mangerud *et al.* (1974) and Kvavadze *et al.* (1992). See Kvavadze and Connor (2005) for further details.

the Blytt-Sernander scheme was developed.[87] And, in supposed glacial refugia such as Colchis, the clear break between the Pleistocene and Holocene may not be obvious in pollen stratigraphy, let alone assumed differences between the early and mid Holocene climate.[88] Independent dating is therefore of paramount importance.

The vast majority of pollen diagrams from the former Soviet Union are unsupported by radiocarbon ages.[89] Even in regional syntheses, where considerable effort has been expended in gathering dated pollen records, undated sections represent 20–30 percent and poorly dated diagrams are in the majority.[90] Where sediments are typically low in organic content, such as in the Caucasus region, conventional ^{14}C dates may not be obtainable in any case. This explains why, amongst the 233 sites included in a comprehensive study by Tarasov *et al.*,[91] Caucasian sites averaged less than two dates per diagram, compared to twice that number across the former Soviet Union generally.

[87] Davis *et al.* 2003.
[88] Mangerud *et al.* 1982.
[89] Grichuk 1984.
[90] Khotinskiy 1984; Tarasov *et al.* 2000; Tarasov *et al.* 1998.
[91] Tarasov *et al.* 1998.

A second important aspect of palynology in Georgia, particularly in regions where peat accumulation is negligible, is the widespread use of soils and archaeological deposits as palaeovegetation archives. Exposed river terraces have often been used for palaeoenvironmental reconstruction,[92] but the pollen in these sediments is often carried fluvially from upstream vegetation, making the source of the pollen difficult to define. The continuity of these largely alluvial deposits may be also questioned; in the absence of strict dating control, it is unclear whether these deposits are affected by pollen redeposition and hiatuses in sedimentation.

Pollen analysis of archaeological strata[93] presents some difficulties for reconstructing vegetation because the source of the pollen varies through time, depending on the form and use of the site's structures.[94] Cereal pollen, for example, is rarely found in archaeological contexts due to its delicate structure, but on former grain-threshing floors it occurs in great abundance.[95] The construction of dwellings from mud bricks introduces secondary and aquatic pollen,[96] and additional pollen may be incorporated during the occupation, destruction and reworking of the site.[97] Pollen taphonomy in these sites often leads to an over-representation of Asteraceae compared to fragile pollen types.[98] These considerable biases make interpretation of archaeological pollen spectra inherently difficult, somewhat subjective, and unsuitable for the reconstruction of past vegetation on any more than a local scale.

Thirdly, the identification of anthropogenic indicators[99] in pollen diagrams would seem to be of prime importance in a landscape so heavily indebted to human activity as Georgia.[100] However, as Kvavadze and Gabashvili note,[101] this has rarely been the case. Only a handful of diagrams make any distinction between cereal and grass pollen. The differentiation of the various diagnostic *Plantago* species is even less common. Many diagrams ignore herbaceous pollen entirely, presenting only data for arboreal taxa and some shrubs.[102] In these circumstances it is very difficult to disqualify human impact as a cause of vegetation change.

A final point relates to the interpretation of pollen data. The almost ubiquitous use of the Blytt-Sernander scheme introduces an element of climatic determinism and argument circularity into any interpretation of vegetation change. The elm decline,

92 *E.g.* Bortenschlager 1987; Gogichaishvili 1984b; Gogichaishvili *et al.* 1977; Kvavadze 1978.

93 *E.g.* Kalandadze *et al.* 1977; Kvavadze *et al.* 2004; Kvavadze and Todria 1992; Martín Sánchez *et al.* 1999.

94 van Zeist and Bottema 1991.

95 Behre 1986.

96 Bottema 1992.

97 Tomescu 2000.

98 Bottema 1975; Havinga 1964; Havinga 1967.

99 Behre 1986.

100 Gogichaishvili 1990.

101 Kvavadze and Gabashvili 1998.

102 Kvavadze and Connor 2005.

for instance, which occurred in Europe with surprising synchronicity around 5800 cal BP, is usually attributed to climate change in Soviet-era publications.[103] This view was shared by Western European scholars some decades ago,[104] but more recent thought posits that the catastrophic nature of the elm decline was probably caused by an outbreak of Dutch Elm Disease, perhaps related to Neolithic forest disturbance.[105] In the Georgian palynological literature, the impacts of disease, human activity, soil weathering and fire are seldom considered to have played a more significant role than climate in governing the timing and direction of past vegetation change.

Palynological Studies in Georgia

The first palynological studies in the Caucasus region were conducted by Russian scientists in the early 1930s.[106] Dokturovskii's work in the peatlands of lowland Colchis and the highland bogs of Bakuriani to a large extent dictated the future direction of Holocene palynological research in Georgia, allowing Neishtadt to enumerate his influential, tripartite scheme of vegetation change in Colchis.[107] In recognition of the interpretive problems associated with undated pollen diagrams, the following discussion will focus on records with at least one radiocarbon date (**Table 2.1**, **Fig. 2.10**).[108]

As **Figure 2.10** shows, there are many more dated pollen diagrams from humid Western Georgia compared to the semi-arid east. The greatest concentration of these is in the autonomous region of Abkhazia, the mountainous strip of land between the Main Caucasus Range and the Black Sea in northwestern Georgia. Interestingly, since this area is considered to have been a glacial refugium for trees, its lowland pollen diagrams are characterised by consistently high arboreal pollen proportions during the Late Pleistocene.

The longest such record is the Sukhumi-36 core (**Fig. 2.11**), collected from the Black Sea coast near the Abkhazian capital of Sukhumi. Although the 45-m core has only one radiocarbon date of 6425±60 ^{14}C bp at a depth of 10 m, it dates back well into the Late Pleistocene (possibly more than 40 kya), as indicated by the pollen of now-extinct *Tsuga, Carya, Cedrus* and Taxodiaceae in the lowermost zones. Late Pleistocene spectra are dominated by *Abies, Pinus* and *Carpinus* amongst arboreal types, and Poaceae, Asteraceae and Polypodiaceae in the non-arboreal group. *Abies* declines during the early Holocene and *Alnus* becomes dominant. The mid Holocene has

[103] Wright and Barnosky 1984.
[104] *E.g.* Tauber 1965.
[105] Girling and Greig 1985; Peglar and Birks 1993.
[106] Dokturovskii 1931; Dokturovskii 1936.
[107] Neishtadt 1957.
[108] A detailed overview of all palynological research in the Caucasus and Central Asia from 1931 to 1975 has been published in Russian: Pakhomov and Klopotovskaya 1980.

Table 2.1. Details of Georgian Pollen Sites Shown in **Fig. 2.10**.

No.	Site name	Sub-region	Lat.	Long.	Elev.	^{14}C	From	To	Source
1	Adange	Abkhazia	43.28	41.28	1750	3	3000	0	Kvavadze and Rukhadze 1989
2	Amtkel	Abkhazia	43.25	41.28	1830	2	2000	0	Kvavadze and Rukhadze 1989
3	Anaklia	Colchis	42.38	41.55	5	5	6000	0	Serebryannyi *et al.* 1984
4	Bathing Chamois	NW Caucasus	43.63	40.53	2490	2	5000	0	Kvavadze and Efremov 1994
5	Bazaleti B	E Georgia	42.03	44.68	875	1	3000	0	Kvavadze and Connor 2005
6	Chamlukh	Abkhazia	43.18	41.03	1650	1	4000	0	Yazvenko 1994
7	Cheliagele	Racha	42.47	43.10	1100	25	7000	0	van Leeuwen, unpubl.
8	Devis Tba	Racha	42.42	43.30	1380	?	4000	300	van Leeuwen, unpubl.
9	Didajaris Tba B	Ajara	41.68	42.50	1850	6	11000	0	van Leeuwen, unpubl.
10	Gagra 471	Abkhazia	43.23	40.27	5	2	7000	0	Kvavadze and Rukhadze 1989
11	Gagra 607	Abkhazia	43.27	40.27	0	2	6000	4000	Kvavadze and Rukhadze 1989
12	Gomnis Tba 1–2	S Georgia	41.77	43.35	1850	2	12000	0	Margalitadze 1971; Gobet, unpubl.
13	Gudauta 120	Abkhazia	43.08	40.62	0	1	11000	0	Kvavadze and Rukhadze 1989
14	Gudauta 55	Abkhazia	43.08	40.62	0	2	7000	5000	Kvavadze and Rukhadze 1989
15	Imnati	Colchis	42.10	41.82	2	4	7000	0	Neishtadt *et al.* 1965
16	Kartsakhi	S Georgia	41.25	43.30	1800	2	8000	0	Margalitadze and Ramishvili 1972
17	Kazbegi	Central C.	42.63	44.55	3360	2	7000	3000	Margalitadze, 1995
18	Kvartsevoe	NW Caucasus	43.63	41.12	2726	1	6000	0	Kvavadze and Efremov 1996
19	Lashkhrashi	Svaneti	43.13	42.43	2000	1	3000	0	Margalitadze and Kimeridze 1985
20	Luganskoe	NW Caucasus	43.38	40.68	2428	2	5000	0	Kvavadze and Efremov 1996
21	Malyi Khodzhal	Abkhazia	42.95	41.92	2030	1	1000	0	Kvavadze and Rukhadze 1989
22	Rybnoe	NW Caucasus	43.65	41.15	2151	2	2000	0	Kvavadze and Efremov 1996
23	Sagarejo	E Georgia	41.68	45.35	550	2	14000	0	Gogichaishvili 1984a
24	Sibista	Abkhazia	43.23	41.43	2160	2	2000	0	Kvavadze and Rukhadze 1989
25	Sukhumi 36	Abkhazia	43.00	41.00	0	1	20000	0	Kvavadze and Rukhadze 1989
26	Sukhumi 721	Abkhazia	43.00	41.00	0	2	10000	0	Kvavadze and Rukhadze 1989
27	Sukhumi 723	Abkhazia	43.00	41.00	0	4	10000	0	Kvavadze and Rukhadze 1989
28	Supsa	Colchis	42.02	41.85	8	4	8000	0	Kvavadze 1982
29	Supsa 2	Colchis	42.03	41.83	10	1	2000	0	Kvavadze and Connor 2005
30	Tseros Tbis	S Georgia	41.77	43.35	1920	2	6000	0	Margalitadze 1967

Key to abbreviations: Central C. – Central Caucasus; Lat. – latitude (decimal degrees); Long. – longitude (decimal degrees); ^{14}C – number of radiocarbon dates (v – varve ages); From – basal age in ^{14}C bp; To – most recent sample age (0 = modern).

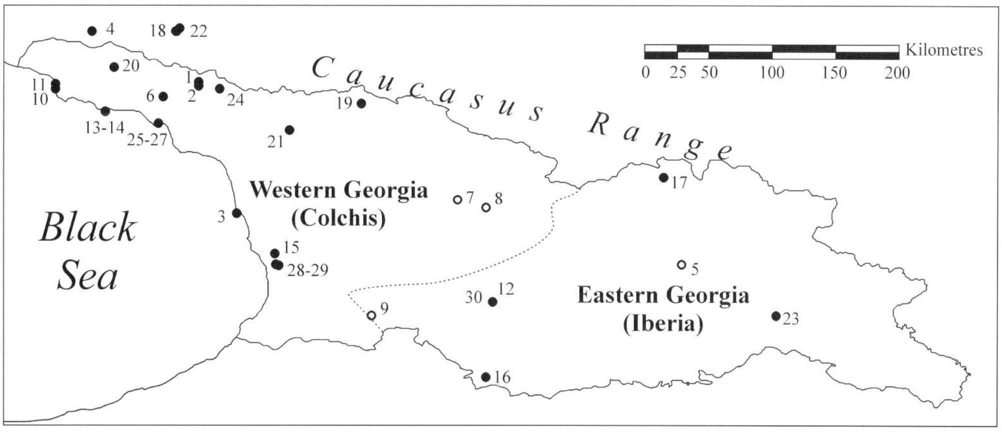

Figure 2.10. Location map of radiocarbon-dated palynological sites in Georgia. See Table 2.1 for site names and other details. Filled circles indicate published pollen data; open circles represent unpublished data. Dashed line indicates the approximate bioclimatic boundary between Western and Eastern Georgia (AN GSSR, 1964).

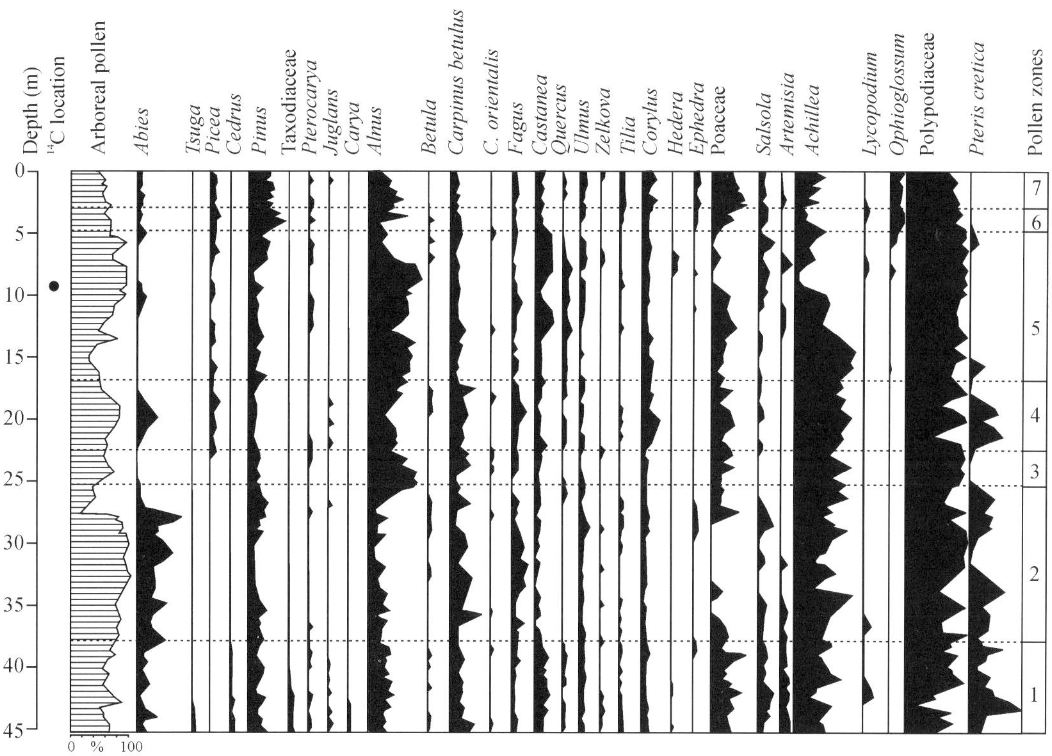

Figure 2.11. The Sukhumi-36 pollen diagram from lowland Abkhazia, Western Georgia. Radiocarbon date is 6425±60 ^{14}C bp. Redrawn and simplified from Kvavadze and Rukhadze (1989, p. 66).

increased proportions of *Fagus, Castanea* and *Quercus*, leading the authors to conclude that this was the time of climatic optimum.[109] Inferred late Holocene spectra have an increase in *Pinus*, thought to indicate a climatic deterioration.[110]

Similar patterns are evident in the Sukhumi-721 and Sukhumi-723 records, collected nearby. These well-dated cores have high proportions of *Alnus* pollen in the early stages of the Holocene, increased proportions of *Abies, Fagus* and *Tilia* in the early-mid Holocene, and a prevalence of *Castanea* from the mid to late Holocene. At both Sukhumi and Gagra, increasing *Alnus* and Cyperaceae proportions in the last 4000 years are explained by climatic deterioration and human impact.[111] Geomorphic evidence, however, suggests an elevated sea level around that time.[112]

In the mountains of Abkhazia, the temporal record of vegetation change is considerably shorter than in the lowlands, as highland peat deposits seem to have begun forming from the mid Holocene onwards. The longest of these records is from Luganskoe, situated at an altitude of 2428 metres in the alpine zone. Pollen spectra from the mid Holocene contain elevated proportions of herbaceous taxa, including Chenopodiaceae, *Cirsium,* Apiaceae and Caryophyllaceae, which the authors think represents a rich herbaceous vegetation developed under humid climatic conditions.[113] Pollen data from Kvartsevoe, Bathing Chamois and Adange also have an increased representation of herbaceous pollen during the mid Holocene.[114] The interpretation of these taxa as indicative of greater humidity and warmth seems difficult to reconcile with the ecology of the plants, which are mostly herbs of dry environments. Since these increases in herbaceous pollen correlate with lithological changes (from minerogenic to organic sediment), it is possible that the herbaceous component is derived from plants growing in the local vegetation.

Opinion is divided on the causes of late Holocene vegetation change in highland Abkhazia. Some have interpreted rapid changes in pollen assemblages as evidence of tree line shifts in response to climatic events such as the 'Little Ice Age'.[115] For others, these excursive and often asynchronous changes could have been initiated by human activities, especially grazing and deforestation.[116] The most problematic of these changes is the recent increase in *Abies* and *Picea* at many sites in highland Western Georgia (*e.g.* Luganskoe, Bathing Chamois, Lashkhrashi, Adange and Chamlukh); the varied timing and rapid rate at which this conifer ingression took place seems difficult

[109] Kvavadze and Rukhadze 1989, p. 67.
[110] Kvavadze and Rukhadze 1989.
[111] Kvavadze and Rukhadze 1989.
[112] Chepalyga 1984; Dzhanelidze 1980; Tvalchrelidze *et al.* 2004.
[113] Kvavadze and Efremov 1996.
[114] Kvavadze and Efremov 1994; Kvavadze and Efremov 1996; Kvavadze and Rukhadze 1989.
[115] Kvavadze and Rukhadze 1989.
[116] Yazvenko 1994.

to explain adequately in climatic terms alone.

Another group of sites is clustered in the swampy lowlands of Colchis. The longest record (**Fig. 2.12**) was obtained from exposed alluvial deposits along the Supsa River, between the towns of Poti and Kobuleti.[117] The early Holocene phase is distinguished by its representation of coniferous taxa (*Picea-Pinus-Abies*). From approximately 7500 cal BP, *Picea* and *Abies* decline and herbaceous pollen increases dramatically. Kvavadze interprets this as a temperature decrease.[118] Around 6000 years ago, Tertiary-relict *Pterocarya* became very important in pollen spectra, indicating an increase in gallery forest on this floodplain and possibly a warmer, wetter climate. The disappearance of *Pterocarya* during the mid-late Holocene is thought to indicate a cooling climate and the intensification of human activity since the fourth century AD.[119]

The next closest site is Imnati, approximately 10 km north of Supsa and part of the extensive *Sphagnum* mires of Central Colchis. In contrast to other diagrams, where *Castanea* becomes prevalent in the mid or mid-late Holocene, at Imnati this taxon declines around 6650 cal BP to be replaced by *Alnus*.[120] Another lowland diagram comes from Anaklia, at the mouth of the Inguri River.[121] The seven-metre core dates back 6500–7000 years and has five radiocarbon dates. The most outstanding feature of the Anaklia diagram is the decline in *Alnus* and coincident increase in *Pinus, Picea, Abies, Carpinus, Fagus, Quercus, Ulmus, Castanea, Corylus* and herbaceous types such as Asteraceae. Dated from 5500–3300 cal BP, this event follows a period of rapid sediment accumulation, and is thought to relate to the reactivation of glaciers on the Main Caucasus and the subsequent depression of tree lines.[122] The problem with this interpretation is that most pollen diagrams from Western Georgia show no evidence for such a change, and it seems strange for tree line depression to manifest at the coast rather than in the mountains.

Moving now to Eastern Georgia, we can see that dated pollen diagrams are particularly lacking (**Fig. 2.10**). This is because the climatic conditions are largely unfavourable for peat formation. Seven radiocarbon-dated diagrams are available, of which five are published. The Gomnis Tba and Tseros Tbis diagrams are both derived from lake-wetland sediments cored on the Dabadzveli Plateau, on the western flank of the Trialeti Range above the spa resort of Borjomi. The climate of this section of the Trialeti Range is wetter than most parts of Eastern Georgia, due to its interception of moist westerly air currents originating over the Black Sea.[123]

[117] Kvavadze 1982.
[118] Kvavadze 1982.
[119] Kvavadze 1982.
[120] Botch and Masing 1983; Neishtadt *et al.* 1965.
[121] Serebryannyi *et al.* 1984.
[122] Margalitadze 1995.
[123] Margalitadze 1998.

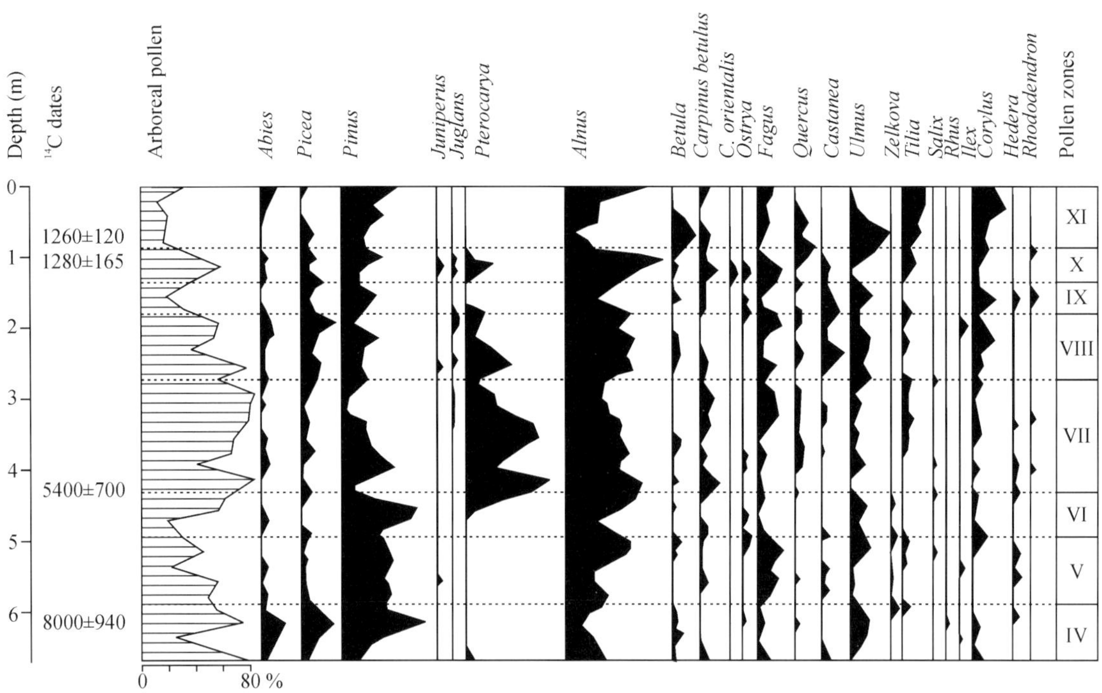

Figure 2.12. Pollen diagram from Supsa, coastal Western Georgia. Redrawn from Kvavadze (1982).

Gomnis Tba is a small lake of volcanic origin set in a herb meadow and surrounded by *Picea orientalis*-dominated forest.[124] The 9.5-metre core consists of sedge peat in the upper four metres, with a black silt layer at 4.0–4.25 m and underlain by lake clays (**Fig. 2.13**). Pollen analysis revealed an early Holocene landscape dominated by herbaceous taxa, although the high proportions of Asteraceae could also suggest post-depositional changes to the pollen assemblage. *Pinus* and *Abies* were the most abundant arboreal taxa. Margalitadze has argued that the landscape of that time was rendered treeless by volcanic activity in tandem with a cooler climate.[125] A more recent, unpublished diagram from the same site indicates that these levels date to the early Holocene prior to 9500 cal BP.[126] In the subsequent period, mixed deciduous forests expanded up the surrounding mountain slopes, changing into monodominant *Fagus orientalis* forest approximately 6000 years ago. In both diagrams there is a rapid reduction in beech pollen and an increase in conifers around 2000 years ago. This shift from deciduous to coniferous is thought to indicate cooler, drier conditions.[127]

[124] Margalitadze 1967.
[125] Margalitadze 1971.
[126] E. Gobet, unpublished data.
[127] Margalitadze 1971.

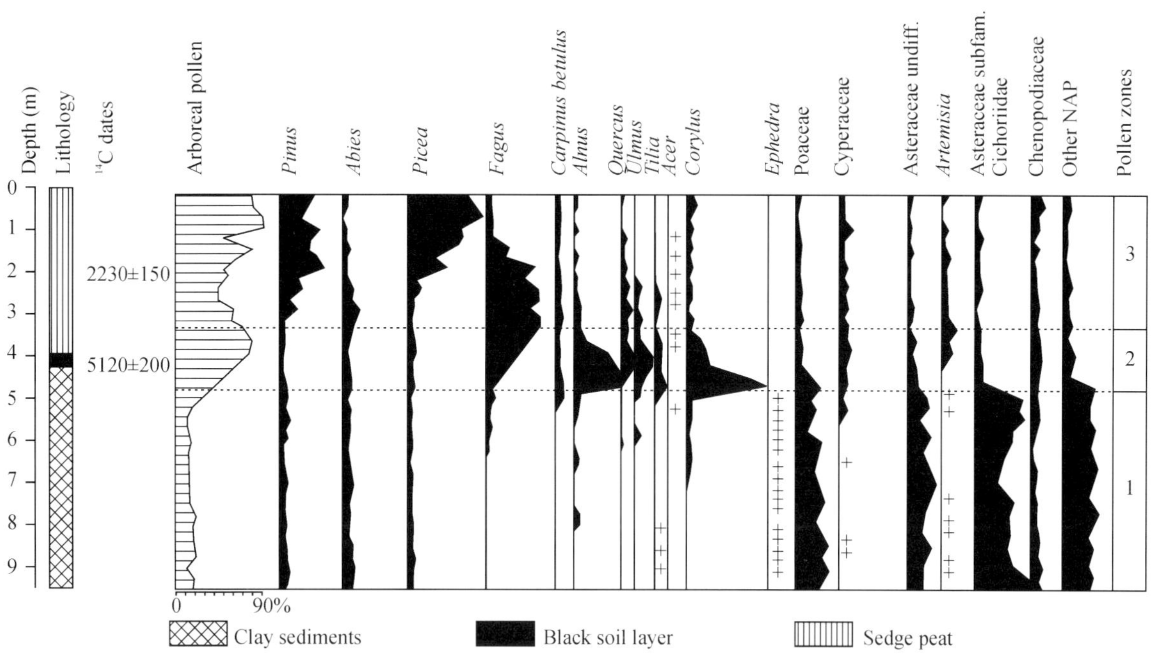

Figure 2.13. Pollen diagram from Gomnis Lake, Dabadzveli Plateau, Southern Georgia. Plus (+) symbol denotes trace proportions. Redrawn from Neustadt (1971).

Tseros Tbis is located near Gomnis Tba in the same volcanic terrain. It is a small, mesotrophic *Sphagnum* bog surrounded by woodlands of *Pinus* and *Betula*. Like the Gomnis core, the sediments were peat underlain by silty sediments with low arboreal pollen percentages. Two radiocarbon dates suggest the 11.5 metres of sediment accumulated during the last 7000 years. The palynological changes are similar to those in the Gomnis record and their interpretation is likewise in climatic terms.[128]

Lake Kartsakhi is located near the Turkish border on the treeless Javakheti Plateau of Southern Georgia. An 8.25-metre core was taken from the lake margins in a *Carex dichoandra-Equisetum fluvatile* wetland association.[129] Sediments of the early-mid Holocene have low arboreal pollen counts and are dominated by Asteraceae, while the more organic sediments in the upper four metres have more arboreal pollen and an increasing proportion of sedge pollen (**Fig. 2.14**). Much of the tree pollen at this site could have come from long-distance pollen transport, and thus the authors interpret the Kartsakhi record as evidence for persistent open landscapes in the South Georgian

[128] Margalitadze 1995.
[129] Margalitadze 1995.

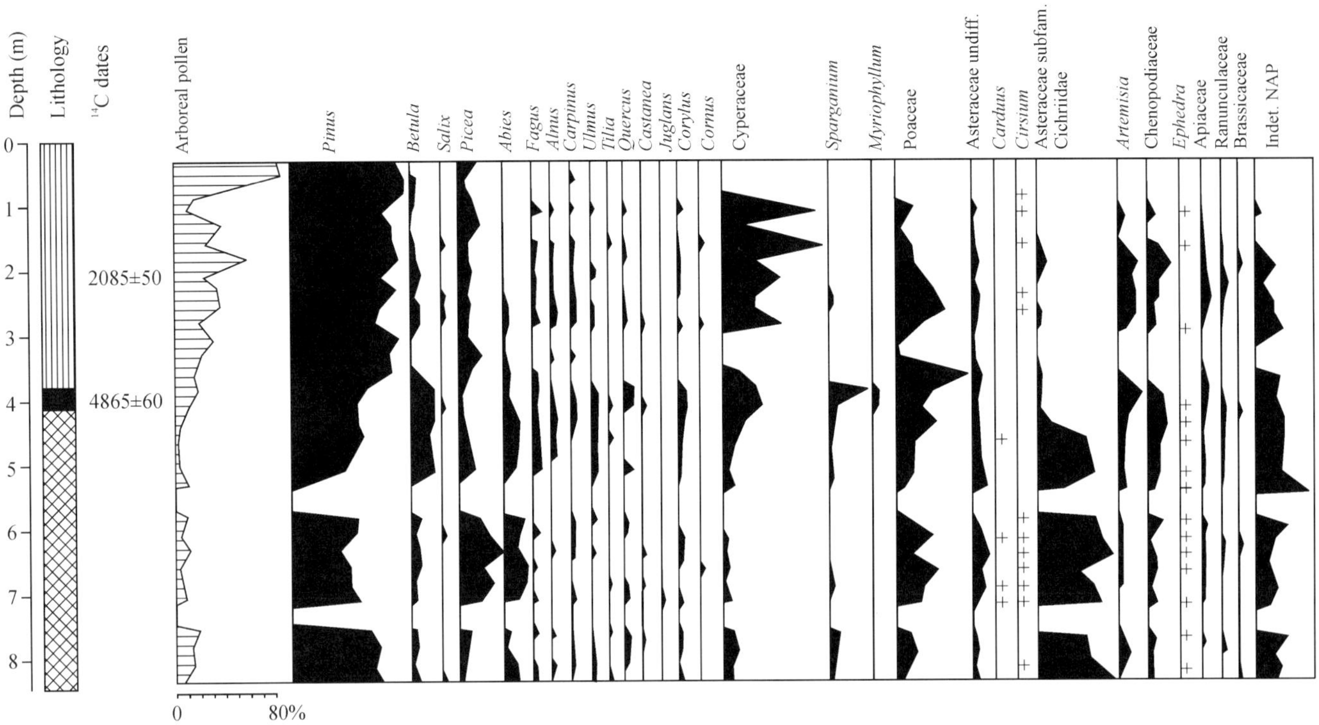

Figure 2.14. Pollen diagram from the Kartsakhi bog, Javakheti Plateau, Southern Georgia. Symbols are the same as those used in Figure 2.13. Redrawn and simplified from Margalitadze (1995).

Uplands during the Holocene.[130] However, the interpretation of these three diagrams (Gomnis, Tseros and Kartsakhi) is influenced by sediment hiatuses and changes in the depositional environment, as reflected in the non-arboreal pollen component.[131]

The only lowland pollen diagram from Eastern Georgia comes from alluvial terraces at Manavi, near the town of Sagarejo in the Iori Valley. The local vegetation at the site is gallery forest, but much of the surrounding vegetation is xerophytic scrub or agricultural land, heavily human-impacted.[132] The pollen diagram (**Fig. 2.15**) contains a record spanning the last 17,000 years. The lowermost levels of the Sagarejo diagram have high proportions of herbaceous pollen and *Pinus*. The landscape of the lowland Iori Valley at the last glacial maximum was inferred as treeless steppe with pines sheltering in less exposed areas.[133]

[130] Gagnidze and Margalitadze 1998; Margalitadze 1977; Margalitadze and Ramishvili 1972.
[131] Dr N. Margalitadze, Botanical Institute, Tbilisi, pers. comm., 2003.
[132] Gogichaishvili 1990; Tumajanov and Gogichaishvili 1969.
[133] Gogichaishvili 1984b.

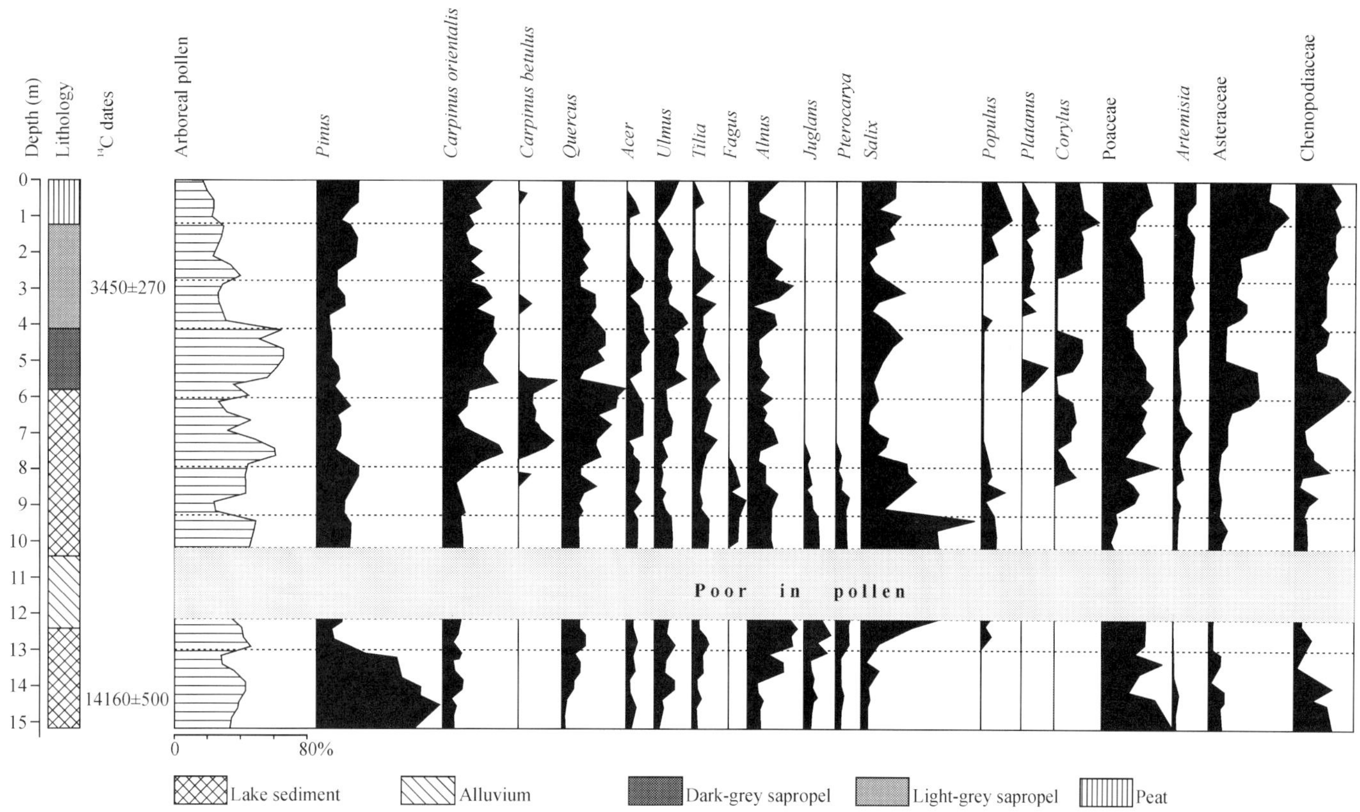

Figure 2.15. Pollen diagram from Sagarejo, lowland Eastern Georgia. Redrawn and simplified from Gogichaishvili (1984).

An interesting aspect of the Late Pleistocene levels from Sagarejo is the presence of *Juglans* and *Pterocarya* pollen. These two, thermophilous, Tertiary-relict trees must have survived the cold and aridity of the glacial period on the lowland floodplains of glacier-fed rivers, becoming locally extinct in these habitats during the early Holocene. Mesophilic tree species expanded during the early Holocene, with a pronounced reduction in forest cover during the mid Holocene that the author interpreted as the result of Neolithic human activities and a 'xerothermic' (hot, dry) climate.[134] Human impacts are thought to have further intensified during the mid-late Holocene throughout Eastern Georgia, maintaining the open, steppic character of the landscape from 3500 years ago until the present-day.[135]

Hence the pollen evidence from Eastern Georgia contradicts the dominant botanical paradigm that Georgia was entirely forested until recently: Margalitadze concluded

[134] Gogichaishvili 1984b; Gogichaishvili 1990.
[135] Gogichaishvili 1984b.

that the highland landscapes of Southern Georgia remained open and grassy throughout the Holocene,[136] and Gogichaishvili argued for an expansion of xerophilous plants coinciding with the adoption of agriculture in the lowlands of Eastern Georgia (approx. 7800 to 6600 cal BP).[137] These interpretations support Avakov's suggestion that non-forest plant communities have a considerable history in Georgia.[138] However, the evidence is not altogether conclusive. Gogichaishvili's argument is based on floodplain sediments and it is difficult to define the source of the pollen or to exclude the possibility that older pollen has been brought in by erosion. It would be preferable to use lake sediments as a means of avoiding these issues. On the other hand, Margalitadze's pollen diagrams from the highland lakes of Southern Georgia seem to be affected by different sedimentation processes, such that herb and grass pollen always prevail in minerogenic sediments, while tree pollen is dominant in peaty sediments. Some form of correction is required to ensure that the apparent lack of trees is not simply the result of differential pollen degradation or changing pollen source-area due to lake level variations.

In conclusion, there is much to be learned about the history of Georgia's vegetation, not least of all in the open landscapes of the semi-arid regions. The aim of this study is to examine the causes of vegetation change in Southern Georgia during the last 14,000 years. Currently, opinion is divided over whether climatic change has been the dominant cause of Holocene vegetation changes,[139] or whether human impacts in fact overrode these climatic influences.[140] Reconciling these divergent causes rests on obtaining carefully dated pollen data with a greater attention to anthropogenic indicators, and, more specifically, formulating an objective method for reconstructing changes in climate, forest cover and land-use. These tasks are described in the following chapter.

[136] Margalitadze 1977.
[137] Gogichaishvili 1984b.
[138] Avakov 1982.
[139] Kvavadze and Rukhadze 1989; Margalitadze 1995.
[140] Gogichaishvili 1984b; Gogichaishvili 1990.

CHAPTER THREE

METHODS

A complete study of vegetation history through pollen analysis is a multifaceted endeavour, involving diverse techniques that complement the elementary method of identifying pollen grains freed from sediment.

Attention must be paid to the selection of sites, as this determines the spatial scale of the investigation.[1] The vegetation growing in and around the study site should be recorded in a way that can be usefully compared to the pollen currently being deposited there.[2] The sediments containing fossilised pollen must be carefully gathered so as not to disturb or contaminate them, and the physical properties of these sediments may reveal important changes in the sedimentary environment through time.[3]

Once pollen has been isolated, identified and counted, the development of the vegetation can only be interpreted in an historical sense with the aid of absolute ages, in the form of radiocarbon dates, marker horizons or other indicators.[4] Finally, underlying patterns in the data can be teased out using statistical analysis and the results presented in a standard format.[5] Each of these stages for the present study is detailed in the sections that follow.

Site Selection

Sites were selected to represent a variety of vegetation units, including semidesert-steppes, coastal swamps, broadleaf forests, tree lines and upland grasslands. The sites are located along a transect running through Southern Georgia, from the high-rainfall coast of the Black Sea to the arid lowlands of Eastern Georgia (**Fig. 3.1**). This transect of sites is designed to give a spatial dimension to the study that furthermore allows local phenomena to be differentiated from a background of regional change: "Experience has proved, what *a priori* considerations should have foreseen, that few problems can be adequately resolved by the investigation of a single site".[6]

[1] Davis 2000; Jacobson and Bradshaw 1981.
[2] Andersen 1970; Sugita 1994, and references therein.
[3] Birks and Birks 1980.
[4] Olsson 1986; Roberts 1998.
[5] Bennett 2002; Moore *et al.* 1991.
[6] Walker 1971, p. 93.

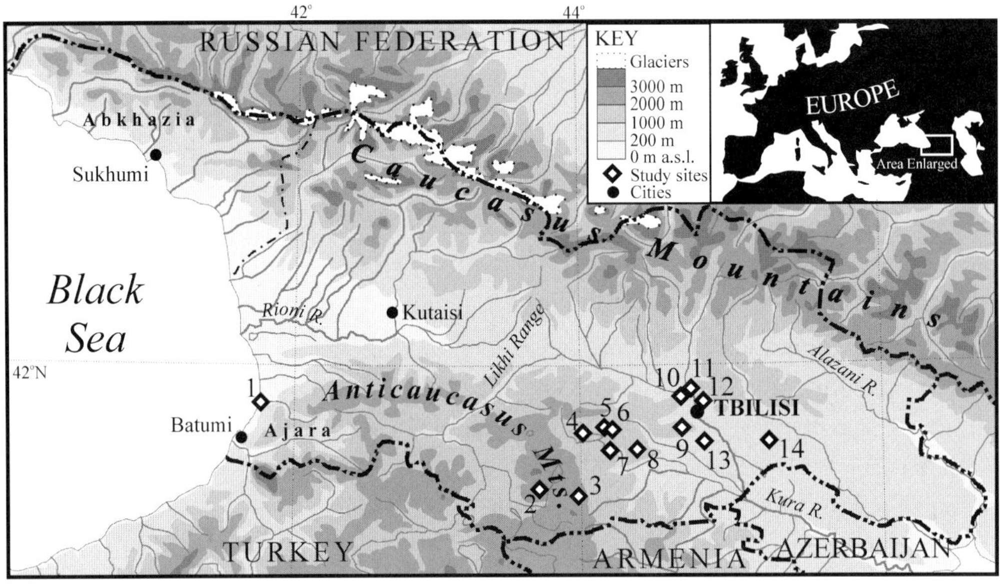

Figure 3.1. Study sites on a topographic map of Georgia. Sites numbered thus – 1: Ispani (see Connor *et al.* 2007); 2: Avchalagöl; 3: Khachalgora; 4: Lake Aligol; 5: Lake Bareti; 6: Lake Imera; 7: Trialeti Wetland; 8: Lake Cherepanov; 9: Tsavkisi Wetland; 10: Chili Lake; 11: Lake Jvari; 12: Patara Gldani Lake; 13: Lake Kumisi; 14: Sakhare Lake.

While site selection was limited to the sedimentary basins occurring in the landscape, the aim of studying vegetation change on a scale comparable to the archaeological record necessitated the selection of *small* lakes and wetlands.[7] Where possible, closed basins (*i.e.* without inflowing streams) were used because their pollen and sediment source-area is greatly simplified.[8] Sites with shallow standing water (0.5 to 1.5 m) were thus favoured, except when a boat or other water craft was available, and sites with very steep basin topography or those that were difficult of access were avoided.

Small lakes and wetlands with the appropriate characteristics were identified from 1:200,000 topographic maps; my colleagues and I then made a field expedition to each site to determine the suitability of its sediments for coring. At the same time, each basin's morphometry was recorded using a Global Positioning System (GPS) receiver. Where appropriate, lake terraces and other obvious geomorphic features were recorded and measured using a clinometer and tape measure.

[7] Jacobson and Bradshaw 1981; Prentice 1985; Sugita 1993; Sugita 1994.
[8] Dearing and Foster 1986.

VEGETATION SURVEYS

To satisfy the aim of comparing pollen and vegetation composition, the local vegetation of each site was described in some detail, because existing data on the study area's vegetation composition is particularly scanty.[9] Dr Giorgi Arabuli of the State Museum of Georgia and Dr Henrikh Avakov of the Palaeobiology Institute described the vegetation and identified plants according to standard nomenclature.[10] Percentage cover of the most common species was estimated subjectively using the ten-point Domin scale,[11] a variant of the Braun-Blanquet scale. The pollen source-area model of Sugita[12] was applied to define the area of dryland vegetation described on a palynologically meaningful scale. I found it useful to approximate the model by defining the 50 percent source area of most common pollen types as ten-times the lake basin radius. Thus for a lake 100 m in radius, the 50 percent pollen source-area encompasses a radius of 1 km.[13] The area of vegetation described was consequently larger for larger lakes and wetlands.

Separate cover abundance estimates were produced for aquatic, wetland, ground layer and forest canopy plants, producing total cover abundances in excess of 100 percent where forest was present. These percentages were then converted to a single total by calculating percentage forest area within each site's 50 percent pollen source-area from 1:100,000 topographic maps (**Fig. 3.2**). In several forested places we also undertook short belt transect surveys[14] to characterise the forest's composition and age structure.

Surveying vegetation for pollen-based studies in a heterogeneous landscape presents particular challenges.[15] Though the study's environmental gradient passes from lowland semidesert-steppes through the forest belt to mountain grasslands, in reality each of these vegetation types may occur within the pollen 'catchment' of any and all of the sites. In the forest belt, for instance, it is not uncommon to encounter considerable grassy clearings and patches of xerophytic vegetation. Cover abundance values are therefore compositional estimates for 'vegetation areas'[16] in the 50 percent pollen source-area (**Fig. 3.2**). This approach was considered the most appropriate and practical solution to the problem of semi-quantitatively relating heterogeneous vegetation to pollen spectra at an equivalent spatial scale.

[9] Prof. R. Gagnidze, Botanical Institute, pers. comm.
[10] Czerepanov 1995. Plant nomenclature throughout follows this scheme.
[11] Kent and Coker 1992.
[12] Sugita 1993.
[13] *Cf.* Sugita *et al.* 1999.
[14] Kent and Coker 1992.
[15] Sugita 1994.
[16] *Sensu* Malmer and Regnell 1986.

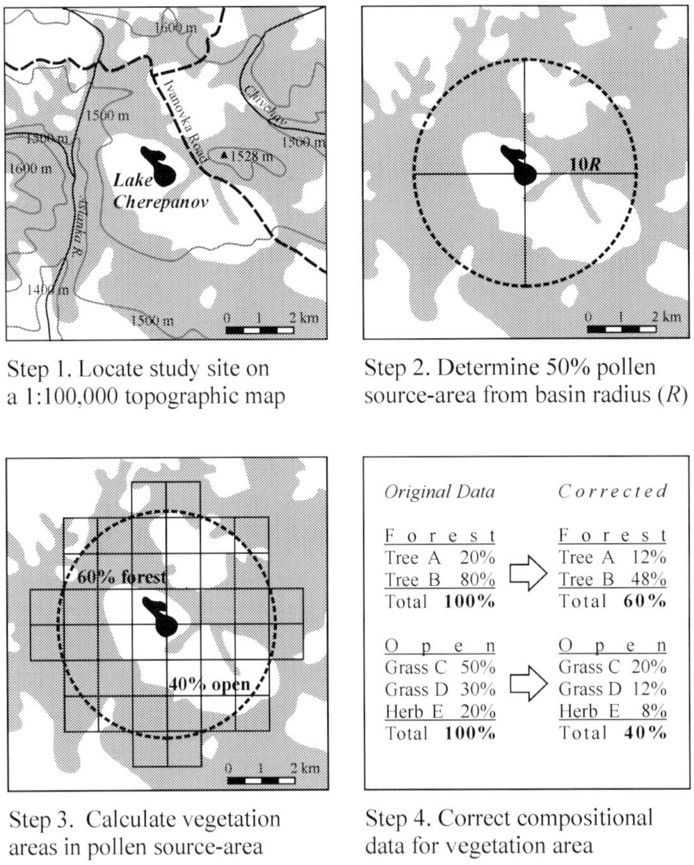

Figure 3.2. An example of the method used in this study to obtain estimates of vegetation cover abundance in the 50% pollen source area. Shading represents forested areas. The site shown is Lake Cherepanov (see Chapter Four).

Sediment Sampling

Heterogeneity of sedimentary environments in the study area necessitated sampling with a variety of hand-held coring equipment. Peats and organic-rich lake sediments were collected with a stainless-steel Russian D-section corer[17] with an internal diameter of 5 cm and a barrel 50 cm long. Where this was insufficient to penetrate consolidated silty sediments, a 'sticky-clay' auger (Dormer Engineering, Australia) was used instead. My Swiss colleagues collected one core (Lake Bareti) using their own piston corer, constructed of a rigid perspex tube 1.2 m in length. Surface sediments were collected either by hand or, where the water depth was too great, using the D-section corer. To avoid sampling bias, each surface sample was an aggregation of several sub-samples collected over a 5 × 5 metre area.

[17] Jowsey 1966.

Each type of coring device has its characteristic advantages and disadvantages.[18] The D-section corer is invaluable because it removes stratigraphically intact sections of sediment with minimal disturbance and contamination, but the corer is limited to soft sediments. The 'sticky-clay' auger penetrates the hardest sediments with ease, but disturbs their stratigraphy and may introduce contamination from younger sediments. Piston corers fall somewhere between the two extremes of the D-section corer and the auger, allowing for the preservation of stratigraphy, but compacting the sediment and introducing the possibility of contamination by repeatedly coring down the same hole.[19]

Cores were collected with a 10-cm overlap when the D-section corer was used. In hard sediments it was unfeasible to obtain such an overlap, so contiguous cores were gathered from the same hole. Great care was taken to avoid contamination during fieldwork. Samples, once collected, were immediately wrapped in plastic film or placed in plastic sample bags, labelled and stored in a cool, dark place until they could be transferred to refrigerated storage in Melbourne. Duplicate samples of 1 cm^3 for physical and microscopic analyses were initially taken from each core at a relatively coarse interval, which was later refined. A compromise had to be reached between spatial and temporal resolution, since this study is primarily concerned with changing small-scale spatial patterns within a larger geographic region, rather than the detailed vegetation history of a single lake or mire.

Physical Analyses

The physical structure of sediments reveals much about their genesis and the environments in which they formed.[20] Sediment colour was defined using Munsell colour charts and sediment description followed the standard system developed by Troels-Smith.[21] This widely-adopted scheme classifies sediments into a number of types based on physical characteristics, degree of humification and sediment composition.[22]

In addition to these descriptive data, the moisture and organic content of the sediments was determined using the method of Bengtsson and Enell,[23] based on overnight drying at 100 °C and ignition for 2 hours at 550 °C. The magnetic susceptibility of 10 cm^3 sediment aliquots from several cores was measured using a Bartington Magnetic Susceptibility Meter (Bartington Instruments MS-2). Unfortunately, this equipment developed a fault before all the cores could be analysed, so magnetic susceptibility profiles are lacking from some sites. Where carbonates were present in the sediment, this was recorded semi-quantitatively based on their reaction to the addition of 10 percent hydrochloric acid (HCl).

[18] Faegri and Iversen 1989.
[19] Aaby and Digerfeldt 1986; Birks and Birks 1980; Wright 1980.
[20] Dearing and Foster 1986.
[21] Troels-Smith 1955.
[22] Birks and Birks 1980.
[23] Bengtsson and Enell 1986.

MICROSCOPIC ANALYSES

Before chemical pretreatment for microscopy, two *Lycopodium* marker spores (University of Lund, batch 938934, n=10,679) were added to each 1 cm^3 sediment sample for the calculation of pollen concentrations.[24]

Most of the samples collected for this study contained a high proportion of silt and other inorganic particles. Hydrofluoric acid digestion[25] of samples for pollen analysis was quickly abandoned because of violent reactions upon addition of HF to some samples. Instead, gravity separation was performed on all samples, using sodium polytungstate ($3Na_3WO_4.9WO_3.H_2O$) solution with a specific gravity of 2.0.[26] Also, boiling samples with low organic content in a 10 percent solution of potassium hydroxide (KOH) proved to be unnecessary and damaging to fragile pollen preserved in compact, silty sediments. Silts and clays were instead deflocculated overnight in a 3–5 percent solution of sodium hexametaphosphate ($Na_2P_2O_7$).

Samples were subjected to standard acetolysis, stained with safranin and mounted in glycerol. I identified all pollen and spores at a magnification of 400× on an Olympus binocular microscope. My primary source for identifications was the excellent pollen key of Moore *et al.*,[27] supported by the collection of Caucasian type material housed in the Palynology Laboratory at the University of Melbourne and several published pollen atlases.[28] In some cases, difficult pollen grains were referred to Dr Eliso Kvavadze (in Georgia) or Jacqueline van Leeuwen (in Switzerland) for identification. An average of 1100 pollen grains were identified in each sample, but this figure varied considerably depending on differences in pollen preservation and concentration from core to core.

In addition to pollen, microscopic charred particles encountered on pollen slides were quantified by the point-count method.[29] This method provides a probability-based estimate of the area of charcoal in the sediment sample and is a proxy for fire. Fire history has received little or no attention in palynological literature from the Caucasus and consequently almost nothing is known of the palaeoecological importance and successional effects of fire, even though fire clearly is commonplace in steppe landscapes in the Caucasus today.[30] Methods for quantifying sedimentary charcoal are many and their selection depends on the questions being asked of the data.[31]

[24] Moore *et al.* 1991. The complete pollen preparation method used here is detailed in Appendix C.
[25] Faegri and Iversen 1989.
[26] Hart 1988.
[27] Moore *et al.* 1991.
[28] Beug 2004; Bobrov *et al.* 1983; Chester and Raine 2001; Gogichaishvili *et al.* 1975; Kuprianova and Aleshina 1972; Kuprianova and Aleshina 1978; van Zeist and Bottema 1977.
[29] Clark 1982.
[30] Connor *et al.* 2004.
[31] Innes *et al.* 2004; Patterson *et al.* 1987; Rhodes 1998, and references therein.

Point-count estimation is distinguished from other methods[32] in that it rapidly quantifies the *area* rather than *number* of charred particles on a pollen slide (**Fig. 3.3**). Hence it is not biased by the fragmentation of charred particles that occurs during pollen extraction or through sediment diagenesis.[33] During counting, microscopic charcoal were grouped into size classes of 5–20, 20–40, 40–60, 60–80 and >80 μm, as charcoal size broadly reflects a fire's proximity to the site of deposition.[34]

Other microscopic plant and animal remains were identified from the descriptions and microphotographs provided by van Geel and co-workers.[35] Not surprisingly, the identity of many such remains in Caucasian deposits could not be found in the published literature, so only common taxa were recorded. Only those taxa that could be confidently identified and have well-defined ecological significance are presented in pollen diagrams. These remains include coprophilous fungi (*Sporormiella* and *Podospora*), thermophilous members of the Zygnemataceae (*Penium, Spirogyra* and *Mougeotia*), green algae (*Botryococcus*), *Ceratophyllum* leaf spines, *Tilletia sphagni* spores, Rhabdocoela flatworm cocoons, *Pediastrum* coenobia, conifer stomata and chironomid remains.[36]

Figure 3.3. Simulated pollen slides with charcoal particles (black). Point-count estimation using the grid gives a result of 33 points for the left-hand graph and 31 points for the right (out of a possible 121 points). In contrast, charcoal counts are vastly different (36 versus 12), even though the charcoal area on both slides is the same. While charcoal counts and charcoal area are positively-correlated (Tinner and Hu, 2003), point-count estimation retains many theoretical and practical advantages over raw charcoal counts.

[32] *E.g.* Tinner and Hu 2003.
[33] Clark 1982.
[34] Blackford 2000.
[35] van Geel 1978; van Geel and Aptroot 2006; van Geel *et al.* 1981; van Geel *et al.* 2003.
[36] Komárek and Jankovská 2001; van Geel and Aptroot 2006; van Geel *et al.* 1981; van Geel and van der Hammen 1978.

Radiocarbon Dating

All of the radiocarbon dates presented in this study were determined through accelerator-mass-spectrometry (AMS). Prior to pre-treatment, samples were picked over for contaminating rootlets and silt-rich samples were gravity separated (see above) to remove excess inorganic material. Most samples were pre-treated using the standard acid-base-acid (ABA) method, which involves carbonate removal with 2 M HCl, then repeated humic acid removal with increasing concentrations of sodium hydroxide (NaOH), followed by a final HCl treatment.

A few samples were not prepared in this way, and were pre-treated to isolate only the pollen fraction of the sediment or using only hydrochloric acid washes. Pretreatment for dating of pollen concentrate involves repeated bleaching and sulphuric acid treatment to remove organic detritus and chitin from the sample,[37] although the absolute ages produced from this method have not always been consistent with those produced by dating macroscopic plant remains.[38] Repeated hydrochloric acid washes are the simplest method and are particularly effective in carbonate-rich sediments, but can render ages that are too recent on account of humic acid infiltration from higher stratigraphic levels.[39] As such, sample pre-treatment can have a strong bearing on the resulting age determination.

Once pre-treated, the samples were combusted to convert the remaining organic carbon to carbon dioxide. This process liberates the carbon from its contaminants and is followed by graphitisation (conversion to a graphite target) for dating by AMS.[40] Samples preceded by codes OZG and OZH were dated at the Australian Nuclear Science and Technology Organisation AMS facility, Lucas Heights, Sydney.

All dates reported in this study are given in calendar years before 1950 (cal BP), determined by calibrating conventional dates to the tree-ring dataset in the computer program CALIB 4.3.[41] Conventional radiocarbon dates, where reported, are always followed by the suffix '^{14}C bp.' An AD – BC chronology is adopted when discussing archaeological data.

Numerical Analyses

Several numerical analyses have been employed in this study and are discussed at greater length in the sections in which they are presented. The principal methods are all multivariate analyses and have been used to identify the prevailing trend or trends in spatial and temporal data.

37 Brown *et al.* 1989.
38 Kilian *et al.* 2002.
39 Olsson 1986.
40 Hua *et al.* 2001.
41 Stuiver and Reimer 1993.

Detrended Correspondence Analysis (DCA),[42] an indirect gradient analysis, has frequently found application in ecological and palaeoecological research because it assumes unimodal species response to environmental gradients,[43] sample scores represent species-compositional turnover (β-diversity)[44] and, perhaps most attractively, it "works well in practice with real-life data".[45] It is a variant of Correspondence Analysis (CA) that uses detrending-by-segments to remove the 'arch effect' (curvilinear distortion) and axis compression associated with CA, PCA and other methods.[46]

Following its introduction, several ecologists and statisticians seriously questioned the robustness of DCA as an ordination method. Criticisms revolved around its sensitivity to rare species, possible distortions introduced by detrending, poor results associated with the second and higher-order gradients, and unstable results depending on sample input order.[47] At least some of these problems were solved more recently by Oksanen and Minchin,[48] who identified the software bug that caused sample order to bear heavily on the ordination results, causing instability in the second and third ordination axes.

Oksanen and Minchin's de-bugged 'super-strict' version of DCA was used in this study to analyse percentage pollen data, through the statistical software package, PC-ORD 4.25.[49] Rare species were down-weighted and axes were rescaled to represent species turnover. The resulting trends can be compared ecologically to climatic, topographic, anthropogenic, edaphic and other data to determine the dominant controls on pollen composition now and in the past. The selection of an *indirect* ordination technique is important where the major environmental controls on species compositional change are unknown or poorly understood.[50]

A second method employed in this study, Two-Way Indicator Species Analysis (TWINSPAN),[51] is a divisive classification method which produces a hierarchy of dichotomies based on a weighted averaging ordination similar to DCA. PC-ORD's de-bugged version was used to classify samples in this study.[52] TWINSPAN works by splitting the entire set of samples into two groups at each division, according to each sample's position on the ordination axis. At each split, the 'pseudospecies' (a quantitative measure of presence/absence) that best differentiates the two groups is given as an

[42] Hill and Gauch 1980.
[43] ter Braak and Prentice 1988.
[44] Gauch 1982.
[45] Birks *et al.* 2004, p. 537.
[46] Hill and Gauch 1980; Kent and Coker 1992.
[47] Dargie 1986; Minchin 1987; Tausch *et al.* 1995; ter Braak 1995; van Groenewoud 1992; Wartenburg *et al.* 1987.
[48] Oksanen and Minchin 1997.
[49] McCune and Mefford 1999; Oksanen and Minchin 1997.
[50] ter Braak and Prentice 1988.
[51] Hill 1979.
[52] McCune and Mefford 1999.

indicator for that group. These indicators can greatly assist the ecological interpretation of the resulting groups. They also allow samples that were not included in the original dataset to be classified when, as in this study, the ordination is performed on untransformed data.

Because both DCA and TWINSPAN are based on the chi-squared (χ^2) dissimilarity co-efficient, these two methods have often been criticized for their greater sensitivity to rare species compared to other methods such as Canonical Correspondence Analysis (CCA), Principal Components Analysis (PCA) and Non-metric Multidimensional Scaling (NMS).[53] The chi-squared dissimilarity coefficient's sensitivity to rare species can be reduced in DCA by down-weighting and in TWINSPAN by selecting cut-levels that give preference to more common taxa.[54] There are also several *a priori* reasons for favouring signal-to-noise dissimilarity coefficients such as chi-squared over others in palaeoecological applications.[55]

Unlike living organisms, palaeoecological data is a fossilised 'death assemblage' that represents the community of living organisms in a way heavily distorted by factors such as the sedimentary environment, fossilization processes, fossil production and dispersion.[56] One of the most significant distortions in any pollen data-set is the background 'noise' created by regional pollen, which effectively obscures changes on a more local scale.[57] The chi-squared dissimilarity coefficient has a noise-reducing capability that helps to isolate a 'signal' that may otherwise be obscured by regional pollen rain.[58]

To demonstrate this capacity, **Figure 3.4** presents the results of NMS and DCA applied to the same pollen dataset and related to environmental gradients. DCA extracted strong gradients on both axes and ordered samples in an ecologically interpretable way, whereas NMS produced only one climatic gradient, with a much noisier sample scatter that is poorly related to biogeographical patterns and much more susceptible to the presence or absence of rare species. Although it has been argued that "DCA is not a statistically elegant technique",[59] it may perform better when faced with noisy or highly heterogenous data and appears to be quite robust even when its assumption of unimodal response to environmental gradients is violated.[60]

[53] See discussions in Clark 1993; Minchin 1987; Peet *et al.* 1988; Wartenburg *et al.* 1987. CCA is not considered appropriate for ordination of data in this study because there are no *a priori* grounds to assume that known environmental variables exert direct control over pollen composition at different sites. PCA can also be dismissed, as there is no evidence that pollen taxa respond to environmental gradients in a linear fashion. The appropriateness of NMS is discussed later.

[54] Vermeersch *et al.* 2003.

[55] Jackson and Williams 2004; Overpeck *et al.* 1985; Prentice 1980; Prentice 1986.

[56] Birks and Birks 1980.

[57] Davis 2000; Jacobson and Bradshaw 1981.

[58] Jackson and Williams 2004; Overpeck *et al.* 1985; Prentice 1980.

[59] Peet *et al.* 1988, p. 924.

[60] Ejrnæs 2000; Ejrnæs and Bruun 2000; Prentice 1986.

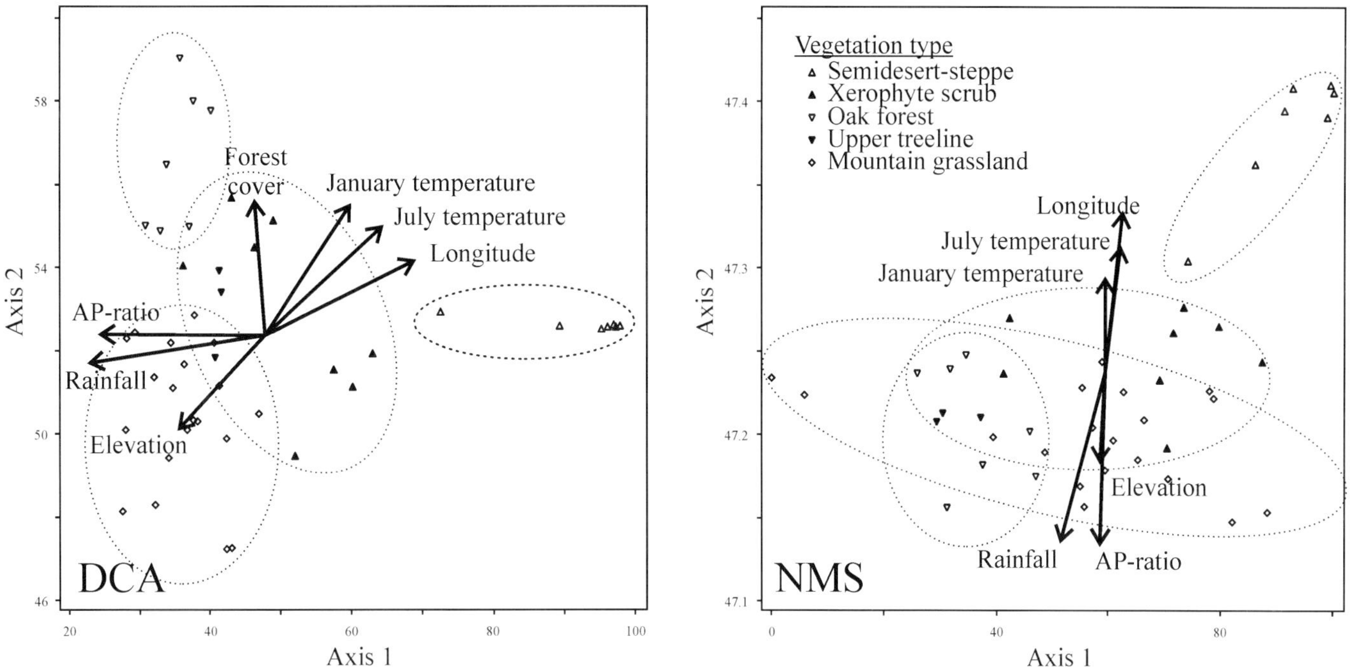

Figure 3.4. Comparison of DCA and NMS results presented as sample scores (points) with correlated environmental variables (arrows). Ellipses enclose samples from semidesert-steppe, treeline (upper and lower), forest and mountain grassland vegetation. Data from Chapter Four. NMS result based on Sørensen distance and a two-dimensional solution.

Most palaeoecological applications of multivariate analyses are, by definition, heuristic, since we have no way of directly measuring the environmental gradients that species were responding to in the past. Indeed, the reason that multivariate analyses are used in this study is to reconstruct past environmental changes from changing species abundances. Circular arguments in interpretation are easily avoided by referring to the present-day ecological tolerances of the species in question and, applying uniformitarian principles, making the assumption that these tolerances were much the same in the past as they are now.

The idea that the 'present is the key to the past' is a theme carried through all the numerical analyses in this study. In Chapter Four, DCA is used to identify the prevailing environmental gradients that shape the composition of pollen samples in the study area presently. Each pollen taxon has a specific relationship to each of these gradients: the stronger the relationship between a pollen taxon and a gradient, the more heavily 'weighted' it will be (though this will also depend on its ubiquity – more common taxa require less weight than rare taxa to have the same effect.) In subsequent chapters these weights are applied to fossil pollen data through Weighted Averaging,[61] allowing environmental gradients to be reconstructed for the past, firmly anchored to present bioclimatic patterns. If particular pollen assemblages in the past have no modern analogue,[62] Weighted Averaging may still reconstruct meaningful, albeit distorted, trends, as it places each sample at the "centre of gravity"[63] of the species in the ordination space. It stands to reason that a comprehensive modern dataset will produce reconstructions of greater certainty.

Presentation of Results

Pollen data are conventionally calculated and graphed as percentages of a sum that includes all trees, shrubs and terrestrial herbs, forbs and grasses.[64] Pollen of wetland and aquatic plants are excluded because these plants often become incorporated into the sediment itself, and hence their pollen is over-represented. For deep lakes and perennially-wet bogs, the calculation of the pollen sum presents no particular problem. In shallow lakes and wetlands prone to desiccation, however, it is possible that terrestrial plants grew very near to or on the sediment surface itself at various times. Taphonomic processes associated with semi-arid environments can lead to pollen degradation and a further distortion of the death assemblage.[65]

Basin size also plays a very important part: small basins have a pollen assemblage dominated by locally produced pollen, while larger basins contain a more regional assemblage.[66] According to the model of Jacobsen and Bradshaw, regional pollen prevails at sites with a diameter of 300 m or more; extra-local pollen dominates sites around 200 m in diameter; local pollen is most prevalent in sites less than 100 m in diameter. Sugita's simulations of pollen representation in heterogenous vegetation suggest that small lakes (radius 50 m) are dominated by a local signal, which is not apparent in larger lakes (radius 750 m).[67] This means that the entire pollen assemblage may be strongly affected by lake level variations related to hydrological change, a process illustrated in **Figure 3.5**.

Because the study area is largely semi-arid, and probably was so in the past,[68] sums for pollen diagrams have been calculated in a variety of ways. Where there appears to have been major pollen source-area or post-depositional changes due to lake level variations recorded lithologically, percentages have been calculated from separate arboreal

[61] Whittaker 1967.
[62] Anderson *et al.* 1989; Birks 1973; Webb 1980.
[63] Prentice 1986, p. 783.
[64] Faegri and Iversen 1989; Wright and Patten 1963.
[65] Havinga 1964; Havinga 1967; Horowitz 1992; Moore and Stevenson 1982.
[66] Jacobson and Bradshaw 1981; Prentice 1985; Sugita 1993.
[67] Sugita 1994.
[68] Wick *et al.* 2003.

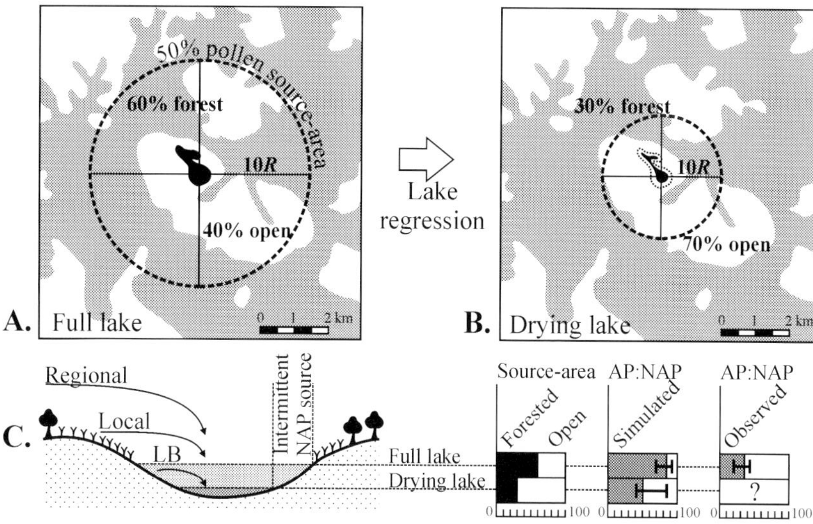

Figure 3.5. A model of pollen behaviour under changing lake levels. In this example (**Figure 3.2**), when the lake is full (A) the 50% pollen source-area contains 60% forest and 40% open landscapes. A lake regression shrinks the source-area (B), which is now dominated by open land. The effect of this change is shown (C). The contracted lake is theoretically more sensitive to changes in local pollen than the full lake (Sugita, 1994). Local pollen may also be contributed by vegetation growing on the exposed lake bed (LB), constituting an ephemeral source of non-arboreal pollen (NAP). Graphs at right compare source-area openness to pollen composition. AP:NAP ratios simulated for southern Sweden (**Figure 4.1**) can be applied to predict that arboreal pollen (AP) will decline at this site as the lake level drops. Observed values (Chapter Four) may indicate that the lake-bed component (LB) is an important pollen source in seasonally fluctuating lakes.

and non-arboreal sums. Where no significant changes in source area have been detected, the pollen sum is calculated conventionally. The particular sum used is reported alongside the results from each site. These adjusted pollen sums are only for the purposes of displaying the data and were not used in Weighted Averaging.

Pollen diagrams were all drawn using Psimpoll.[69] The pollen diagrams have been subdivided stratigraphically into pollen assemblage zones based on optimal splitting by information content.[70] The number of zones was determined by the reduction in variance produced by each split, where the variance reduction exceeded that anticipated by the 'broken stick' statistical model.[71] This means that the placement and number of pollen assemblage zones has a robust statistical basis, independent of any regional stratigraphic, palaeoclimatic or chronological schemes. Palaeoecological research in the study area must move beyond such schemes if it is to truly analyse past vegetation changes in an objective manner.

[69] Bennett 2002.
[70] Bennett 1996; Birks and Gordon 1985.
[71] Bennett 1996.

CHAPTER FOUR

MODERN POLLEN-VEGETATION RELATIONSHIPS

> Know the changes of the times, remember the proverbs, and liken the future to the past.[1]

INTRODUCTION

The aim of palaeoecology is to reconstruct past ecosystems. As in many of the natural sciences, this aim can only be fulfilled by making certain assumptions in relation to the physical environment, assumptions collectively known as uniformitarian principles. In part, these state that natural processes are governed by the laws of physics, which do not change with the passing of time, even though the processes themselves and the rate at which they occur may vary.[2] On this premise, palaeoecological data represent past ecology in the same way that equivalent data collected today represent the ecology of organisms living (or dying) now.

To reconstruct the environmental conditions that led to certain assemblages of organisms existing in the past, it is most feasible to infer these conditions from a solid knowledge of how they operate on the same organisms in the present-day. Furthermore, because of processes that shape and distort fossil assemblages, the abundance of fossils in a sedimentary deposit does not represent the organisms that produced them in any simple, straightforward manner.[3]

Pollen analysis is used to reconstruct the composition and character of plant communities in the past. With a firm understanding of the ecology of the vegetation in the present-day and the pollen it produces, past pollen assemblages may be matched to modern analogues, lending greater certainty to the reconstruction.[4] The environmental conditions of the past can be inferred if the conditions associated with this assemblage are known in the present. And while there are errors inherent in this process, a study of modern pollen-vegetation relationships is a necessary prelude to the interpretation of fossil pollen data.

[1] *The Georgian Chronicles*, translation by Thomson 1996, pp. 341–342.

[2] Birks and Birks 1980.

[3] Birks and Birks 1980.

[4] Anderson *et al.* 1989; Birks 1995; Jackson and Williams 2004; Overpeck *et al.* 1985; Wright 1967.

Traditionally, studies of how modern pollen is represented in sediments have made use of moss cushion or surface soil samples and related these to the surrounding vegetation.[5] Several such studies have been previously undertaken in the Caucasus region, using soil or moss samples collected over long altitudinal transects.[6] While they reveal important broad-scale changes in pollen composition, the application of these studies to the interpretation of fossil pollen is limited. This is because few lake and wetland sites usually used in palaeoenvironmental reconstructions have a pollen source area that is comparable to soil or moss samples.[7] Indicator pollen types derived from soils and mosses cannot be strictly applied to the interpretation of pollen spectra in lake and wetland sediments.

The pollen source-area of soils, mosses, wetlands and lakes can be strikingly different. Moss 'polsters' beneath forest canopies are dominated by pollen originating in the surrounding 20–30 m; small forest openings by pollen from the surrounding 50–100 m; while sediments in the centre of a large lake are typically dominated by a regional component with a source area of many hundreds of kilometres.[8]

These source-area concepts are based on conditions in Northern Hemisphere temperate forests and their applicability to areas where semidesert-steppes, scrub, savanna and mountain grasslands are characteristic vegetation types may be questioned.[9] Sites in open vegetation have a greater potential source area than sites in forest, due to the removal of the 'trunk space' or 'gravity' component and the absence of filtration by arboreal vegetation.[10]

Highland sites are often dominated by arboreal pollen generated by forests growing at lower altitudes,[11] so determining the position of tree lines palynologically is difficult and potentially flawed.[12] In mountain environments, joint analyses of conifer stomata and pollen have shown that arboreal pollen percentages and influx rates increase as forests advance toward a study site, but often decline once forests are present *in situ*.[13] Human-induced forest disturbance can also cause a greater influx of tree pollen, increasing arboreal pollen percentages, paradoxically, during deforestation events.[14]

[5] *E.g.* Bradshaw 1981; Bunting 2002; Davis 1980; Elliot 1999; Gajewski *et al.* 2002; Islebe and Hooghiemstra 1995; Paez *et al.* 2001; Tinsley and Smith 1974; Wright *et al.* 1967.

[6] *E.g.* Klopotovskaya 1973; Kvavadze 1993; Kvavadze 1999a; Kvavadze and Stuchlik 1990; Kvavadze and Stuchlik 1991; Kvavadze and Stuchlik 1993; Peterson 1983; Stuchlik and Kvavadze 1993; Stuchlik and Kvavadze 1998; Yazvenko 1991.

[7] Hicks 2001; Kvavadze and Efremov 1995; Seppä and Bennett 2003.

[8] Andersen 1970; Calcote 1995; Jacobson and Bradshaw 1981; Janssen 1966; Janssen 1986; Prentice 1985; Sugita 1993.

[9] Davies and Fall 2001.

[10] Jacobson and Bradshaw 1981; Tauber 1965; Tauber 1967.

[11] Fall 1992; Frei 1997; Haberle and Bennett 2001; Islebe and Hooghiemstra 1995; Markgraf 1980.

[12] Birks and Birks 2000.

[13] Ammann and Wick 1993; David 1997; Pisaric *et al.* 2000.

[14] Aaby 1988.

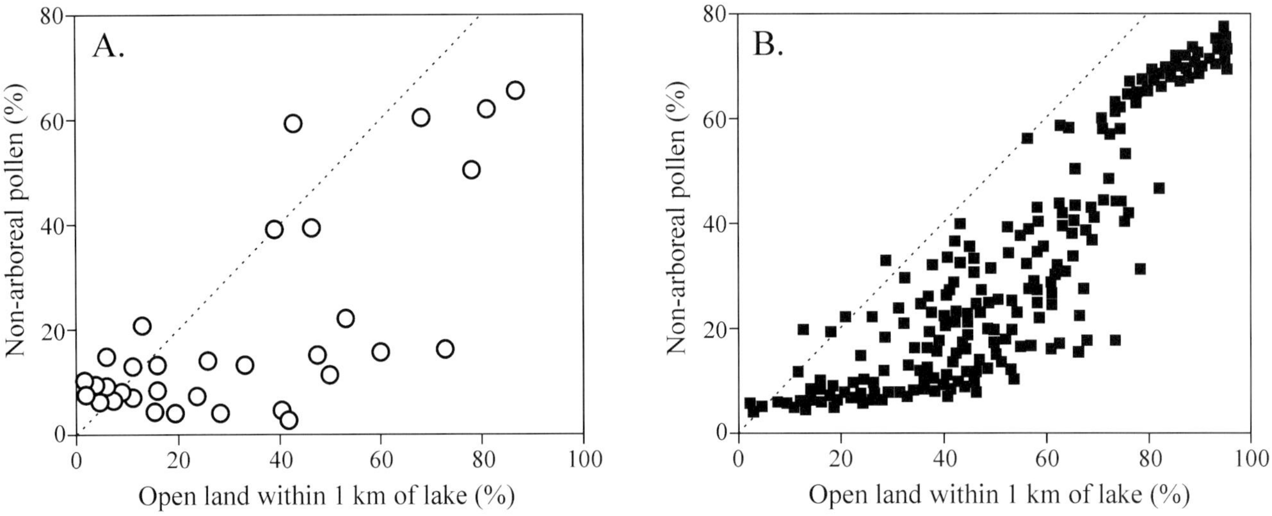

Figure 4.1. Non-arboreal pollen (NAP) percentages plotted against the percentage of open land within a kilometre of a one-hectare lake. A: observed relationship; B: simulated relationship; dashed line: direct linear relationship (simplified from Sugita *et al.*, 1999).

The relationship between arboreal pollen abundance and forest abundance is therefore non-linear (**Fig. 4.1**), influenced heavily by how pollen disperses in various environments and the density and composition of the vegetation itself.

In the semi-arid parts of the mountainous Caucasus region, accurate differentiation of open versus forested landscapes from pollen data is crucial for reconstructing past vegetation, contextualising a rich archaeological record and developing a model of Holocene palaeoclimatic fluctuations. Hence the aims of this chapter are:

To study the relationships between the present-day vegetation of the study area and the pollen that it produces, by comparing vegetation survey data with modern pollen spectra from lakes and wetlands.

To determine whether treeless and forested environments can be differentiated using pollen data, and if there are pollen indicator taxa specific to certain vegetation types.

To examine statistical relationships between environmental gradients and pollen spectra that may assist interpretation of palaeoecological data in terms of past climate and land-use.

My approach to satisfying these aims is both descriptive and experimental. Much of the following chapter is descriptive, in terms of data presentation and statistical analyses. The experimental part is essentially a model-testing exercise, in which a classification of the pollen data is validated using additional data. The results are then synthesised into a discussion on pollen representation and environmental gradients.

Study Sites and their Pollen-Vegetation Relationships

The study sites are located along an altitudinal transect through Southern Georgia and include places surrounded semidesert-steppe, xerophytic scrub, oak-hornbeam forest and mountain grassland vegetation (**Fig. 4.2**). This transect does not attempt to capture the full range of vegetation types in Georgia since these are discussed in Chapter Seven. Instead, it is a small-scale exploration of pollen-vegetation relationships within the specific study area based on original data.

From the outset it must be said that the vegetation that surrounds each of the study sites is and has been heavily influenced by human activities. So, while the aim of analysing surface samples is to obtain modern analogues for past pollen spectra, this aim cannot always be fulfilled with precision in the present-day vegetation.

Eight sites were initially chosen for study. Their locations are shown in **Figure 4.2**, and some associated characteristics are provided in **Table 4.1**. Forest area was calculated using the method described in Chapter Three. Forest area data in **Table 4.1** represent forest cover within each site's 50 percent pollen source-area.

Pollen data from each of the eight sites is shown in **Figure 4.3** along with major altitudinal, temperature and precipitation gradients. It is immediately apparent that these gradients have a profound influence on pollen spectra. In the hot, dry lowlands

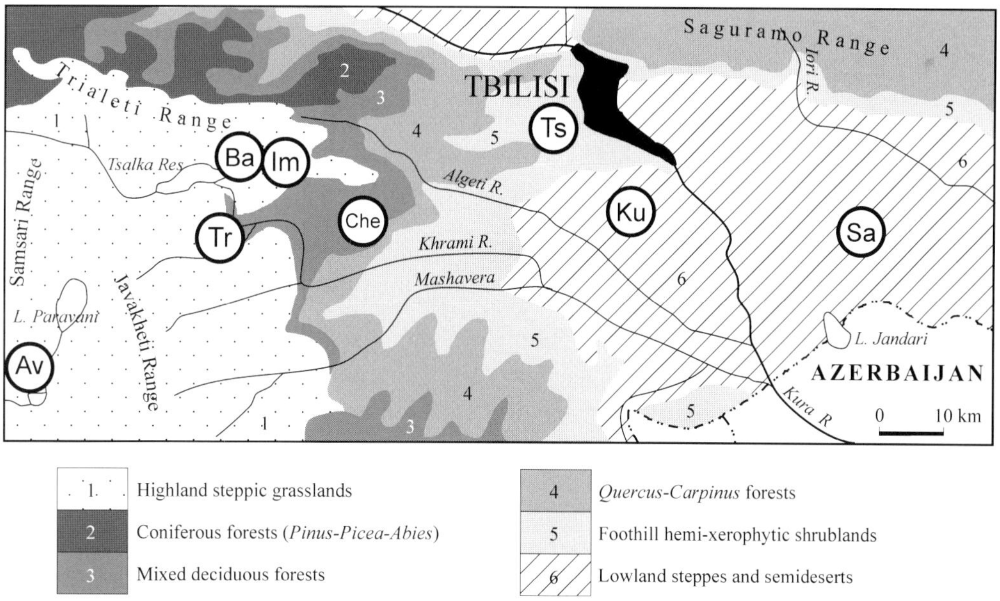

Figure 4.2. Study site locations on a map of principal vegetation zones in the study area. Site names given in Table 4.1. Adapted from Ketskhoveli (1959) and 'Landscape Map' in AN GSSR (1964).

of the eastern part of the transect, Chenopodiaceae pollen is absolutely dominant. This changes as one moves upslope, where the pollen of various deciduous trees (*Quercus, Carpinus, Fagus*) and grasses (Poaceae) become important. Finally, in the westernmost part of the transect, pine (*Pinus*) pollen prevails over the deciduous types.

Each of the study sites is discussed in more detail in the following sections, beginning with the lowland semidesert-steppe area.

Table 4.1. Site locations and associated environmental variables. See Chapter Three for an explanation of forest area calculations. Type: L- lake site, W- wetland site.

Site code	Site name	Type	Location (Lat.; Long.)	Elevation (m a.s.l.)	Basin avg. radius (m)	Forest area (%)	Vegetation type
SA	Sakhare	W/L	41°35′N; 45°19′E	800	190	0	Semidesert-steppe
KU	Kumisi	L	41°35′N; 44°50′E	469	1070	<1	Semidesert-steppe
TS	Tsavkisi	W	41°41′N; 44°45′E	1100	100	60	Steppe-forest
CHE	Cherepanov	L/W	41°34′N; 44°22′E	1400	210	70	Oak forest
TR	Trialeti	W	41°32′N; 44°07′E	1520	140	45	Upper tree-line
IM	Imera	L/W	41°39′N; 44°12′E	1610	140	0	Mountain grassland
BA	Bareti	L/W	41°39′N; 44°10′E	1630	620	<5	Mountain grassland
AV	Avchalagöl	W	41°20′N; 43°42′E	2050	417	<1	Mountain grassland

Sakhare Lake

Sakhare Lake is a small, saline wetland situated on the Iveria Plateau of Eastern Georgia (**Fig. 4.4**). The wetland is one of several salt lakes nestled in the striking, 'cuesta-like'[15] folds of the Udabno-Gareji landscape (**Fig. 4.5**). Its basin is around 600 m long and 200 m wide, but the lake's size varies according to the seasons. Geologically, the area is predominantly Mio-Pliocene sediment that has been tilted and subsequently heavily eroded.[16] Lake basins have formed through preferential erosion of soft, sandy strata interbedded between more resistant conglomerates (**Fig. 4.6**).

[15] Kvavadze and Todria 1992.
[16] Tsereteli 1971.

The wetland surface, when it is not seasonally inundated, is mainly vegetated with *Salicornia europaea* and *Phragmites australis*. Dense swathes of *Bolboschoenus maritimus* occur around the wetland margins along with other semidesert plants. The surrounding landscape is covered with semidesert-steppe vegetation, dominated by *Festuca valesiaca, Stipa pulcherrima, Artemisia lerchiana* and *Salsola nodulosa*. In describing the Udabno-Gareji landscape in the early 18th century, the Georgian geographer Prince Vakhushti Bagrationi observed that, 'there is no forest, if we do not count the Christ-thorn [*Paliurus*], sumac and suchlike; no rivers flow, and springs and lakes are rare, besides being hard and salty. Grassy even in winter…the summer is oppressively hot'.[17] His description is as valid today as it was then.

A comparison of the site's vegetation composition and pollen spectra (**Table 4.2**) demonstrates that Chenopodiaceae pollen are over-represented palynologically, that Poaceae pollen are well-represented, whereas many other herbaceous plants are severely under-represented, including *Artemisia*. Consistent with the predominantly treeless landscape, arboreal pollen is represented only in tiny proportions. *Polygonum aviculare*-type and unidentified Polygonaceae pollen occur frequently in the Sakhare Lake pollen spectra, even though no source plants were identified in our surveys. A previous botanical description of this site records the presence of the halophytic annual, *Polygonum argyrocoleum*, along with various Chenopodiaceae.[18]

The lands around Sakhare Lake are used predominantly as winter pasture. Cattle, goats and fat-tailed sheep graze here and their herders periodically burn the vegetation to promote fresh growth for their stock. The high charcoal concentrations in the Sakhare Lake sediments probably reflect this practice (**Fig. 4.3**).

Lake Kumisi

Lake Kumisi is also situated in the semidesert-steppe area, but it is a much larger lake than Sakhare Lake, with a maximum diameter of 2.6 km (**Fig. 4.7**). Kumisi is a salt lake impounded by the tectonic uplift of the surrounding Teleti and Yagluja Ranges and the suffosion of gypseous strata in the underlying sedimentary strata of Oligocene-Miocene age (**Fig. 4.8**).[19] Because the lake has no outlet and is fed only by intermittent streams, its water level fluctuates considerably from year to year. In the last two decades the lake water has been supplemented with water diverted from the Iori River, such that the water depth in the centre of the lake is now about 3 metres.[20] Local villagers recall the lake being much smaller and salt flats much more extensive prior to this irrigation scheme.

[17] Maruashvili 1956, p. 74.
[18] Ketskhoveli 1959, p. 297.
[19] Janelidze 1955; Tsereteli 1971.
[20] Kvavadze and Connor 2005.

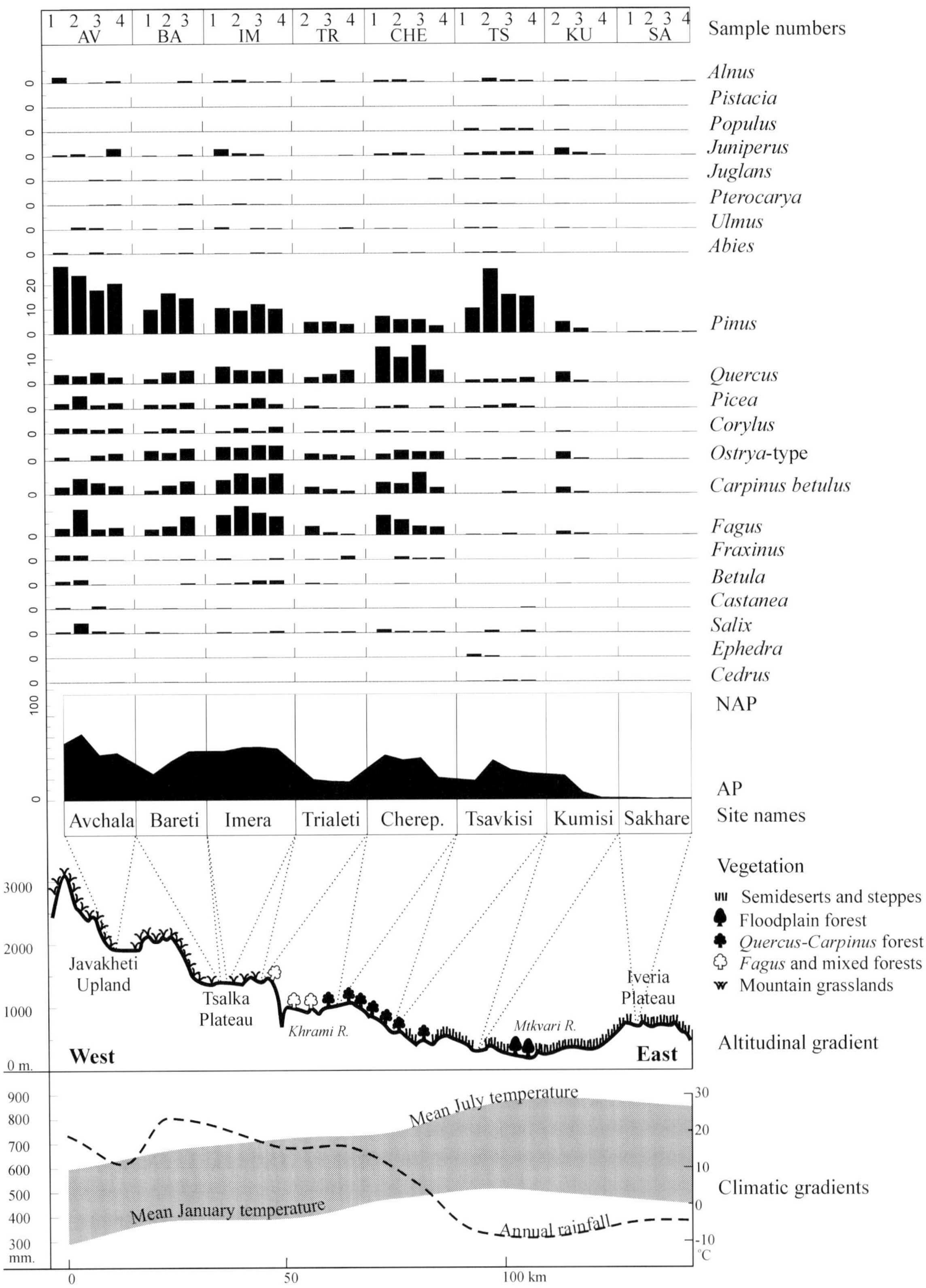

Figure 4.3 (a). Diagram of altitudinal and climatic gradients, AP-to-NAP ratios and percentages of selected AP taxa, calculated from the dryland pollen sum. Taxa are ordered according to TWINSPAN ordination. Climatic data from AN GSSR (1964). See Appendix E for the complete data-set.

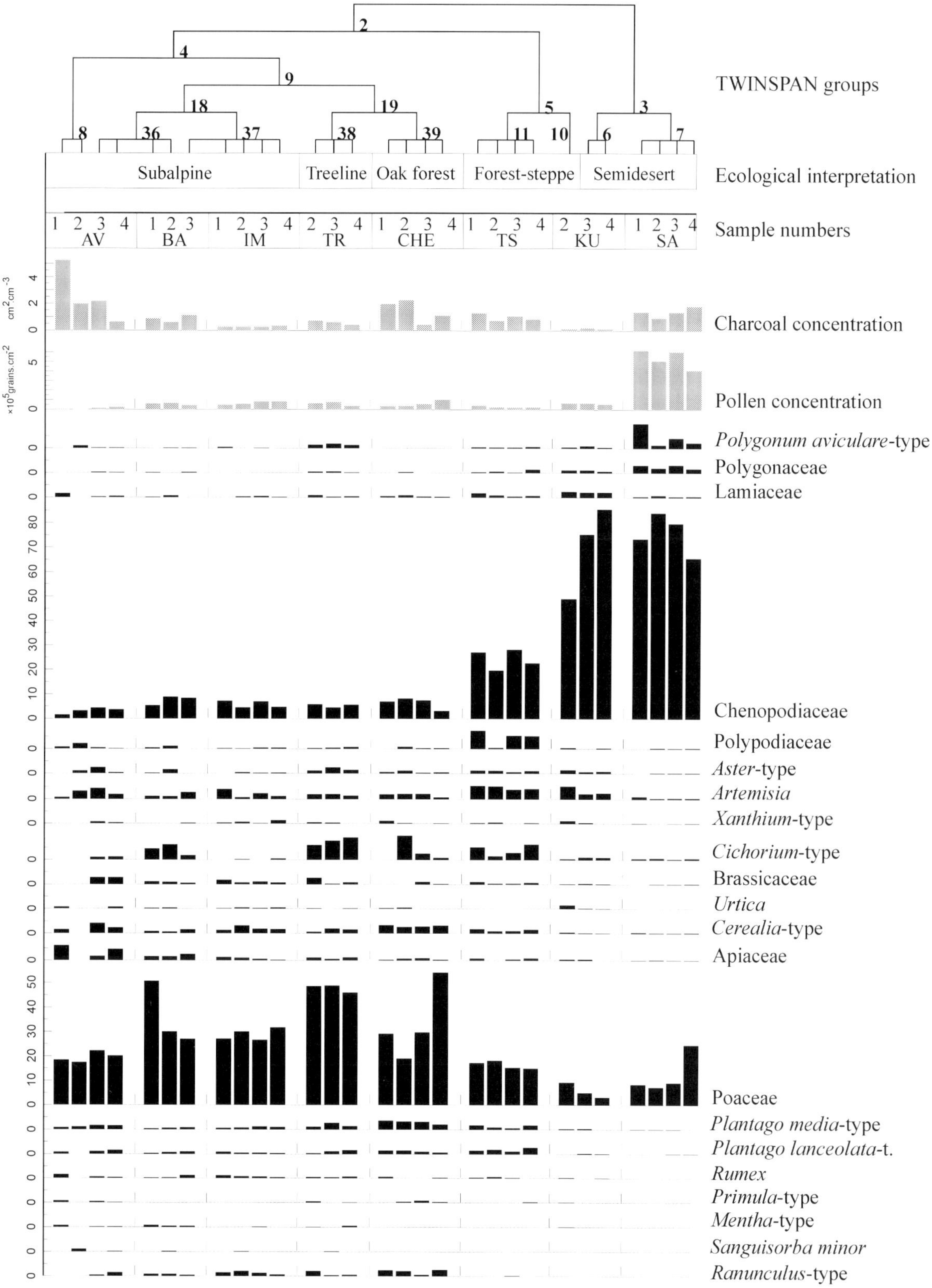

Figure 4.3 (b). Diagram of selected NAP percentages. Pollen and charcoal concentrations, as well as the TWINSPAN result, are shown at the top of the diagram. TWINSPAN indicators are given in Table 4.8.

Table 4.2. Comparison of plant community composition and pollen spectra at Sakhare Lake. The table compares cover abundance ('% cover') of plant species to the abundance of corresponding pollen types ('% pollen'), the values indicated by shading (see legend below). Pollen percentages are a mean value for the four samples studied.

Herbaceous plants:	% cover	% pollen	Pollen type represented
Achillea micrantha			-
Artemisia fragrans			*Artemisia*
Carduus albidus			*Aster*-type
Carduus seminudus			*Aster*-type
Cousinia macroptera			-
Cousinia orientalis			-
Onopordum acanthium			-
Senecio vernalis			*Aster*-type
Anchusa italica			Boraginaceae undiff.
Camelina microcarpa			-
Rorippa prostrata			-
Thlaspi arvense			*Hornungia*-type
Chenopodium album			Chenopodiaceae undiff.
Salsola nodulosa			Chenopodiaceae undiff.
Astragalus stevenianus			*Astragalus*
Lathyrus hirsuta			-
Medicago orbicularis			-
Trigonella orthoceras			-
Erodium cicutarium			*Erodium*
Salvia nemorosa			Lamiaceae undiff.
Salvia viridis			Lamiaceae undiff.
Papaver fugax			*Papaver*
Agropyron cristatum			Poaceae undiff.
Alopecurus pratensis			Poaceae undiff.
Bromus japonicus			Poaceae undiff.
Dactylis glomerata			Poaceae undiff.
Eragrostis starosselskyi			Poaceae undiff.
Festuca valesiaca			Poaceae undiff.
Hordeum leporinum			Poaceae undiff.
Lolium rigidum			Poaceae undiff.
Puccinellia gigantea			Poaceae undiff.
Stipa pulcherrima			Poaceae undiff.
Adonis aestivalis			-
Reseda lutea			-

Herbaceous plants:	% cover	% pollen	Pollen type represented
Verbascum phoeniceum			-
Wetland plants:			
Salicornia europaea			Chenopodiaceae undiff.
Bolboschoenus maritimus			Cyperaceae undiff.
Phragmites australis			Poaceae undiff.

Legend for **Tables 4.2** to **4.6**

	absent
	<1%
	1–10%
	11–33%
	34–75%
	76–100%

Figure 4.4. Photograph of Sakhare Lake in the Udabno region of Eastern Georgia (May 2002).

Figure 4.5. The semidesert-steppe landscape of Udabno Gareji. The tower of the Davit Gareji monastery is seen in the foreground, and exposures of Mio-Pliocene sedimentary rock in the background.

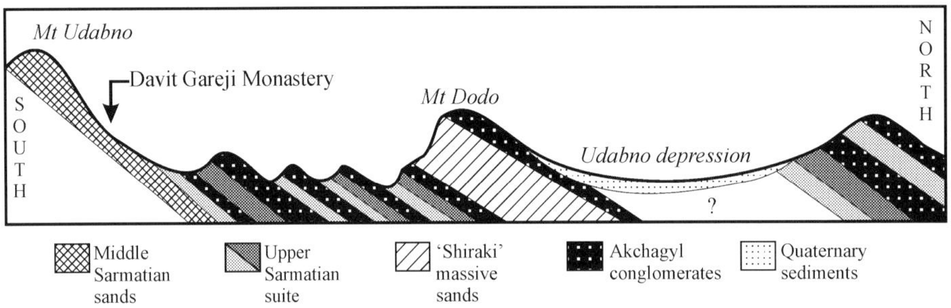

Figure 4.6. Idealised geological cross-section from the Davit Gareji Monastery to the Udabno Depression, Eastern Georgia (after Tsereteli, 1971, p. 297).

Ruppia spiralis grows in the saline water and *Phragmites australis* beds are scattered round the lake margins. The surrounding steppes are 'without forest and without water'.[21] This site and Sakhare Lake are notable for the extensive representation of halophytic semidesert plants on the saline soils (solonchaks) that surround both basins.[22] These soils are given over to pastures while more elevated, less salty soils are

[21] Maruashvili 1956, p. 59.
[22] Ketskhoveli 1959, pp. 296–7.

Figure 4.7. Lake Kumisi in the semidesert-steppe lowlands. Dr Maia Jejelava at left (May 2002).

Figure 4.8. Exposed sediments of Oligocene-Miocene age on the shores of Lake Kumisi (May 2002).

used for irrigated agriculture. The only trees in the catchment are a few planted pines (*Pinus eldarica*) and poplars, though oak-hornbeam forests grow in the upper reaches of the Nagubi Creek, the main tributary of Lake Kumisi.

It proved impossible to survey the vegetation of so large a pollen source area as Lake Kumisi's. Nevertheless, several points may be made about the relationship between its pollen spectra and the surrounding vegetation. Chenopodiaceae are dominant, but not quite as prevalent as at Sakhare Lake. This deficit is counteracted by higher levels of *Artemisia* and Poaceae at Lake Kumisi. Arboreal pollen is also more important, and especially so in surface samples gathered from the lake centre where the pollen source area is greatest. Samples from the centre contain 23 percent arboreal pollen compared to just 2 percent in samples taken nearer the lake edge. Most of this arboreal pollen must arrive as long-distance-transported pollen from trees growing outside the lake catchment, though some of the *Pinus, Quercus* and *Juniperus* pollen could come from more local sources. The Lake Kumisi pollen spectra demonstrate with clarity the strong influence of lake area and water depth on pollen source area in semi-arid lakes.

Tsavkisi Wetland

Tsavkisi Wetland is an infilled lake situated at 1110 m elevation in the rugged foothills above the Georgian capital, Tbilisi (**Fig. 4.9**). The wetland is around 400 m long and 100 m wide, contained within a steep-sided basin composed of Upper Eocene sedimentary rocks, which have been warped and folded into a series of abrupt ridges and steep ravines. Preferential weathering of anticlines, occurrence of landslips and suffosion of gypsum layers within the sedimentary units have led to the formation of a number of closed lake basins near Tbilisi, occupied by spring-fed, alkaline lakes.[23]

Tsavkisi Wetland is located at the very border of forest and steppe. Forests of *Quercus iberica* and *Carpinus orientalis* occur extensively to the north and west of the wetland (**Fig. 4.10**), while scrub vegetation, steppe and phrygana occur in various associations to the south and east. Plantations of pine (*Pinus kochiana*), apple (*Malus domestica*) and walnut (*Juglans regia*) occupy some of the site's catchment (**Fig. 4.11**). The wetland itself is densely clothed with *Phragmites australis* and *Typha latifolia*, and fringed with sedges (*Carex vesicaria*).

Vegetation survey data are compared to pollen spectra from the wetland edge and wetland centre in **Table 4.3**. In general, the agreement between the three sets of data is good. Amongst the arboreal species, pollen from shrubs is much under-represented compared to tree pollen. This is especially the case for shrubs in the Rosaceae family, which are palynologically 'silent' because their pollen are insect-dispersed.

[23] Dzhanelidze 1980; Tsereteli 1971.

Figure 4.9. Tsavkisi Wetland, between the villages of Tsqneti and Tsavkisi, looking eastward (August 2003).

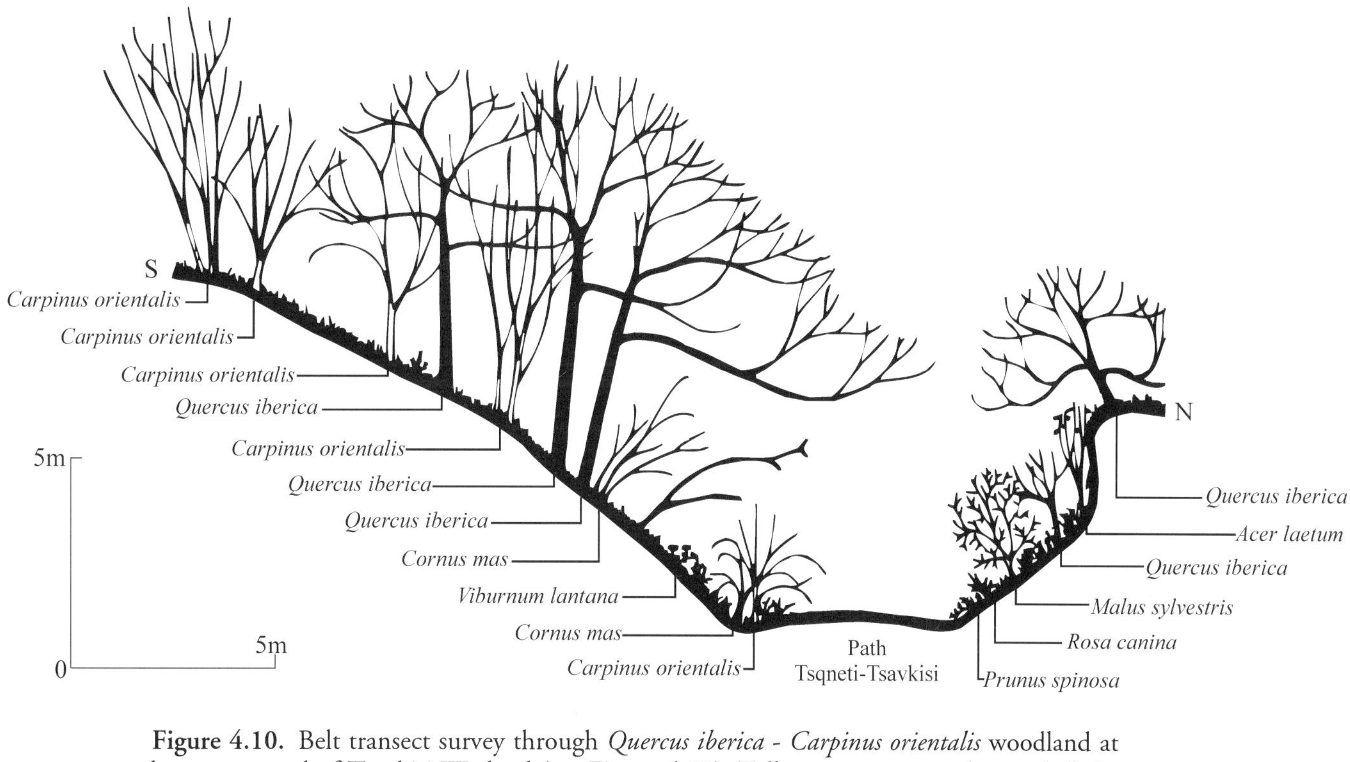

Figure 4.10. Belt transect survey through *Quercus iberica - Carpinus orientalis* woodland at the western end of Tsavkisi Wetland (see **Figure 4.11**). Taller trees grow on the north-facing slope, more xeric species on the south-facing slope.

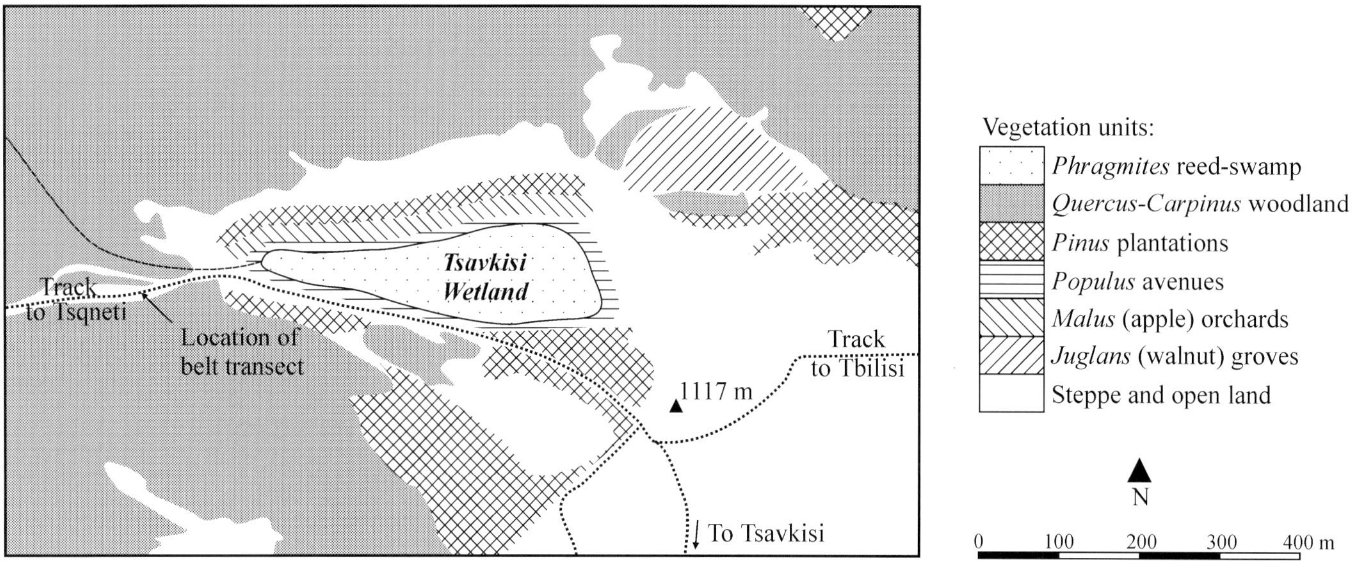

Figure 4.11. Map of Tsavkisi Wetland, showing principal vegetation units. Compiled from aerial photographs, topographic maps and ground surveys.

Although *Quercus iberica* and *Carpinus orientalis* are the most common trees in the catchment, this is not reflected in pollen spectra from the wetland edge, where *Pinus* pollen from the adjacent plantation prevails. *Carpinus orientalis* is very much under-represented in pollen spectra; oak is better represented. Spectra from the lake centre more accurately reflect the surrounding forest composition, but fail to adequately represent the substantial areas of steppe and shibliak vegetation nearby. Again, this is the result of poor pollen production and dispersion amongst most steppe herbs, with the notable exception of Chenopodiaceae, Poaceae and *Plantago* species.

The Tsavkisi area is mainly used as rough pasture for cattle, as is the wetland's sedge vegetation (**Fig. 4.12**). High proportions of *Plantago lanceolata* pollen are probably indicative of this activity.[24]

[24] *Cf.* Behre 1986.

Table 4.3. Comparison of plant community composition and pollen spectra at Tsavkisi Wetland. **See Table 4.2** for explanation and legend. Asterisks (*) denote plantations.

Trees and shrubs:	% cover	% edge	% centre	Pollen type represented
Acer laetum				*Acer*
Lonicera caucasica				-
Viburnum lantana				*Viburnum*
Cornus mas				*Cornus*
Carpinus orientalis				*Ostrya-type*
Fagus orientalis				*Fagus*
Quercus iberica				*Quercus*
*Juglans regia**				*Juglans*
Fraxinus excelsior				*Fraxinus*
Ligustrum vulgare				*Ligustrum*
*Pinus kochiana**				*Pinus*
*Malus domestica**				-
Malus orientalis				-
Prunus spinosa				-
Rosa canina				-
Rubus caesius				*Rubus-type*
*Populus canadensis**				*Populus*
Herbaceous plants:				
Alisma arcuatum				-
Crepis micrantha				*Cichorium*-type
Senecio vernalis				*Aster*-type
Lappula barbata				-
Hesperis matronalis				Brassicaceae undiff.
Campanula collina				*Campanula*
Silene viscosa				-
Euphorbia helioscopia				-
Astragalus viciifolius				-
Onobrychis transcaucasica				-
Scutellaria galericulata				Lamiaceae undiff.
Aegilops cylindrica				Poaceae undiff.
Bromopsis riparia				Poaceae undiff.
Koeleria cristata				Poaceae undiff.
Melica picta				Poaceae undiff.
Melica taurica				Poaceae undiff.
Nardurus krausei				Poaceae undiff.
Rumex acetosella				*Rumex*

Herbaceous plants:	% cover	% edge	% centre	Pollen type represented
Asplenium trichomanes				Polypodiaceae undiff.
Polypodium vulgare				Polypodiaceae undiff.
Polystichum braunii				Polypodiaceae undiff.
Trollius ranunculinus				-
Geum urbanum				-
Asperula arvensis				*Galium*-type
Digitalis ferruginea				-
Veronica jacquinii				-
Veronica serphyllifolia				-
Hyosciamus niger				-
Wetland plants:				
Carex vesicaria				Cyperaceae undiff.
Phragmites australis				Poaceae undiff.
Typha latifolia				*Typha*

Figure 4.12. A cow grazing on sedges (*Carex vesicaria*) at the edge of Tsavkisi Wetland (July 2005).

Lake Cherepanov

Lake Cherepanov is a small lake situated in the Nadarbazevi Basin, approximately 8 km WNW of the township of Tetri Tsqaro (**Figs 4.13, 3.2**). The relief here is muted on account of the basin's volcanic origin. Both Cainozoic and Plio-Pleistocene basalts occur in the lake's catchment,[25] and a small-scale quarry operates near the south side of the lake. The basalt was previously mined in the Mediaeval Period for the construction Queen Tamar's summer residence, now ruined (**Fig. 4.14**). The water depth in the sections of open water was only 1–2 m; numerous dams and dykes have been built to artificially elevate the water level, which fluctuates from year to year.[26]

Lake Cherepanov is located in the forest belt proper. *Quercus iberica* and *Carpinus betulus* are the most common forest trees, with an understorey of rosaceous shrubs (**Table 4.4; Fig. 4.15**). In this part of Georgia, wild pears (*Pyrus caucasica*) often form part of the forest canopy or occur solitarily in park-like landscapes.[27] Small fragments

Figure 4.13. Dr Henrikh Avakov poses beside an oak (*Quercus iberica*) at Lake Cherepanov, Southern Georgia.

[25] Maruashvili 1971.
[26] Dr Eristo Kvavadze, Zoological Institute, Georgian Academy of Sciences, pers. comm.
[27] Ketskhoveli 1959.

Figure 4.14. 'Nadarbazevi' – part of the ruined summer residence of the mediaeval monarchs of Georgia (May 2002).

of pure beech forest (*Fagus orientalis*) are found on moist, north-facing slopes. Another feature of this region is the occurrence of polydominant forests, in which, along with the above species, *Acer campestre, A. laetum, Carpinus orientalis, Cerasus avium, Crataegus curvisepala, C. pentagyna, Fraxinus excelsior, Pyrus caucasica, Ulmus glabra* and other trees are admixed more-or-less equally. In clearings, such as the lake terrace of Lake Cherepanov (**Fig. 4.13**), meadow vegetation occurs, dominated by *Agrostis planifolia* and other grasses. Lake Cherepanov's wetland has extensive stands of *Phragmites australis, Typha latifolia* and *Carex vesicaria*, whilst *Nymphaea candida, Potamogeton lucens* and *Lemna trisulca* grow in the water itself.

Table 4.4 shows the relationship between vegetation composition and the pollen spectra from Lake Cherepanov, moss polsters from the oak forest floor and soil samples from the same forest. Rosaceae are all but absent in all three pollen spectra even though they are very important in the vegetation. Pollen spectra from lake sediments are dominated by Poaceae and *Quercus* (arboreal pollen: 35 percent), the moss polsters by *Quercus, Carpinus betulus* and *Pinus* (AP: 67 percent), and the soil samples by *Pinus*, Chenopodiaceae and *Quercus* (AP: 43 percent). These significant differences in pollen composition highlight the difficulties in using soil samples and moss polsters as modern analogues for pollen records based on lake sediments.

Table 4.4. Comparison of plant community composition and pollen spectra at Lake Cherepanov. Pollen samples from lake sediments ('% pollen'), moss polsters ('% moss') and soil samples ('% soil'). See **Table 4.2** for further explanation and legend.

Trees and shrubs:	% cover	% pollen	% moss	% soil	Pollen type represented
Carpinus betulus					*Carpinus betulus*
Carpinus orientalis					*Ostrya*-type
Fagus orientalis					*Fagus*
Quercus iberica					*Quercus*
Cerasus avium					-
Crataegus curvisepala					-
Crataegus pentagina					-
Malus orientalis					-
Prunus divaricata					-
Pyrus caucasicus					-
Rosa canina					-
Populus tremula					-
Salix caprea					Salix
Herbaceous plants:					
Astrantia maxima					Apiaceae undiff.
Chaerophyllum aureum					Apiaceae undiff.
Lapsana grandis					*Cichorium*-type
Sonchus palustre					-
Arnebia pulchra					-
Lithospermum sibthorpianum					-
Scleranthus annuus					Caryophyllaceae undiff.
Galega orientalis					-
Trifolium pratense					-
Hypericum perforatum					-
Phlomis pungens					Lamiaceae undiff.
Salvia nemorosa					Lamiaceae undiff.
Marrubium goktschaicum					Lamiaceae undiff.
Ajuga genevensis					Lamiaceae undiff.
Dactylorhiza euxina					-
Agrostis planifolia					Poaceae undiff.
Deschampsia parviflora					Poaceae undiff.
Poa palustris					Poaceae undiff.
Dactylis glomerata					Poaceae undiff.
Erempoa persica					Poaceae undiff.
Koeleria cristata					Poaceae undiff.
Poa sylvatica					Poaceae undiff.

Herbaceous plants:	**% cover**	**% pollen**	**% moss**	**% soil**	**Pollen type represented**
Polygala transcaucasica					-
Adonis vernalis					-
Pedicularis comosa					-
Veronica beccabunga					-
Wetland plants:					
Carex vesicaria					Cyperaceae undiff.
Phragmites australis					Poaceae undiff.
Typha latifolia					*Typha*
Aquatic plants:					
Lemna trisulca					*Lemna*
Nymphaea candida					-
Potamogeton lucens					*Potamogeton*-type

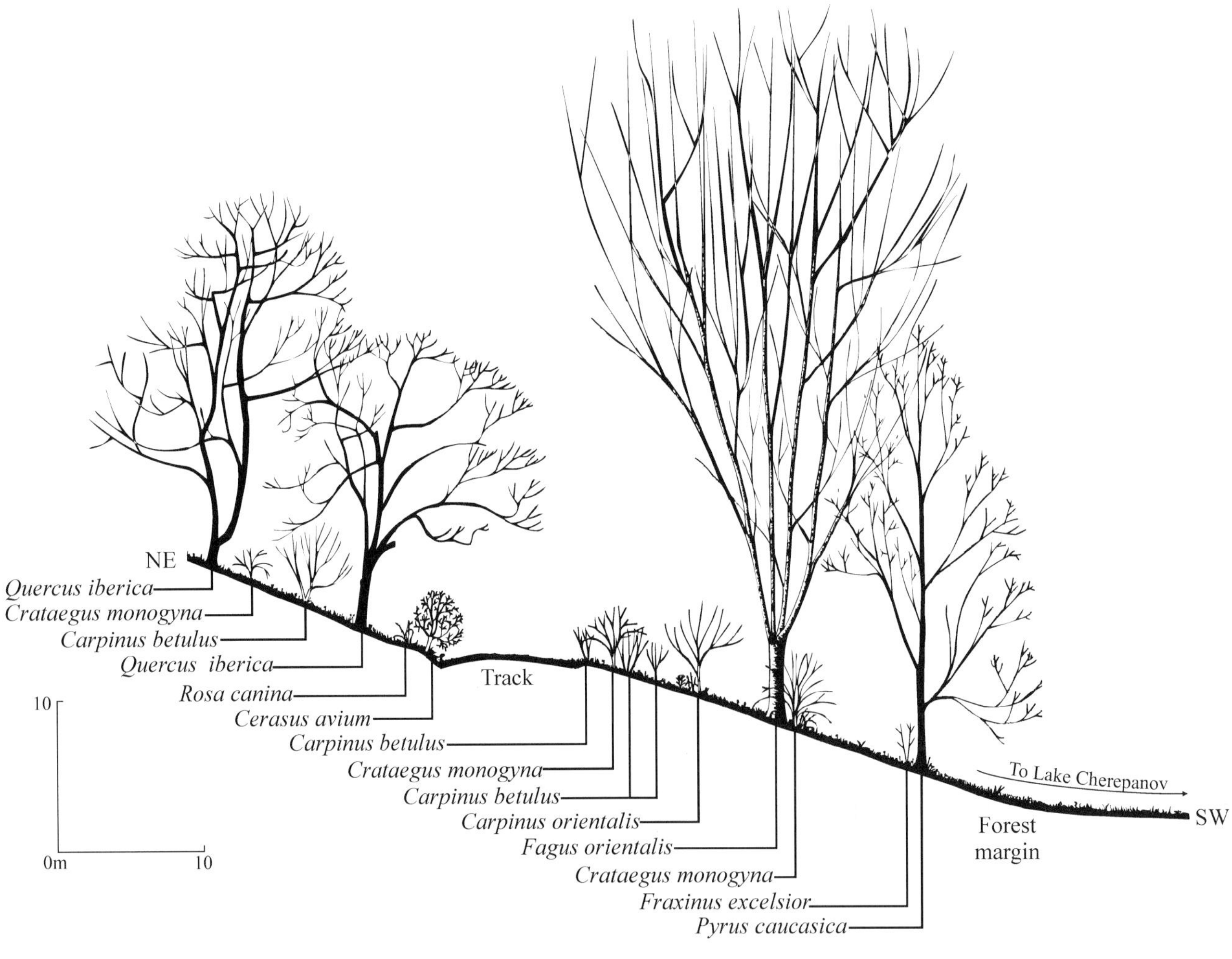

Figure 4.15. Belt transect survey through the forest beside Lake Cherepanov. The large *Fagus orientalis* was the only beech tree observed in the vicinity and appears be an old pollard. South-facing slope.

Trialeti Wetland

Trialeti Wetland is located 1 km south of the town of the same name on what is known to geomorphologists as the Chochiani Plateau. The Plateau is composed of Upper Pliocene andesite-basalts and is punctuated by a number of small, shallow lake basins. These depressions are thought to have formed at the same time as the lava flows through the subsidence of the underlying rocks.[28]

The wetland is approximately 300 metres in diameter with a water depth of around 1 metre (**Fig. 4.16**). It is situated on a residual andesite-basalt cap directly above the Khrami River Canyon and the Chochiani Gorge. The steep slopes of the river gorges are densely forested, while the basalt plateau is treeless. This site is where the narrow forest belt of Southern Georgia yields to extensive mountain grasslands. On the southern slope immediately adjacent the wetland is an oak forest of *Quercus macranthera* and *Q. iberica* (**Fig. 4.17**). On the northern slope is a secondary woodland dominated by *Corylus avellana, Carpinus betulus* and *Fraxinus excelsior* (**Fig. 4.18**). Grassy steppic meadows prevail to the west and around the lake, which is vegetated with *Phragmites australis, Equisetum palustre, Juncus articulatus* and *Potamogeton nodosus.*

Figure 4.16. Trialeti Wetland at the oak forest border. Boulders of andesite-basalt visible in the foreground; the forested slopes of the Chochiani Gorge and snow-capped mountains of the Javakheti Range are seen in the background (May 2002).

[28] Maruashvili 1971.

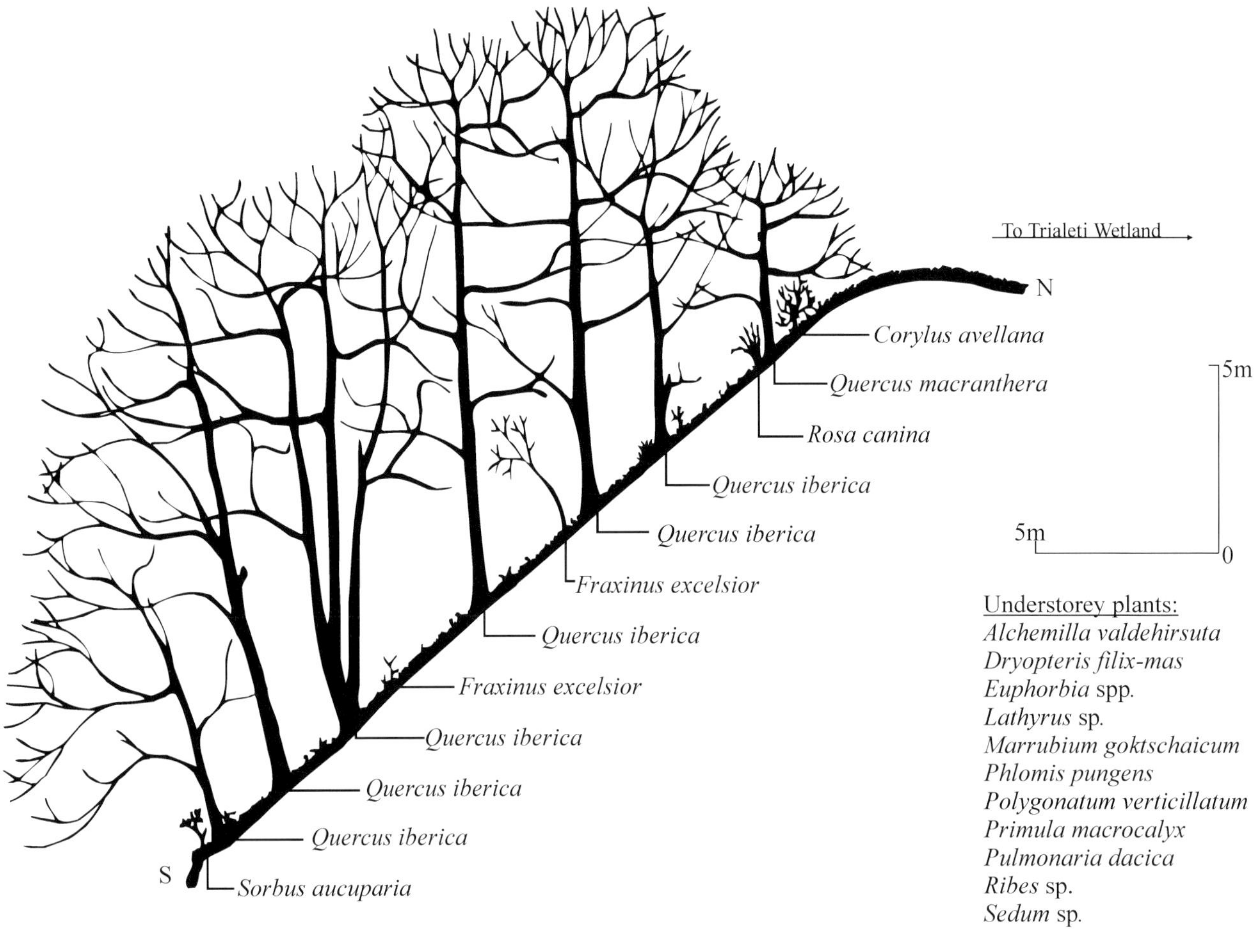

Figure 4.17. Belt transect through oak woodlands on the side of the Chochiani Gorge, near Trialeti Wetland. South-facing slope.

Arboreal pollen is poorly represented in the surface sediments from this site, masked by overwhelming proportions of Poaceae (**Table 4.5**). *Corylus* is particularly under-represented, even though dominant in the secondary woodland canopy. Grasses and herbs are much better represented palynologically than the trees. This could be a source-area effect related to the small size of Trialeti Wetland, or over-representation due to the abundance of Poaceae in the wetland flora. It is also possible that the site's topography and its location at the forest margin cause tree pollen to be carried away by strong wind updraughts. Price and Moore showed that pollen spectra in steep-sided uplands were influenced by anabatic winds,[29] and it is not uncommon for tree line situations in the Caucasus Mountains to have low arboreal pollen percentages, probably for this reason.[30]

[29] Price and Moore 1984.
[30] Kvavadze and Rukhadze 1989.

Table 4.5. Comparison of plant community composition and pollen spectra at Trialeti Wetland. Cover abundance and pollen spectra from lake sediments ('% pollen') and moss polsters from a mixed deciduous forest ('% moss 1') and a *Fagus*-dominated forest ('% moss 2') in the nearby Khrami Canyon are compared. See **Table 4.2** for legend.

Trees and shrubs:	% cover	% pollen	% moss 1	% moss 2	**Pollen type represented**
Lonicera caucasica					-
Lonicera steveniana					-
Carpinus betulus					*Carpinus betulus*
Carpinus orientalis					*Ostrya*-type
Corylus avellana					*Corylus*
Corylus colurna					*Corylus*
Quercus iberica					*Quercus*
Quercus macranthera					*Quercus*
Fagus orientalis					*Fagus*
Fraxinus excelsior					*Fraxinus*
Rhamnus catharticus					-
Crataegus curvisepala					-
Rosa canina					-
Sorbus boissierii					*Sorbus*-type
Herbaceous plants:					
Sonchus palustris					-
Pulmonaria dacica					-
Capsella bursa-pastoris					*Hornungia*-type
Scleranthus annuus					Caryophyllaceae undiff.
Marrubium goktschaicum					Lamiaceae undiff.
Phlomis pungens					Lamiaceae undiff.
Asphodeline lutea					Liliaceae undiff.
Polygonatum verticillatum					Liliaceae undiff.
Alopecurus sericeus					Poaceae undiff.
Bromopsis riparia					Poaceae undiff.
Dryopteris filix-mas					Polypodiaceae undiff.
Trollius ranunculinus					-
Alchemilla valdehirsuta					*Alchemilla*-type
Asperula taurina					*Galium*-type
Veronica arvensis					-
Wetland plants:					
Equisetum palustre					*Equisetum*
Juncus articulatus					-
Phragmites australis					Poaceae undiff.
Potamogeton nodosus					*Potamogeton*-type

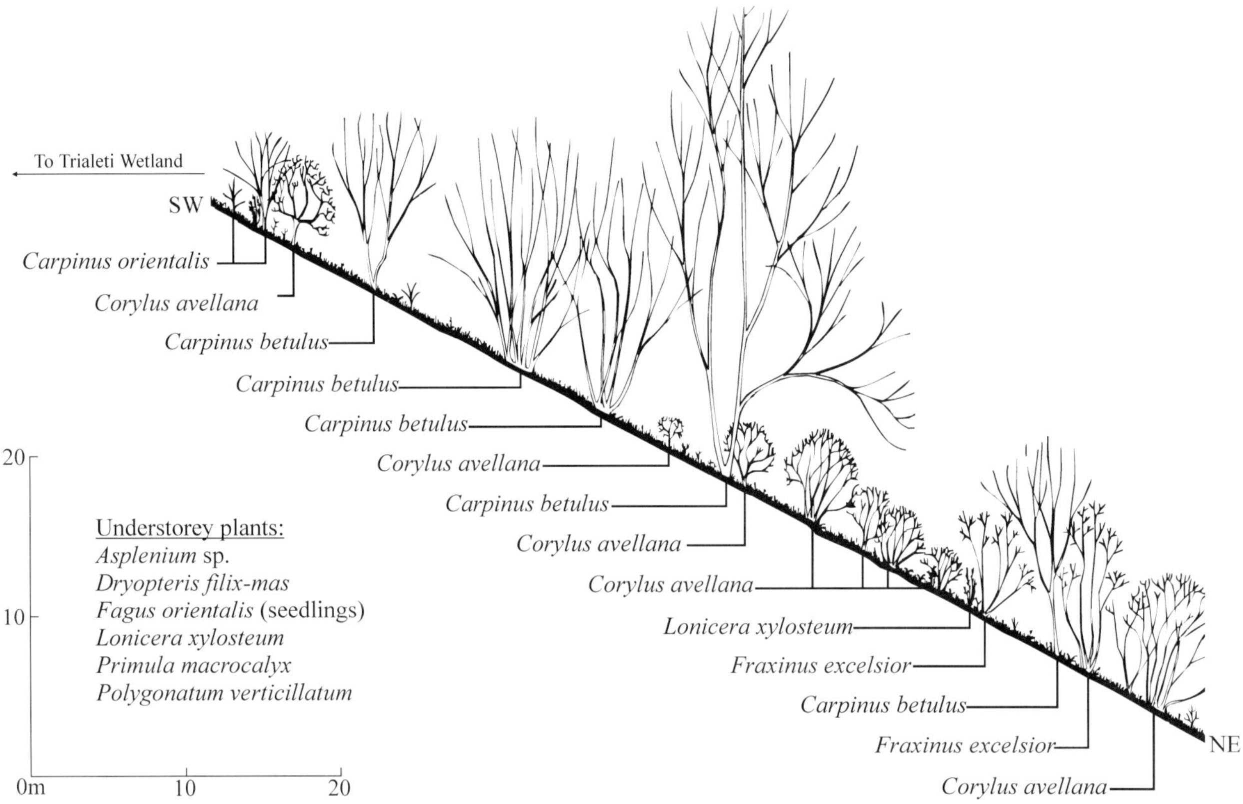

Figure 4.18. Belt transect through secondary woodlands on the edge of the Khrami Canyon, near Trialeti Wetland. North-facing slope.

Like Lake Cherepanov, herders have attempted to manipulate the water level of Trialeti Wetland by building small dams around it, perhaps to encourage sedges for cattle. The wetland vegetation has been burnt in the past, leaving a few hummocks of the former peat surface protruding from the water. Charred particles are thus well represented in samples from this site.

Lake Imera

Lake Imera is situated on the Tsalka Plateau at an elevation of 1610 m. It occupies a small synclinal depression in Plio-Pleistocene basalt, andesite and dolerite substrate, formed more recently than Trialeti Wetland, but in much the same way.[31] The 12-hectare lake has a closed basin and is quite shallow. Its water depth depends heavily on atmospheric precipitation: in May 2002, after a dry spell, the water depth at the coring site (**Fig. 4.19**) was 0.5 m, while in August 2003 it was over 2 metres, dropping to 1.5 m a month later. Villagers observed that the lake level was often low in spring and early summer, but was usually full by mid-summer.

[31] Maruashvili 1971.

Table 4.6. Comparison of plant community composition and pollen spectra at Lake Imera.

Herbaceous plants:	% cover	% pollen	Pollen type represented
Achillea setacea			-
Arctium lappa			-
Centaurea salicifolia			*Centaurea*
Cichorium intybus			*Cichorium-type*
Grossheimia macrocephala			*Centaurea phrygia*-type
Inula aspera			*Aster*-type
Taraxacum praticola			*Cichorium*-type
Anchusa gmelenii			-
Capsella bursa-pastoris			*Hornungia*-type
Clypeola jonthlaspi			Brassicaceae undiff.
Erophila verna			-
Scleranthus annuus			Caryophyllaceae undiff.
Cephalaria gigantea			-
Euphorbia villosa			-
Trifolium pratense			*Trifolium*-type
Vicia purpurea			Fabaceae undiff.
Vicia variabilis			Fabaceae undiff.
Gentiana gelida			-
Anacamptis pyramidalis			-
Agrostis planifolia			Poaceae undiff.
Phleum phleoides			Poaceae undiff.
Phleum pratense			Poaceae undiff.
Dactylis glomerata			Poaceae undiff.
Deschampsia parviflora			Poaceae undiff.
Poa pratensis			Poaceae undiff.
Alopecurus aequalis			Poaceae undiff.
Polygala transcaucasica			-
Rumex acetosella			*Rumex*
Trollius ranunculinus			-
Alchemilla erythropoda			-
Sanguisorba officinalis			*S. officinalis*
Asperula caucasica			*Galium*-type
Galium verum			*Galium*-type
Rubia tinctorum			*Galium*-type
Digitalis ferruginea			-
Pedicularis comosa			-
Pedicularis condensata			-
Rynchocorys elephas			-
Veronica anagallis-aquatica			-

Herbaceous plants:	% cover	% pollen	Pollen type represented
Viola montana			-
Wetland plants:			
Alisma plantago-aquatica			-
Sagittaria sagittifolia			-
Carex dichroandra			Cyperaceae undiff.
Scirpus lacustris			Cyperaceae undiff.
Carex vesicaria			Cyperaceae undiff.
Juncellus pannonicus			Cyperaceae undiff.
Juncus effusus			-
Mentha arvensis			*Mentha*-type
Stachys palustris			Lamiaceae undiff.
Epilobium hirsutum			-
Phragmites australis			Poaceae undiff.
Scolochloa festucacea			Poaceae undiff.
Calamagrostis neglecta			Poaceae undiff.
Poa palustris			Poaceae undiff.
Typha latifolia			*Typha*
Aquatic plants:			
Myriophyllum spicatum			*Myriophyllum*
Nymphaea candida			*Nymphaea alba*-type
Polygonum amphibium			*P. amphibium*
Potamogeton gramineus			*Potomogeton*-type
Potamogeton lucens			*Potomogeton*-type
Potamogeton natans			*Potomogeton*-type
Sparganium minimum			*Sparganium*
Sparganium neglectum			*Sparganium*

The lake has a distinct wave-cut terrace on its eastern side, positioned about 2 metres above the wetland's floating sedge mat (**Fig. 4.20**). Behind the terrace is a low, rounded, lunette-like feature composed of fine sediments that may have been built by deflation of the wetland sediments during low lake level phases.

The vegetation of the Tsalka Plateau is treeless mountain steppe. Although situated above the present tree line, these grasslands are not regarded as truly subalpine.[32] Ketskhoveli argues that these plains were once forested.[33] Their characteristic species

[32] Nakhutsrishvili 1999.
[33] Ketskhoveli 1959.

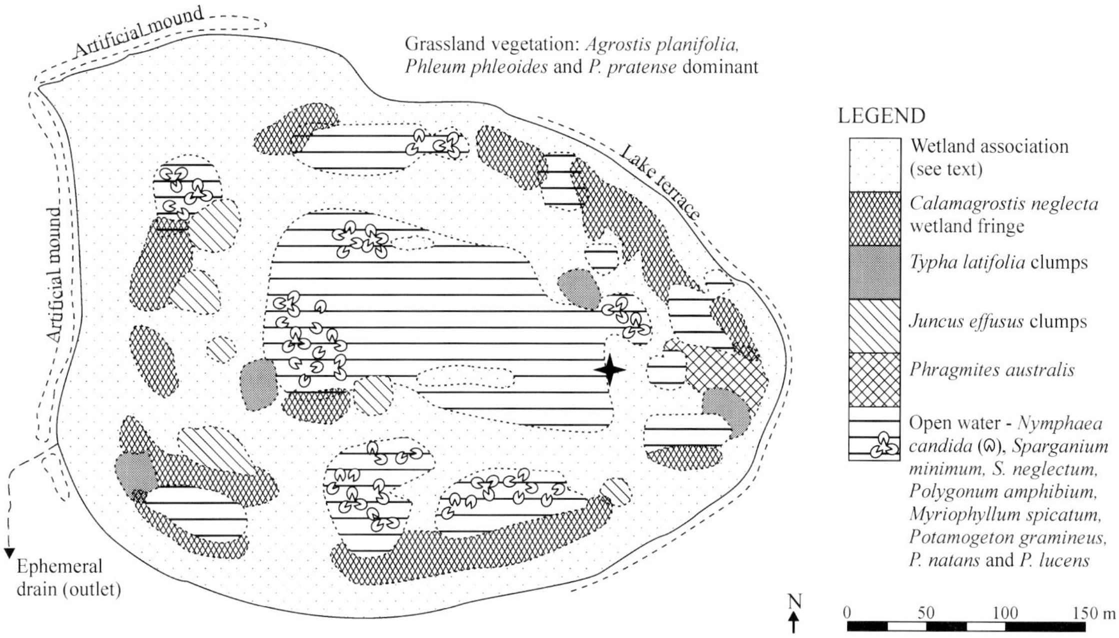

Figure 4.19. Map of Lake Imera, Tsalka Plateau, showing major wetland vegetation units. Coring location is shown by a star. Map constructed by GPS-referenced ground survey.

Figure 4.20. Photograph from the edge of Lake Imera looking northward. The light-coloured plants in the very foreground are *Calamagrostis neglecta*, followed by the lake's floating sedge mat, and areas of open water colonised by *Nymphaea candida* (May 2002).

are *Stipa tirsa, Festuca valesiaca, F. ovina, Falcaria vulgaris, Filipendula vulgaris* and others.[34] Around Lake Imera, the dominant grassland species are *Agrostis planifolia* (60–70 percent), *Phleum phleoides* (40–50 percent) and *P. pratense*. Many other species were encountered in the survey, and are listed in **Table 4.6**. Most of the plants are represented in pollen spectra, with the exception of some of the insect-pollinated herbs, such as *Polygala, Cephalaria, Gentiana* and the various Scrophulariaceae.

Large numbers of Poaceae in the wetland flora of Lake Imera (*Scolochloa festucacea, Phragmites australis, Calamagrostis neglecta, Poa palustris*) present a special problem for interpretation of fossil pollen spectra. Grasses are dominant in both the wetland and the mountain steppes that surround it. Any change in either steppe or wetland vegetation will affect Poaceae pollen percentages, so interpretation of landscape openness in the fossil context must rest on other subalpine herbs and forbs.

Pinus is the most abundant arboreal pollen type encountered in surface sediments from Lake Imera. The nearest sources of *Pinus* pollen are pine plantations established in the 1950s approximately 3 km from the site. Most of the pine pollen, however, must come from more extensive forests bordering the Tsalka Plateau. The same is true of taxa such as *Pterocarya, Buxus, Olea europaea, Zelkova, Pistacia* and *Picea*, which do not currently occur on the Plateau, but are nevertheless represented in pollen spectra. *Quercus, Fagus, Carpinus, Corylus* and *Betula* do grow on the Plateau, but have such a restricted distribution there that much of their pollen must also be carried by the wind from more extensive forests at lower altitudes.

Lake Bareti

Lake Bareti (or Bashkoi) is situated only 3 km WNW of Imera Lake in a synclinal depression.[35] The terrain here is also Plio-Pleistocene volcanics, but is slightly hillier on account of E-W anticlinal ridges running north and south of the lake (Mt Bareti and Mt Elibaba). Lake Bareti is a freshwater lake of moderate size, approximately 1300 m in diameter with a water depth of 1.5 metres (**Fig. 4.21**).

Lake Bareti is surrounded by mountain steppic grassland vegetation, similar in composition to the Lake Imera flora (**Table 4.6**). Wetland vegetation at Lake Bareti is also diverse. Areas of open water are populated by *Myriophyllum* and *Potamogeton* spp. A thick sward of *Phragmites australis* borders the open water. Between the *Phragmites* beds and the lake shoreline grow various associations of *Sparganium, Typha, Phragmites, Carex, Scirpus, Equisetum*, wet meadow plants and a few *Salix caprea* shrubs. Not surprisingly, the pollen spectra of Lake Bareti are very similar to those

[34] Ketskhoveli 1959.
[35] Maruashvili 1971.

Figure 4.21. Lake Bareti on the Tsalka Plateau. In the foreground (beneath the fishing boat) is *Equisetum palustre*, while the reeds in the middle ground are *Phragmites australis*. The rounded hill to the left is Mt Elibaba and the snow-capped peaks in the background are part of the Trialeti Range.

from Lake Imera (**Fig. 4.3**). *Pinus* pollen percentages are slightly higher, perhaps because of the pine plantation nearby on the slopes of Mt Bareti. The basin's larger pollen source area could also account for this difference. Again, the diversity of grasses in the wetland and grassland flora makes interpretation of Poaceae pollen difficult.

Avchalagöl Mire

Avchalagöl is situated on the Javakheti Plateau at the foot of Mt Patara Abuli (2800 m), the southernmost volcanic cone in the Samsari Range. The mire is large (1.25 km across) and occupies a basin impounded by Upper Pliocene andesite-basalt hills. Despite such extensive vulcanism, the mire is situated in a graben that runs from Avchalagöl through Lake Saghamo to Lake Madatapa in the south, so a tectonic genesis cannot be discounted.

The mire is vegetated with sedges growing in small hummocks, except for small sections of open water. Its surrounds are steppic grasslands, with a few shrubs growing on steep, rocky slopes on the north side of the basin. These shrubs include *Viburnum lantana, Rosa spinossisima* and *Rubus*, species most often associated with deciduous

Figure 4.22. Avchalagöl Mire on the Javakheti Plateau, looking southwards. The darker areas on the wetland surface are fire scars. The fearsome Caucasian Sheepdog in the foreground proved to be friendly, and his owner, an Armenian shepherd, told of how the wetland had been burnt the previous October (Photo: May 2002).

woodlands.[36] Recent pine plantations also occur nearby. *Pinus* and Poaceae dominate pollen spectra from this site. Arboreal pollen percentages are higher in samples from this site than at any of the sites described above, despite the nearest forests being some 50 kilometres away.

The Avchalagöl sediments are exceedingly rich in charred particles. A significant area of this shallow, *Carex*-dominated wetland appeared burnt in late 2001 (**Fig. 4.22**). A local shepherd told of how the entire landscape of this area, including the wetland and surrounding slopes, had been burnt the previous October. This is an annual event, he said, deliberately lit by herders to boost spring grass growth when flocks and herds return to the highland pastures. High charcoal concentrations in sediments are therefore a good indication of local fires.

[36] Ketskhoveli 1959.

Other sites

Several other sites were added to the original eight: Lake Patara Gldani on the outskirts of Tbilisi, Lake Jvari near the old Georgian capital of Mtskheta, Chili Lake in the oak-forested mountains south of Mtskheta, Lake Aligol on the western part of the Tsalka Plateau, and the Khachalgori Wetlands on the Javakheti Range (**Table 4.7**; **Fig. 4.23**).

Data from these sites are given in **Figure 4.24** and will be used to test the statistical model developed in a later section.

Table 4.7. Additional site locations and associated environmental variables. See text for an explanation of forest area calculations. Type: L- lake site, W- wetland site. Sites marked with an asterisk (*) are discussed in greater detail in subsequent chapters.

Site code	Site name	Type	Location (Lat.; Long.)	Elevation (m a.s.l.)	Basin avg. radius (m)	Forest area (%)	Vegetation type
PG	Patara Gldani	W/L	41°48′N; 44°50′E	535	100	5	Agricultural land
JV	Jvari*	L	41°50′N; 44°44′E	570	100	50	Xerophyte scrub
CHI	Chili	L/W	41°48′N; 44°40′E	1120	150	90	Oak forest
AL	Aligol*	L/W	41°38′N; 44°01′E	1540	100	1	Mountain grassland
KH	Khachalgori	W	41°18′N; 43°56′E	2650	20	0	Mountain grassland

POLLEN REPRESENTATION

The relationship between the vegetation and pollen spectra of the study sites (**Tables 4.2–4.6**, above) permits some discussion of the different pollen productivities of various plants. According to these data, *Pinus* and *Quercus* are abundant pollen producers and are always well represented in pollen spectra. *Pinus* pollen, because of its aerodynamic characteristics, is transported far from source trees so that it is dominant in treeless environments. *Fagus* and *Carpinus betulus* are moderate pollen producers, occurring in all spectra but in subordinate abundance to *Pinus* and *Quercus*, a reflection of their subordinate role in the forest vegetation of the transect. *Carpinus orientalis* and *Corylus* appear to be poor pollen producers, and *Fraxinus* and the various Rosaceae are very poor (**Fig. 4.25**). These observations are in accord with the representation factors calculated by Andersen.[37]

The pollen representation of herbaceous plants seems to depend largely on their

[37] Andersen 1970.

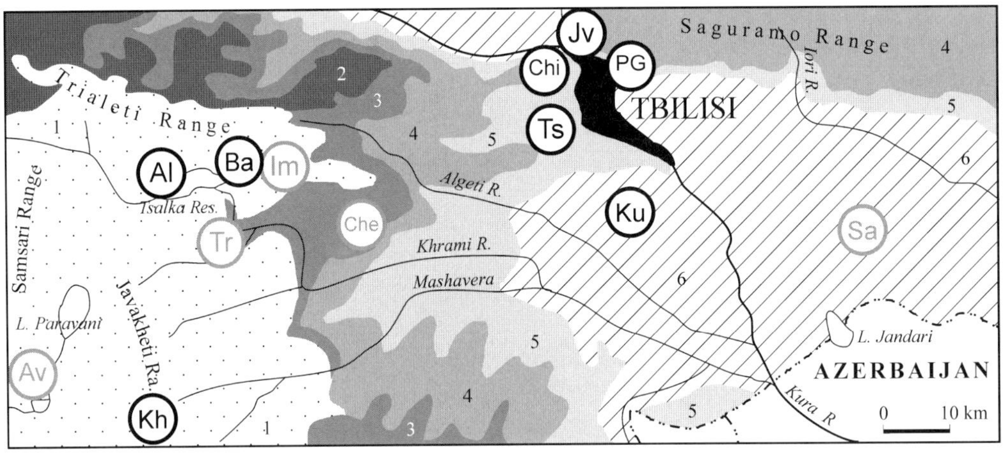

Figure 4.23. Locations of additional study sites (in black). See **Table 4.7** for site names. Vegetation units are the same as those in **Figure 4.2.**

pollination strategy. Wind-pollinated genera, such as the Poaceae, Brassicaceae and Polypodiaceae, are well-represented, whereas insect-pollinated Scrophulariaceae, Boraginaceae, Caryophyllaceae and Fabaceae are generally under-represented. Others, including the Chenopodiaceae, Plantaginaceae and Polygonaceae, are clearly over-represented, occurring in all pollen spectra but in few vegetation surveys.

Since the use of the arboreal to non-arboreal pollen (AP:NAP) ratio is unsuitable for differentiating forested from open vegetation types in this study, the 'background' pollen level must be estimated.[38] This is illustrated in **Figure 4.26**, which shows proportions of selected tree pollen along the altitudinal transect. In general, *Pinus* pollen comprises up to 35 percent of the arboreal pollen sum even when it is absent from the extra-local vegetation, 35–50 percent when it occurs nearby, and over 50 percent when present and locally abundant. In treeless environments outside the forest belt, however, *Pinus* percentages may be elevated above 50 percent by long-distance pollen transport.

Further difficulties arise deriving similar thresholds for *Quercus, Fagus, Carpinus betulus, Ostrya*-type, *Corylus* and *Juniperus*. When oaks are present in the surrounding vegetation, *Quercus* pollen percentages vary between 5 and 50 percent, but in the forest belt proper the figure is consistently above 15 percent. It can be seen from **Figure 4.26** that the relationship between source plants and pollen percentages is exceedingly complex. This pattern perhaps suggests that pollen production is influenced by factors other than the proximity and abundance of source plants in

[38] *Cf.* Sugita *et al.* 1999.

the surrounding vegetation. Factors such as pollen aerodynamics, the velocity and direction of wind, local variations in the vegetation mosaic, the pollen source area of individual study sites and even the use of percentage calculations themselves can all influence pollen percentages.

Pollen production is another factor that may vary in relation to environmental gradients. Different species of plants have "unique ecological responses to the multivariate environment",[39] pollen production being one such response. In **Figure 4.26**, for instance, *Quercus* constitutes less of the tree pollen percentage at the upper and lower limits of its distribution. At the lower tree line, *Quercus* percentages are diluted by *Pinus* and *Juniperus* pollen, even though pine and juniper trees are considerably less abundant in the vegetation here than *Quercus iberica*. A similar *Quercus* dilution occurs at the upper tree line where *Quercus macranthera* trees grow. As a second example, *Ostrya*-type pollen appears to be almost inversely related to the occurrence of its source (*Carpinus orientalis*) in the vegetation. *Carpinus orientalis* trees are co-dominants at the lower tree line, but the pollen data reflect this very poorly (**Fig. 4.26**). Some other factor apart from vegetation cover abundance is influencing pollen representation in the study area; it is possible that individual species' pollen productivity and cover abundance respond to climatic gradients in slightly different ways.

The foregoing shows that pollen representation is influenced by pollination strategies, vegetation cover abundance and environmental gradients. The latter stands as an impediment to determining the background pollen level, since this level changes according to environmental factors. Multivariate numerical analyses may be more useful than simple pollen thresholds in determining which environmental gradients have the greatest bearing on pollen data.

Numerical Analyses

Classification

A recent paper presented an analysis of modern pollen data from eight sites in Southern Georgia.[40] TWINSPAN was used to classify the data into two main groups – a semidesert-steppe group and a forest-upland group – on the basis of the greater abundace of *Pinus* and *Quercus* pollen in forest and upland spectra. The forest and upland group could be further sudivided into 'subalpine', 'treeline', 'oak forest' and 'forest-steppe' classes since *Picea* pollen is most prevalent (on this transect) in treeless grasslands, *Cerealia*-type pollen in mixed oak forest, and Chenopodiaceae pollen at the border between forest and steppe.[41] Because each TWINSPAN division is based

[39] Jackson and Williams 2004, p. 522.
[40] Connor *et al.* 2004.

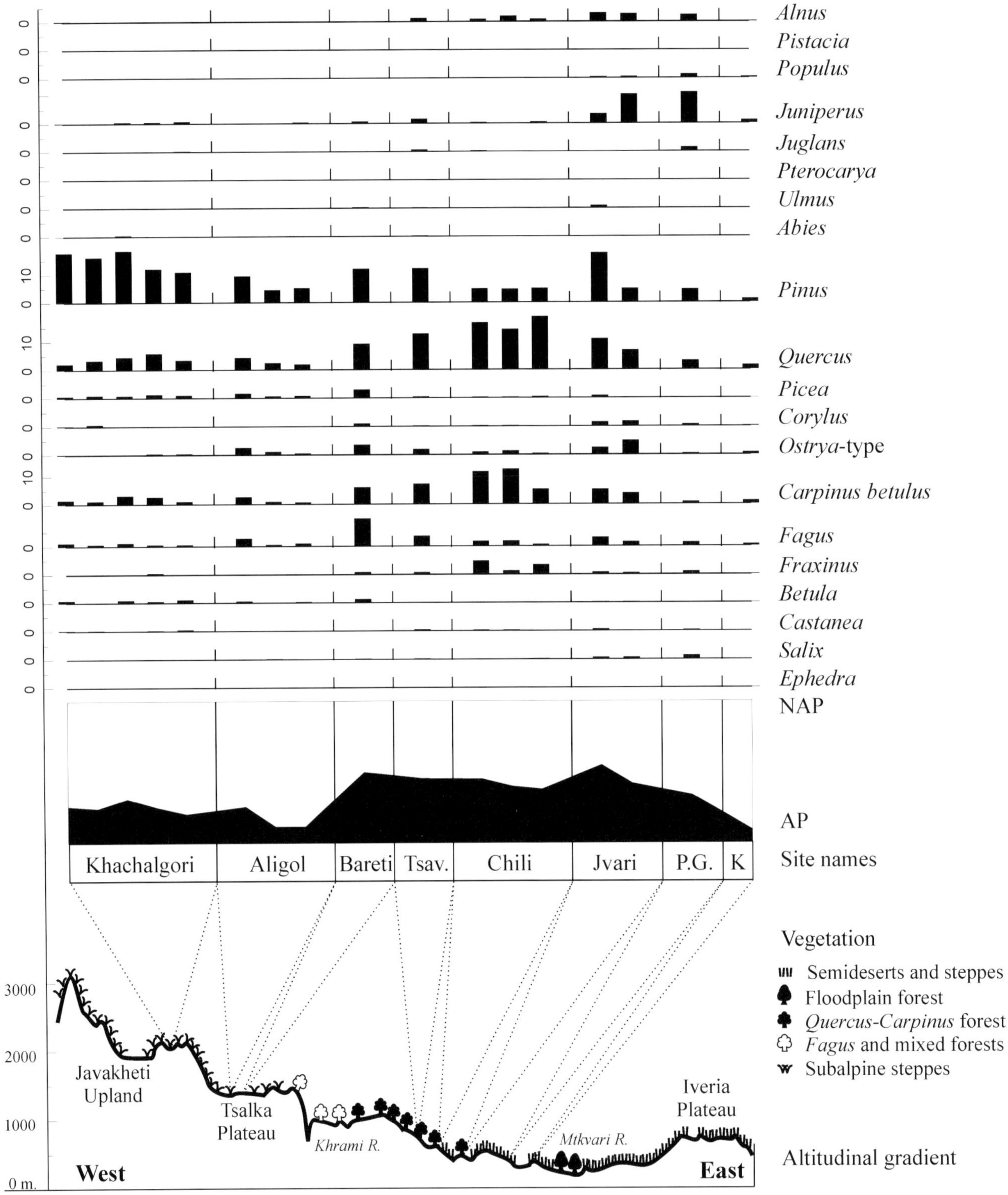

Figure 4.24 (a). Percentage arboreal pollen data from the additional sites listed in **Table 4.7**, shown in relation to the altitudinal gradient from **Figure 4.3**. See Appendix E for the complete data-set.

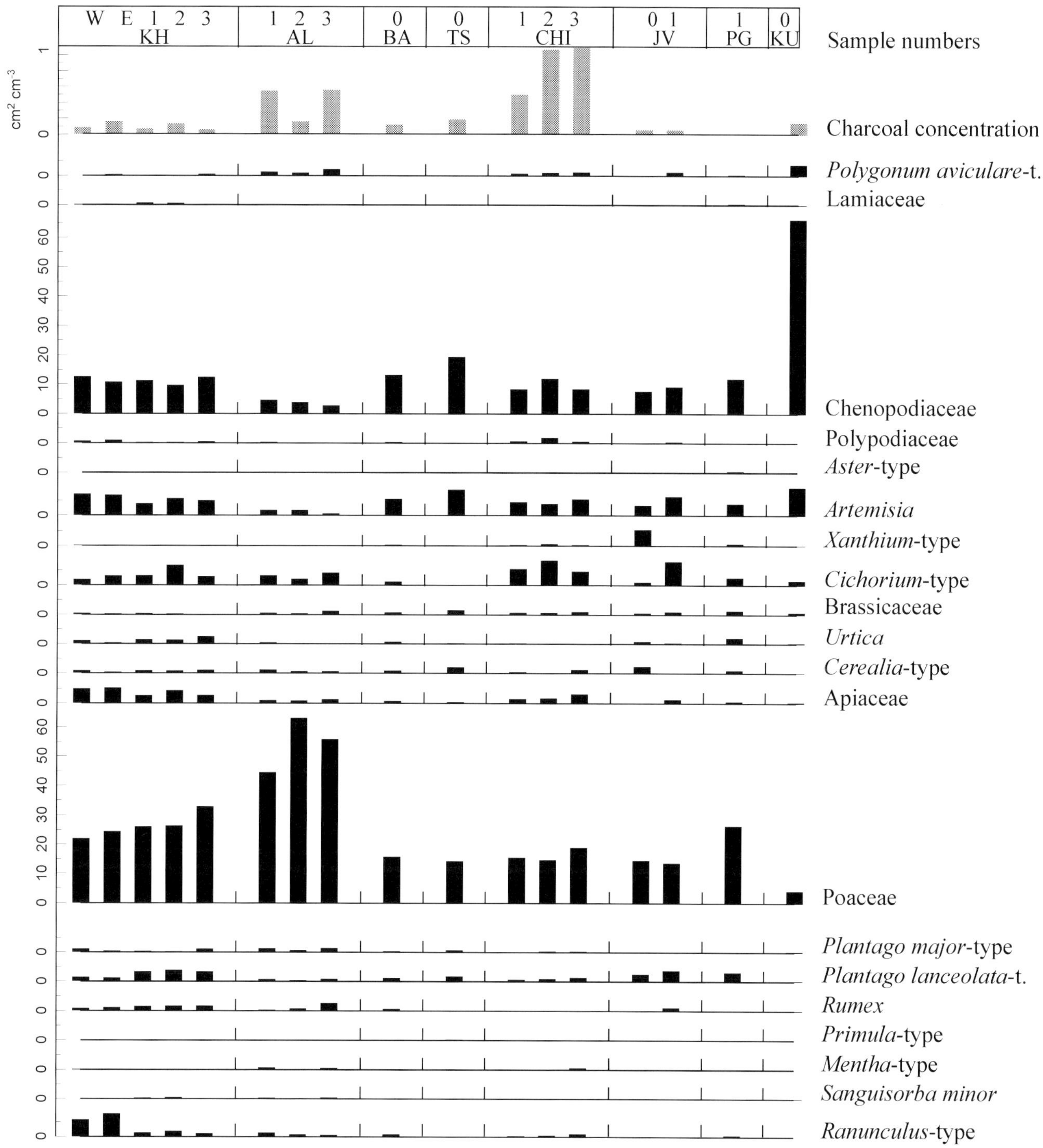

Figure 4.24 (b). Percentages of non-arboreal pollen and charcoal concentrations from additional sites listed in **Table 4.7**. Taxa displayed are the same as those in **Figure 4.3**.

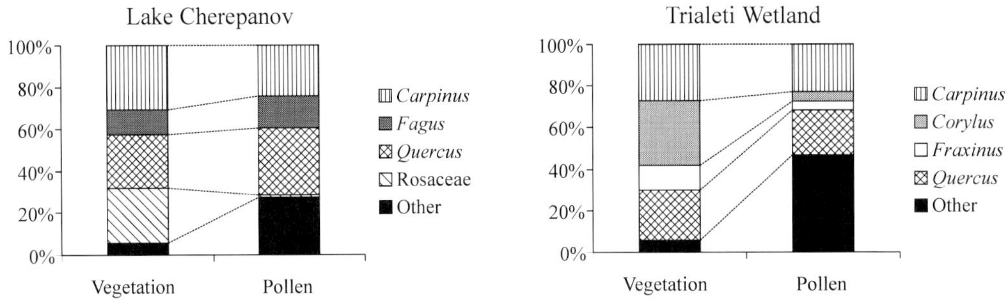

Figure 4.25. Comparison of tree cover abundance versus pollen percentages at Lake Cherepanov and Trialeti Wetland.

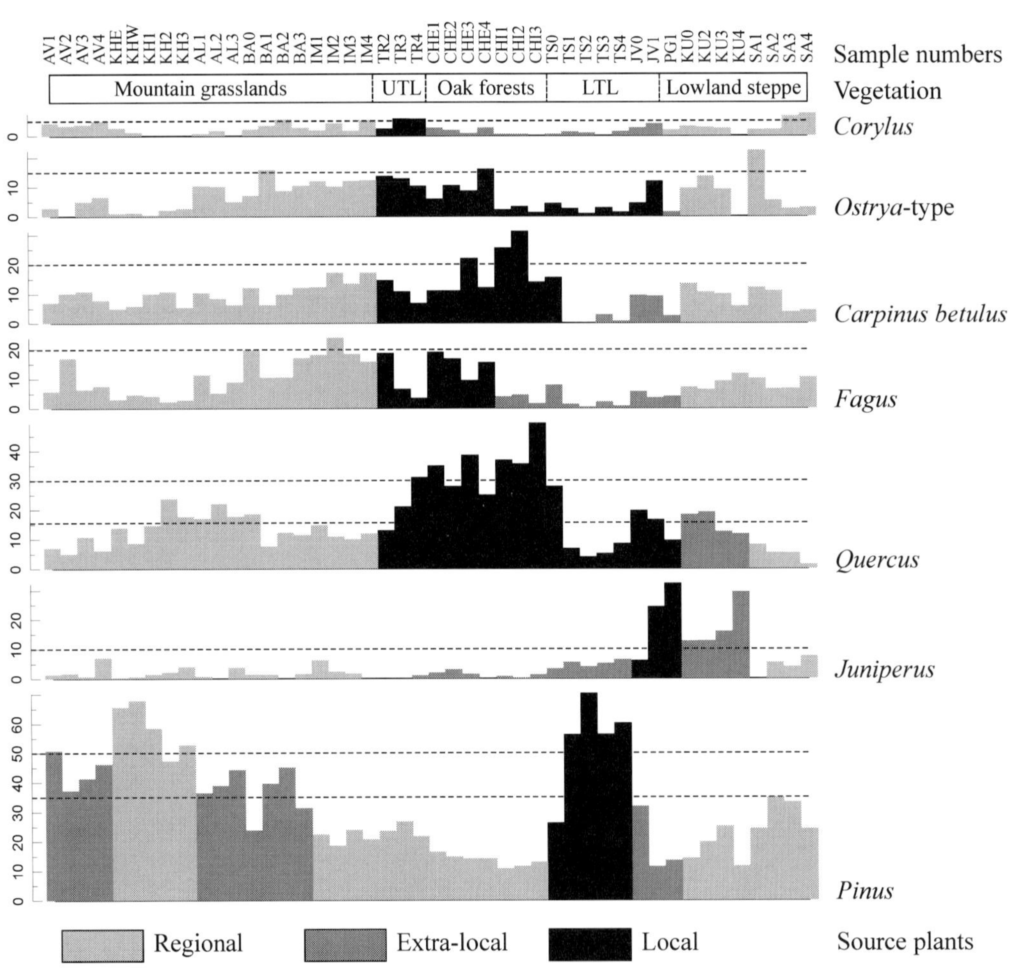

Figure 4.26. Pollen representation of the dominant forest trees of the study area. Percentages calculated from the arboreal pollen sum. Samples arranged in geographic order from W to E. Histograms are shaded according to the tree's occurrence in the vegetation and the dashed lines are thresholds discussed in the text. 'UTL': upper tree line; 'LTL': lower tree line.

on the presence and abundance of these indicator species, it is possible to use them to classify samples that were not included in the original analysis, and, in doing so, test the original classification.

Samples from Lake Kumisi, Lake Patara Gldani, Lake Jvari, Chili Lake, Tsavkisi Wetland, Lake Bareti, Lake Aligol and Khachalgora Wetlands (**Fig. 4.23**) were classified according to the TWINSPAN classification shown in **Figure 4.3** and the accompanying indicators in **Table 4.8**. The results are shown in **Table 4.9**.

Of the 17 additional samples, only 6 were correctly classified. Although the TWINSPAN classification presented previously[42] is consistent with biogeographical patterns, the model is susceptible to the over-representation of Poaceae pollen in some samples, no doubt due to grass species being dominant in both wetland and terrestrial vegetation at some sites (*i.e.* Lake Imera, Lake Bareti, Trialeti Wetland). To correct for this, Poaceae was excluded and the ordination repeated. The results are essentially the same as the previous classification (**Fig. 4.3**), although samples from the upper tree line and mountain grasslands are intermixed (**Fig. 4.27**). TWINSPAN indicators are also slightly different (**Table 4.10**) and the new result has the advantage of being simpler than the original.

When the additional samples are run through this classification, a much better fit is obtained (**Table 4.11**). Only two samples were incorrectly classified and both of these from lower tree line sites where the vegetation has been greatly modified by agricultural activities and the establishment of pine plantations (which, in the lowlands, mimic the effect of long-distance pollen transport in upland environments).

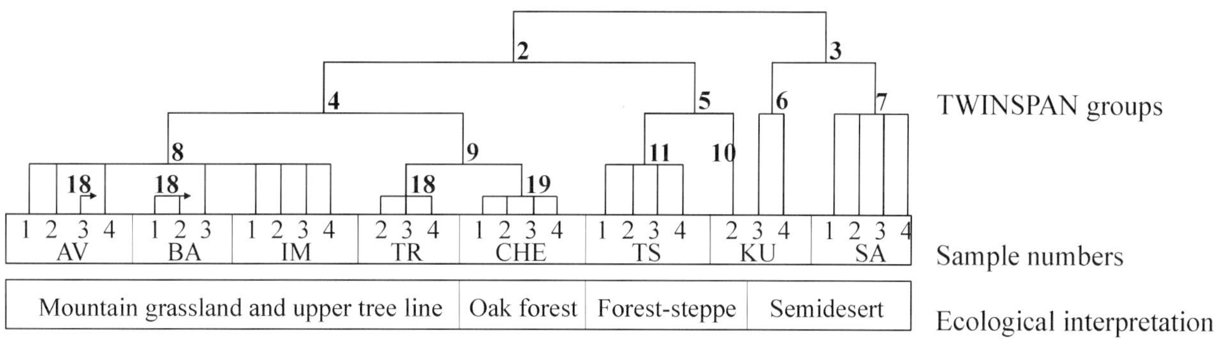

Figure 4.27. New TWINSPAN classification of pollen data from the eight original sites with Poaceae excluded. Indicator species for each group are given below in **Table 4.10**.

41 Connor *et al.* 2004.
42 Connor *et al.* 2004.

Table 4.8. Positive 'perfect' indicators for TWINSPAN groups in **Figure 4.3**. Each indicator occurs in that group at the stipulated level. Groups not listed have no indicators.

Group hierarchy	Vegetation (inferred)	TWINSPAN indicators and cut levels
2	Forest and upland	*Quercus* 1%; *Pinus* 2–3%
5	Steppe-forest	Chenopodiaceae 10%; *Populus* (presence)
6	Lowland steppe	*Artemisia* 2%; *Aster*-t. 0.5%; Lamiaceae 2%; *Juglans* (presence)
7	Semidesert-steppe	Polygonaceae 1.5%
8	Subalpine (burnt)	*Fraxinus* 2%
9	Subalpine and forest	*Ranunculus* (presence)
11	Steppe-forest	*Cedrus* (presence); *Pinus* 5%; *Picea* 0.5%; Poaceae 10%; *Plantago lanceolata*-t. 0–1%
18	Subalpine steppe	*Picea* 1.5%
37	Subalpine steppe	*Fagus* 5%; *Quercus* 5%; *Carpinus betulus* 5%
38	Upper tree line	*Polygonum aviculare*-t. 0–1%; *Hornungia*-t. (presence)
39	Oak forest	*Cerealia*-t. 2%

Table 4.9. Classification of additional surface pollen spectra using the TWINSPAN indicators presented in **Table 4.8** and the groups in **Figure 4.3**. Sample names followed by a zero are core-top samples from the initial group of sites.

Sample name	Actual vegetation	TWINSPAN group (No.)	Goodness-of-fit
Kumisi 0	Semidesert-steppe	'Forest-steppe' (10)	+
Patara Gldani	Lowland agricultural	'Forest-steppe' (10)	++
Jvari 0	Oak-xerophyte scrub	None – poor fit	–
Jvari 1	Oak-xerophyte scrub	'Forest-steppe' (10)	+++
Chili 1	Oak forest	'Upper tree line' (38)	–
Chili 2	Oak forest	'Forest-steppe' (10/11)	–
Chili 3	Oak forest	'Upper tree line' (38)	–
Tsavkisi 0	Forest-steppe	'Forest-steppe' (11)	+++
Bareti 0	Mountain grassland	'Forest-steppe' (11)	–
Aligol 1	Mountain grassland	'Subalpine steppe' (36)	+++
Aligol 2	Mountain grassland	'Upper tree line' (38)	–
Aligol 3	Mountain grassland	'Upper tree-line' (38)	–
Khachalgori 1	Mountain grassland	'Forest-steppe' (11)	–
Khachalgori 2	Mountain grassland	'Subalpine steppe' (36)	+++
Khachalgori 3	Mountain grassland	'Forest-steppe' (11)	–
Khachalgori E	Mountain grassland	'Forest-steppe' (11)	–
Khachalgori W	Mountain grassland	'Forest-steppe' (11)	–

Table 4.10. TWINSPAN indicators for the new classification (**Fig. 4.27**). N.B. There is no 'perfect' indicator for groups 8 and 9, which are differentiated based on four taxa together.

Group hierarchy	Vegetation (inferred)	TWINSPAN indicators and cut levels
2	Forest and upland	*Pinus* 2%
5	Forest-steppe	Chenopodiaceae 20%
6	Artemisia steppe	*Artemisia* 2%
8	Mountain grassland	*Pinus* 10% + *Picea* 2% – group 9 indicators
9	Grassland-forest border	*Plantago maritima*-t. 2% + *Cichorium*-t. 3% – group 8 indicators
10	Forest-steppe	*Fagus* 1%
19	Oak forest	*Quercus* 10%

Table 4.11. Classification of additional surface pollen spectra using the new TWINSPAN indicators presented in **Table 4.10** and the groups in **Figure 4.27**.

Sample name	Actual vegetation	New TWINSPAN group	Goodness-of-fit
Kumisi 0	Semidesert-steppe	Semidesert-steppe	+++
Patara Gldani	Agricultural/urban	Grassland-forest border	–
Jvari 0	Oak-xerophyte scrub	Oak forest	+
Jvari 1	Oak-xerophyte scrub	Grassland-forest border	–
Chili 1	Oak forest	Oak forest	+++
Chili 2	Oak forest	Oak forest	+++
Chili 3	Oak forest	Oak forest	+++
Tsavkisi 0	Forest-steppe	Forest-steppe	+++
Bareti 0	Mountain grassland	Mountain grassland	+++
Aligol 1	Mountain grassland	Mountain grassland	+++
Aligol 2	Mountain grassland	Mountain grassland	+++
Aligol 3	Mountain grassland	Mountain grassland	+++
Khachalgori 1	Mountain grassland	Grassland-forest border	++
Khachalgori 2	Mountain grassland	Mountain grassland	+++
Khachalgori 3	Mountain grassland	Grassland-forest border	++
Khachalgori E	Mountain grassland	Grassland-forest border	++
Khachalgori W	Mountain grassland	Grassland-forest border	++

This second TWINSPAN classification suggests that the various vegetation types encountered along the altitudinal transect in Southern Georgia can be differentiated with relative confidence based on palynospectra. Semidesert-steppes are differentiated from the forests and highlands by the paucity of tree pollen in their sediments. Forest-steppe at the lower tree line can be distinguished from mid-altitude forests and the treeless highlands by an abundance of Chenopodiaceae pollen. Highland sites dominated by long-distance-transported pollen have higher proportions of *Pinus* and *Picea* than sites with a more local pollen signal, including oak forests, in which *Quercus* is the dominant arboreal pollen type. Although the result differs slightly from the one presented in Connor *et al.*,[43] the major biogeographical patterns revealed by both analyses are the same.

Environmental Gradients

Major trends in the pollen data were identified using Detrended Correspondence Analysis (DCA) and the full surface pollen data set of eight original and five additional sites. DCA assumes no direct relationship between the data and its associated environmental variables. Where there is a relationship, however, this is identified indirectly by correlation. The resulting ordination is presented in **Figure 4.28** with vector lines representing correlations with a matrix of environmental variables.

The prevailing trend in the pollen data is, judging from these correlations, a rainfall gradient from the arid semidesert-steppes to the moister forests and mountains (DCA axis 1). The strongest correlation of this axis is with annual rainfall (r^2=0.63), followed by the AP:NAP ratio (r^2=0.59), longitude (r^2=0.53), July temperature (r^2=0.42), elevation, January temperature and summer rainfall (r^2=0.3).

The secondary trend in the data appears to be one of landscape openness (DCA axis 2). The correlation with forest cover (r^2=0.47) is followed by correlations with January temperature (r^2=0.45), July temperature (r^2=0.38), elevation (r^2=0.32), longitude (r^2=0.26), and summer rainfall (r^2=0.25). The third DCA axis (**Fig. 4.29**) has weak negative correlations with the AP:NAP ratio (r^2=0.28) and annual rainfall (r^2=0.12). Interestingly, charcoal concentrations had little correlation with any of the measured variables, suggesting that fire occurrence may be largely independent of physical environmental factors.

These results indicate that the composition of pollen assemblages is strongly influenced by climatic factors, especially rainfall. This is hardly surprising in a semi-arid region, but it is perhaps surprising that this characteristic is so plainly revealed through pollen spectra. Responses to temperature generally follow the altitudinal trend and indicate that the forested sites have slightly warmer, drier summers than the treeless mountain grasslands, even though their annual rainfall is similar.

[43] Connor *et al.* 2004.

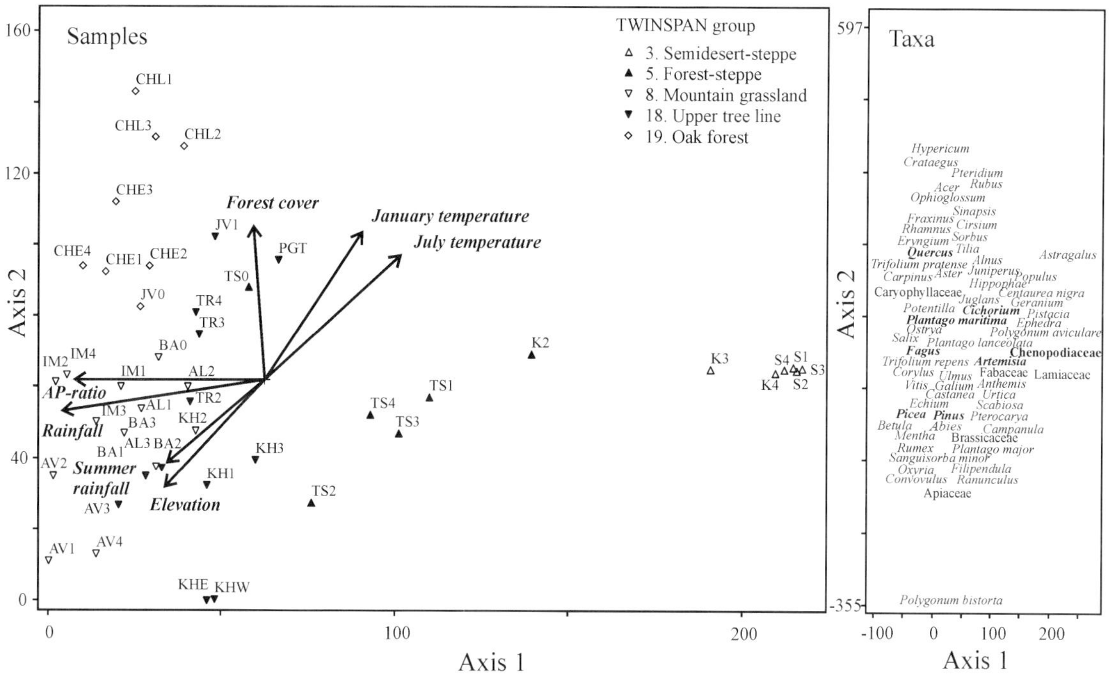

Figure 4.28. DCA axes 1 and 2 biplot of sample scores (points), new TWINSPAN groups (symbols) and correlated environmental variables (vector lines). Samples are labelled by site and sample number (see **Figures 4.3** and **4.24**). Taxa correlations are shown at right (on different axis scale), with TWINSPAN indicators in bold typeface. Axis units are standard deviations × 100.

Of interest is the relationship between rainfall and arboreal pollen representation. In semi-arid regions such as the study area, this relationship suggests that basin morphometry, lake depth and water level variability exert a much greater influence on AP:NAP ratios than does forest cover. In fact, forest cover appears to have little relationship to AP:NAP ratios in this study.

An ecological dimension can be added to the DCA results by examining the taxa most responsible for the patterns observed. The relationships between DCA axes and pollen taxa are shown in **Table 4.12**. These taxa have the advantage of being relatively common types that can be expected in most samples from the study area.

Consistent with the rainfall gradient identified in DCA axis 1, the taxa positively correlated with this axis are all xerophilous herbs, while negative correlates are moisture-loving trees (**Table 4.12**). The second axis, a gradient of forest cover and temperature, puts oaks and oak forest associates together against pines and herbs found mainly in treeless environments. The third axis is difficult to interpret, but may have something to do with local land use. Positive correlates are species often associated with heavy grazing and pasturage, while negative correlates are associated with grain agriculture (*Triticum* and Caryophyllaceae) and secondary woodland vegetation (*Fagus* and *Carpinus orientalis*).

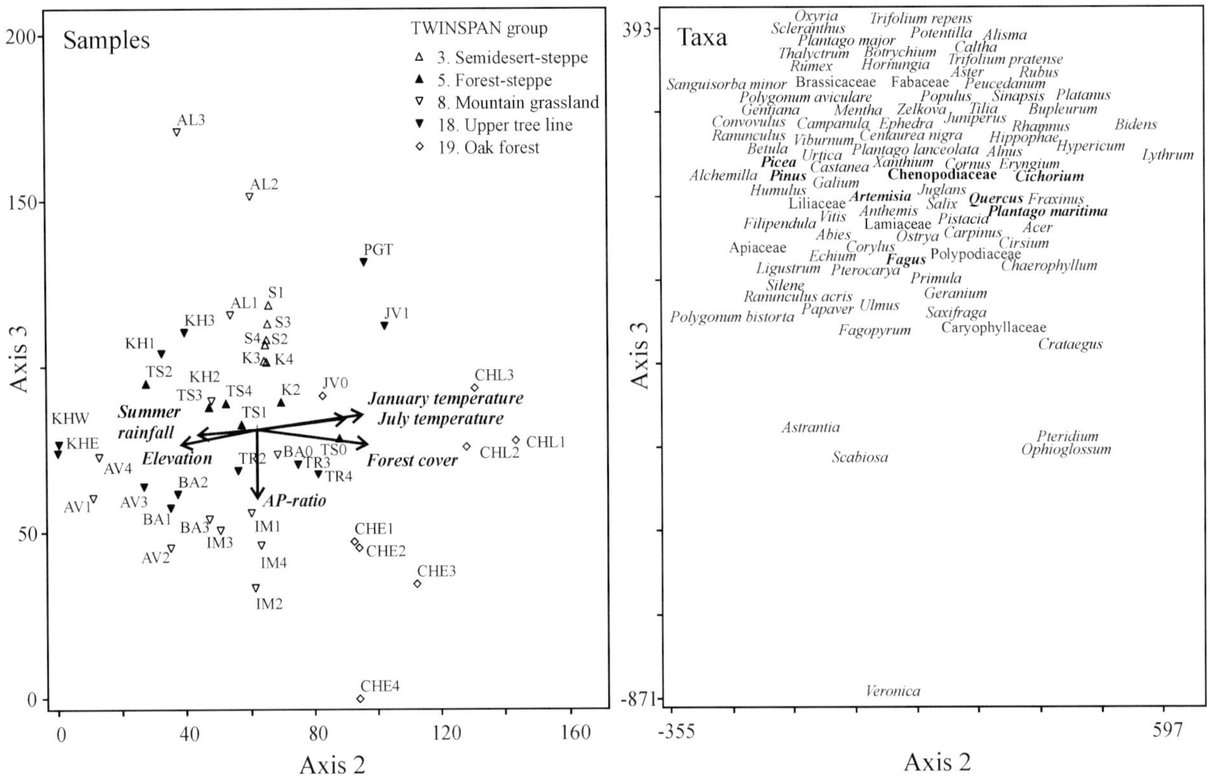

Figure 4.29. DCA result, axes 2 and 3.

Table 4.12. Pollen taxa most strongly correlated with each of the DCA axes. Correlation coefficients (r^2) are given in parentheses.

Axis 1	+	Chenopodiaceae (0.95), *Polygonum aviculare*-t. (0.46), Lamiaceae (0.22)
	–	*Carpinus betulus* (0.57), *Fagus* (0.29), *Picea* (0.28), *Corylus* (0.27)
Axis 2	+	*Quercus* (0.48), *Ophioglossum* (0.29), *Hypericum*-t. (0.28), *Fraxinus* (0.26)
	–	*Pinus* (0.39), Apiaceae (0.35), *Ranunculus* (0.30), *Rumex* (0.20)
Axis 3	+	*Trifolium repens*-t. (0.32), *Secale*-t. (0.27), *Oxyria*-t. (0.26), *Plantago major*-t. (0.25)
	–	*Triticum*-t. (0.42), *Fagus* (0.38), Caryophyllaceae (0.36), *Ostrya*-t. (0.31)

Based on the strength of correlations with the first two DCA axes, it seems possible to reconstruct gradients of rainfall, temperature and forest cover from fossil data using these indicator taxa. These and TWINSPAN indicators provide the means to develop a detailed picture of past climate, vegetation and landscape openness from pollen data.

Conclusion

This analysis of modern pollen-vegetation relationships along an environmental gradient has shown that pollen assemblages in the study area change in accordance with broad-scale vegetation patterns. Chenopodiaceae pollen dominates in the semi-desert-steppe lowlands. Pollen produced by deciduous trees (particularly *Quercus*) becomes important in the forest belt. And, above the tree line, *Pinus* and Poaceae pollen are dominant.

The relationship between vegetation and pollen is not straightforward, however. A most significant problem for interpretation is the dominance of grasses (Poaceae) in both wetland and terrestrial floras from many of the sites. Because wetland and terrestrial grasses cannot be distinguished by pollen morphology, Poaceae pollen percentages cannot be interpreted in any meaningful way and constitute a major source of 'noise' in pollen data from this region. The problem could be overcome by sampling deep lakes without wetland vegetation, but this would restrict analysis to larger lake basins that are a poor source of information on local vegetation history. Neither can the Poaceae problem be resolved with reference to pollen assemblages from terrestrial soils and moss polsters: in this area these are fundamentally different to lake sediment spectra and have limited value as modern analogues.

Certain plants, because of their pollination strategy, their abundance in the vegetation and environmental factors that limit pollen production, are under-represented in pollen spectra from the study area. These include *Fraxinus, Corylus* and most trees that inhabit the lower tree line (*Prunus spinosa, Paliurus spina-christi, Spiraea hypericifolia, Carpinus orientalis*, etc.). Xerophyte scrub communities appear to be almost invisible palynologically and perhaps can be identified only by the abundance of associated taxa like *Plantago* and *Artemisia*. In addition, the pollen production of some species does not relate to their abundance in vegetation in a linear fashion. This is exemplified by *Quercus iberica* and *Carpinus orientalis*, the two dominant trees of the lower tree line (forest-steppe zone); both appear to have decreased pollen production at the lower limits of their distribution, perhaps as a physiological response to aridity.

Other pollen taxa are over-represented and these include *Pinus*, Chenopodiaceae, Plantaginaceae and Polygonaceae. All of these plants appear to have very high pollen production and dispersion. *Pinus* pollen is an indicator of treeless environments in the study area, as it is carried from lower altitudes by the wind. The same process introduces pollen of *Pterocarya, Buxus, Juglans, Castanea, Olea europaea* and others from distant sources. The presence of these pollen taxa does not necessarily indicate the presence of source plants in the surrounding vegetation.

[44] Kvavadze 1993; Stuchlik and Kvavadze 1995.

An aim of this study was to determine whether forested and open vegetation types could be differentiated using pollen data, a question that has thus far proved difficult to answer in the mountainous Caucasus region.[44] The arboreal to non-arboreal pollen ratio, the traditional yardstick of landscape openness, has no relationship with forest cover in the sites and samples studied here. In Southern Georgian lakes, the AP:NAP ratio is influenced much more by basin size, water depth and hence pollen source-area than it is by forest cover. AP:NAP ratios may therefore prove useful for reconstructing changing lake levels within particular sites.

Forest cover or landscape openness can be gauged indirectly by numerical analysis. Taxa such as *Quercus, Ophioglossum, Hypericum*-type and *Fraxinus* are characteristic of forested vegetation, whereas *Pinus*, Apiaceae, *Ranunculus* and *Rumex* distinguish treeless environments. Using these and other indicators, ordination scores may be calculated for new samples through weighted averaging to detect changes in forest cover from fossil pollen data. The same is true for rainfall, which is the dominant environmental gradient affecting pollen assemblages in the study area, as well as temperature and land-use, which are also identifiable palynologically. Despite pollen assemblages from this semi-arid region being subject to distortion by changing lake levels and highly localised differences in vegetation, broad-scale environmental gradients are detectable using numerical analyses that isolate a signal from background noise.

Some of the observations arising from this study may have applicability beyond the realm of Southern Georgia. For instance, the clear relationship between AP:NAP ratios and rainfall could help explain, in part, reduced AP levels during the last glacial maximum as lake levels receded in response to aridity. Likewise, the apparent reduction in some trees' pollen production at the forest-steppe border suggests that different species respond to environmental gradients by varying pollen production, a factor that may be responsible for no-analogue pollen spectra in late glacial-aged lake deposits around the world.

CHAPTER FIVE

THE TBILISI AREA

Introduction

With a knowledge of the relationship between Southern Georgia's present-day vegetation and the pollen produced by it, the following chapters move to the question of vegetation history and apply the knowledge gained in the previous chapter to the analysis of fossil pollen preserved in lake deposits. The results are grouped into two geographic areas: the Tbilisi area and the Tsalka Plateau. It is the first of these that is presented here.

Tbilisi is nowadays surrounded by cultural landscapes. Like most major cities in Western Asia, its vegetation has been strongly modified by past human activities. Yet the Georgian capital differs from many others in retaining considerable tracts of forest in its immediate surrounds. The city itself is located on the very border of forest and steppe, and thus its vegetation reflects not only human influences, but also climatic ones. Because variations in the position of the lower tree line can be traced using pollen analysis, the Tbilisi area provides an opportunity to examine these changes in the light of past climatic fluctuations and a long, well-documented, socio-economic history.

The Lost Forests of Tbilisi

It is widely believed that Tbilisi was surrounded almost entirely by forests up until the nineteenth century. Various botanists have concluded that these original forests were of beech or oak,[1] although the available historical evidence for this is fragmentary and often ambiguous. When Tbilisi was founded as the Georgian capital, King Vakhtang Gorgasali was out hunting pheasants, which, being birds of the forest, seem to indicate that Tbilisi was a forest in the sixth century AD.[2] Curiously, though, the same king soon issued an edict for his subjects to quit the city and flee to the distant forests of Kakheti:

> He dispatched heralds so that they might abandon the villages and cities which were not fortified and flee to the people of the Caucasus and Kakheti, because Kakheti was forested and was impenetrable by enemies.[3]

Had Tbilisi's forests vanished in the few years since the city's establishment?

[1] Ketskhoveli 1959; Sosnovskii 1915.
[2] Ketskhoveli 1959; Sosnovskii 1915.
[3] *The Georgian Chronicles*: translation by Thomson 1996, p. 219.

Georgia's eighteenth century geographer, Prince Vakhushti Bagrationi, describes places around the city good for hunting, taken by some to signify a forested landscape.[4] But Vakhushti never once mentions the word 'forest' in his *Description of Tbilisi.*[5] Instead, he notes that, 'from Tbilisi to Mtskheta is waterless, yet fruitful,' including the plain of Didi Dighomi: 'dry, but grassy, and burgeoning with sheep and cattle in winter'. The hills above, 'from the Vere River up to Tsqneti are all orchards'.[6]

Vakhusti's contemporary, and botanist to Louis XIV, Joseph de Tournefort, visited Tbilisi in 1702 and described the surrounding mountains as being "quite bare" and a population that relied on dung fuel rather than wood for cooking.[7] Old photographs show that the hills were still "quite bare" during the early twentieth century (**Fig. 5.1**).

Because of these contradictions, the question of Tbilisi's original vegetation seems impossible to answer on the basis of historical evidence. On botanical grounds, Ketskhoveli came to the following reconstruction:

> Around Tbilisi, the plains and hills were once covered with beautiful forest. Here, many forest types were distributed: gallery forest of poplars and oak along the Kura floodplain; on the hills grew forests of Georgian Oak, with maples, ash, hornbeam, wild pears, wild apples and rowan trees. Forests of oak and elm grew on the plains; savanna on south-facing slopes. But where are these forests today? Only fragments can be found, high in the hills, where they escaped destruction.[8]

When this reconstruction is compared to the only pollen record from the Tbilisi area, Lake Lisi,[9] there is little evidence for forests. The Lake Lisi record is dominated by *Pinus*, Chenopodiaceae and Asteraceae, and suggests that steppe landscapes prevailed around the lake through much of the period represented by the 10-metre-long core. However, because this very detailed record has no independent radiocarbon chronology and appears to be affected by post-depositional pollen degradation,[10] it is difficult to interpret the results in any historical sense.

The significance of Tbilisi's past vegetation extends well beyond questions of historical geography. The city's location at the border of forest and steppe means that aridity exerts a major influence on the vegetation; because of steep topography and sharp environmental gradients, the lower tree line here could be expected to respond with great sensitivity to climatic changes. The area is also known for its high plant biodiversity, especially in phrygana and xerophyte scrub communities. Little is known about the history of these important semi-arid communities because they are so poorly expressed in pollen spectra.[11]

[4] Sosnovskii 1915.
[5] Vakhushti Bagrationi 1983.
[6] Vakhushti Bagrationi 1983, p. 439. Tsqneti is a town less than 2.5 km NW of Tsavkisi Wetland.
[7] Tournefort 1718, p. 230.
[8] Ketskhoveli 1959, p. 344.
[9] Kvavadze 1999b.
[10] Dr Eliso Kvavadze, pers. comm.
[11] Bottema 1974.

Figure 5.1. Two photographs of the Mtatsminda (Holy Mountain) funicular railway. The right-hand photograph was taken in 1905, and the left-hand photograph in 2003 next to the Palaeobiology Institute. In a century forests have expanded to such an extent that the Church of St David so prominent in the older photograph is scarcely visible in the present-day (lower left). (Source: Gersamia, 1984, and author's collection).

The Tbilisi landscape is highly heterogenous. This is the product of the rugged terrain, steep environmental gradients, and, significantly, the varied impact of humans. Along with rainfall and temperature variations, disturbances such as fire, deforestation and grazing must have moulded the vegetation in the past. The relative contribution of each of these factors to the evolution of heterogenous vegetation is unknown. It is clear that these issues can only be resolved with reference to multiple sites in different plant communities.

Using palaeoecological data from four sites along the altitudinal transect presented in Chapter Four – two sites in the semidesert-steppe, one in xerophyte scrub and one in oak-hornbeam forest – this chapter seeks to answer the following questions:

- What was the past character of Tbilisi's vegetation?
- How did it change through time?
- Were these changes related to climatic or anthropogenic factors?

Topography and Climate

The Tbilisi landscape is exceedingly varied, owing to its dissected topography and complex history of land-use. Tbilisi's hilly surrounds are composed mostly of Upper Eocene and Miocene sedimentary rocks that have been heavily warped and folded to create a rugged topography of jagged peaks and steep ravines. As mentioned in Chapter Four, this landscape is punctuated by a number of small lake basins (**Fig. 5.2**), formed through suffosion, landslips and preferential weathering.[12]

'The climate of Tbilisi is excellent and invigorating… the summer is hot, but not unbearably so; winter is cold; and both spring and autumn are wonderful'.[13]

Tbilisi's climate varies considerably according to topography. Low-lying areas southeast of city are very hot and dry in summer, with an average annual temperature range of 2–23 °C (winter-summer) and annual precipitation less than 500 mm. In contrast, the mountains around the city have a temperature range of 0 to 20 °C and rainfall up to 800 mm.[14] Two precipitation minima occur annually: during winter and late summer (**Fig. 5.3**).

Vegetation

> The 'Tbilisi landscape' therefore, determined mainly by shrub, steppe and partly by mountain xerophile vegetation, has rather a dismal aspect.[15]

Complex topography, microclimates and land-use patterns enable a great variety of vegetation types to coexist within a relatively small area. Far from being dismal, Tbilisi's vegetation is extraordinarily species-rich, with some 1643 or so species — about 25 percent of the total number of plant species found in the Caucasian flora and 40 percent of all species represented in Georgia.[16] A few species are local endemics, including *Primula saguramica* and *Anthemis saguramica*. Six major vegetation groups can be enumerated:[17]

[12] Dzhanelidze 1980; Janelidze 1948.
[13] Vakhushti Bagrationi, quoted in Maruashvili 1956, p. 47.
[14] Svanidze and Papinashvili 1992.
[15] Sakhokia 1961, p. 18.
[16] Gagnidze 2005; Sakhokia 1961.
[17] Sakhokia 1958.

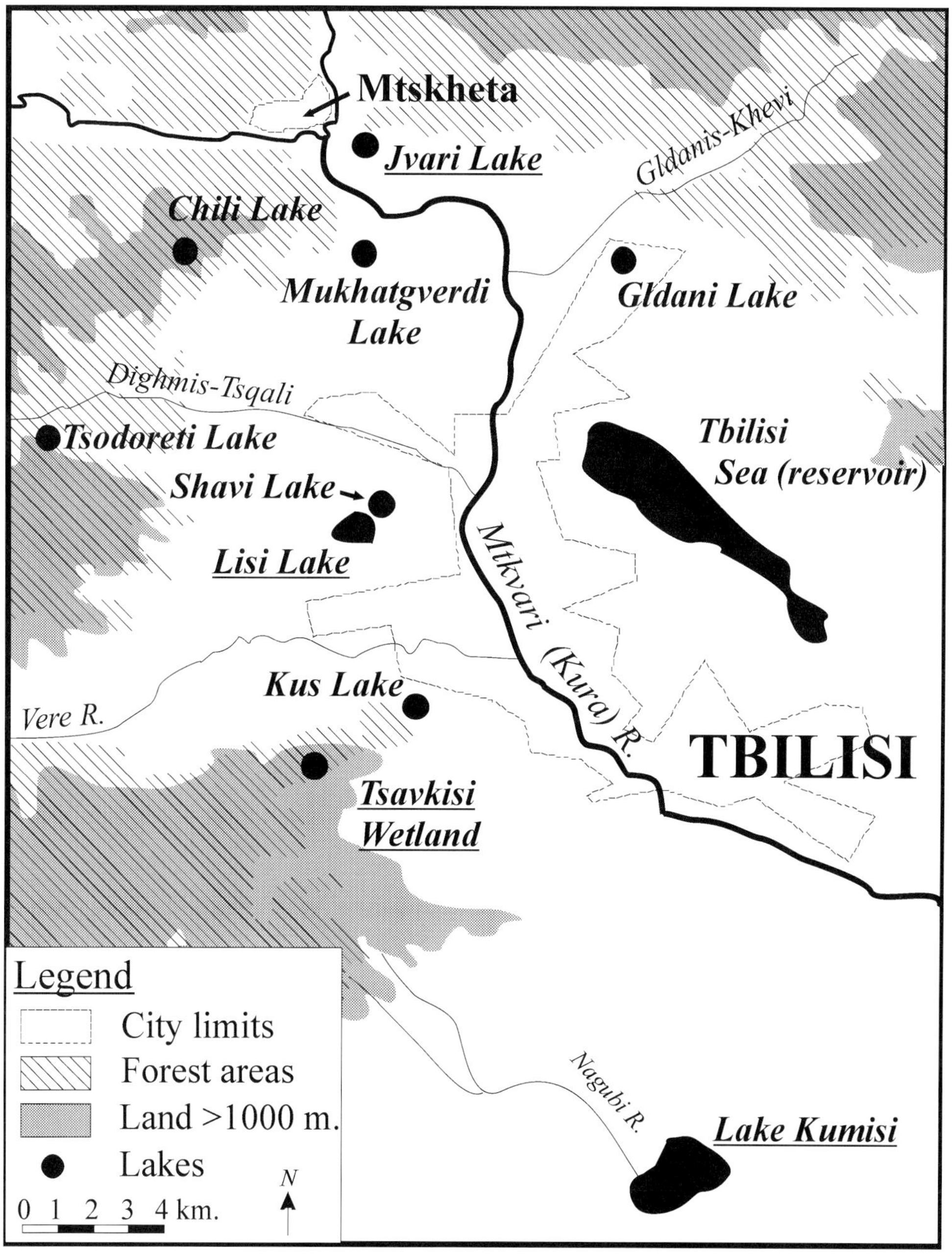

Figure 5.2. Map of the Tbilisi area showing the location of lakes, forests and land over 1000 metres in elevation (source: GUGK 1:200,000 topographic map series, 1984). Pollen diagrams are available for sites with underlined names.

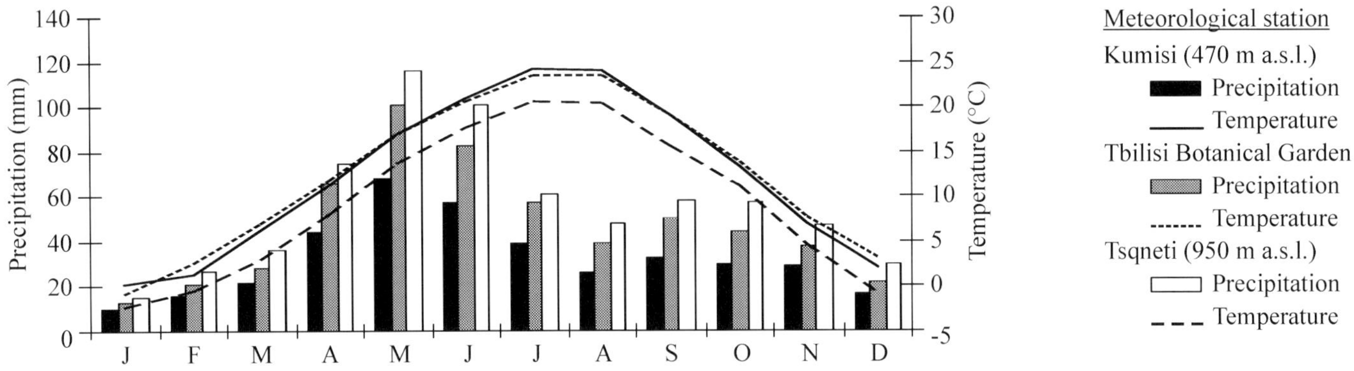

Figure 5.3. The variety of Tbilisi climates
(average monthly temperature and precipitation data compiled from Gulisashvili, 1964).

Semidesert — Composed largely of *Artemisia fragrans*-Chenopodiaceae (*Salsola, Suaeda, Petrosimonia* spp.) associations. This community is found only in the driest areas of the city, often in association with salt lakes.

Steppe — Dominated by *Bothriochloa ischaemum, Festuca valesiaca* and other grasses, this community is widespread on the left bank of the Mtkvari (Kura) River and in some places where woodland or shrubland has been cleared.

Scrub — Composed of *Juniperus*, xerophile *Pyrus* species and deciduous maquis (forest species assuming shrubby form, *e.g. Quercus iberica, Carpinus orientalis, C. betulus* and *Fraxinus excelsior*). This community is found on lower mountain slopes and in some cleared areas, especially near Mtskheta.

Forest — Most frequently an association of *Quercus iberica* and *Carpinus orientalis*, with a lesser role played by *Fagus orientalis* (on north slopes), *Carpinus betulus* and *Pinus kochiana*. Found on most of the mountains surrounding the city.

Shibliak — This is a shrubby community whose indicator species is the Christ Thorn (*Paliurus spina-christi*). It often takes the place of forest where clearing has taken place and under the continual influence of livestock grazing. It may be found in association with steppe vegetation or with species typical of the forest understorey (maquis). Other typical species include *Spiraea hypericifolia, Lonicera iberica, Rhamnus pallasii, Carpinus orientalis, Crataegus, Cotoneaster* and *Rosa* spp.

Phrygana — A community of widespread but fragmentary distribution, occurring chiefly amongst scrub and steppes. It is characterised by low, exceedingly prickly subshrubs (*Astragalus, Astracantha* and *Acantholimon* spp.) or perennials possessing pungently scented foliage (*Thymus tiflisiensis, T. transcaucasicus, Teucrium polium* etc.). They occur particularly in places where livestock grazing is so heavy that no other plants can establish there – even the grasses amongst individual phrygana plants are often chewed down to the soil.

Alongside these six groups are azonal plant communities found along rivercourses and in proximity to salt lakes,[18] as well as other, more obviously anthropogenic vegetation types, such walnut groves, orchards, vineyards, croplands and pine plantations.

The following sections describe pollen records from four sites that have yielded new data on the history of vegetation in the Tbilisi area.

The Sakhare and Kumisi Records

Lakes Sakhare and Kumisi are salt lakes situated in the lowland semidesert-steppe zone. Their vegetation is detailed in Chapter Four, along with an analysis of their modern pollen spectra.

Sedimentology and Dating

Dr Ian Thomas (University of Melbourne) collected cores from both lakes in May 2002. The Sakhare Lake core was 90 cm long, composed of sticky black muds with occasional inclusions of sand-sized particles and decayed organic matter (**Table 5.1**). These sediments are low in organic content, with loss on ignition values in the lower part of the core of around 10 percent, increasing slightly in the upper part (**Fig. 5.4**). Magnetic susceptibility values fluctuate considerably, but again are generally higher in the upper part of the Sakhare core. The one radiocarbon date from 83–84 cm depth yielded an age of 1450±80 ^{14}C bp (OZG-614), which calibrates to calendar years between 425–693 AD (1525–1257 cal BP).[19]

Over the past 1500 years or so, sedimentation in this small lake basin has been very slow. The rate of sedimentation (approximately 0.6 mm per annum) and low organic content are attributable to small catchment size and the aridity of the climate. If hydrological conditions have changed substantially at Sakhare Lake in recent millennia, this is not clearly reflected in its sediments.

The second core, from the centre of Lake Kumisi, is composed of unconsolidated sulphide muds in the upper part, and laminated silts in the lower part (**Table 5.2**). These laminations are usually 2–5 mm in thickness and are most apparent at 90–110 cm, 150–155 cm, and 168–177 cm depth. Macrofossil remains of *Chara* and *Ruppia* occur in some levels; carbonates are consistently present throughout the profile. Organic content ranges between 10 and 17 percent, the higher values occurring between 190–120 cm in the weakly laminated sediment unit (**Fig. 5.5**). A radiocarbon

[18] Ukleba 1989.

[19] Based on highest probability 2-sigma (95%) result; *i.e.* Method B of Stuiver and Reimer 1993. Please see Appendix D for further details on radiocarbon dates.

Table 5.1. Sediment description of the Sakhare Lake core.

Depth (cm)	Colour (Munsell)	Description
0–45	10YR2/1	Black clay/silt with some reed fragments and sand particles
45–70	10YR2/1	Black clay/silt with sand inclusions
70–80	10YR2/2	Dark grey clay/silt with abundant sand inclusions and mottling
80–90	10YR2/3	Compact, dark grey silt

Table 5.2. Sediment description of the Lake Kumisi core.

Depth (cm)	Colour (Munsell)	Description
0–70	10YR3/3	Massive yellowish sulphide muds
70–115	10YR4/2	Strongly laminated silts with *Ruppia* and *Chara* macrofossils
115–180	10YR4/2	Laminated silts with *Ruppia* and some macroscopic charcoal
180–250	10YR2/2	Massive dark grey silt with *Chara*, algal (?) filaments and charcoal

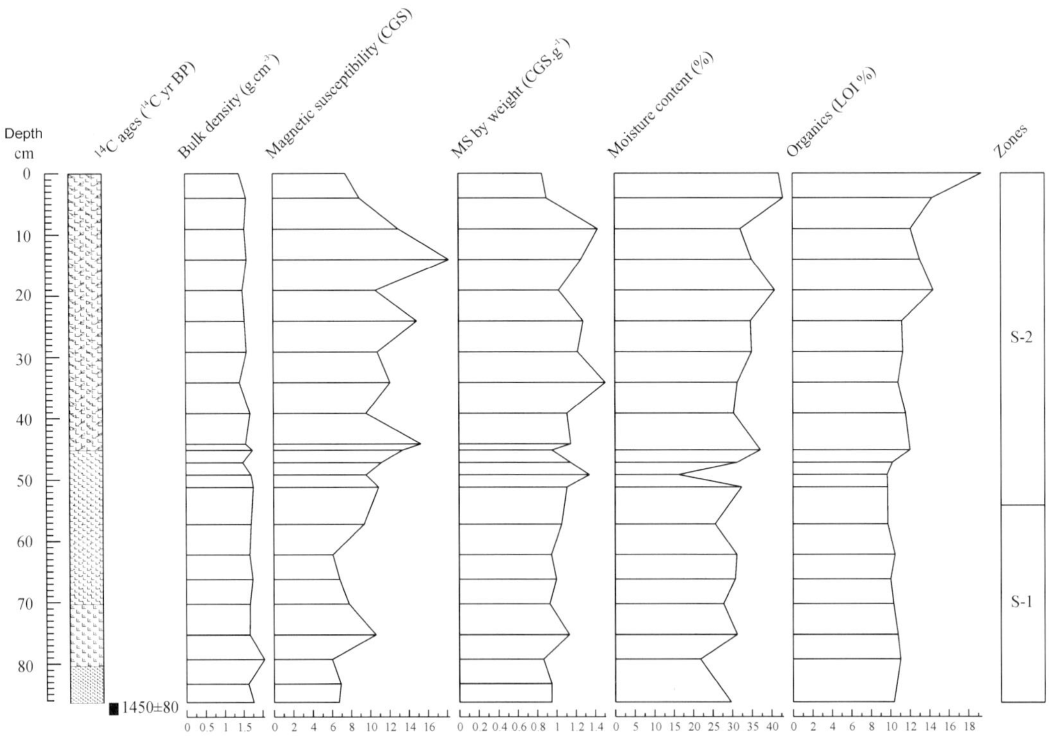

Figure 5.4. The Sakhare Lake core: lithology, radiocarbon dating, bulk density, magnetic susceptibility, moisture and organic content determinations. Pollen zones at right.

date from bulk organics at 246 cm depth produced an age of 1150±40 ^{14}C bp (OZG-616), corresponding to calendar ages between 771 and 904 AD (1179–1046 cal BP). This date is cautiously accepted, while acknowledging that uncertainty remains (*cf.* Fig. 5.22).

Given the sedimentological changes in the Kumisi core, it is probably unreasonable to assume constant sedimentation. The increased abundance of redeposited pollen in the upper metre suggests that erosional inputs have increased in recent centuries. On average, the Kumisi sediments accumulated at a rate of 2.1 mm per annum, approximately 3.5 times faster than the Sakhare material. This is due to Lake Kumisi's much larger, steeper catchment and the additional sediment inputs provided by the Nagubi Creek, the tributaries of which rise in the Teleti Range. Each tributary begins in a deeply incised ravine and joins with others to form the Nagubi on the flat, alluvial plain NW of Lake Kumisi. The name 'Nagubi', and that of its main tributary, the 'Chanchrebi', suggests that this alluvial plain was a reed swamp in the historical past.[20] Thick Quaternary lacustrine and alluvial sediments underlie the whole Kumisi Basin and ought be the subject of more detailed study.

In the 250 cm profile studied here, the lowermost and uppermost sediments are most similar in sedimentological and textural characteristics. The sediment units between 70 and 180 cm have, in contrast, higher organic content, more plant macrofossils and obvious laminations. These laminations are made up of light-coloured carbonate layers deposited during summer and darker layers of organic detritus and inorganic material deposited in cooler, wetter months. Summer evaporation concentrates the dissolved carbonates, which in turn precipitate out of the water column and are incorporated into the sediment. This process is not occurring in Lake Kumisi today, but is observed in Lake Van in Eastern Anatolia.[21] The formation of laminated sediments is usually associated with lakes that are relatively deep compared to their surface area and where bioturbation is negligible.[22] If these conditions were to be met in Lake Kumisi in the past, the lake level must have been considerably higher than it is now. This would suggest that the period of laminated sediment accumulation occurred under a climate of higher rainfall or reduced evaporation.

[20] In Georgian, Nagubi means 'former pond' and Chanchrebi means 'swampy'.

[21] Lemke and Sturm 1997.

[22] Saarnisto 1986.

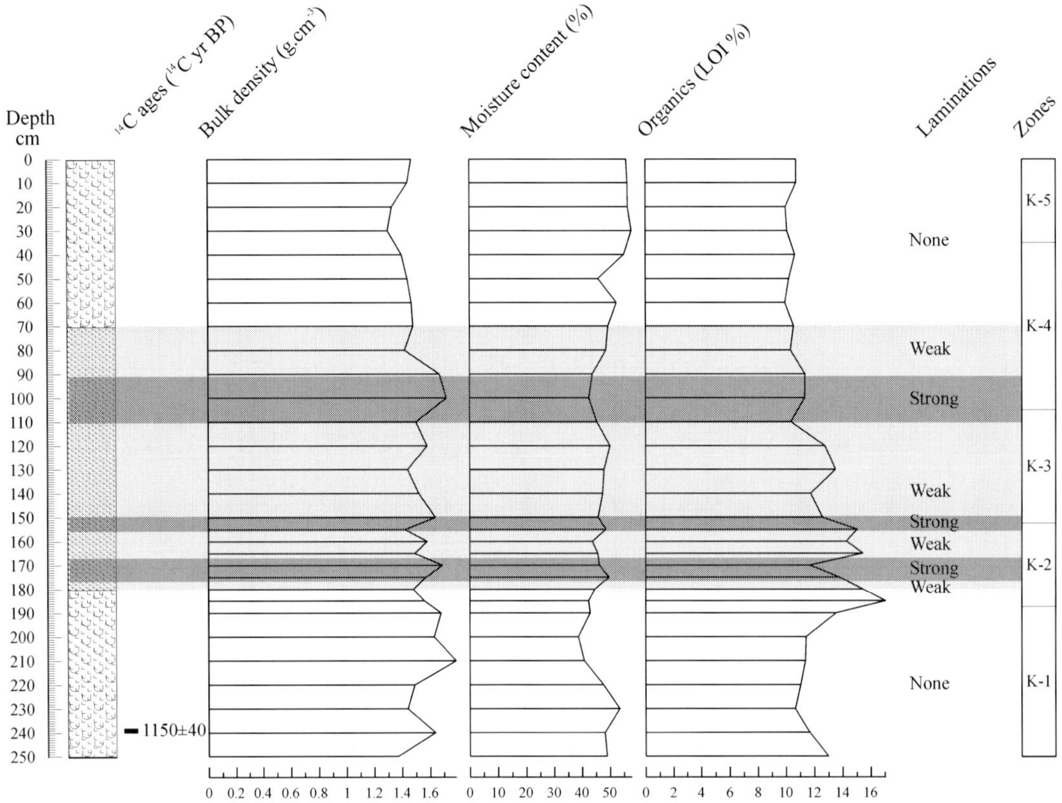

Figure 5.5. The Lake Kumisi core: lithology, radiocarbon dating, bulk density, moisture and organic content determinations. Shaded areas indicate laminated sediments. Pollen zones at right.

Pollen Diagrams

The Lake Kumisi pollen diagram is presented in **Figure 5.6**. Prior to the artificial manipulation of the lake's level two decades ago, Chenopodiaceae plants were common in the vegetation of the saline flats of the Kumisi Basin.[23] As well as being part of the broader semidesert flora, Chenopodiaceae can be regarded as part of the wetland vegetation. Chenopodiaceae pollen have been excluded from the dryland pollen sum for that reason. The diagram is subdivided into five pollen assemblage zones:

Zone K-1. Typical semidesert-steppe pollen taxa — Chenopodiaceae, *Artemisia*, Poaceae and *Polygonum aviculare*-type — dominate the first and every zone. Amongst arboreal pollen, *Pinus, Quercus* and *Carpinus betulus* are most abundant. Some shrub

[23] Ketskhoveli 1959.

taxa typical of scrub vegetation, such as *Acer, Juniperus, Rhamnus*-type and *Hippophaë*, are relatively important during this zone, despite their poor pollen representation. Amongst wetland and aquatic plants, *Ruppia* pollen dominates and is accompanied by the brackish-water *Potamogeton* subgenus *Coleogeton*, as well as *Typha*, Cyperaceae and *Phragmites*-type. Charcoal proportions increase through this zone, declining towards the end.

Zone K-2. Arboreal taxa increase in the second zone, especially *Carpinus betulus, Quercus* and *Alnus*. The olive (*Olea europaea*), which was present in Zone K-1, vanishes in Zone K-2. Two other economic plants, *Vitis* and *Morus*, also decline somewhat. Amongst non-arboreal taxa, Chenopodiaceae and *Polygonum aviculare*-type percentages decline and fluctuate considerably throughout the zone. Most other herbaceous types remain more or less stable. *Ruppia* becomes dominant in the aquatic flora along with *Botryococcus* green algae. Charcoal concentrations are higher in this zone than any of the others and dung fungi (*Sporormiella*) are all but absent.

Zone K-3. Arboreal pollen decline and herbaceous taxa expand. Foremost amongst the latter are semidesert taxa such as Chenopodiaceae, *Polygonum aviculare*-type, Lamiaceae and various Asteraceae. *Ruppia* dominates the aquatic component in the lower part of this zone, replaced by *Potamogeton* subgenus *Coleogeton* in the upper part. Charcoal concentrations are relatively low.

Zone K-4. This fourth zone differs little from the preceding zone, except that Chenopodiaceae and *Polygonum aviculare*-type decline, whilst Poaceae and *Artemisia* expand. *Juniperus, Ostrya*-type and *Cannabis*-type also increase through this zone. *Potamogeton* subgenus *Coleogeton* replaces *Ruppia*, as in the previous zone.

Zone K-5. The final zone sees a further, slight decline in forest taxa (*Quercus, Carpinus betulus, Corylus, Ulmus*) and a modest expansion of *Juniperus, Rhamnus*-type, *Cotinus* and *Platanus*. At the same time, Chenopodiaceae and *Polygonum aviculare*-type increase, and *Ambrosia*, introduced in the 1930s from North America, appears for the first time. *Ruppia, Potamogeton* subgenus *Coleogeton* and *Botryococcus* all decline through this zone.

The pollen diagram from Sakhare Lake is shown in **Figure 5.7**. Unlike the well-preserved Kumisi material, pollen in the Sakhare Lake sediments were affected by oxidation. The diagram is split into two pollen assemblage zones:

Zone S-1. The dominant pollen types in this zone are Chenopodiaceae, *Artemisia*, Poaceae and *Polygonum*. These are set against a diverse backdrop of herbaceous pollen types. The role of arboreal pollen is minor, suggestive only of long-distance pollen transport. Mid-way through the zone, Chenopodiaceae and *Artemisia* percentages drop suddenly and Poaceae increase. This trend is reversed toward the end of the first zone, when charcoal concentrations also begin to rise.

Zone S-2. The transition to the second zone is accompanied by the brief appearance of *Potamogeton* and *Sparganium* pollen and a persistent increase in Cyperaceae.

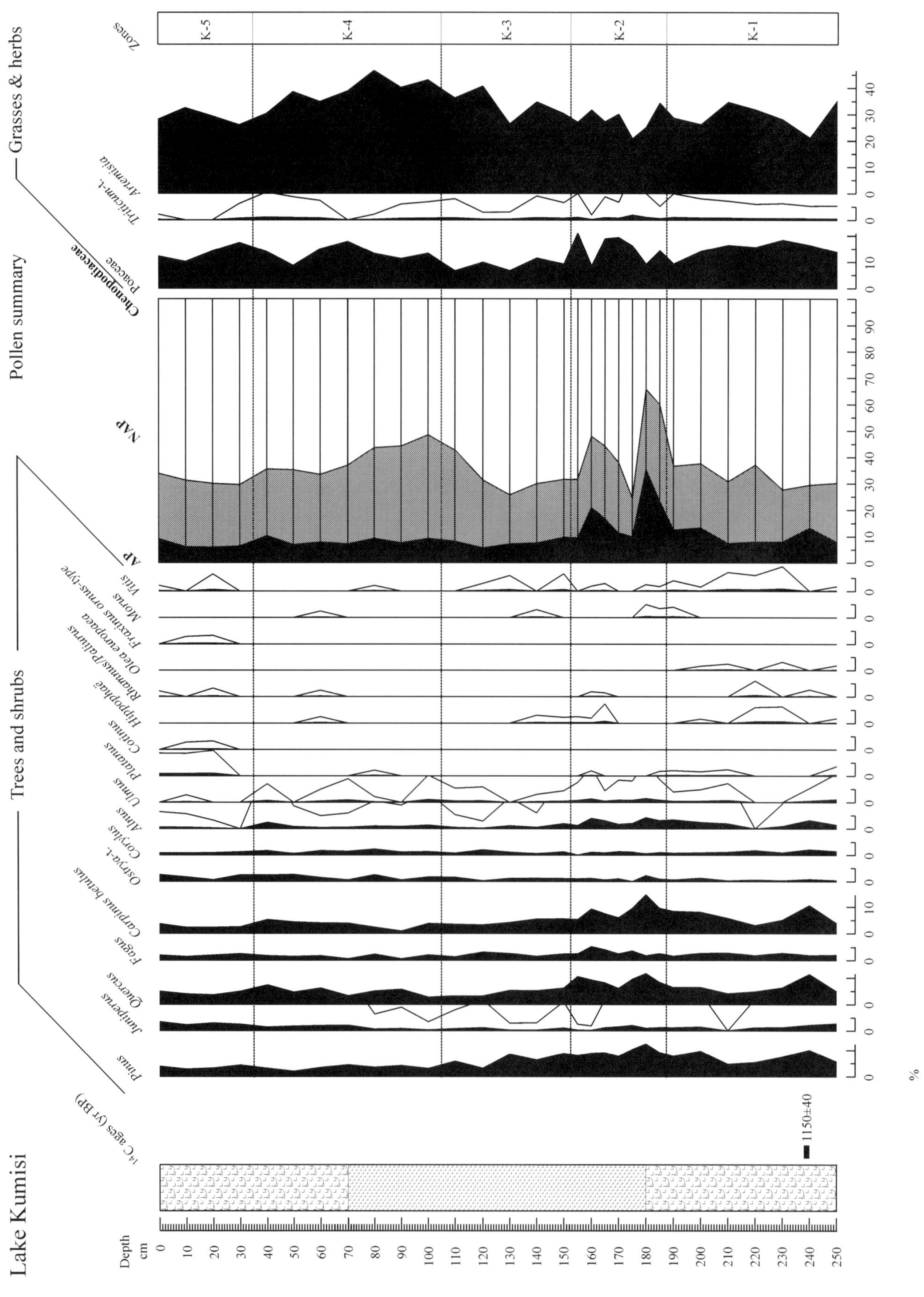
Lake Kumisi
Trees and shrubs
Pollen summary
Grasses & herbs
Depth cm
14C ages (yr BP)
1150±40
Pinus
Juniperus
Quercus
Fagus
Carpinus betulus
Ostrya-t.
Corylus
Alnus
Ulmus
Platanus
Cotinus
Hippophaë
Rhamnus/Paliurus
Olea europaea
Fraxinus ornus-type
Morus
Vitis
AP
NAP
Chenopodiaceae
Poaceae
Triticum-t.
Artemisia
Zones
K-1
K-2
K-3
K-4
K-5
%

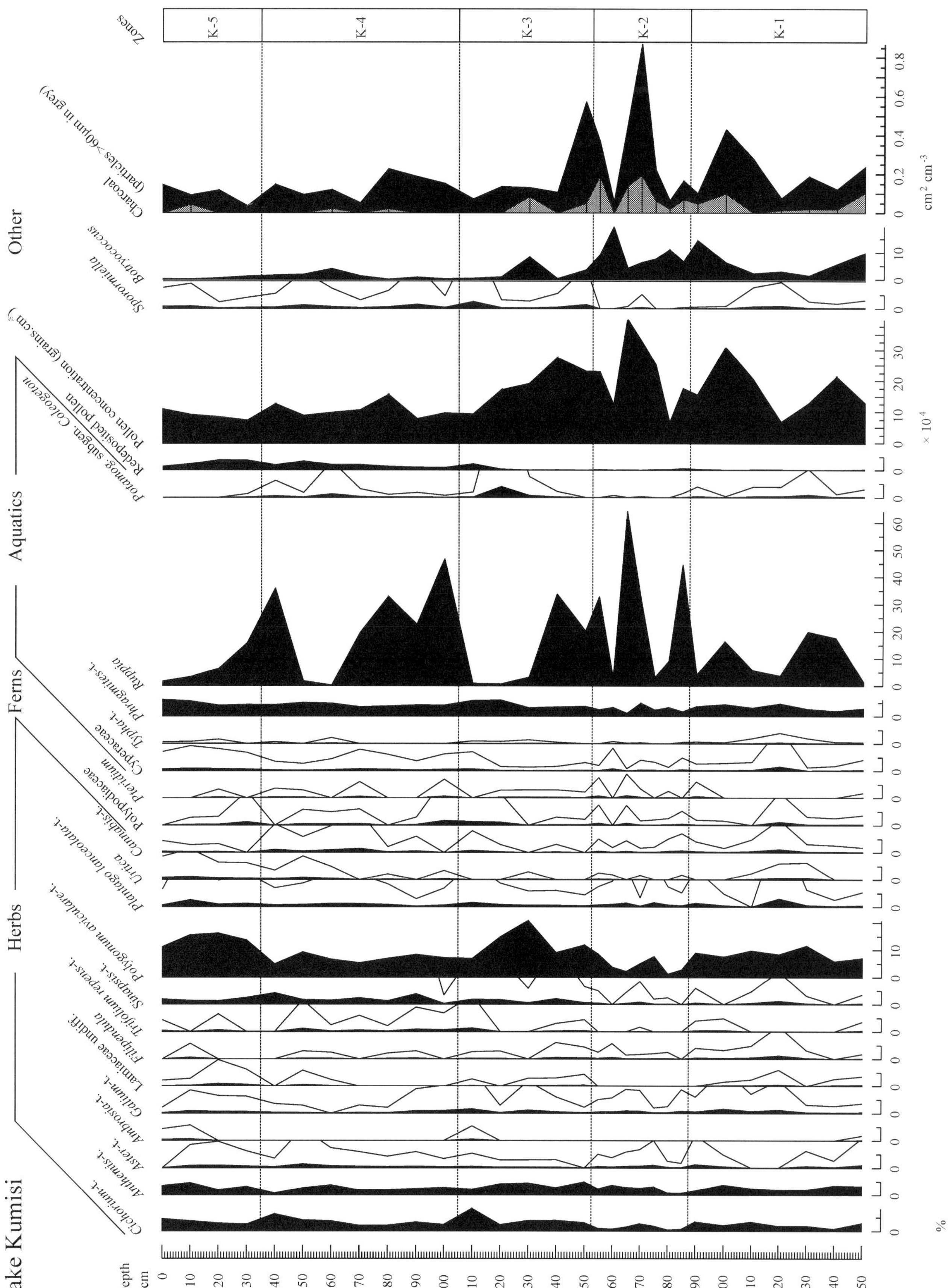

Figure 5.6. The Lake Kumisi pollen diagram. Dryland pollen sum excludes Chenopodiaceae. See Appendix E for the complete pollen diagram.

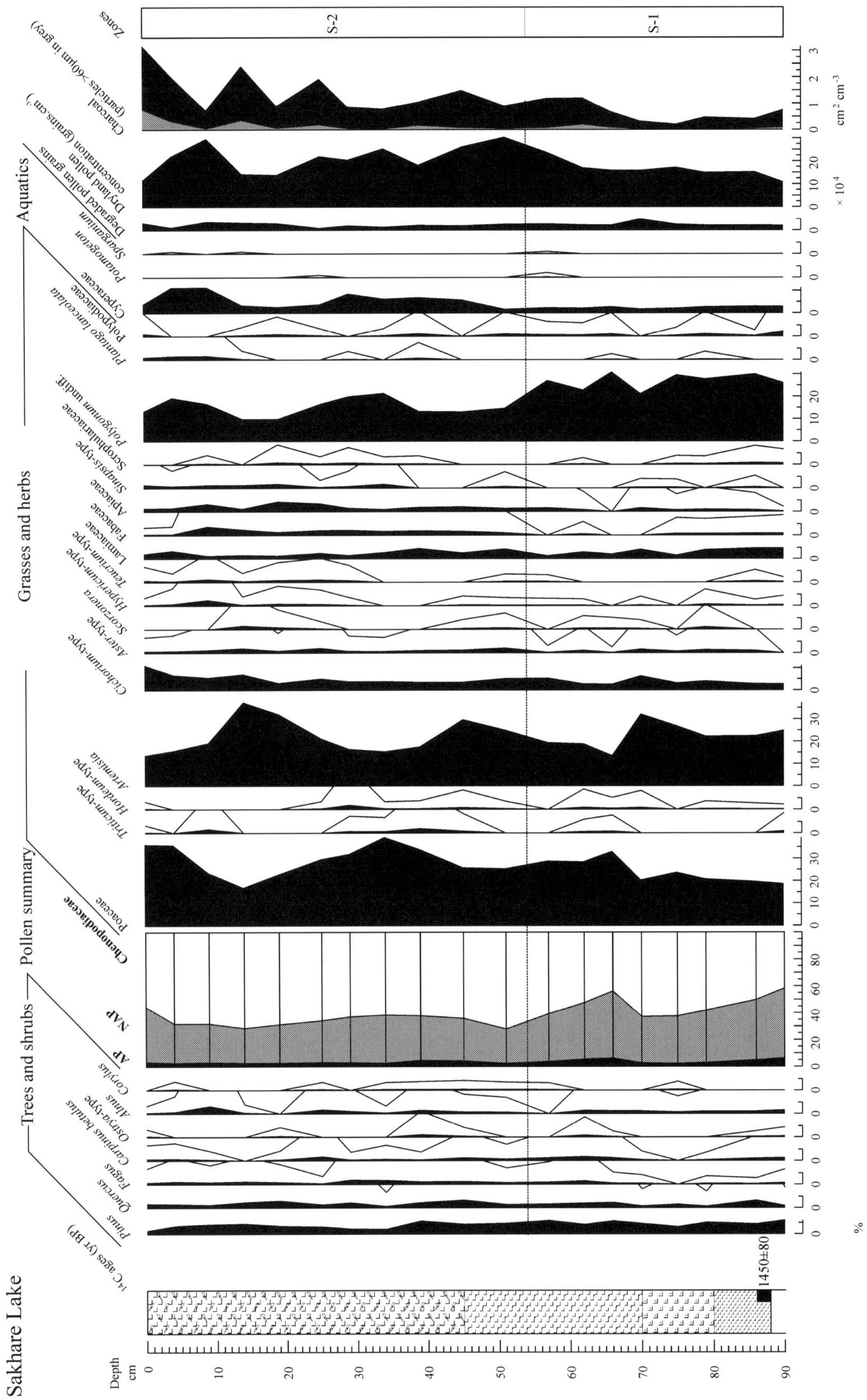

Figure 5.7. The Sakhare Lake pollen diagram. Dryland pollen sum excludes Chenopodiaceae. See Appendix E for the complete diagram.

This suggests a freshening of the water in Sakhare Lake. At the same time, *Polygonum* declines and Chenopodiaceae increases. The changing abundances of major pollen taxa indicate a relationship between Chenopodiaceae and *Artemisia* on the one hand, and Poaceae, *Polygonum* and Cyperaceae on the other. Charcoal concentrations are higher through this second zone than the first.

Numerical analyses

In the previous chapter, pollen data from a range of environments were analysed using detrended correspondence analysis (DCA). This analysis showed that rainfall, temperature and forest cover were the most important environmental gradients related to the composition of pollen spectra in Southern Georgia. Each of these gradients is associated with varying ratios of different pollen types, *e.g.* pollen spectra with high proportions of Chenopodiaceae compared to *Carpinus betulus* are almost certainly from the arid zone, just as a high ratio of *Quercus* to *Pinus* pollen is characteristic of forest landscapes. By applying the same ratios to fossil pollen data through weighted averaging, it is possible to reconstruct these environmental gradients in the past. These gradients can then be compared to a separate DCA ordination of the site's pollen data to see whether the dominant trends in the data correspond to the main environmental gradients in the present-day landscape. This may provide some indication of the major causes of vegetation change.

The results of this procedure applied to the Sakhare Lake record are shown in **Figure 5.8.**[24] Environmental changes around this site have been relatively subdued through the course of the last 1500 years. The first gradient, one of rainfall, shows that conditions were slightly moister 1500 years ago. Other phases with a climate similar to today's occurred around 1000, 700 and 300–200 years ago.[25] These periods were rhythmically interspersed with drier times, but these changes had a minor impact on the vegetation.

The second gradient is a measure of forest cover and partly temperature. It shows that these parameters around Sakhare Lake have changed very little during the last 1500 years. The third gradient also demonstrates remarkable stability. It is the only gradient to be correlated with the DCA ordination (**Fig. 5.8**; Appendix B), indicating, most probably, that variations in land-use are responsible for vegetation changes around this site.

[24] Throughout, weighted averaging has been applied to conventional, dryland pollen percentages, rather than the adjusted percentages shown in the pollen diagrams (see Chapter Four), while each site-specific DCA is based on pollen percentages as presented in the diagrams in Appendix E.

[25] Assuming uniform sedimentation rates.

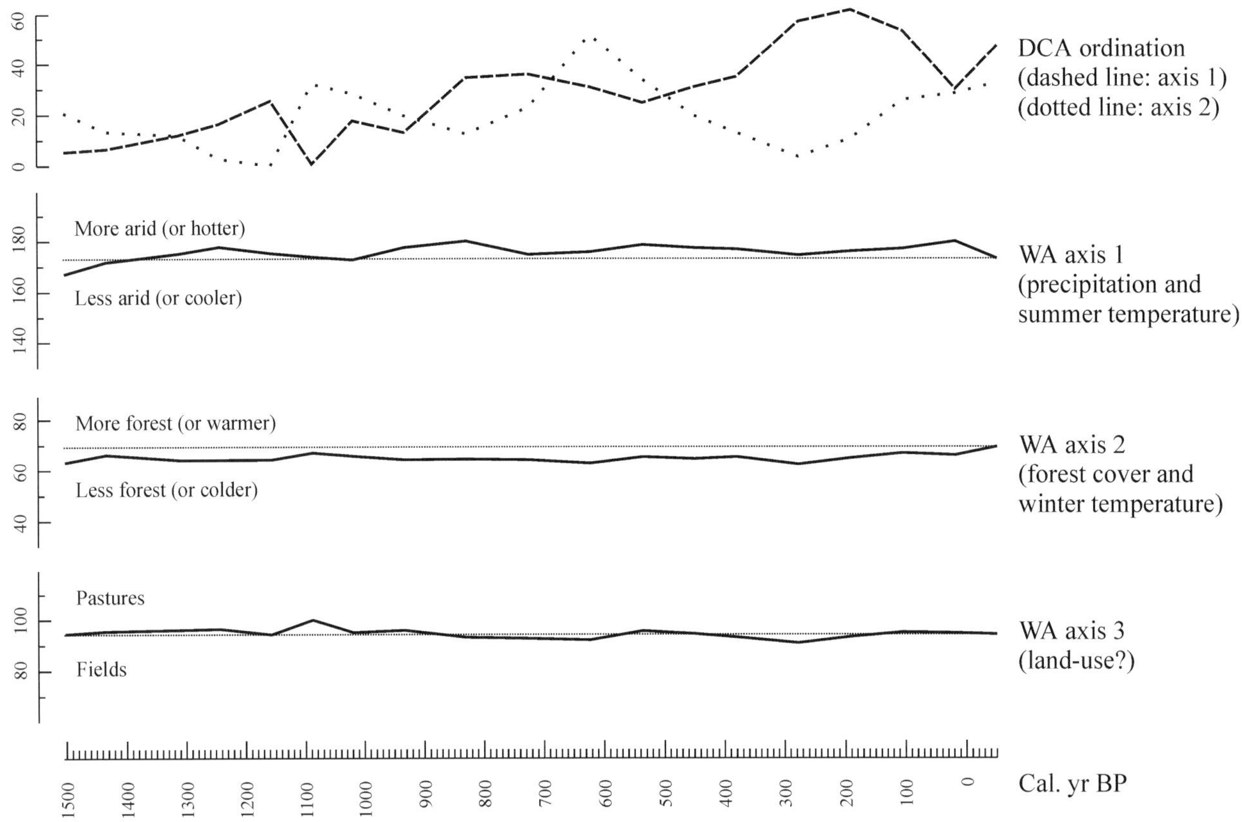

Figure 5.8. Environmental changes at Sakhare Lake: rainfall, forest cover and land-use gradients produced by weighted averaging. Dominant trends in the pollen data (DCA ordination) are shown at the top for comparison.

According to the TWINSPAN classification presented in the previous chapter, the vegetation surrounding Sakhare Lake changed from 'steppe-forest' in pollen zone S-1 to '*Artemisia* steppe' in pollen zone S-2, suggesting a minor reduction in tree cover through the last millennium.

The history of Lake Kumisi appears to have been more changeable (**Fig. 5.9**). The primary rainfall gradient shows that the climate 1100 years ago was, in palynological terms, quite similar to today's. At some time around 900 years ago, the climate suddenly became much wetter, reverting to a drier state about 700 years ago. It became wetter again 500–400 years ago and dried during the last few hundred years. This climatic model is substantiated by the sedimentology of the Lake Kumisi core – the laminated sediments coincide perfectly with these periods of hypothesised wetter climate. In those times, Lake Kumisi must have been considerably larger and deeper than it is now as a result of increased stream and rainwater influx or reduced evaporation.

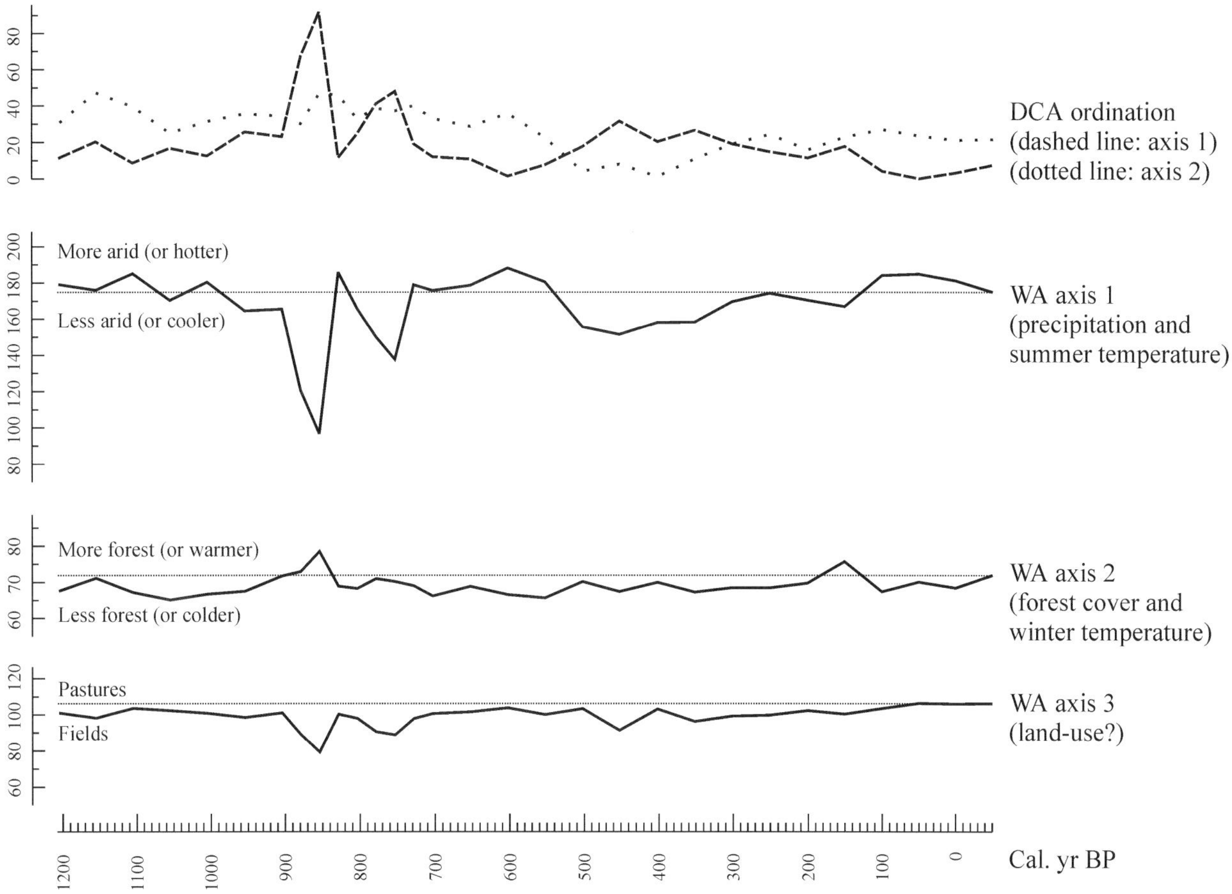

Figure 5.9. Environmental changes at Lake Kumisi: rainfall, forest cover and land-use gradients produced by weighted averaging. Dominant trends in the pollen data (DCA ordination) are shown at the top for comparison.

The second gradient shows that although arboreal pollen percentages fluctuate considerably in the Lake Kumisi pollen diagram, the changes in forest cover or temperature were relatively insignificant. The only possible increase occurs at the same time as the first wet phase, between 900 and 800 years ago, but was very short-lived and of low magnitude. Importantly, the lack of variation in the second gradient demonstrates that fluctuating arboreal pollen percentages during the earlier wet phase are merely the result of increased pollen source area as rainfall increased and the lake became larger.[26] The second wet phase is accompanied not by arboreal pollen, but by *Artemisia* and *Ruppia*. Because *Artemisia* pollen percentages are positively correlated with atmospheric

[26] Compare surface samples K2 and K3 in Chapter Four.

moisture and *Ruppia* pollen is most abundant in shallow saline waters,[27] this second phase may be the result of decreased evaporation rather than increased rainfall.

If the third gradient can be interpreted in terms of land-use, it may be that during the wetter phases some farmers put aside the cattle prod and took up the plough.

The strong relationship (r^2=0.93) between the rainfall gradient and the first axis of the DCA ordination (**Fig. 5.9**; Appendix B) indicates that the vegetation history of this large lake is primarily a reflection of climatic or lake-level factors.

Like Sakhare Lake, TWINSPAN classification of Lake Kumisi's pollen zones indicates the surrounding vegetation changed from 'steppe-forest' in zones 1–3 to '*Artemisia* steppe' in zones 4–5. This apparent change in the vegetation is not clearly portrayed by any of the trends in **Figure 5.9**, so one must assume that it was of relatively small magnitude.

The Lake Jvari Record

Lake Jvari is situated at an altitude of 570 m in a steep-sided depression next to the 6th century Jvari Monastery, one of Georgia's oldest cruciform churches and perhaps the country's most iconic building. The geology of this area is very diverse, consisting of marine sand, marl, conglomerate and tufa of Eocene, Oligocene and Miocene age. These sedimentary rocks have been tilted, faulted and eroded to form a WSW–ENE-trending valley occupied by Lake Jvari on one side of the Kura River and the deep ravine of Karsanis-Khevi on the other. Like most of the salt lakes around Tbilisi, Lake Jvari was probably formed through a combination of tectonic activity and the suffosion of underlying gypsum.[28] Upwellings on the lake surface suggest that the lake is largely fed by subterranean springs.

Lake Jvari is surrounded by a mixture of forest, shibliak scrub and open steppes (**Fig. 5.10**). The dominant species in the forest are *Carpinus orientalis, C. betulus, Quercus iberica* and *Fraxinus excelsior*, with an understorey of *Helleborus caucasicus,* various grasses and *Echinops sphaerocephalus*. At the forest edges, species such as *Spiraea hypericifolia, Crataegus curvisepala, Cotoneaster integerrimus, Prunus spinosa, Lonicera iberica* and *Acer laetum* are more or less common.

Paliurus spina-christi dominates in areas of shibliak vegetation, along with *Rosa canina, Spiraea hypericifolia* and *Genista flagellaris*. The open, semidesert-steppe vegetation that grows on the lake terrace is dominated by grasses, with *Artemisia chamaemelifolia, Cirsium arvense, Eryngium campestre, Echium vulgare, Euphorbia squamosa* and many other species. The lake itself is so salty and alkaline that few vascular plants grow in or near the water. The exception is *Ruppia spiralis*, which grows in the water when the lake is full. Once the water evaporates, only halophilous algae remain.

[27] El-Moslimany 1990; Vance and Mathewes 1994.
[28] Dzhanelidze 1980.

Figure 5.10. Photograph of Lake Jvari, with the Monastery of the Cross in the background (September 2003).

The land around Lake Jvari is mostly used as rough pasture for cattle. Pine plantations are also common in the area, as are plantings of cedar and poplar. Hunters roam through the neighbouring forests, which, along with the lake and its sediments, are the property of the Patriarch of Georgia.[29]

Sedimentology

The 2.5-metre-long Lake Jvari core was collected in September 2003 from the centre of the then dry lake. It consists of black, unconsolidated, sulphide muds in the upper 40 cm, changing to soft, homogenous, dark-grey marl in the lower part. The sediments are highly alkaline (pH >10) and carbonate rich. Sediment moisture and organic content exhibit considerable variation, but generally decline with increasing depth (**Fig. 5.11**). The highest loss-on-ignition value (28 percent) occurs at the surface, and the lowest (9 percent) at 150 cm. Macrofossils, mostly *Chara* oogonia, *Ruppia* seeds and filaments, occur throughout the core.

[29] Connor 2003.

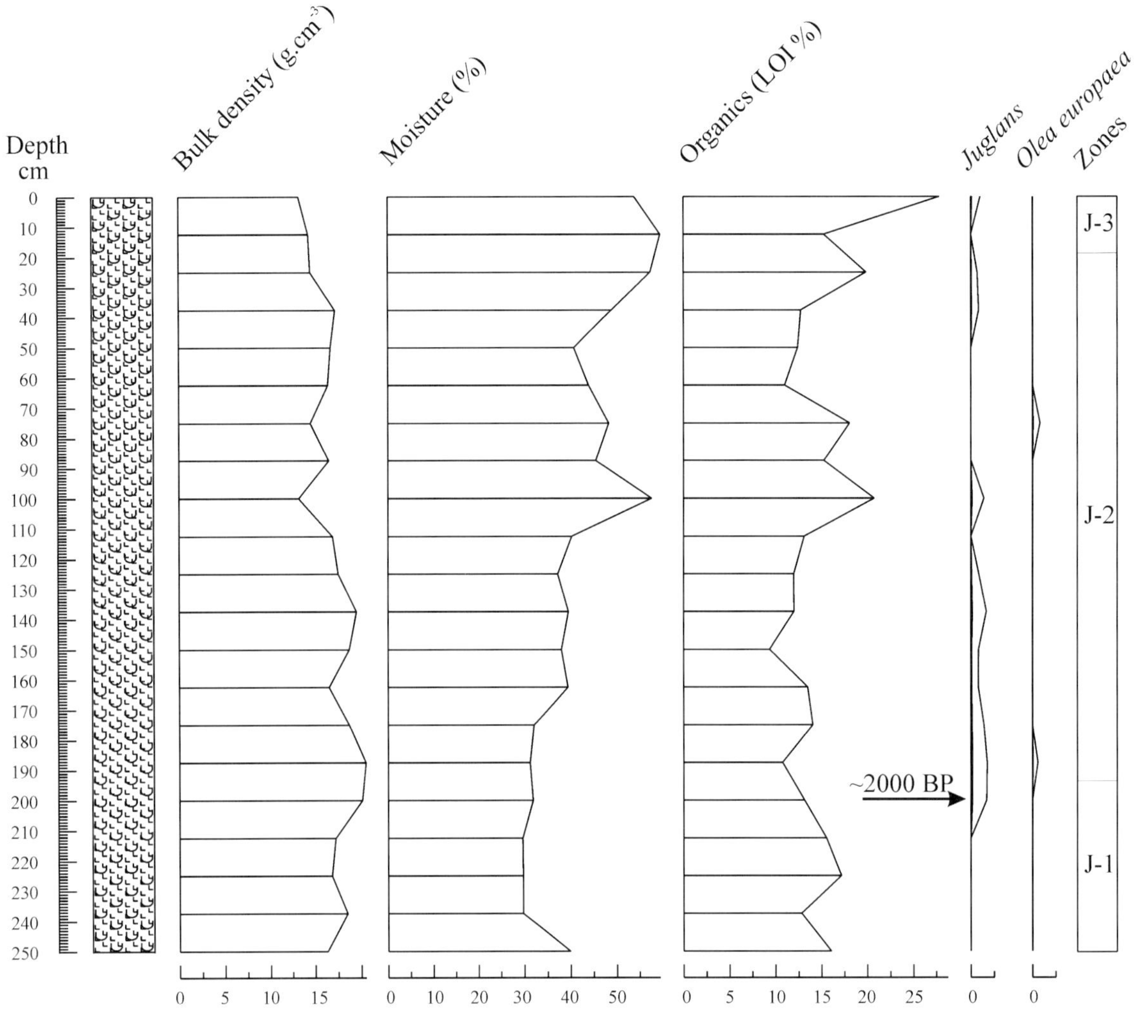

Figure 5.11. The Lake Jvari core: lithology, bulk density, moisture and organic content determinations. Biostratigraphic markers used for dating purposes are shown. Pollen zones at right.

The extreme salinity and alkalinity of Lake Jvari is a function of its considerable age, its water source in alkaline springs, the lack of inflowing or outflowing streams, and intense evaporation during hot, dry summers.

No radiometric dates were obtained for the Lake Jvari core, but on the basis of pollen markers and correlation with other records (see below) it seems to span the last 2000–2500 years. If this is correct, sedimentation has occurred at an average of 1.0-1.3 mm per year. This rate falls somewhere between that of Lake Kumisi and Sakhare Lake, consistent with the lake's catchment size, topography and climate.

Pollen Diagram

Pollen preservation throughout the Jvari material is excellent, despite the high pH of the sediments. This is because the strongly reducing conditions protect pollen from oxidation. Good preservation also implies that this lake has not desiccated completely in recent centuries, even though the sediment surface may be exposed in dry years. The Jvari pollen diagram (**Fig. 5.12**) is divided into three pollen assemblage zones:

Zone J-1. The lowermost zone is characterised by relatively high arboreal pollen proportions, particularly *Pinus, Juniperus* and *Carpinus betulus*. Amongst non-arboreal taxa, Poaceae dominates, followed by *Artemisia*, Chenopodiaceae and indicators of open pastures – *Galium*-type, *Plantago lanceolata, P. major*-type and *Trifolium repens*-type. An abundance of *Sporormiella* dung fungal spores confirm the existence of grazing, and the presence of *Paliurus* (*Rhamnus*-type) pollen indicates that shibliak vegetation was already present at that time. The landscape during this first zone was probably a mixture of forest, shibliak and open pastures, much as it is presently. *Ruppia* is the most important aquatic taxon, followed by *Potamogeton* subgenus *Coleogeton*, indicating that the lake water was brackish to saline.

Zone J-2. Arboreal pollen declines at the beginning of the second zone, compensated by a slight increase in steppic herbs, especially *Centaurea nigra*-type. Amongst economic plants, pollen of the walnut (*Juglans*) appears for the first time, and both wheat (*Triticum*-type) and grape (*Vitis*) pollen become more abundant. *Sporormiella* proportions reach very high levels: 8.5 percent of the total pollen sum. Following a charcoal peak mid-way through this zone, *Carpinus betulus, Quercus* and *Pinus* percentages increase, but non-arboreal pollen percentages do not change appreciably. Only *Marrubium* and *Polygonum aviculare*-type pollen appear to rise after this temporary increase in AP. Charcoal concentrations peak in the upper part of the zone and then decline steadily through the remainder of the record. In the lake itself, *Ruppia* and *Botryococcus* populations fluctuated considerably and were at times replaced by *Potamogeton* subgenus *Coleogeton*. Presumably this occurred during higher lake phases when the water was less saline.

Zone J-3. The uppermost zone comprises only two samples, distinguished from the previous zone by an increase in arboreal pollen proportions (especially *Juniperus* and *Pinus*) and a decline in *Artemisia* and Chenopodiaceae. The appearance of *Ambrosia*-type and *Zea mays* pollen in this zone place it within the 20th century. *Sporormiella* declines to its lowest levels in the entire Jvari record, suggesting that grazing intensity has decreased in the last 100 years. Charcoal concentrations follow the same trend.

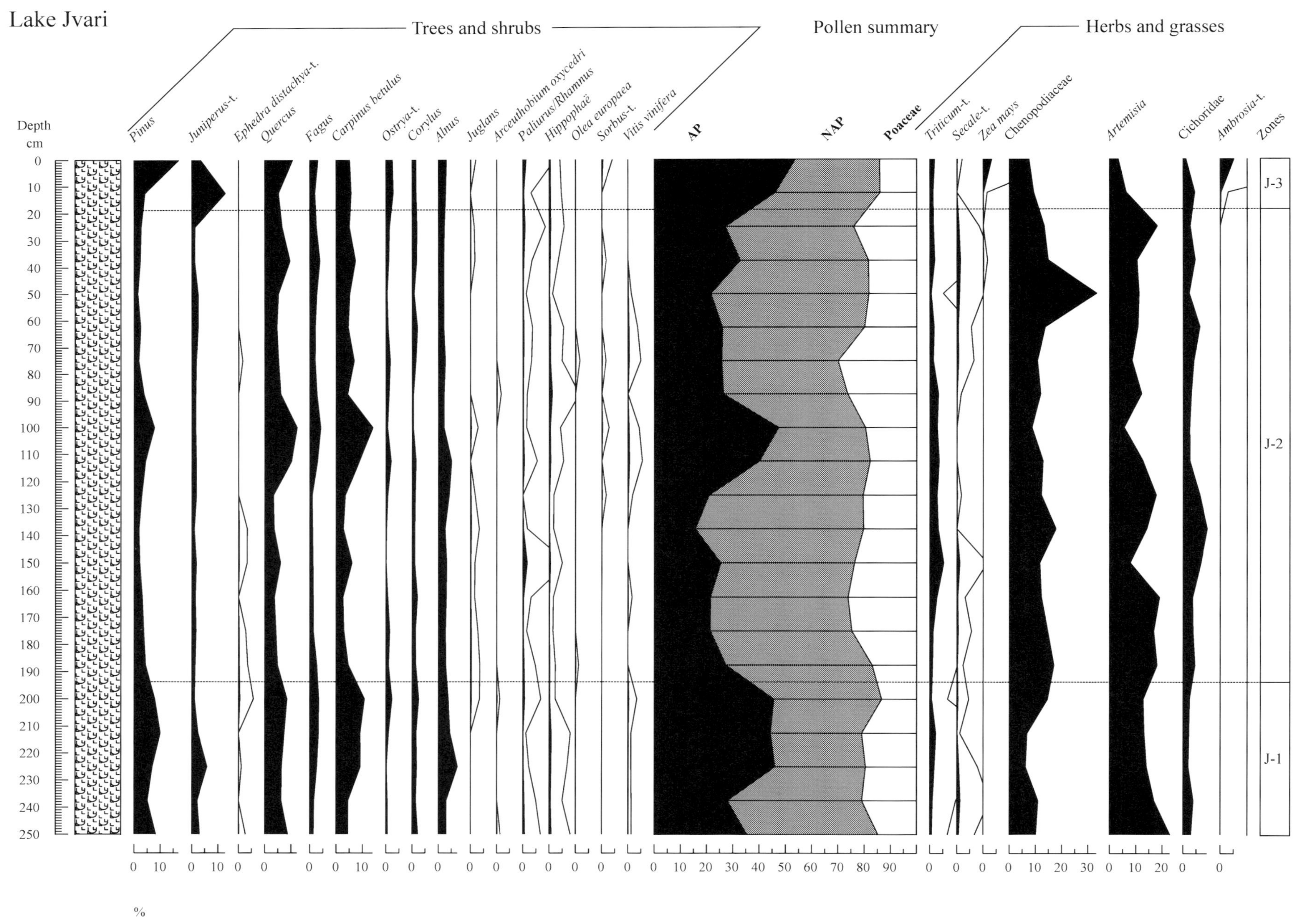

Lake Jvari
Trees and shrubs
Pollen summary
Herbs and grasses
Depth cm
Pinus
Juniperus-t.
Ephedra distachya-t.
Quercus
Fagus
Carpinus betulus
Ostrya-t.
Corylus
Alnus
Juglans
Arceuthobium oxycedri
Paliurus/Rhamnus
Hippophaë
Olea europaea
Sorbus-t.
Vitis vinifera
AP
NAP
Poaceae
Triticum-t.
Secale-t.
Zea mays
Chenopodiaceae
Artemisia
Cichoridae
Ambrosia-t.
Zones
J-3
J-2
J-1
%

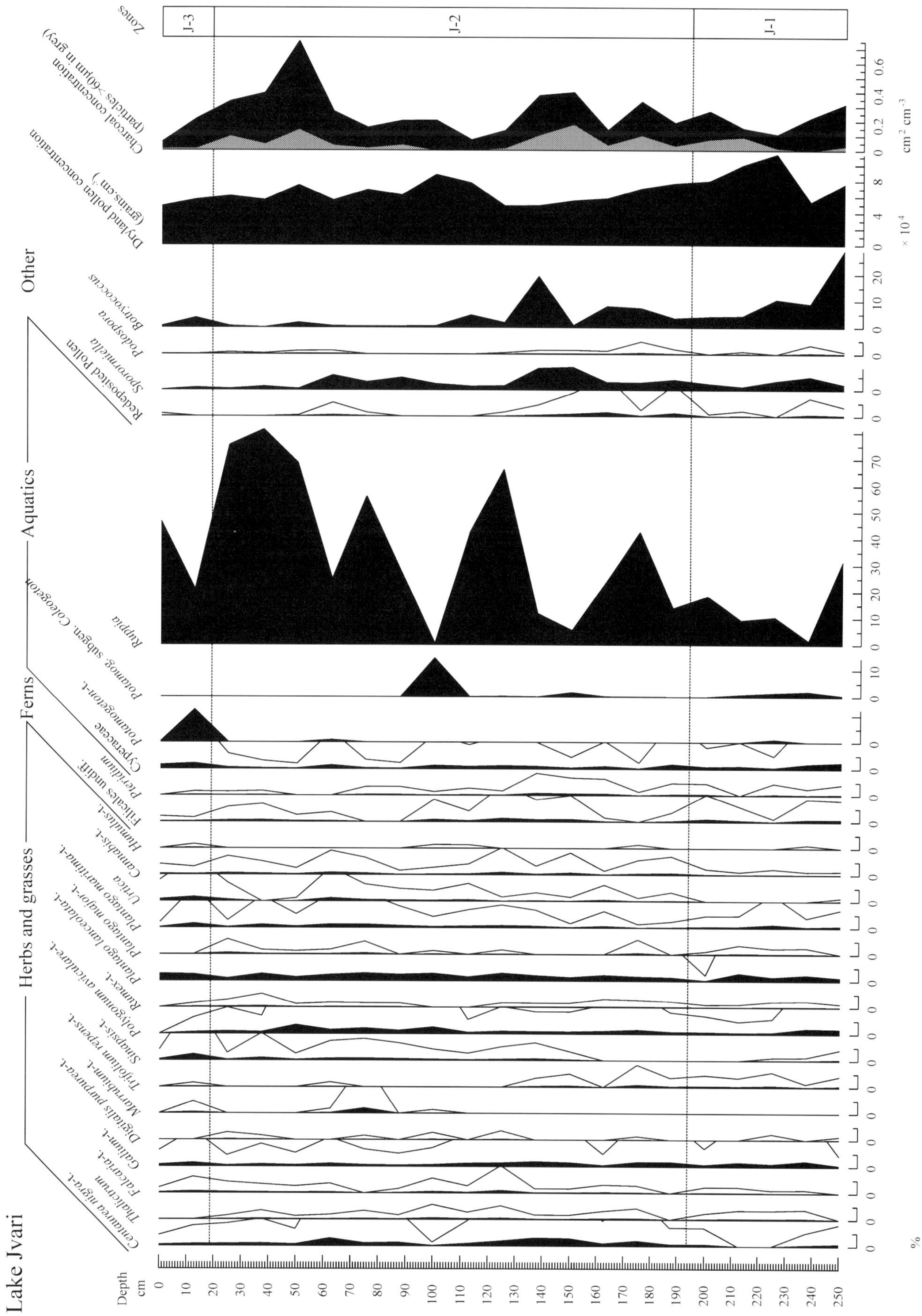

Figure 5.12. The Lake Jvari pollen diagram. Percentages calculated conventionally. See Appendix E for the complete diagram.

Numerical Analysis

The results of applying the weighted averaging technique to the Lake Jvari material are shown in **Figure 5.13**. The primary, climatic gradient follows the *Ruppia* pollen curve very closely, indicating that the changes in atmospheric moisture were reflected with exceptional clarity in both dryland and aquatic pollen spectra. *Ruppia* pollen was most abundant in relatively dry periods, when lake levels were consequently low. When the lake level was higher, *Potamogeton* species became more abundant and the accumulation of organic matter and arboreal pollen in the sediments increased.

The second gradient represents changes in forest cover. It shows that the extent of woodland has oscillated throughout the record, but that the changes were relatively minor. The sample scores here do not indicate that forest cover was ever continuous, nor do they suggest that forest was ever completely absent. All of the Jvari pollen zones were classified as 'forest-steppe' by TWINSPAN. Comparison of the first two gradients permits the conclusion that forest area expanded when the climate was wetter, and retracted either as a result of fire or when the climate became drier. Grazing pressure also had a strong influence on the vegetation.

The third axis seems to indicate that grazing was more widely practiced during drier times, and cropping during wetter phases, but this relationship is far from clear or straightforward. It also seems that relaxation of grazing pressure aided the spread of (secondary) woodland in some instances.

The main trend identified by DCA ordination is correlated with the first WA axis (r^2=0.71; **Fig. 5.13**; Appendix B), indicating that climatic variations are responsible for the major changes in the pollen record at this site.

Dating these changes relies on the appearance of *Juglans* and *Olea europaea* pollen at a depth of 200 cm. These pollen taxa appear consistently in pollen records from the study area from the Roman Period onwards (see Chapter Eight). The rainfall gradient shown in **Figure 5.13** also correlates well with the same curves for the last 2000–2500 years from other sites (Tsavkisi, Imera and Aligol – this chapter and Chapter Six).

While some dating uncertainties remain, in broad terms both landscape and climate at Lake Jvari 2000 years ago were similar to today's. Around 1500 years ago the climate became drier, and forest cover retracted somewhat as grazing pressure increased. Similarly dry spells recurred about 1000 years ago and 500 years ago, interrupted by periods with cooler or wetter conditions. In essence, though, the landscape changed remarkably little through the last 2500 years. The Lake Jvari pollen diagram, like those from Lake Kumisi and Sakhare Lake, is a good example of vegetation and land-use stability in the face of climatic variability.

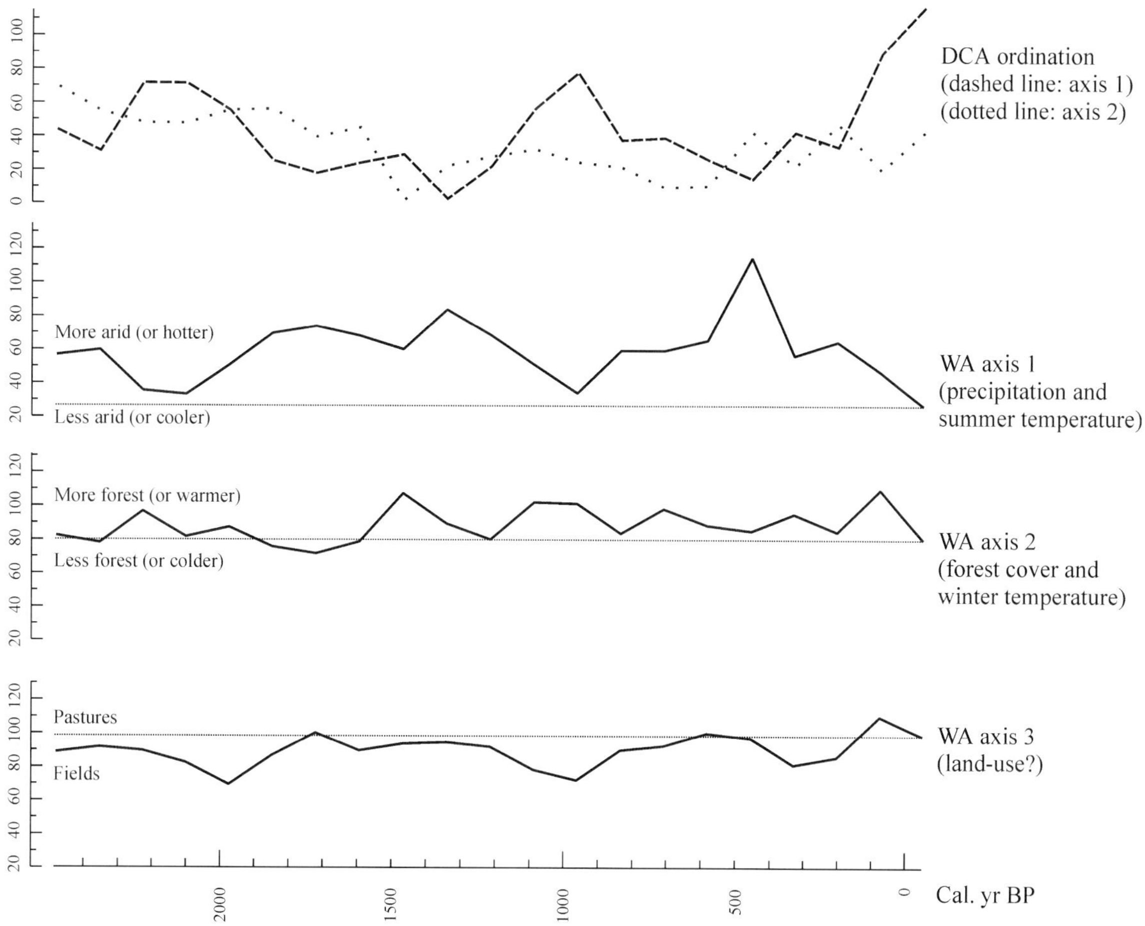

Figure 5.13. Environmental changes at Lake Jvari: rainfall, forest cover and land-use gradients produced by weighted averaging. Dominant trends in the pollen data (DCA ordination) are shown at the top for comparison. Dating relies on the appearance of *Juglans* pollen, ~2000 cal BP.

The Tsavkisi Wetland Record

Tsavkisi Wetland is a reed-dominated mire located in the hills six kilometres west of Tbilisi, mid-way between the villages of Tsqneti and Tsavkisi, at an elevation of 1110 metres. Within an anticlinal fold of Upper Eocene sedimentary units (tufogenetic sands, clays, marls and schists) in the eastern Trialeti Range, it occupies a triangular depression that was probably created by the faulting, metamorphosis and subsequent erosion of interbedded saline sedimentary strata.[30] This tectonic genesis is

[30] Dzhanelidze 1980; Tsereteli 1971.

reflected in the steepness of the basin sides, the wetland's lateral constriction, and the straight, geologically-determined basin sides. The wetland surface covers an area of approximately 32 ha. A map of the geomorphology (**Fig. 5.14**) highlights the complexity of the surrounding relief. Data on the site's vegetation and modern pollen representation are given in Chapter Four.

Whilst the wetland has no visible inflow (it is at least partially spring-fed), there is a small outlet that flows westward into the Ukanakhevi (Back Creek). This outlet was inactive during all visits to the site and is probably intermittently active only after heavy rain.

Sedimentology and Dating

Dr Ian Thomas (University of Melbourne) and Dr Henrikh Avakov (Georgian Institute of Palaeobiology) collected the first core from Tsavkisi Wetland in May 2002. The core was one metre in length and was obtained with a Russian corer at a place between the southern shore and the centre of the wetland. At a metre's depth, the corer repeatedly struck an obstruction, so further penetration was not possible. The wetland sediments were cored twice more: in September 2003 and July 2005. In 2003, Michael Fletcher and I cored to a depth of 7.8 m using a 'sticky clay' auger. The length of this core was limited by a lack of equipment rather than a lack of

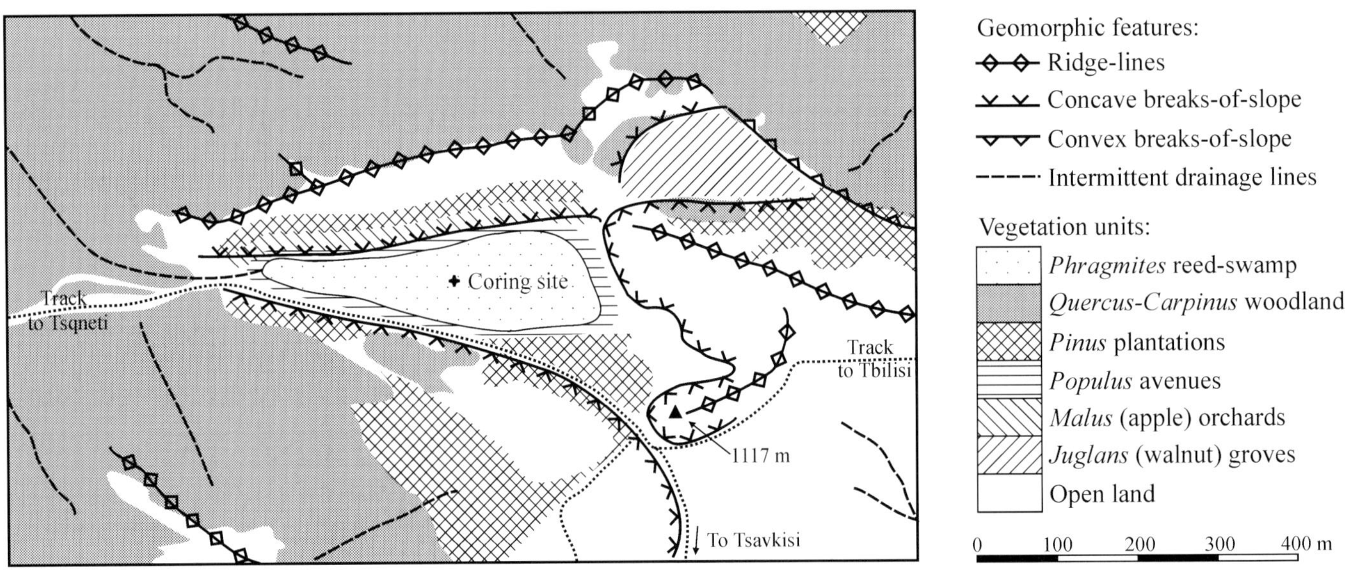

Figure 5.14. Map of Tsavkisi Wetland, showing principal geomorphic features, vegetation units and coring location. Compiled from aerial photographs, topographic maps and ground surveys.

sediment below 7.8 metres. In 2005, the record was extended to a depth of 11.1 m. Hydrostatic pressure at this depth was too great to permit further boring with hand-operated equipment. Both of these cores were obtained from the centre of the wetland (see **Fig. 5.14**).

The sediments of Tsavkisi Wetland are extremely varied, consisting predominantly of lake marl. They alternate between silty and more organic strata, and contain macrofossils including gastropod shells, molluscs, *Chara* oogonia, *Phragmites* fragments and *Potamogeton* seeds.

The lowermost section of the core consists of massive, dark grey, silty sediment. Layers with a more peaty texture occur between 1080–1045 cm and 960–915 cm. Above 665 cm the sediments are altogether more organic, first dark brown, then yellow and unconsolidated. The sediment changes sharply to crumbly, dark brown silty peat above five metres, then changes again to dark grey silts that closely resemble those from the bottom of the core. Around three metres' depth, the sediments yet again become peatier. Brown crumbly peat appears at 2.25 m and the upper part of the core is characterised by cappuccino-coloured silty gyttja with peat layers, abundant gastropods, charcoal bands and macrofossil remains of wetland plants. A detailed description is given in **Table 5.3**.

Moisture content and loss on ignition curves are given in **Figure 5.15**. The sediments can be classed broadly into three phases based on organic content: silt-rich basal sediments (1110–640 cm), silts with a higher organic content (630–190 cm) and the peaty upper layers (185 cm to the surface). Every sample from this core contained carbonates. In the silty sediments, this was held in sedimentary carbonates; in the peat layers, it resided in gastropod remains.

The sedimentological changes in the Tsavkisi core can be interpreted with reference to analogous lakes in the study area. The dark-grey lake marl sediments resemble those from salt lakes such as Jvari, Sakhare and Shavi Lakes (**Fig. 5.2**), which are situated at lower altitudes than Tsavkisi Wetland, experience a more arid climate, and have less vegetation cover. Tsavkisi Wetland's dark-grey sediments were probably deposited in similar conditions, of greater aridity and intense summer evaporation. However, the laminations visible in parts of the Tsavkisi core are absent from these other lakes, suggesting that although the past climate was more arid, lake levels in Tsavkisi Wetland were still relatively high, with intense summer evaporation.

The organic layers in the Tsavkisi core were probably deposited when the water level was stable or aquatic vegetation was able to colonise the waters due to a dilution of dissolved salts. This could have occurred with increased precipitation, less droughty summers or reduced evaporation. Lakes with peat accumulation in the present-day include those on the Tsalka Plateau (see Chapter Six), which experience high lake levels in summer when most aquatic plants are growing rapidly.

Table 5.3. Sediment description of the Tsavkisi Wetland cores.

Depth (cm)	Colour (Munsell)	Description
0–30	10YR4/3	Soft, brown detritus peat with gastropod remains
30–35	10YR2/1	Black, elastic peat layer containing abundant rootlets
35–50	10YR3/2	Dark-brown peaty sediment with gastropod remains
50–54	10YR2/1	Black peaty silt with abundant reed fragments
54–117	10YR3/2	Brown organic-rich marl with plant remains, charcoal and gastropods
117–142	10YR5/2	As above, only lighter in colour
142–159	10YR3/1	Crumbly, peaty marl with abundant gastropod and mollusc remains
159–170	10YR2/1	Black, elastic peat, well decomposed with some gastropod remains
170–185	10YR3/2	Brown marl sediments with laminations, charcoal bands and gastropods
185–188	10YR2/1	Distinct black band (charcoal?)
188–225	10YR3/2	Brown peat layer with abundant gastropod remains
225–295	10YR3/2	Brown lake marl
295–335	10YR3/2	Dark-brown, organic-rich marl
335–440	10YR2/1	Soft, dark-grey marl
440–445	10YR3/2	Brown marl sediments
445–505	10YR3/2	Crumbly, brown peaty sediment (with gastropod layer at 455 cm)
505–600	10YR4/4	Soft, yellow, organic-rich marl
600–625	10YR3/2	Dark-brown, peaty marl
625–665	10YR3/2	Brown marl sediments
665–915	10YR2/1	Soft, dark-grey marl with laminations
915–960	10YR3/2	Hard, crumbly, dark-brown marl
960–1045	10YR2/1	Very soft, dark-grey marl
1045–1080	10YR3/2	Compact, brown peaty sediment (gastropod layer at 1065 cm)
1080–1110	10YR2/1	Soft, dark-grey marl

Because Tsavkisi Wetland has no inflowing streams and its small geologically-determined outflow is stable and has only recently been met by the sediment level, these observations can be used to provide a model of lake level change through the period represented by the core. I think it most likely that the lithological changes observed in the Tsavkisi Wetland core were driven by variations in precipitation, evaporation and sedimentation. The influence of tectonic movements or anthropogenic impacts

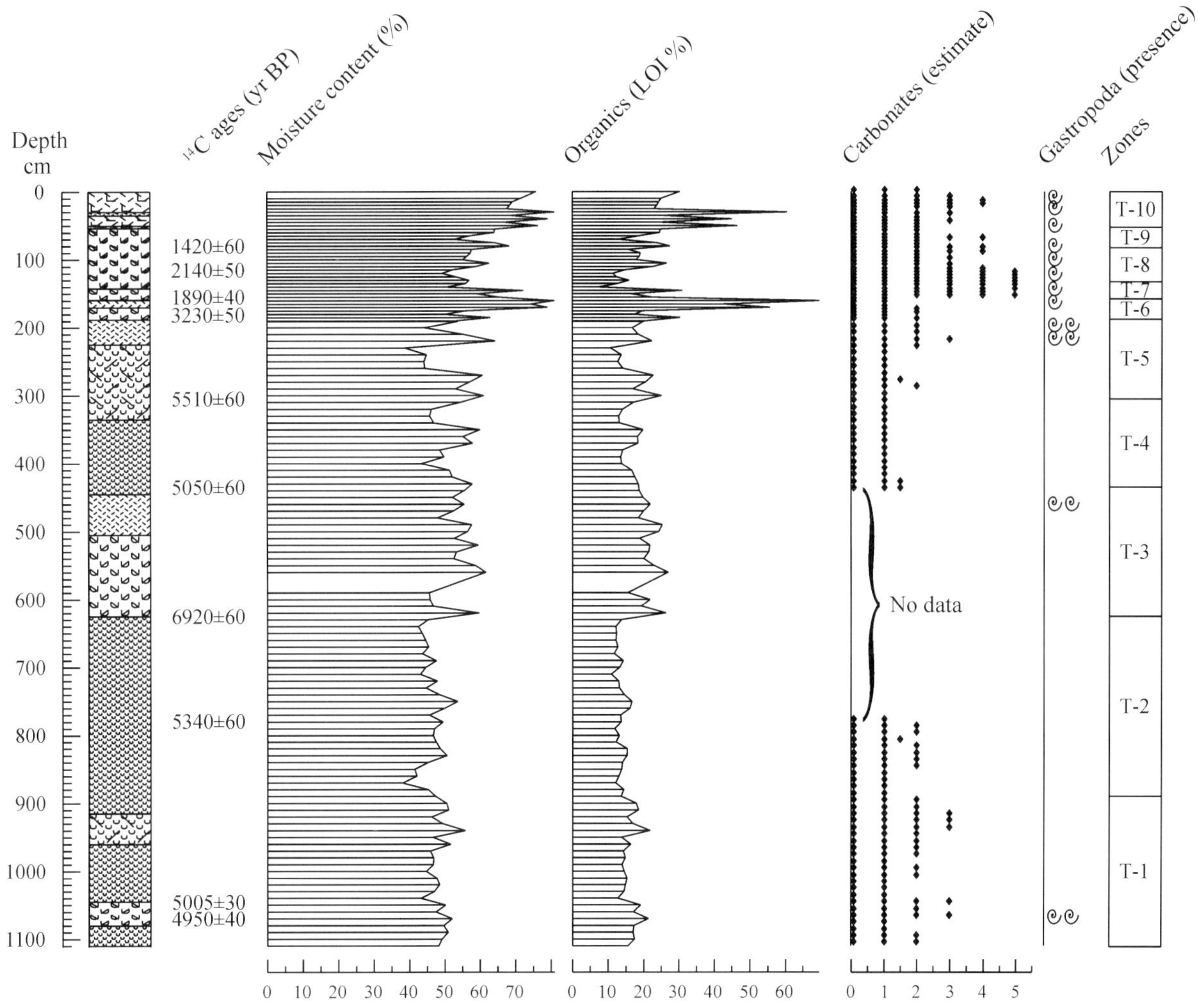

Figure 5.15. The Tsavkisi Wetland core: lithology, moisture content, loss-on-ignition and carbonate determinations. Presence of gastropod remains and pollen zones shown at right.

cannot be discounted, but it seems more parsimonious to regard the changes as reflective of Holocene climatic variability, given that the various sediments can be tied to modern analogues in other lakes that differ primarily in their altitudinal and hence climatic characteristics.

Before these data can be incorporated into a lake-level history for Tsavkisi Wetland, dating of the sediments must be first considered. Radiocarbon dating of the Tsavkisi sediments has proved exceedingly problematic because of contamination by older carbonates and younger humic material. Old carbonates are brought into the basin from the erosion of the surrounding Eocene sedimentary rocks and also from alkaline aquifers that provide some of the wetland's water. Three different methods were employed

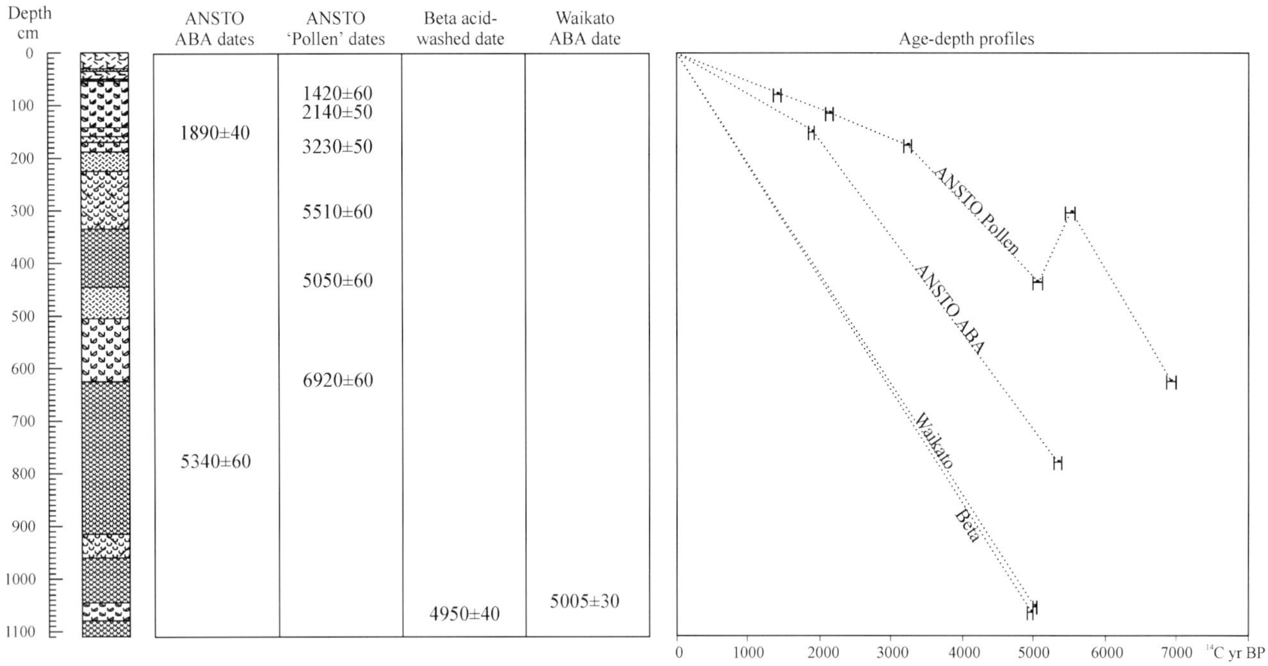

Figure 5.16. Problematic dates for Tsavkisi sediments: discrepancies between AMS radiocarbon dates pre-treated by different methods, presented in stratigraphic order.

in the pre-treatment of the samples – acid-base-acid (ABA), pollen isolation and acid leaching. The discrepancies between the dates are shown in **Figure 5.16** (see also Appendix D).

The two samples pre-treated by traditional ABA methods at the Australian Nuclear Science and Technology Organisation (ANSTO) gave the most convincing results, based on the appearance of dating markers (*Juglans* and *Olea* pollen) and biostratigraphic correlation with other records from the study area.

Dating the isolated pollen fraction using the bleaching and oxidation method of Brown *et al.*[31] produced radiocarbon ages that appear to be much too old. In some cases the pollen dates' divergence from the ABA-derived chronology is 3000 years. Kilian *et al.* have studied this ageing phenomenon in detail using the varved sediments of Lake Gosciaz in Poland.[32] They showed that pollen concentrates produced dates that were approximately 700 years too old when compared to the lake's varve chronology. In contrast, macrofossils preserved in the same sediments were around 100 years too old. They suspect that the large surface-to-volume ratio and high porosity of pollen grains allows 'dead' carbon to adhere to pollen exines.[33]

31 Brown *et al.* 1989.
32 Kilian *et al.* 2002.
33 Kilian *et al.* 2002.

Two additional dates, from Beta Analytic and the University of Waikato, are more difficult to dismiss. They are younger than either the ABA or pollen-fraction chronology. Both dates overlap temporally, and differ stratigraphically by 10 cm. These dates may be erroneously young due to contamination during coring (by use of an auger) or through infiltration by younger humic acids, which may have been incompletely removed by pre-treatment. While these possibilities could also apply to the ANSTO ABA dates, contamination during coring is less likely because the samples were taken from higher in the profile (hence less hydrostatic pressure). The ANSTO samples were also gravity separated before ABA pre-treatment, eliminating contamination from humic material bound to mineral particles, a potential source of erroneous ages in clay- and silt-rich samples.[34]

I have adopted the ANSTO ABA chronology as the best representation of the sediments' age based on three factors: the advantages of the pre-treatment method (complete removal of humic acids), vegetation changes of probable regional extent (a sharp rise in *Quercus* ~5000 years ago) and reference to sedimentation rates in comparable sites. Uncertainty remains, however, and is not likely to be resolved without age-determinations for various sediment components within each dated horizon.

Using the ANSTO ABA chronology, the Tsavkisi Wetland record encompasses the last 8000 years. Sedimentation rates between 8000 and 2000 years ago were, on average, 1.6 mm per annum. In the last 2000 years, the rate of sedimentation has slackened somewhat (0.8 mm per annum), possibly as the sediment level rose to meet the basin's intermittent outlet and inhibited further infilling, or as local vegetation changes reduced sediment supply.

Aquatic Pollen

In addition to the sedimentological data already discussed, changes in lake levels can be studied with reference to the aquatic pollen types preserved in the various sedimentary layers (see **Fig. 5.17**, in the following section).

The lowermost, silty sediments (1110–640 cm) contain large amounts of *Myriophyllum* pollen. Most other aquatic and wetland types are poorly represented during this stage, although *Typha* and Cyperaceae pollen occur consistently and there are a number of peaks in *Pediastrum* algae.

Myriophyllum pollen are difficult to interpret – milfoil may live in still water bodies for long periods of time without flowering, propagating entirely by vegetative means. Even where the plant is abundant, pollen percentages are often low.[35] Some species can be forced into flower by fluctuating water levels.[36] Wasylikowa interpreted high

[34] Dr G. Jacobsen, ANSTO, pers. comm.
[35] Huntley and Birks 1983.
[36] Orchard 1985.

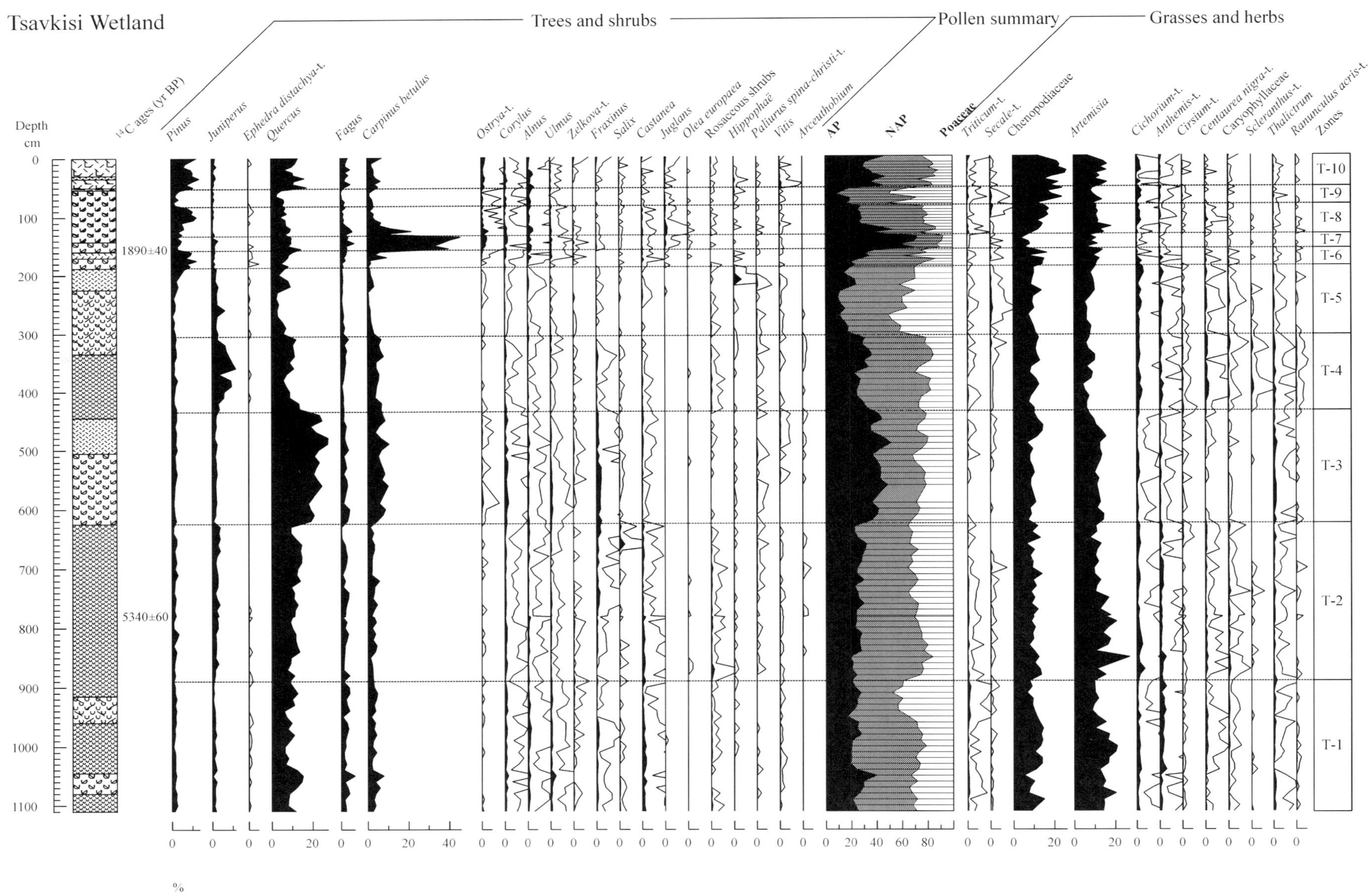

Tsavkisi Wetland
Trees and shrubs
Pollen summary
Grasses and herbs
Depth cm
¹⁴C ages (yr BP)
1890±40
5340±60
Pinus
Juniperus
Ephedra distachya-t.
Quercus
Fagus
Carpinus betulus
Ostrya-t.
Corylus
Alnus
Ulmus
Zelkova-t.
Fraxinus
Salix
Castanea
Juglans
Olea europaea
Rosaceous shrubs
Hippophaë
Paliurus spina-christi-t.
Vitis
Arceuthobium
AP
NAP
Poaceae
Triticum-t.
Secale-t.
Chenopodiaceae
Artemisia
Cichorium-t.
Anthemis-t.
Cirsium-t.
Centaurea nigra-t.
Caryophyllaceae
Scleranthus-t.
Thalictrum
Ranunculus acris-t.
Zones
T-10
T-9
T-8
T-7
T-6
T-5
T-4
T-3
T-2
T-1
%

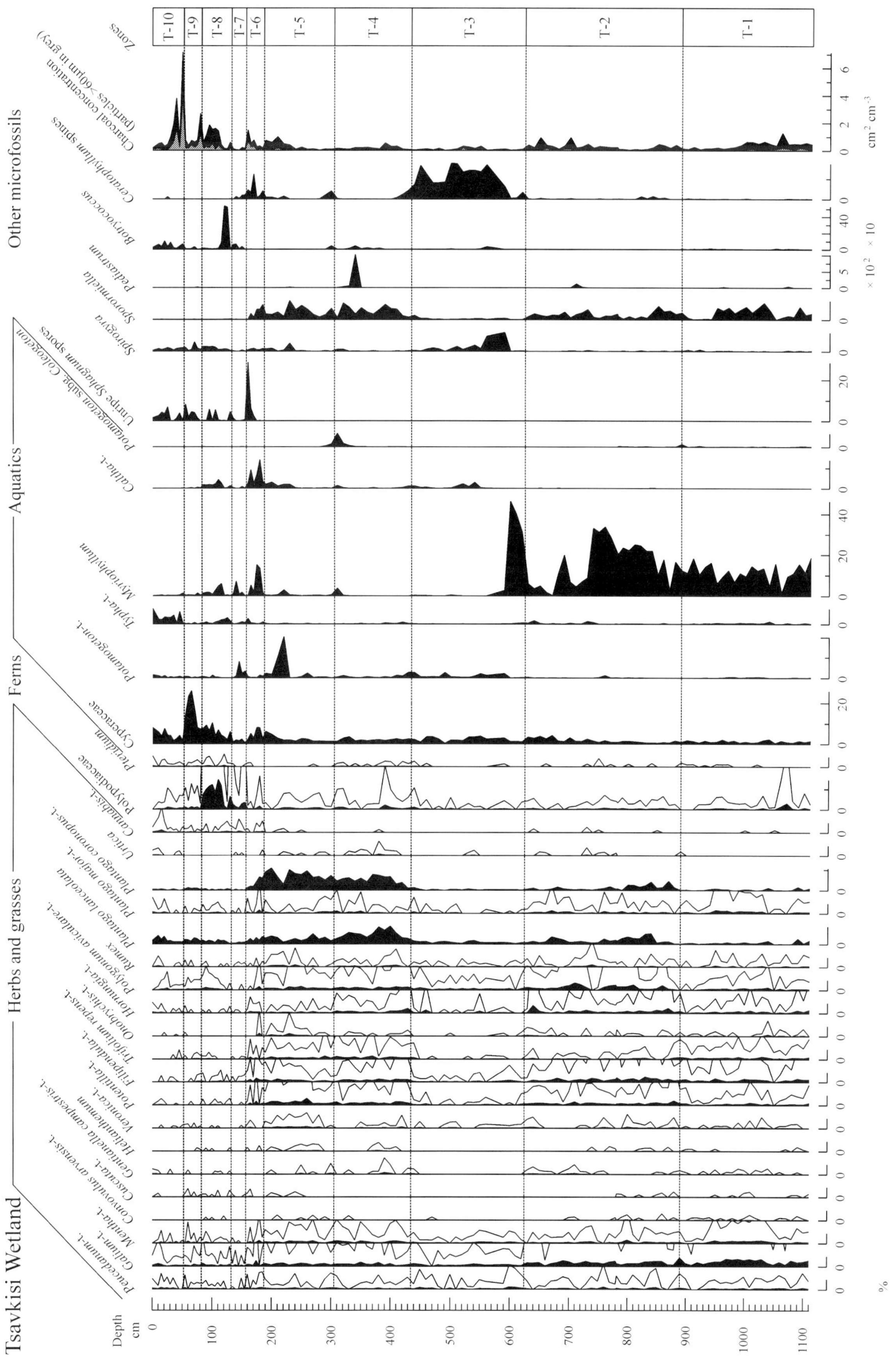

Figure 5.17. The Tsavkisi Wetland pollen diagram. Pollen percentages calculated from a conventional dryland pollen sum. See Appendix E for the complete diagram.

and rapidly varying concentrations of *Myriophyllum spicatum* macrofossils in mid Holocene sediments from Lake Zeribar, western Iran, as indicative of fluctuating water levels and a warm climate.[37] After 6000 cal BP, *Myriophyllum* in Lake Zeribar declined to be replaced by *Ceratophyllum*, indicating a more stable water level.[38]

This succession is echoed so precisely in the Tsavkisi sediments that it seems plausible its cause was the same: a climate with seasonal drought gave way to conditions of relative hydrological stability after 6000 cal BP. Water fluctuations within this phase can be traced through *Potamogeton* pollen (indicative of a depth of standing water) and, on the other hand, *Spirogyra* (shallow warm water) and *Caltha* (boggy conditions).[39] So, although the water level may have been more stable between 5000 and 4000 cal BP, it may have been lower than previously due to rapid infilling of the water column by *Ceratophyllum*.[40]

After 4000 cal BP, *Ceratophyllum* and most other aquatic species declined and the sediments became increasingly silty. The only pollen taxon to increase then was *Potamogeton* subgenus *Coleogeton*, which is able to tolerate brackish water,[41] and suggests that the lake water became more saline. Toward the end of this stage, *Potamogeton* (subgenus *Potamogeton*) began to rise, signalling the return of a substantial depth of fresh water. It then declined to the benefit of *Ceratophyllum* and *Caltha* as the lake paludified, beginning around 2000 cal BP. Water levels remained low, but free of summer desiccation, during the phases of peat deposition. These are notable for the presence of abundant unripe *Sphagnum* spores[42] and fungal remains indicative of shallow water.[43] In recent centuries the wetland has changed from a sedge swamp to a *Typha-Phragmites* association.

In summary, Tsavkisi Wetland existed as a lake with fluctuating water levels and moderate salinity up until 5000 cal BP. After this time, water levels stabilised and salinity was generally lower. Within this phase was a period of low lake levels and higher salinity between 3900 and 2400 cal BP. Around 2000 cal BP, the lake surface was colonised by sedges and other plants and became a wetland.

Pollen Diagram

The sediments of Tsavkisi Wetland are exceedingly rich in pollen. In the 129 stratigraphic levels analysed, a total of 250 different types of pollen, fungi and other microscopic remains were identified. The pollen diagram (**Fig. 5.17**) is divided into ten palynological assemblage zones based on statistical splitting:

[37] Wasylikowa 1967.
[38] Wasylikowa 1967; Wasylikowa 2005.
[39] Ketskhoveli 1973; van Geel *et al.* 1981; van Geel and van der Hammen 1978.
[40] Hejný 1960; Wasylikowa 1967.
[41] Tutin *et al.* 1980.
[42] Identification confirmed by J. van Leeuwen, Institute of Plant Sciences, Bern.
[43] *Cf.* van Geel *et al.* 1981; van Geel and van der Hammen 1978.

Zone T-1. The first zone has high proportions of herbaceous pollen (especially Chenopodiaceae, *Artemisia* and Poaceae). Many of these species are, additionally, indicators of open vegetation, *e.g.* *Secale*-type, *Arctium*-type, *Centaurea nigra*-type, Caryophyllaceae, *Convovulus arvensis*-type, *Cuscuta*-type, *Gentianella campestris*-type, *Phyteuma*-type, *Helianthemum, Prunella*-type and *Plantago* spp. The dominant arboreal pollen type is *Quercus*, followed by *Carpinus betulus* and *Fagus*. Poorly-represented pollen types include *Ephedra distachia*-type, *E. fragilis*-type, *Salix, Castanea, Pistacia, Tilia, Viscum, Juniperus* and many Rosaceae. The mere presence of these pollen types suggests that their parent plants were quite prevalent in the surrounding vegetation. Charcoal particles are consistently represented through this zone, decreasing in abundance in the upper part. Of interest is the high proportion of *Sporormiella* dung fungal spores, indicative of grazing herbivores.[44] These decline in the upper part of the zone, when Poaceae and *Triticum*-type increase.

Zone T-2. The second zone is similar to the first, except that Poaceae declines and various pasture indicators increase (*Potentilla*-type, *Filipendula, Polygonum aviculare*-type and *Plantago* species). *Sporormiella* dung spores and charcoal particles also increase, as does *Juniperus*, suggesting more intensive pastoral activity compared to the first zone. Pollen of juniper's parasite, *Arceuthobium oxycedri*, appears in this zone, confirming the local occurrence of *Juniperus* shrubs or trees. As in the first zone, *Myriophyllum* dominates the aquatic pollen group.

Zone T-3. The transition to the third zone is accompanied by a rapid increase in arboreal pollen, especially *Quercus, Carpinus betulus* and *Fraxinus*. With the important exception of *Thalictrum*, a plant of forest edges, all herbs and grasses decline. Charcoal concentrations and dung fungal spores also diminish and the aquatic vegetation changes from *Myriophyllum*-dominated to *Ceratophyllum*-dominated. This zone records the expansion of deciduous (oak-ash) woodland in the Tsavkisi area.

Zone T-4. At the beginning of the fourth zone, *Fraxinus, Quercus* and *Corylus* decline. They are replaced by herbs such as *Hornungia*-type, *Potentilla*-type, *Filipendula, Polygonum aviculare*-type, *Gentiana*, Apiaceae, Asteraceae and rosaceous shrubs. In the middle of this zone, *Juniperus* increases in tandem with *Plantago, Cirsium, Centaurea, Galium*-type, *Rumex, Urtica, Gentianella* and *Helianthemum*. These taxa indicate open vegetation, probably juniper scrub with dry meadow herbs. The climate must have been dry and droughty for juniper to become so important. Although fire-loving bracken, *Pteridium*, is well represented in this zone, charcoal concentrations are not significantly higher than in Zone T-3 and large charcoal particles (representing local fires) are rare. Brackens also spread under the influence of heavy grazing,[45] a more plausible explanation for their representation throughout this zone, given that

[44] van Geel and Aptroot 2006.
[45] Moore and Stevenson 1982.

quantities of *Sporormiella* dung spores reach up to 8 percent of the total pollen sum. *Sporormiella* values of 4 percent are considered a reliable indication of high livestock densities;[46] hence the Tsavkisi record indicates very high densities. Perhaps intense grazing pressure played a leading role in maintaining the openness of the vegetation.

Zone T-5. In the fifth zone, grass pollen (Poaceae) increases dramatically and arboreal pollen all but disappear. Amongst herbaceous pollen otherwise dominated by Poaceae, *Plantago, Potentilla* and *Filipendula*, some relatively rare pollen types appear – *Scorzonera, Echinops, Cuscuta, Melampyrum, Polygala, Sideritis* and *Verbena*. Many of these are characteristic phrygana species of the dry grasslands and pastures on the hills around Tbilisi.[47]

Zone T-6. *Plantago* species and *Sporormiella* decline in the sixth zone, at the same time as charcoal concentrations and arboreal pollen begin to increase. *Pinus* becomes important for the first time in the history of the site and the pollen of *Juglans* and *Olea europaea* appears for the first time.

Zone T-7. This zone includes the most remarkable change in the Tsavkisi pollen record. After the resurgence of arboreal pollen in the previous zone, here *Carpinus betulus* comes to dominate suddenly and absolutely, reaching up to 47 percent of dryland pollen. Poorly-represented pollen of *Carpinus orientalis* (*Ostrya*-type) also peaks at this time, as dense forests overtook the Tsavkisi landscape.

Zone T-8. This zone witnesses the decline of the short-lived hornbeam forests. It occurs as charcoal concentrations become much higher. Monolete fern spores, of the Aspleniaceae and Polypodiaceae families, maintain values of between 8 and 16 percent throughout this zone, compared to values generally <1 percent in the remainder of the record. By and large, these represent ferns of the forest floor that proliferated after the withdrawal of *Carpinus* forests. Increasing percentages of *Pinus* and various herb pollen types indicate a return to a more open landscape in this zone.

Zone T-9. The vegetation became even more open and steppic in the ninth zone, as ferns and arboreal pollen decline, and Poaceae, Chenopodiaceae and *Plantago* increase. Charcoal particles indicate that fire retained an important impact on the vegetation. Cyperaceae briefly becomes the most important wetland taxon in this zone.

Zone T-10. The role of grasses diminishes in the final zone, countered by the re-emergence of arboreal types (*Pinus, Quercus, Fagus*) at the same time as *Vitis* pollen increases and *Picea* and *Cannabis* achieve their highest representation in the record. This event was coincident with very high charcoal concentrations and followed by a proliferation of Chenopodiaceae. The decreased importance of fire in the latter part of the zone is echoed in a Chenopodiaceae decline and increased arboreal pollen percentages. Importantly, dung fungi (*Sporormiella* and *Podospora*) and grazing indicator pollen (*Potentilla*-type, *Ranunculus acris*-type, *Plantago major*-type and Fabaceae) are

[46] Davis and Shafer 2006.
[47] Ketskhoveli 1959.

much reduced through Zones 7–10, suggesting that grazing pressures were lighter through the last 2000 years than during prior millennia.

Numerical Analyses

The great complexity of the Tsavkisi pollen diagram can be reduced to a few major trends with the aid of statistical methods. The results of weighted averaging based on modern pollen data are shown in **Figure 5.18**. The climatic gradient seems to indicate that precipitation fluctuated considerably from the beginning of the record, stabilising after 6000 years ago with higher rainfall. The climate became gradually drier between 4000 and 2000 years ago, but shifted back during the expansion of *Carpinus betulus* around 1900 years ago. Conditions became again drier in more recent times. Correlations between this reconstructed trend and aquatic pollen (**Table 5.4**) show that *Ceratophyllum*, an aquatic plant requiring between 60 and 200 cm water depth,[48] is related to phases of wetter climate; while sedges, shallow water fungi and pollen degradation occurred when water levels were lower.

The gradient of forest cover exhibits a trend that is inversely related to the first gradient (r^2= 0.75). Since forest cover is positively correlated with winter temperature in the study area (see Chapter Four), this second gradient is also a proxy for temperature variations. Forest cover was most extensive between 5200 and 3800 cal BP (oak-ash forest) and 1900 and 1600 cal BP (hornbeam forest). It was more restricted during all other times.

The third gradient shows that land-use in the earliest part of the Tsavkisi record oscillated between livestock herding and grain agriculture as climatic conditions varied. A comparison of the *Sporormiella* (grazing indicator) and *Triticum*-type (agriculture indicator) curves strongly supports this assessment (**Fig. 5.19**). After 6800 cal BP, grazing appears to have occupied a primary place near Tsavkisi Wetland. This changed with the advance of forests 5200 to 3800 cal BP, but afterward grazing returned and became extremely important. Pastures went into decline around 2000 cal BP, followed by the expansion of hornbeam forest and the establishment of walnut groves in the Tsavkisi area. Since that time, grazing has been reintroduced, but at levels far less intense than in the past.

Without doubt, the dominant impetus for vegetation shifts in the Tsavkisi record has been land-use change. This is demonstrated by the strong relationship (r^2=0.81) between the third WA axis and the first axis of the separate DCA ordination of the Tsavkisi data (**Fig. 5.18**; Appendix B). The second Tsavkisi DCA axis is related to both the forest-cover and climatic gradients, indicating a subordinate role for these factors in directing the course of Tsavkisi's vegetation history.

[48] Korelyakova and Gorbik 1989.

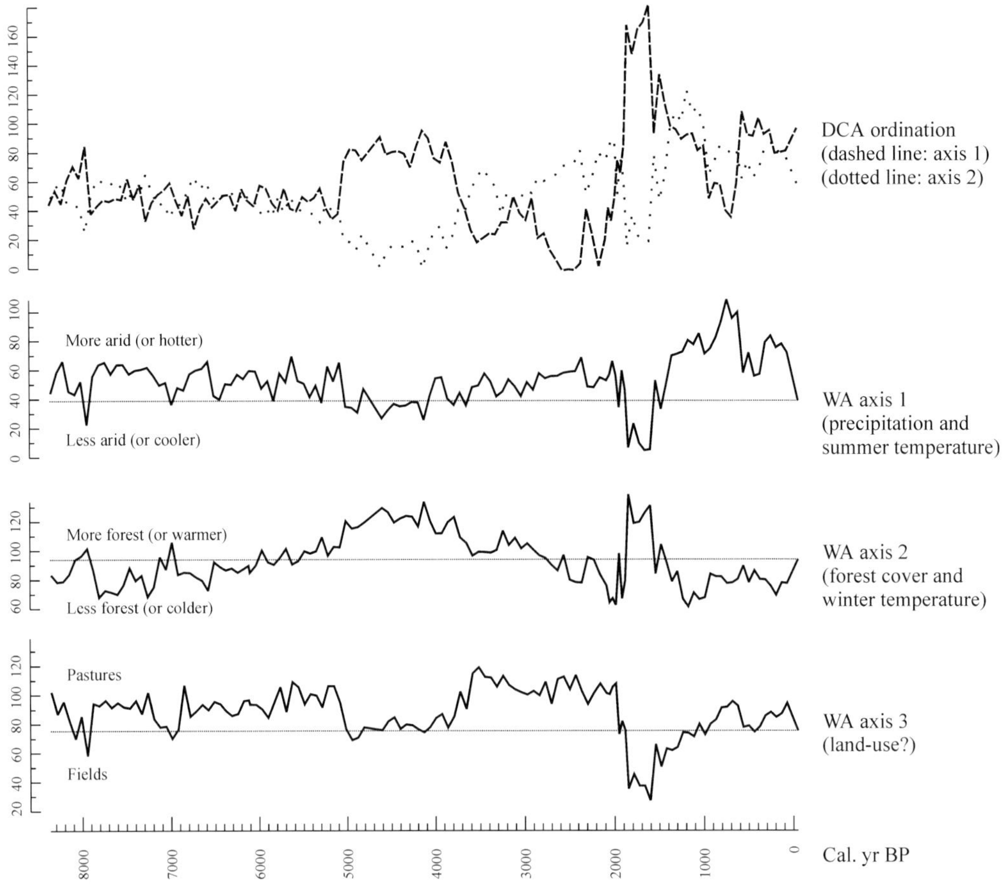

Figure 5.18. Environmental changes at Tsavkisi Wetland: rainfall, forest cover and land-use gradients produced by weighted averaging. Dominant trends in the pollen data (DCA ordination) are shown at the top for comparison.

Table 5.4. Wetland pollen taxa and other microfossils most strongly correlated with each of the WA axes, and all pollen types correlated with charred particles (CPs) and loss on ignition (LOI). Correlation coefficients (r^2) are given in parentheses.

Axis 1	+	Cyperaceae (0.37), degraded pollen (0.29), *Mougeotia* (0.19), *Closterium* (0.13)
	–	Dryland pollen concentration (0.16), *Ceratophyllum* (0.12)
Axis 2	+	*Ceratophyllum* (0.30)
	–	Degraded pollen (0.22), *Closterium* (0.10)
Axis 3	+	*Sporormiella* (0.38)
	–	*Botryococcus* (0.09)
CPs	+	*Pinus* (0.23), *Alnus* (0.19), Chenopodiaceae (0.11), *Vitis* (0.11)
	–	*Filipendula* (0.05), *Fraxinus* (0.05), Caryophyllaceae (0.04), *Juniperus* (0.04)
LOI	+	Unripe *Sphagnum* spores (0.31), *Pinus* (0.30), Charcoal concentrations (0.21)
	–	*Polygonum aviculare* (0.13), *Galium* (0.12), *Filipendula* (0.11), *Sporormiella* (0.09)

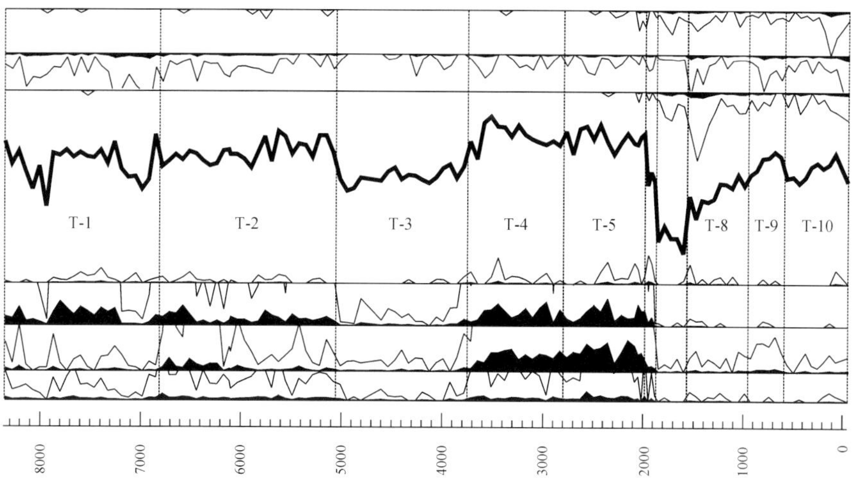

Figure 5.19. The Tsavkisi land-use trend compared with palynological indicators of cultivation (top) and pasturage (bottom). Pollen zones T-3 and T-7 were forested stages with diminished local agricultural or pastoral activity.

Vegetation History

While these trends are useful for interpreting changes in climate, forest cover and human impact, they provide little data on how the *composition* of the vegetation changed through time. Because the Tsavkisi pollen record was so rich, containing many taxa not used in the TWINSPAN classification in Chapter Four, a separate analysis was performed so the two sets of results could be compared side-by-side.

The relevant TWINSPAN groups and their indicators appear in **Table 5.5**. The classification splits the Tsavkisi pollen record into two 'stages'. The earlier stage encompasses the six millennia prior to the Classical Period. It occupies pollen zones T-1 to T-5 and is indicated by relatively high proportions of *Juniperus*. For this reason it is referred to as the '*Juniperus* stage'. The second stage commenced in the Classical Period and continues through to the present day. This period is referred to as the '*Pinus-Juglans* stage'. Each of the two stages is divided further into 'phases', discussed below.

Table 5.5. TWINSPAN indicators for each of the classified stages and phases at Tsavkisi.

Stage/phase	Indicator species (and cut levels)
Juniperus	*Juniperus* (2%)
Phase 1	*Artemisia* (10%)
Phase 2	*Plantago coronopus*-type (presence)
Pinus-Juglans	*Pinus* (5%), *Juglans* (presence), *Ostrya*-type (presence), *Cannabis*-type (presence)
Phase 3	*Polygonum aviculare*-type (presence), *Cirsium*-type (presence)
Phase 4	*Carpinus betulus* (5%), *Fraxinus* (presence), *Zelkova*-type (presence)

The *Juniperus* stage (8000–2000 cal BP) – Juniper is a shrub or tree characteristic of Caucasian scrub and savanna vegetation. It was more clearly much more prevalent in the Tsavkisi area prior to the Classical Period than it is at present. It must have been quite abundant locally, as the pollen of its associated parasite, *Arceuthobium oxycedri* attests. Junipers are extremely sensitive to fire,[49] a characteristic exploited by graziers in the Alps today to prevent junipers invading mountain pastures. Although the charcoal record from Tsavkisi demonstrates that fires did occur through the *Juniperus* stage, their frequency and intensity appear to have been low compared to the subsequent *Pinus-Juglans* stage. Grazing was almost certainly the dominant land-use during the *Juniperus* stage, as indicated by high proportions of *Sporormiella* dung fungal spores throughout. The *Juniperus* stage can be divided into two major phases:

Phase 1 (pollen zones T-1 to T-3): This phase lasted from 8000 until about 3800 cal BP. It is characterised by high proportions of *Artemisia* pollen and is associated with *Quercus* and Rosaceae. These species are found together in both xerophile woodland and scrub communities in this part of Georgia.[50] Taking into consideration the forest cover gradient discussed in the previous section, it appears that the vegetation during pollen zone T-1 was open scrub, gradually becoming more wooded through zone T-2 and forming a forest during T-3.

Phase 2 (pollen zones T-4 and T-5): The indicator species for Phase 2 is *Plantago coronopus*-type. This pollen type includes several *Plantago* species that grow primarily in grasslands and pastures. Associates of *Plantago* during this phase include *Juniperus, Agrimonia eupatoria, Helianthemum, Aster, Campanula,* Liliaceae, *Scleranthus, Geranium* and *Ranunculus acris*-type. *Plantago* and these associated species occur in phrygana communities around Tbilisi.[51] This suggests that grazing pressure during pollen zones T-4 and T-5 was sufficiently intense to mould vegetation composition. Increased proportions of *Juniperus, Acer* and *Carpinus betulus* during pollen zone T-4 indicate that the vegetation changed from scrub with phrygana elements to a landscape that was much more open and steppic in zone T-5.

The *Pinus-Juglans* stage (2000 cal BP to the present) – This stage, beginning some time in the Classical Period, has *Pinus, Juglans, Ostrya*-type and *Cannabis*-type as its indicator species. Pines are usually found on poor, rocky soils where more competitive deciduous trees will not grow, but also compete with these trees in less marginal habitats by promoting fire through flammable leaf litter and resinous bark.[52] The correlations in **Table 5.4** demonstrate that *Pinus* and fire are interrelated. Both *Ostrya*-type and *Cannabis*-type represent plants that favour various kinds of disturbance.

[49] Grove and Rackham 2001.
[50] Ketskhoveli 1959.
[51] Ketskhoveli 1959.
[52] Grove and Rackham 2001; Rackham 1990.

The appearance of *Juglans* pollen, alongside *Olea europaea*, strongly suggests that these economically-important plants were introduced to the Tbilisi area by Greek or Roman contacts. While *Juglans* is cultivated near the Tsavkisi Wetland today, the climate in the hills above Tbilisi is too cold for olives, which are not cultivated in Georgia at present. The 18th century Georgian geographer, Vakhushti Bagrationi, describes places on the lower Khrami River which, 'more than other places, produce pomegranates, figs, olives and other fruits' because, 'in winter it is warm and very hot in summer'.[53] The low, but consistent percentages of *Olea* pollen in both the Tsavkisi and Kumisi records demonstrate that olives were grown in the lowlands of Eastern Georgia from the Hellenistic Period until relatively recent times.

Phase 3 (pollen zones T-6, T-8, T-9 and T-10): This phase appears several times through the *Pinus-Juglans* stage, namely during much of pollen zone T-6, the upper part of T-8, all of T-9 and sporadically through T-10. Its indicators are *Polygonum aviculare*-type and *Cirsium*-type. These types represent various species that colonise open, often disturbed ground. Other taxa associated with phase 3 include *Plantago, Potentilla, Cuscuta, Sanguisorba minor* and *Gentianella campestris*-type, all of which are found in meadows and steppes.

Phase 4 (pollen zones T-6, T-7, T-8 and T-10): The final phase in Tsavkisi's vegetation development occurs momentarily during pollen zone T-6, through the whole of T-7 and the earlier part of T-8, and finally at various times during T-10. The uppermost sample from the record, representing the present-day vegetation, falls into this phase. Within the context of the *Pinus-Juglans* stage, indicators of the last phase are *Carpinus betulus, Fraxinus* and *Zelkova*-type, all representing deciduous forest trees. This phase occurred at times of greater forest cover, interspersed with the more open and steppic vegetation of Phase 3. This analysis indicates that the vegetation is more forested now than it was 100 years ago, a fact confirmed by early postcards of Tbilisi and aerial photographs taken in the mid 20th century (**Fig. 5.20**).

Based on these results, the following stages of vegetation development are proposed for the Tsavkisi area (**Table 5.6**). The conclusion to be drawn from this vegetation history is that the landscape near Tsavkisi Wetland was, for much of the last 8000 years, much less forested than it is today.

Discussion

Finding Tbilisi's Lost Forests

The vegetation history of Tbilisi and its surrounds has attracted the attention of a number of botanists and other workers over the years. The general opinion is that the vegetation of Tbilisi was thickly forested until quite recently, having been destroyed

[53] Vakhushti Bagrationi 1983, p. 428.

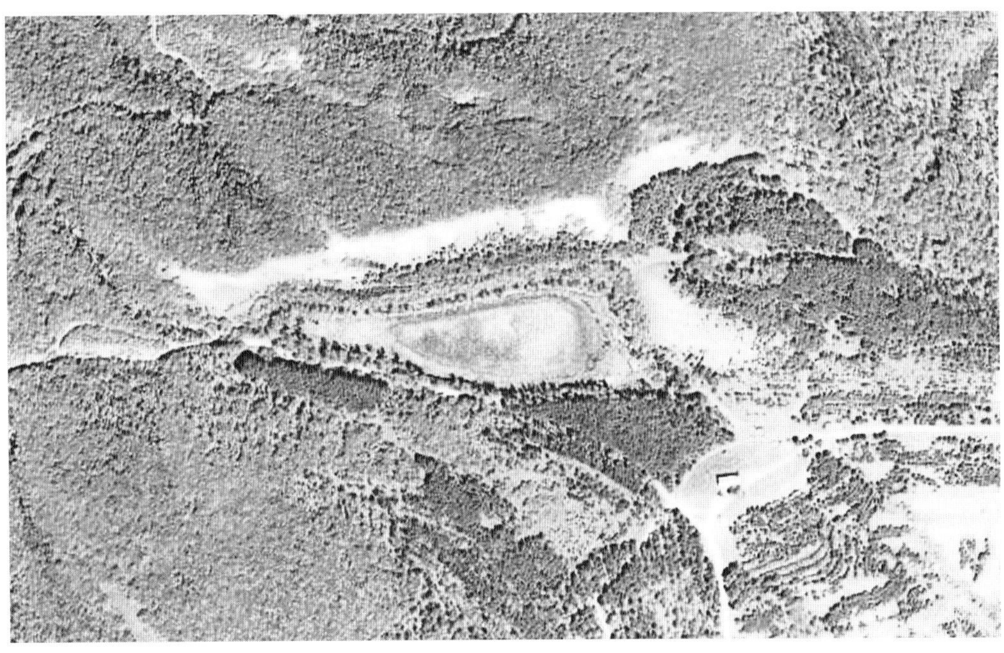

Figure 5.20. Aerial photograph of Tsavkisi Wetland taken shortly after World War II (top), showing open areas that were later planted with pines, apples, walnuts and poplars – see recent satellite image (below). Source: Declassified US Military Aerial Photography and © 2008 Google Image, © 2008 Digital Globe, © 2008 Europe Technologies

Table 5.6. Stages of vegetation development at Tsavkisi Wetland.

Zone	Time (cal BP)	Inferred vegetation	TWINSPAN (this chapter)	TWINSPAN (Chapter 4)
T-1	8000–6800	Open rosaceous scrub	Juniperus Phase 1	Grassland-forest border
T-2	6800–5100	Shibliak vegetation	Juniperus Phase 1	Oak forest
T-3	5100–3800	Oak-ash woodland	Juniperus Phase 1	Oak forest
T-4	3800–2800	Juniper scrub and phrygana	Juniperus Phase 2	Oak forest
T-5	2800–2000	Steppe-phrygana	Juniperus Phase 2	Grassland-forest border
T-6	2000–1900	Steppic fallow land	Pinus-Jugl. Phase 3	Grassland-forest border
T-7	1900–1600	Hornbeam secondary forest	Pinus-Jugl. Phase 4	Oak forest
T-8	1600–1000	Ferny meadow (deforested)	*Pinus-Jugl.* Phases 3 and 4	Grassland-forest border
T-9	1000–600	Steppe	Pinus-Jugl. Phase 3	Forest-steppe
T-10	600–today	Forest-steppe	*Pinus-Jugl.* Phases 3 and 4	Forest-steppe

Timeline based on linear interpolation.

by human activity in recent centuries.[54] This is a view that now seems difficult to reconcile with the long histories of steppe and xerophyte communities yielded by the pollen records, but it is nonetheless a view long held and widely accepted.

An alternative analysis of Tbilisi's vegetation history is presented by the palaeobotanist Avakov.[55] He notes that, in the panoramic drawings made by the French botanist Joseph Tournefort during his visit to Tbilisi in 1702, the hills around Tbilisi appear quite deforested. Likewise, the 1673 engraving by Jean Chardin, although depicting lush avenues of trees and gardens in the city, illustrates the surrounding hills apparently devoid of forest (**Fig. 5.21**).

Any trees around the city were, according to Avakov,[56] probably burnt for charcoal production, and the open areas thus created could then be used economically for grazing, gardens and agriculture. The same author has even observed areas near the city where thick-trunked forest trees (*Fagus orientalis*) have invaded former agricultural terraces, indicating that a relaxation of human impact has resulted in reafforestation in some places. The pollen records strongly support Avakov's alternative view.

Forests are, of course, represented in the Tsavkisi pollen record, but their appearance is of far shorter duration than that of xerophyte communities. And any notion that the lost forests were of primaeval character can be dispensed with completely.

[54] Badenkov *et al.* 1990; Ketskhoveli 1959; Sakhokia 1961; Sosnovskii 1915; Tumajanov 1971.
[55] Avakov 1982.
[56] Avakov 1982.

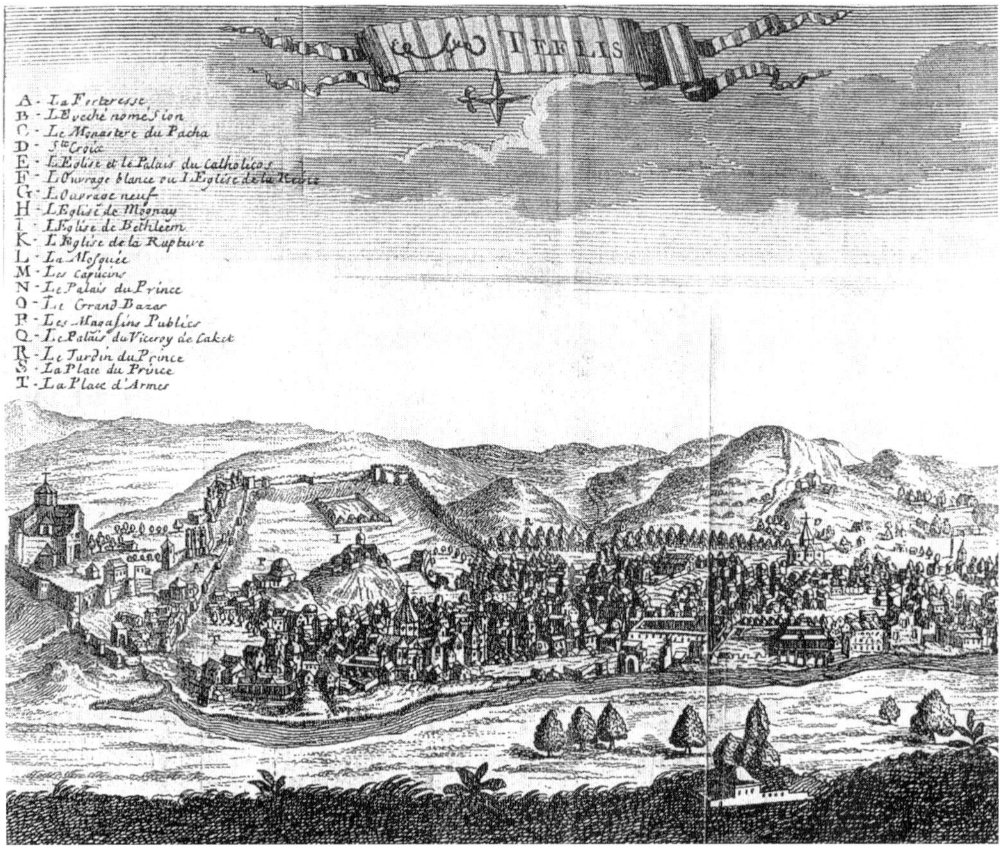

Figure 5.21. Engraving of the city of Tbilisi in 1673 by Jean Chardin. Source: National Library of Georgia.

Forests throughout the Tsavkisi pollen record were as much the product of deliberate human intervention as any other plant community represented. Unfortunately the record does not extend back sufficiently far to see whether primaeval forests existed at any stage during the earliest Holocene, but given the aridity of the climate at that time[57] and the long history of human habitation in Eastern Georgia,[58] it would be surprising if dense, tall-timbered forest was widespread about Tbilisi prior to the Neolithic, except in the most sheltered of situations.

This raises the question of the relative importance of climatic and anthropogenic factors in determining the course of Tbilisi's vegetation through the Holocene.

[57] Lemke and Sturm 1997.
[58] Kiguradze 1986.

Climatic Changes and Human Impacts

The climatic trends distilled from the pollen data in this chapter demonstrate that forest cover expanded during phases of warmer and/or wetter climate. Perhaps the best evidence for Holocene climatic change in the Caucasus region is the oxygen isotope record from Lake Van, in eastern Anatolia.[59] This record has the advantage of coming from a very large lake (thus reflecting changes on a regional scale), and being independent of its pollen-based vegetation history (which is subject to many influences other than climate).[60] Van shows that the climate became gradually wetter through the course of the early Holocene, peaking in relatively moist and warm conditions between 6000 and 4000 years ago. Then the climate became briefly drier and probably cooler, and this cooler climate appears to have lasted until about 2000 years ago. After that time the climate seems to have been quite variable, but not substantially different to what we experience today.

The oxygen isotope curve from Lake Van sits alongside the climatic and forest cover trends from Sakhare, Kumisi, Jvari and Tsavkisi in **Figure 5.22**. The correspondence between the early part of the record from Tsavkisi and the Van isotopic curve is extraordinarily good. Clearly forest cover in the Tbilisi area at this time was responding very faithfully to variations in temperature and precipitation, perhaps caused by changes to the Atlantic and Mediterranean weather systems during the mid Holocene.[61] These climatic shifts are also reflected in the changing textural characteristics of the Tsavkisi sediments.

The close relationship between the Van and Tsavkisi records breaks down during the *Carpinus* expansion at Tsavkisi during the Classical Period. The Tsavkisi climatic record would seem to suggest a dramatic increase in temperature and rainfall at this time. The rainfall trend may be evident also in the Jvari and Kumisi records if their respective radiocarbon chronologies are ignored (**Fig. 5.22** inset). What changed in the Tbilisi environment two thousand years ago to cause its climatic signal to deviate so considerably from the regional one?

Several possibilities could be entertained. One is that Tbilisi's climatic characteristics are inherently different to Lake Van's and hence their history cannot be expected to accord in all aspects. A second possibility is that the isotopic record is less sensitive to certain types of climate change than pollen. A third possibility is that human impacts on the vegetation have mimicked the influence of climate, quite independently of regional climatic changes.

[59] Lemke and Sturm 1997.
[60] Wick *et al.* 2003.
[61] Wick *et al.* 2003.

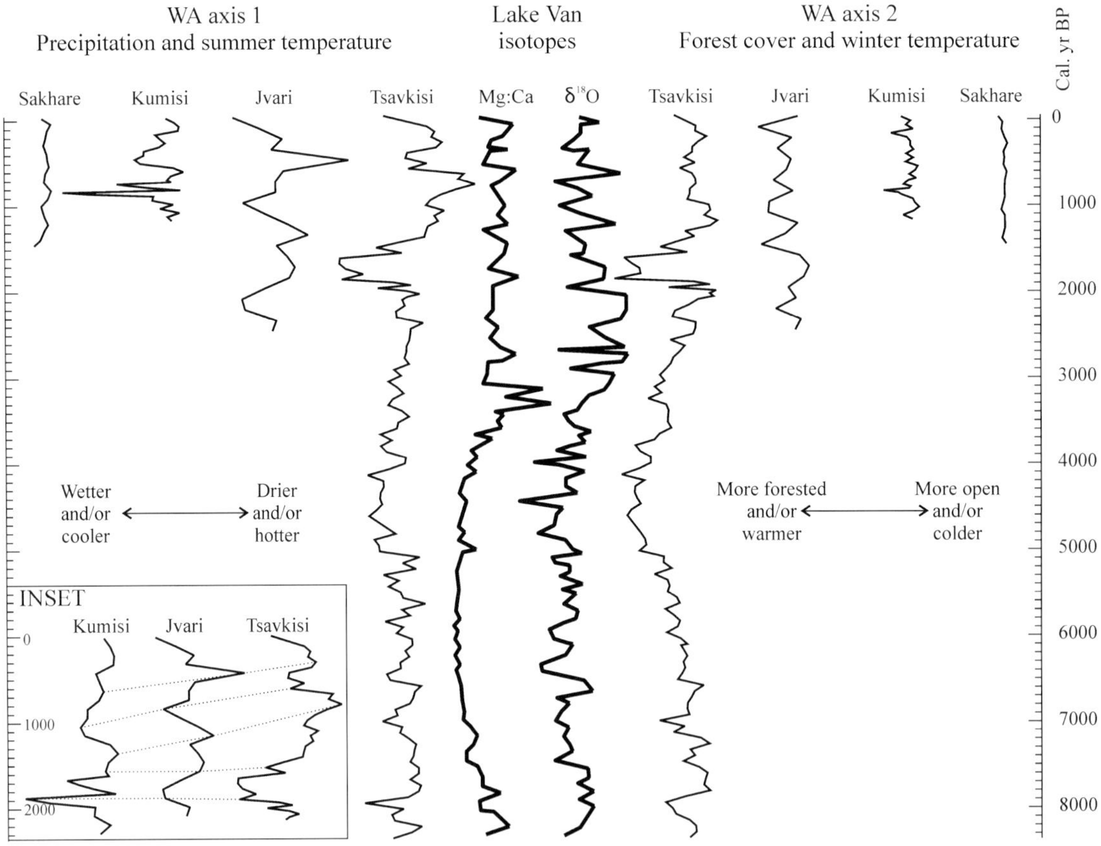

Figure 5.22. Environmental trends from the Tbilisi area compared to a regional climatic trend. Comparison of the Lake Van isotopic records (bold curves at centre) with the results of weighted averaging applied to pollen records from the Tbilisi area (shown at the same scale). INSET: An attempt to match the Kumisi and Jvari records to the Tsavkisi record using environmental trends. The result could suggest that the ^{14}C date for the Kumisi record is 1000 years too young, while the adopted Jvari chronology is 500 years too old. It may also indicate that Tsavkisi pollen zone T-8 was of shorter duration than the linear age-depth model suggests, which would be in harmony with the ephemeral plant community represented in T-8.

The Tbilisi and Lake Van areas differ in their climatic characteristics, but also have a number of similarities. Tbilisi is situated at an altitude of 490 m, compared to Lake Van's 1650 m. Orography is important in the distribution of rainfall, which in both areas varies from less than 500 to over 800 mm per annum.[62] Both experience a period of moisture deficit in summer. This is more pronounced at Van, whereas

[62] Schweizer 1975; Svanidze and Papinashvili 1992.

Tbilisi also has dry winters brought on by persistent anticyclonic activity in lowland Eastern Caucasus.[63] Maritime influence on both areas is minor because the mountain ranges of Western Georgia and Northern Anatolia intercept rainfall originating over the Black Sea.

If, however, maritime influence were greater in the past, this might be expected to manifest in Tbilisi to a greater extent. Atlantic low-pressure cells track across the Black Sea from the Mediterranean, bringing heavy rain to the Colchis lowland during winter.[64] An intensified westerly influence, or a weakened high-pressure cell over Central Asia, could allow moisture-laden, relatively warm air-masses to enter Eastern Georgia during winter, reducing the continentality of the climate.

Two aspects of the Tsavkisi pollen record may support this theory: the appearance of *Olea europaea* pollen and the sharp rise of *Carpinus betulus* in the Roman Period. It seems that *Olea* was grown in the lowlands near Tbilisi in the past when winters were less severe (see Appendix A). An "explosive expansion"[65] of *Carpinus betulus* in some parts of Europe during the Roman Period has been ascribed to the same diminished continentality that allowed grapes to be grown widely in England.[66] A problem with this theory is that it is not recorded universally in palaeoecological records.[67] Nor can it account for the extraordinary rate at which the open landscapes of Tsavkisi were converted from open pastures to dense forest during the Roman Period.

The second hypothesis – that the isotopic record is insensitive to climatic changes that pollen data may record – can be dismissed by comparing the Lake Van isotopic and pollen records.[68] It is true that isotopic and pollen data record certain climatic events differently,[69] but the Lake Van pollen record exhibits remarkable stability throughout the Roman Period, with no evidence for the forest expansion so apparent at Tsavkisi.

The answer may instead lie in the human history of the area. From the abundance of *Sporormiella* dung fungal spores in the Tsavkisi sediments, we know that livestock grazing was a very significant, and perhaps dominant, activity in the surrounding land. In statistical terms, changes in land-use are the most important trend reflected in the Tsavkisi pollen diagram. The impact of grazing on the vegetation was to suppress the growth of forest and to allow scrub, shibliak and phrygana communities to gain the upper hand. This relationship between grazing and vegetation persisted until it was broken down in the Roman Period. After the massive expansion of hornbeam

[63] Svanidze and Papinashvili 1992.
[64] Lydolph 1977.
[65] Godwin 1975, p. 266.
[66] Huntley 1988; Huntley and Birks 1983.
[67] *E.g.* Davis *et al.* 2003.
[68] Lemke and Sturm 1997; Wick *et al.* 2003.
[69] Siegenthaler and Eicher 1986.

forests during that period, fire thenceforth played a defining role in the ecology of forest and steppe communities.

On the Mediterranean island of Crete, a recent increase in pine in the vegetation has been attributed to a change from a grazing-dominated landscape to a fire-dominated one.[70] If one compares the charcoal (fire indicator) and *Sporormiella* (grazing indicator) curves from Tsavkisi, they suggest that this same process occurred in the mountains around Tbilisi in Classical times. The economy of the Tbilisi area changed considerably as Graeco-Roman cultural influences were first felt. Land-use of the hills around the city changed from pastures to orchards.

Forest composition changed as well. The importance of *Quercus, Juniperus* and *Fraxinus* declined, and forests of *Carpinus betulus* and *C. orientalis* expanded suddenly and rapidly. The causes for this event in Europe have been debated for many years.[71] Around Tbilisi, milder winters certainly could have aided hornbeam's spread. But to disqualify human intervention would be to ignore the appearance of olive and walnut pollen, the decline of livestock grazing and the increasingly intensive fire regime that led up to the rapid spread of these pure *Carpinus* forests.

The extraordinary rate at which these trees enveloped the landscape seems difficult to explain in terms of ecological succession. Instead, it seems more probable that hornbeam were already present in the vegetation prior to their expansion. Today, stunted copses of hornbeam grow in many areas that are heavily grazed by livestock (**Fig. 5.23**). These 'cow coppices' are so bitten-down that they seldom develop into mature trees. Their pollen production must also be severely restricted. Once livestock are removed, cow coppices of hornbeam grow rapidly, overtopping all other shrubs, herbs and grasses, and turning into forest within a few years. I have little doubt that this process occurred on a large scale at Tsavkisi during the Roman Period. Just why it occurred is a matter for speculation, but as Lang puts it:

> Once Roman hegemony was established, however, the Iberians rapidly adapted themselves to Roman ways, which had far-reaching effects on Georgia's social and economic life. The building of roads gave the country access to markets in Asia Minor and other parts of the Roman Empire.[72]

The Romans were doubtlessly beguiled by Strabo's descriptions of the country's almost incredible bounty. Here, the fruits "spring up for them without sowing and ploughing", grains "when sown only once produce two crops or even three", and the vines "yield so much that the people leave a large part of the fruit on the branches".[73] Once Georgia was swept forcibly into the Roman Empire's sphere of trade, it seems that some villagers established orchards and vineyards in suitable places with alacrity,

[70] Rackham 1990.
[71] Godwin 1975; Huntley 1988; Huntley and Birks 1983.
[72] Lang 1966, p. 86.
[73] Strabo 1928, 11: 4: 3.

Figure 5.23. Photograph of bitten-down *Carpinus* plants in a grassy clearing near the Mukhatgverdi cemetery, 15 km NW of Tbilisi, 2005. The forest behind also contains *Carpinus*, but these have escaped the cattle and developed into trees.

and introduced new plants like the olive and walnut. Grazing was forced into areas not so valued: the lowland steppes and highland grasslands, as well as continuing in places unsuited to horticulture (*e.g.* Lake Jvari.)

Presumably the utilisation of forests changed as well. *Carpinus* wood burns hot and is particularly prized for the manufacture of charcoal.[74] Removing livestock from the vegetation could allow coppicing practices to be introduced. Coppices can provide substantial quantities of small-diameter wood for charcoal making on a very short rotation.[75] Perhaps fire became part of the maintenance and management of these woodlands once grazing livestock were no longer present to control the vegetation. Most of the woodlands around Tbilisi show clear signs of having been coppiced in the recent past.

What the Tsavkisi pollen record suggests is that humans have played a pre-eminent part in shaping Holocene vegetation *composition* on a local scale. Climate has had more bearing on the *density* of cover, a relationship that was accentuated by land-use changes during the Roman Period. Tbilisi's vegetation exists in a state of dynamic

[74] Huntley and Birks 1983.
[75] Rackham 1988; Rackham 2001.

equilibrium between land-use, forest-cover and climate. In the vicinity of Tsavkisi Wetland, oak forests spread widely only during the early-middle Bronze Age, while since the Roman Period various forms of secondary forest, forest-steppe and steppe prevailed. Vegetation types such as shibliak and phrygana existed since the early Holocene under the influence of grazing, while steppe and secondary forest are moulded to a greater degree by fire (**Fig. 5.24**). At many times during the past 8000 years the vegetation was much more open than at present, a fact that may help account for the area's high botanical diversity.

Concluding Remarks

In this chapter, I have presented a pollen-based reconstruction of vegetation changes in the Tbilisi area over the past 8000 years. In specific terms, the development of the vegetation in the hills can be traced from scrub and shibliak in the Neolithic and Chalcolithic Periods (8000–5000 years ago), oak-ash woodland in the Bronze Age (5000–3500 cal BP), through to more steppic phrygana communities in the Iron Age (*ca* 3000 cal BP), to dense secondary forests and cultural vegetation in the Roman Period and finally to a stage where forests were destroyed by fire, recovered and destroyed again, resulting in the present-day landscape of forest and steppe. At a lower altitude (Lake Jvari), the xerophytic vegetation since the Roman Period has been a model of stability. And, under changing climates, the vegetation of the lowland steppes has remained open and grassy through the past millennium.

Significantly, analysis of microscopic charcoal and dung fungal spores has shown that land-use patterns at one site (Tsavkisi) changed dramatically as the Iberian Kingdom was subsumed by the Roman Empire. Up until the Roman invasion, livestock grazing was of great importance in the hills above Tbilisi, leading to the preservation of open vegetation from the earliest times. Grazing maintained xerophytic scrub, shibliak and phrygana communities and prevented the encroachment of forest. But, around 2000 years ago, intensive grazing appears to have declined in favour of orchards, vineyards, gardens and forests. This means that the economy of the hills around the city changed dramatically as cultural and economic influences arrived from the West, influences that included the cultivation of the olive and walnut.

Climate has always played a role in determining the past character of Tbilisi's surrounding vegetation, as shown by the expansion of woodland during a warmer, wetter phase 6000–4000 years ago. Rainfall variations have allowed xerophytes to expand in drier phases and grasses under wetter conditions. However, human activity has also been a dominant force on the vegetation around Tbilisi. If fire and grazing are removed, forest rapidly expands from its refuges in the gorges and gullies to cover areas formerly occupied by steppe, scrub and agricultural land.[76]

[76] Avakov 1982.

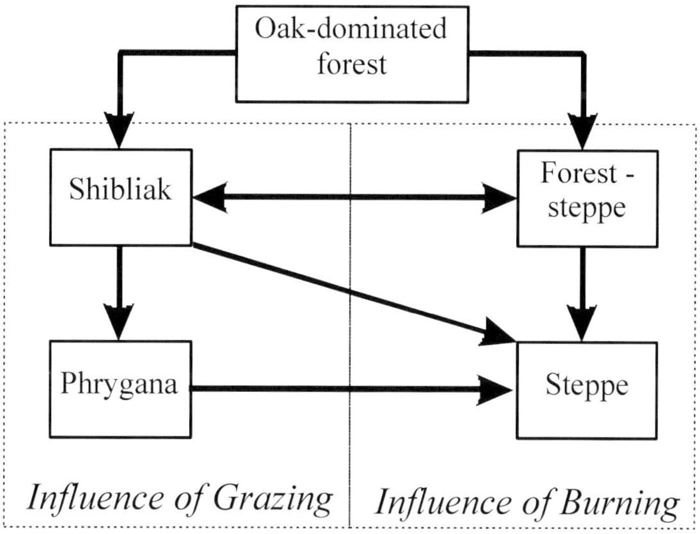

Figure 5.24. A schematic model for plant community succession in the Tbilisi area under different and increasing human impacts. The ability of this process to operate in reverse (*i.e.* reafforestation) is determined largely by topography, since rugged terrain provides ample refuge for forest trees, whilst plains and plateaux, once deforested, are unlikely to recover quickly.

Since the most ancient of times, farmers and graziers have wittingly curtailed the expansion of forests onto agricultural and pastoral lands, and in doing so have created a varied cultural landscape with its own unique flora. Such a heterogenous landscape has existed near Tbilisi for at least 8000 years, and it could be argued that the entire course of Holocene vegetation development has been shaped, to a degree, by human hands.

CHAPTER SIX

THE TSALKA PLATEAU

INTRODUCTION

Let us now turn our attention westward and upward, to the mountain pastures of Southern Georgia and the question of their vegetation history. The Tsalka Plateau is the northernmost part of the South Georgian Uplands, a vast expanse of volcanic terrain that stretches southward from the Trialeti Range to the Georgian border, where it intergrades with the even more extensive highland plateaux of Armenia and northeastern Anatolia. The Tsalka Plateau is hemmed-in to the west by the Samsari Range, rising to 3300 m; and to the east by the steep, forested slopes of the Algeti Valley. Its landscape is a striking combination of flat, grassy plains punctuated by the abrupt peaks of volcanic cones and craters (**Fig. 6.1**).

All this land is devoid of forest. It is not true to say, however, that the landscape is devoid of trees. The stunted beech trees on Mt Tavkvetili, the pine savannas of remote parts of Javakheti, the occasional birch above Nariani Vale, the fragments of mountain oak woodland near Tsalka, wild pears on the Ortashua Plateau: these and other examples have suggested to botanists for many years that this landscape was not always without forest.[1]

Historical evidence suggests that at least part of these highlands was forested even as recently as the 16th century. The *Register of the Gurjistan Vilayet*, in which the Ottomans recorded all the towns within their territory, speaks of the "forested Javakheti region in Khertvisi Liva".[2] Vakhushti Bagrationi, the 18th century geographer, mentions pines and spruces growing near Lake Tabisquri, west of the Tsalka Plateau, where no such trees exist today.[3]

Archaeological evidence may also point to a forested landscape: Bronze Age grave goods depicting deer-hunting scenes in forests, the extensive use of timber in Bronze Age burial chambers and other constructions, and the presence of forest fauna in archaeological vertebrate assemblages.[4] These disparate lines of evidence point to the intriguing possibility that past landscapes were radically different to those we see today.

[1] Grossgeim 1948; Gulisashvili 1964; Ketskhoveli 1959; Nakhutsrishvili 1999; Sosnovskii 1933.

[2] Jikia 1941, p. 191. Khertvisi is an impressive mediaeval fortress at the confluence of the Kura and Paravani Rivers.

[3] Vakhushti Bagrationi 1983.

[4] Bendukidze 1979; Burney and Lang 1971; Howell-Meurs 2001; Kuftin 1941; Kushnareva 1997.

Figure 6.1. Photograph of the Tsalka Plateau at sunset. Near Lake Bareti (August 2003).

But evidence that these highland plateaux were never forested is equally abundant. It includes the palynological investigations of Margalitadze,[5] who argued on the basis of four pollen diagrams that grassy vegetation had dominated the Javakheti and Tsalka Plateaux throughout the Holocene. Faunal assemblages from archaeological sites, although containing some forest species, also include fauna more commonly associated with open landscapes.[6] The geographical reports of Vakhushti Bagrationi and Joseph Tournefort suggest that the southern highlands of the Caucasus were mostly treeless by the early 18th century.[7] And a few botanists have suggested that the mountain xerophilous plants observed in these highlands indicate that steppic vegetation is the original cover.[8]

The question of the Tsalka Plateau's past vegetation is important for understanding the present-day vegetation of the region and for contextualising a very rich archaeological record. It is also of value in estimating how long present-day pastoral activities in the region, such as transhumant grazing, have been practised. The flat, homogenous

[5] Margalitadze 1977; Margalitadze 1995; Margalitadze and Ramishvili 1972.
[6] Bendukidze 1979.
[7] Tournefort 1718; Vakhushti Bagrationi 1983.
[8] Kuznetsov 1909.

landscapes of the Tsalka Plateau provide an ideal counterpoint to the extreme variety of the Tbilisi area. While topography and aspect surely have a strong bearing on the direction of vegetation changes around Tbilisi, these factors cannot be considered important on the Tsalka Plateau. This important difference enables the vegetation histories of the two areas to be compared to identify changes driven by regional climate change versus more localised events.

Topography, Climate and Vegetation

The Tsalka Plateau has a muted relief ranging from 1500–1700 m elevation and relatively few surface streams. The geology of the Plateau is entirely Plio-Pleistocene basalts, dolerites and andesite-basalts.[9] Its geomorphology is somewhat more complex, owing to tectonic movements beneath the basalt mantle. Maruashvili[10] has distinguished seven geomorphological zones (**Fig. 6.2**).

The climate is cold and continental. In Tsalka township the average annual temperature is 5.8 °C. The summers are mild, with an average August temperature of 15.8 °C and an absolute maximum of 30 °C (**Fig. 6.3**). The winters are harsh: their average January temperature is –5.3 °C, and the absolute minimum is –32 °C.[11] Rainfall in Tsalka averages 638 mm per annum, most of which falls in May-June (**Fig. 6.3**). The winters are quite dry, but patchy snow cover lasts around 75–100 days per year, arriving at the end of November and remaining until the end of March.[12]

The vegetation of the individual study sites is described in Chapter Four. The mountain steppic grasslands of the Tsalka Plateau are very rich in species. The dominant species in every location are grasses: *Agrostis planifolia, Festuca valesiaca, F. ovina, Phleum phleiodes, P. pratense, Dactylis glomerata, Deschampsia parviflora, Poa pratensis, Alopecurus aequalis, Agropyron repens, Koeleria cristata, Melica taurica, Nardus stricta* and several others. Tall herbaceous vegetation elements grow amongst grassy meadows in moister places, and include giants such as *Heracleum sosnovskii, Cephalaria gigantea* and *Grossheimia macrocephala* (**Fig. 6.4**). Drier places, such as the rocky slope beside Lake Aligol, have a sparser cover of xeric herbs, such as *Hypericum perforatum, Thymus rariflorus, Allium albidum, Sedum oppositifolium*, and rosaceous shrubs. The diversity of *Centaurea, Cirsium* and other Asteraceae in these highlands is quite extraordinary.

Apart from the occasional *Salix caprea, Quercus macranthera, Betula litwinowii* or *Pinus kochiana* tree, the Tsalka Plateau is treeless. According to Ketskhoveli,[13] only

[9] Gudzhabidze 2003.
[10] Maruashvili 1971.
[11] Kordzakhia 1961, quoted in Margalitadze 1977.
[12] AN GSSR 1964.
[13] Ketskhoveli 1959.

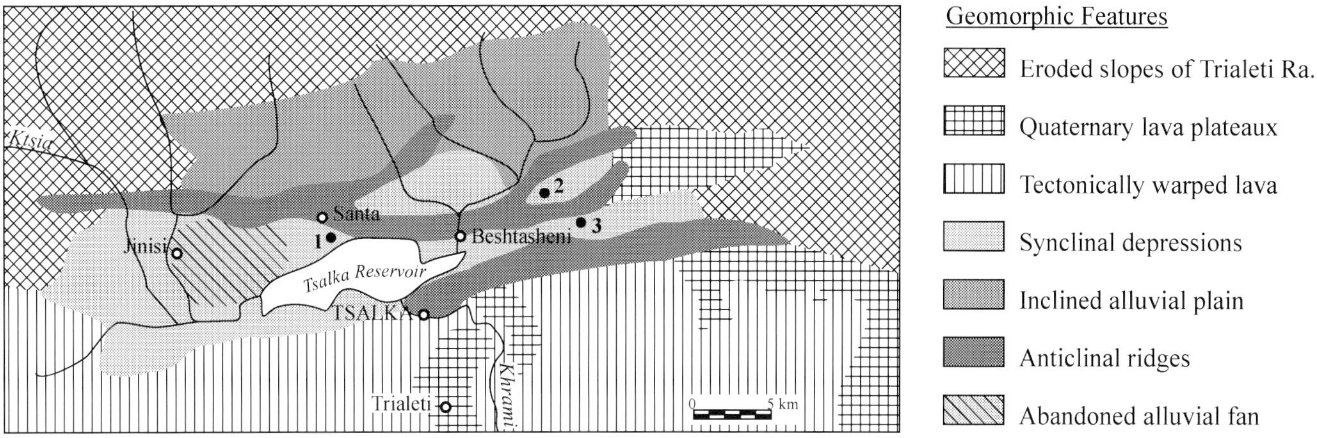

Figure 6.2. Geomorphological scheme for the Tsalka Plateau (Maruashvili, 1971). Study sites are numbered thus: 1. Lake Aligol; 2. Lake Bareti; 3. Lake Imera.

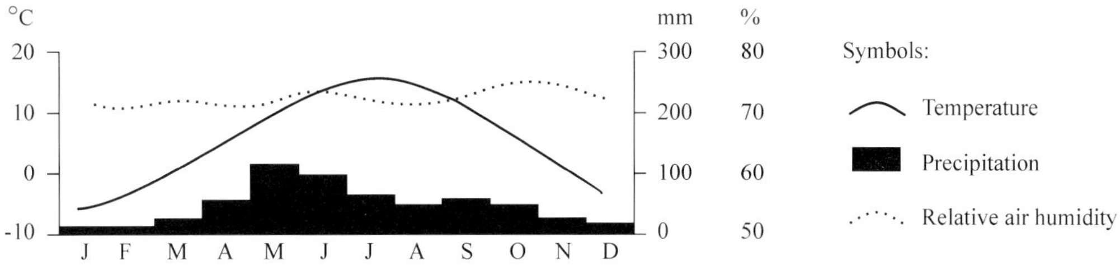

Figure 6.3. Climate diagram, Tsalka, showing monthly averages for temperature, precipitation and relative humidity (AN GSSR, 1964).

2 percent of the Tsalka region can be classified as forest, and most of this grows in the ravines bordering the Plateau itself. Pine plantations were established on the Tsalka Plateau during the Soviet period and now constitute the dominant tree cover in many places.

The Lake Imera Record

Lake Imera is a small, shallow wetland located about two kilometres ENE of Imera village, just north of the Manglisi-Tsalka Road. Its vegetation has been described in detail in Chapter Four and consists of a diverse wetland flora surrounded by steppic grasslands. Although the lake is shallow and a floating sedge mat is gradually covering its surface, it is still visited by fishers, both human and avian. The surrounding meadows are mostly used for haymaking, and their stubble is burnt in autumn. The lake water is quite clean and very low in mineral cations compared to other similar lakes in the region (**Table 6.1**). Lake Imera has little minerogenic input as a consequence of its small catchment size and gently sloping relief.

Figure 6.4. Two Caucasian endemic tall herbs, *Grossheimia macrocephala* (left) and *Heracleum sosnovskii* (right), growing near the Nariani Wetlands of the Tsalka Plateau (August 2003).

Table 6.1. Analysis of lake waters from sites in the Southern Georgian Volcanic Uplands.

	Imera	Bareti	Aligol	Cherepanov	Gomnis	Tabisquri
Chloride ($mg.l^{-1}$)	<0.5	0.7	3.4	0.6	1.1	4.9
Nitrate ($mg\ N.l^{-1}$)	<0.25	<0.25	<0.25	<0.25	<0.25	-
Sulphate ($mg.l^{-1}$)	<5.0	<5.0	<5.0	<5.0	<5.0	3.2
Sodium ($mg.l^{-1}$)	<2.5	3.2	<2.5	<2.5	3.2	9.3
Potassium ($mg.l^{-1}$)	<1.0	<1.0	19.4	<1.0	1.0	-
Calcium ($mg.l^{-1}$)	7.1	14.8	30.7	5.6	11.9	8.6
Magnesium ($mg.l^{-1}$)	<2.5	4.4	8.1	<2.5	3.6	0.8
pH	6.18	6.65	7.37	6.01	6.63	7.40
Alkalinity ($mmol.l^{-1}$)	0.49	1.26	2.83	0.45	1.04	-
Hardness ($mmol.l^{-1}$)	0.29	0.61	1.18	0.22	0.50	-

Data courtesy Michael Sturm, Federal Centre for Water Research, Switzerland. Samples collected by S. Connor and B. Ammann, August 2003. Lake Tabisquri data from Arnol'di (1933).

Sedimentology and Dating

The Lake Imera cores were collected from the floating sedge mat, as close to the centre of the lake as possible. The first core was collected in May 2002 by Dr Ian Thomas (University of Melbourne). This core was 198 cm long. Michael Fletcher and I made a second expedition to Lake Imera in September 2003 and, with the aid of a 'sticky clay' auger, reached weathered bedrock at 393 cm from the same coring location (**Fig. 4.19**).

The sediments of the Imera cores consist of alternating peat and silt layers in the upper part, and compact grey silt in the lower section. The major changes in lithology are shown in **Figure 6.5**. Detailed information on the Imera sediments is provided in **Tables 6.2** and **6.3**.

The two cores correlate well stratigraphically, and for this reason, when constructing a continuous record, depths for the 2003 core were corrected with reference to lithological changes in the 2002 core; specifically, the 41-cm difference between the upper boundary of black silty sediments at 146 cm in the 2002 core and 105 cm in the 2003 core.

Apart from depth, the major difference between the two cores is the depth of the uppermost peat layer. The reason the 2003 core is truncated is that this core was taken through the sedge mat floating on the lake, whereas the 2002 core was taken 2 metres to the west, from the edge of the mat in open water. A period of sediment deposition is therefore missing from the top of the 2003 core. It also must be noted that the greenish sandy sediments between 276 and 393 cm in the 2003 core (equivalent to 317–434 cm, 2002 core) contained negligible amounts of pollen, and probably correspond to weathered basalt or a proto-lacustrine palaeosol, rather than wetland or lake sediment.

Table 6.2. Sediment description of the 2002 Lake Imera core.

Description	Colour (Munsell)	Description
0–47	10YR2/1	Sedge peat with increasing decomposition and silt content with depth
47–61	10YR2/1	Black silt with organic inclusions
61–92	10YR1.7/1	Peat with silt in upper and lower sections
92–108	10YR3/1	Black silt with organic inclusions
108–146	10YR2/1	Peat with variable silt content, *Potamogeton* and *Ceratophyllum* macrofossils
146–152	10YR1.7/1	Sharp transition (146 cm) to black silt with organic inclusions and charcoal
152–198	10YR4/1	Compact grey silts

Table 6.3. Sediment description of the 2003 Imera core. There are some differences in colour between the 2002 and 2003 cores because the 2002 core was described in the laboratory after sampling, while the 2003 core was described in the field during sampling.

Depth (cm)	Colour (Munsell)	Description
0–10	10YR2/2	Dark brown detritus peat
10–13	10YR3/1	Grey clay-silt band
13–30	10YR2/2	Silty peat
30–60	10YR2/2	Dark brown crumbly peat
60–63	10YR3/1	Dark grey silt band
63–105	10YR2/1	Decomposed sedge peat, with many black bands (charcoal particles)
105–115	10YR1.7/1	Black humic silty peat
115–190	10YR4/1	Compact grey silts
190–206	10YR1.7/1	Blackened silts. Abundant macroscopic charcoal particles
206–240	10YR4/1	Compact grey silt with small white mineral concretions at 215cm-depth
240–276	10YR4/1	Compact grey silt with many small, angular pebbles and sand
276–370	10YR4/1	Compact grey sandy silt with no pebbles
370–393	10YR4/1	Compact grey sandy silt with many small pebbles. Weathered bedrock at 393cm-depth.

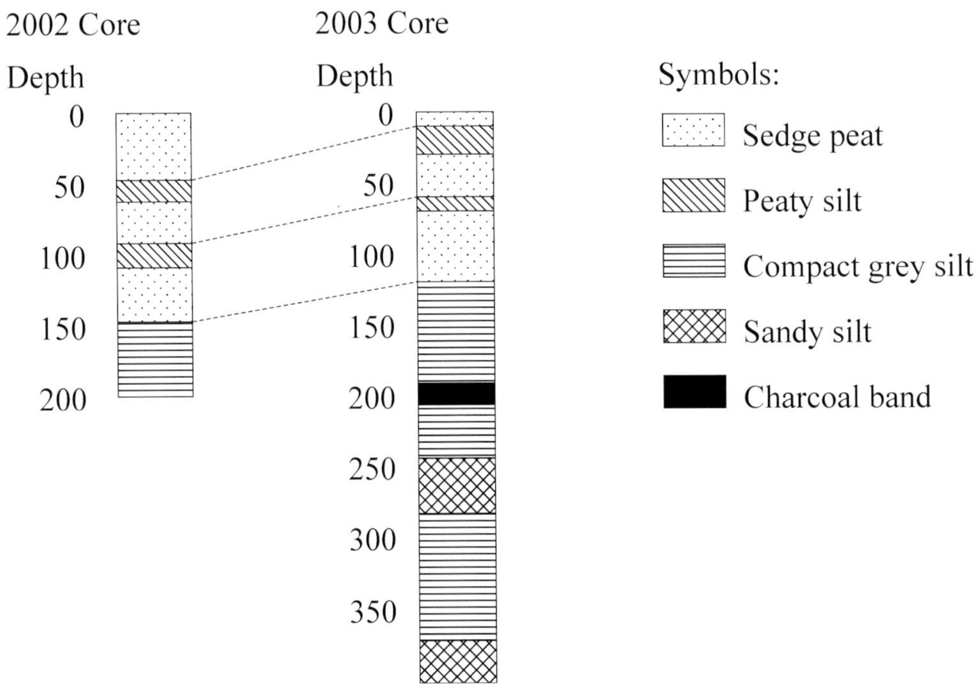

Figure 6.5. Simplified graphic comparison of Lake Imera sediment cores from 2002 and 2003.

Magnetic susceptibility and loss-on-ignition measures are negatively-correlated throughout the Lake Imera core (**Fig. 6.6**). This indicates that magnetic minerals are held in the inorganic fraction, which is probably enriched with basaltic titanomagnetite particles.[14] There is no evidence in the charcoal record (see below) for catchment fires causing magnetic susceptibility variations.[15]

Perhaps the most striking relationship in the Imera record is between organic content and arboreal pollen percentages (**Fig. 6.6**). The peaty sediments contain high proportions of arboreal pollen, whereas non-arboreal pollen dominates in the hard, silty sediments. All previous palynological investigations on the plateau have observed this pattern.[16] As noted in Chapter Four, AP:NAP ratios in the study area reveal much more about the sedimentary environment than they do about forest cover. This is because small, shallow lake basins tend to desiccate under the semi-arid climate. In the Tsalka region, lake levels are usually lowest in spring, when most trees are releasing pollen. On this basis, it is suggested that the reductions in the AP:NAP ratio represent low lake levels (at least seasonally) and reduced pollen source area, rather than having any palaeoecological significance. Lake Imera is particularly sensitive to changes in precipitation and evaporation as it is a closed lake that relies completely on rainfall and groundwater to sustain its water level.

The six radiocarbon dates from the Imera material fit a consistent chronological series (**Fig. 6.6**, Appendix D). Sedimentation rates have varied quite considerably through the last 10,000 years. The highest sedimentation rates are recorded between 1530–930 cal BP (1 mm per annum), followed by the period of 4850–4470 cal BP (0.8 mm p.a.). The slowest rates were in the early part of the record (7471–4850 cal BP: 0.02 mm p.a.). In general, peat layers accumulated at least twice as quickly as the silty sediments.

The sedimentological data can be brought together with the radiocarbon chronology to provide a sequence of lake-level fluctuations over the past 10,000 years. The lowermost levels, of probable late Pleistocene age, accumulated when the lake was dry. Low lake levels continued through the early Holocene, but gave way to slightly moister conditions about 8000 cal BP. The lake began to dry again beginning 5000 years ago and low lake levels continued until about 2300 cal BP. After this the lake filled completely and peat began to form. Peat formation progressed until interrupted by two drying phases: a short spell of perhaps a century's duration around 1500 years ago, followed by a longer droughty period lasting from about 800 to 1200 AD. Lake levels reacted to increased evaporation, reduced rainfall or changes in seasonality during all low phases.

[14] Thompson *et al.* 1975.

[15] *E.g.* Gedye *et al.* 2000; Rummery 1983.

[16] Margalitadze 1977; Margalitadze 1995.

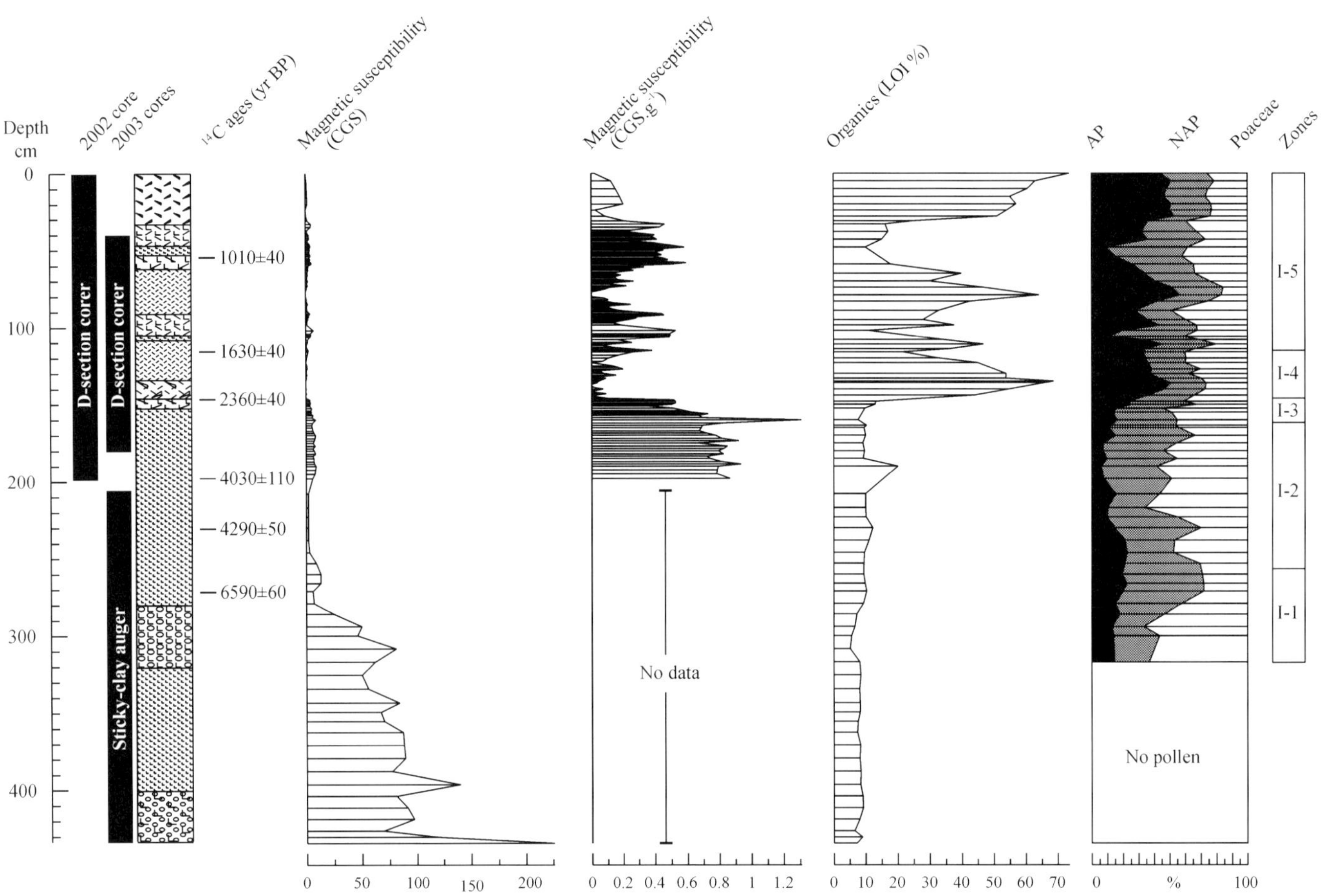

Figure 6.6. The Lake Imera core: lithology, core overlap, radiocarbon dating, magnetic susceptibility, loss-on-ignition and arboreal pollen ratios. Pollen zones shown at right.

Pollen Diagram

Lake Imera's pollen diagram (**Fig. 6.7**) has been constructed using percentages calculated from separate arboreal and non-arboreal pollen sums. This is to account for the changes in source area that occurred during low lake levels. The silt-rich sediments contain abundant pollen of local types from the Poaceae, Asteraceae, Lamiaceae, Brassicaceae and Polygonaceae families. Under the pollen source-area model proposed by Jacobson and Bradshaw,[17] Lake Imera's pollen spectra will be dominated by 'regional' pollen when the lake is full. Any reduction in water level, and hence site diameter, will result in 'extra-local' and 'local' pollen flooding the lake's pollen spectra (**Fig. 6.8**).

[17] Jacobson and Bradshaw 1981.

Pollen preservation in the peat sediments was very good. The same could not be said of the silty strata, in which a small proportion of pollen grains were corroded. Pollen degradation occurred when the lake dried and its sediment surface was exposed to oxidation. Preservation was not so bad, however, as to prevent confident pollen identification. The pollen diagram is divided into five stratigraphic zones:

Zone I-1. The first zone, spanning the period from approximately 10,000 to 5000 cal BP, has high proportions of *Carpinus betulus, Fagus, Zelkova*-type and *Corylus* amongst the arboreal pollen, and non-arboreal pollen dominated by Poaceae, *Cichorium*-type, *Ranunculus*-type, Apiaceae and *Polygonum aviculare*-type. A striking feature of this zone is the large amount of *Dipsacus*-type, *Campanula, Thalictrum, Filipendula* and pollen of other subalpine herbs. Some of these insect-pollinated plants are very much under-represented palynologically, such that their abundance in pollen zone I-1 suggests that subalpine herbaceous plants were widespread, if not dominant, in the early-mid Holocene vegetation around Lake Imera. The wetland vegetation of that time consisted mostly of grasses and sedges, consistent with low lake levels. Charcoal concentrations are high through the middle and upper parts of the zone.

Zone I-2. The second zone begins with a substantial peak in charcoal concentrations, which suggest that fires were commonplace on the Tsalka Plateau between 6000 and 5000 years ago. As the fires became increasingly frequent or intense, fire-adapted taxa such as *Pinus* and *Alnus* increased in importance, while *Carpinus betulus* and *Zelkova*-type declined. The cessation of fires after 5000 cal BP led to the expansion of *Quercus* (probably *Q. macranthera*). The ground flora changed at the same time. Most of the subalpine herbs so prevalent in the first zone yield to Lamiaceae, Fabaceae, Caryophyllaceae, Polygonaceae, *Papaver* and *Hypericum*-type. These are types more characteristic of dry grasslands than of herbfields. However, since Chenopodiaceae and *Artemisia* percentages are reduced through this zone, there is no evidence for the expansion of true semidesert-steppe vegetation.

Zone I-3. Towards the end of the second zone and beginning of the third, *Quercus* begins to decline. Between 3500 and 2500 cal BP, the contraction of oaks and their replacement by pine (*Pinus kochiana*) and fir (*Abies nordmanniana*) could be linked to cooler climatic conditions. *Triticum*- and *Secale*-type pollen suggest that grain was being cultivated in close proximity to Lake Imera at this time. Like the previous zone, charcoal concentrations are low and the wetland vegetation was composed of grasses and sedges.

Zone I-4. Beginning some time around 2400 cal BP, this zone sees the complete transformation of the vegetation. The most significant aspect of this zone is the extraordinary abundance of charcoal in the sediments. Numerous charcoal peaks suggest that fires ravaged the Tsalka Plateau frequently between 2400 and 1700 cal BP. Their impact on the vegetation was profound. *Abies,* a tree easily killed

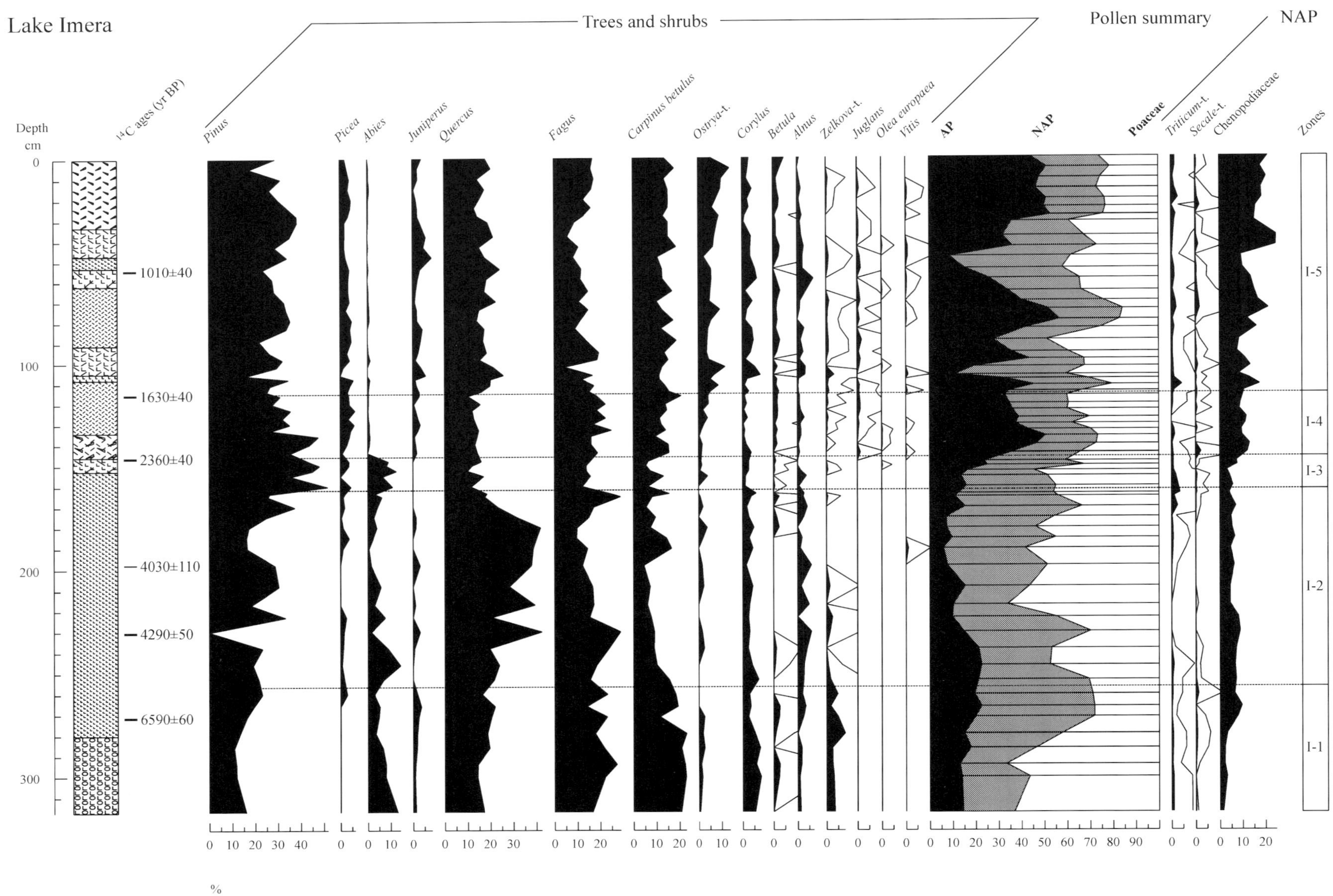

Lake Imera
Trees and shrubs
Pollen summary
NAP
Depth cm
14C ages (yr BP)
Pinus
Picea
Abies
Juniperus
Quercus
Fagus
Carpinus betulus
Ostrya-t.
Corylus
Betula
Alnus
Zelkova-t.
Juglans
Olea europaea
Vitis
AP
NAP
Poaceae
Triticum-t.
Secale-t.
Chenopodiaceae
Zones
1010±40
1630±40
2360±40
4030±110
4290±50
6590±60
I-5
I-4
I-3
I-2
I-1
%

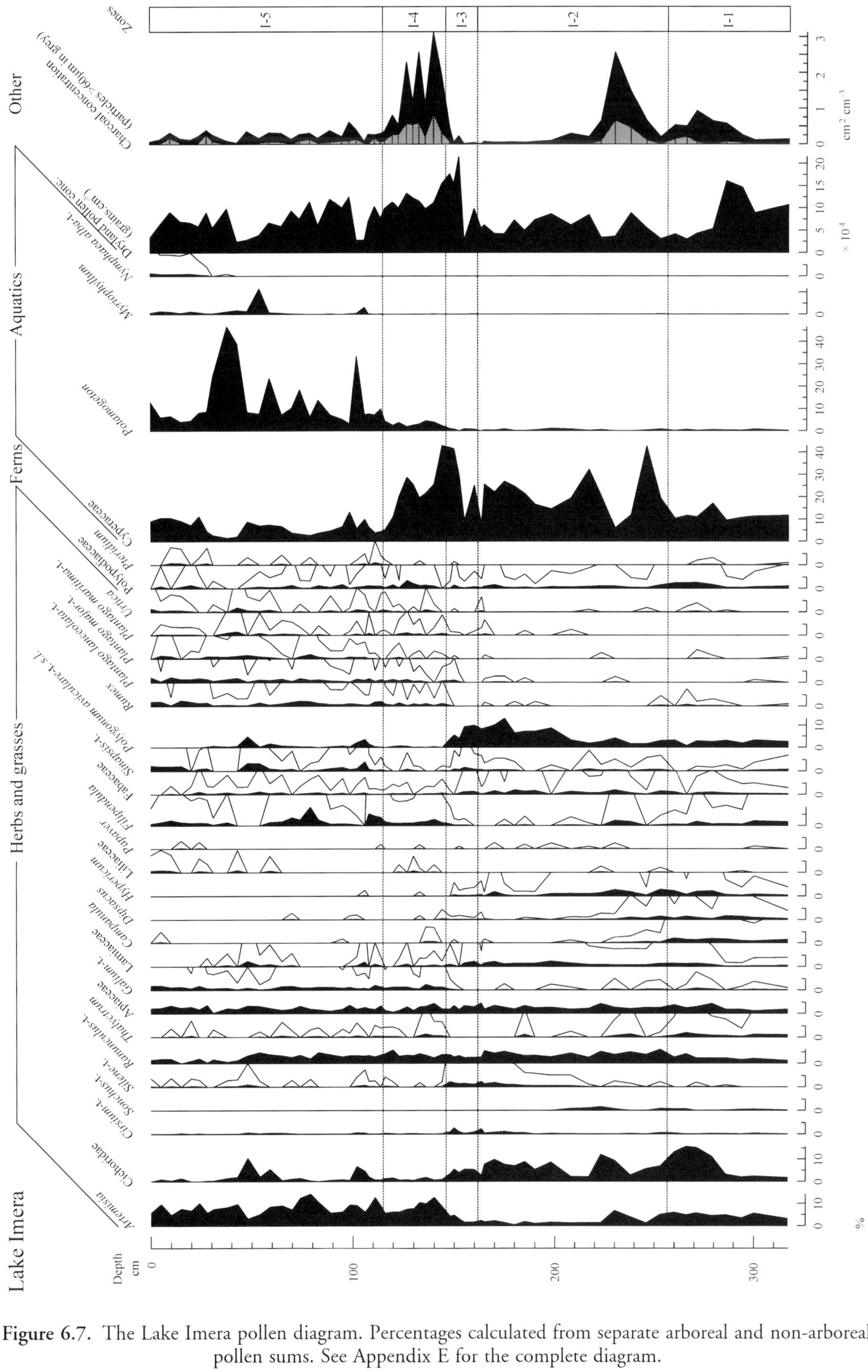

Figure 6.7. The Lake Imera pollen diagram. Percentages calculated from separate arboreal and non-arboreal pollen sums. See Appendix E for the complete diagram.

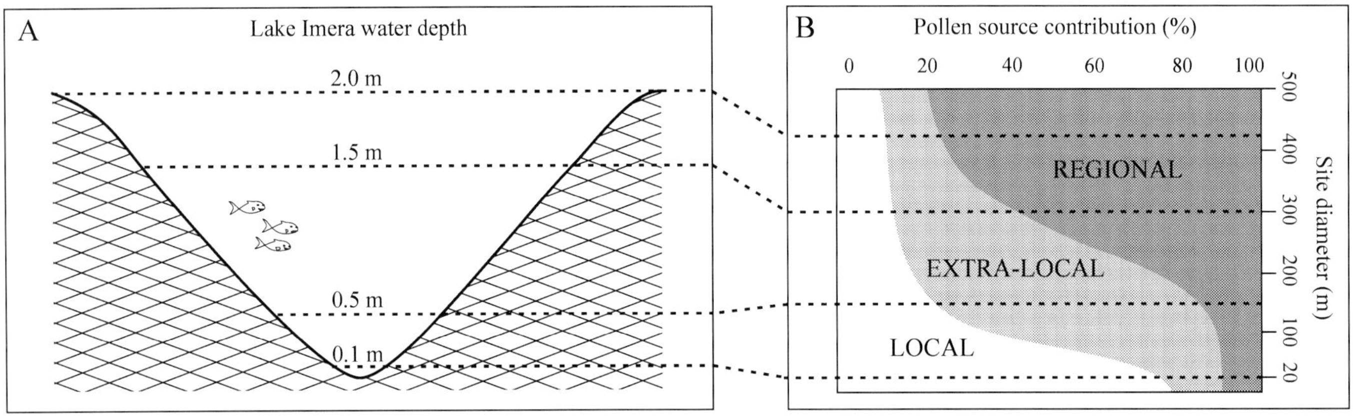

Figure 6.8. Theoretical relationship between lake levels and pollen source-area at Lake Imera. A. Schematic cross-section of Lake Imera, with water depth and basin diameter measurements based on field observation. B. Pollen source-area versus basin size (after Jacobson and Bradshaw, 1981); dashed lines indicate the relative contribution of regional, extra-local and local pollen at varying lake levels. Depths as measured at the coring site.

by fire,[18] ceased to exist in the vicinity of Lake Imera. Even pine was reduced through successive fires. Grassland herbs such as Chenopodiaceae, *Artemisia*, *Galium*, Liliaceae, *Urtica*, *Rumex* and *Plantago* species increased dramatically, signalling the arrival of more open vegetation. This zone also sees the arrival of *Juglans* and *Olea europaea*. After deforestation, Lake Imera began to fill with water, which was readily colonised by *Potamogeton*. Sedges and grasses were pushed to the lake fringes by elevated water levels.

Zone I-5. The final zone is characterised by relative vegetation stability. The destructive fires of zone I-4 gave way to a cycle of low-intensity burning in zone I-5, which may have helped to maintain the grassland vegetation in a state of 'plagioclimax'. Grazing appears to have been an important land-use during the past 2000 years, indicated by rising percentages of *Plantago major*-type, *P. lanceolata*-type, *Rumex* and *Pteridium*. Two lake regressions, at 1500 and 1000 cal BP respectively, allowed the plants represented by *Polygonum aviculare*-type, Brassicaceae and *Cichorium*-type pollen to colonise the drying lake-bed, just as they had done through much of the early and mid Holocene.

[18] Tinner *et al.* 2000.

Numerical Analyses

As in the previous chapter, the approach to analysing these data with multivariate methods relates back to the environmental gradients produced in Chapter Four. The result of applying weighted averaging to the Lake Imera pollen data is given in **Figure 6.9**. According to the first trend, effective moisture was moderate during the early and middle part of the Imera record, with a distinct 'dry spell' during the oak expansion of 5000–3000 cal BP. Then, it seems, rainfall increased and peaked around 2000 years ago. Two dry spells occurred in the late Holocene: one at 1500 cal BP, and a later, longer one centred on the Mediaeval Period.

It is difficult to determine whether this trend is indeed a measure of actual precipitation, or reflects changes in local vegetation due to lake level fluctuations or soil moisture availability. Evidence for a dry spell at Lake Imera between 5000–3000 cal BP conflicts with evidence from Lake Van that this period was one of high lake levels and an 'optimum' in temperature and precipitation.[19] However, the hydrology of Lake Imera is very different to Van, driven by direct precipitation, groundwater fluctuations and evaporation. High rainfall in spring and early summer now sustains Lake Imera through the hottest and driest months of late summer. Low lake levels could be expected if rainfall were concentrated in winter. An hypothesised mid-Holocene decrease in continentality[20] seems to fit with this explanation. On the other hand, the reductions in spring-flowering tree pollen compared to summer-flowering grasses during the Imera dry phases could suggest that lake levels were low in spring, consistent with an increase in continentality (*i.e.* summer-dominated precipitation) as proposed by Stevens *et alia*.[21] Irrespective of the amount of precipitation, hotter summers could cause Lake Imera's level to drop, and, because groundwater levels in a plateau landscape can be strongly influenced by the presence or absence of trees, which act as groundwater 'pumps',[22] any tree cover in the past could have drawn down Lake Imera considerably.

When the gradient of forest-cover and temperature is applied to the Imera material, three phases with increased values are apparent. The first occurred during the expansion of *Carpinus betulus* and *Zelkova*-type around 7500 years ago; the second during the oak expansion of 5000–3000 cal BP; and the third in recent decades.

The third trend may pertain to land-use. If so, it indicates that grazing intensified during the Mediaeval Period. Secondary forest succession was greatest following the cessation of fires 5000 years ago and after the demise of oak woods 3500–3000 years ago.

[19] Wick *et al.* 2003.
[20] Wick *et al.* 2003.
[21] Stevens *et al.* 2006.
[22] Moore 1986; Moore and Evans 1991.

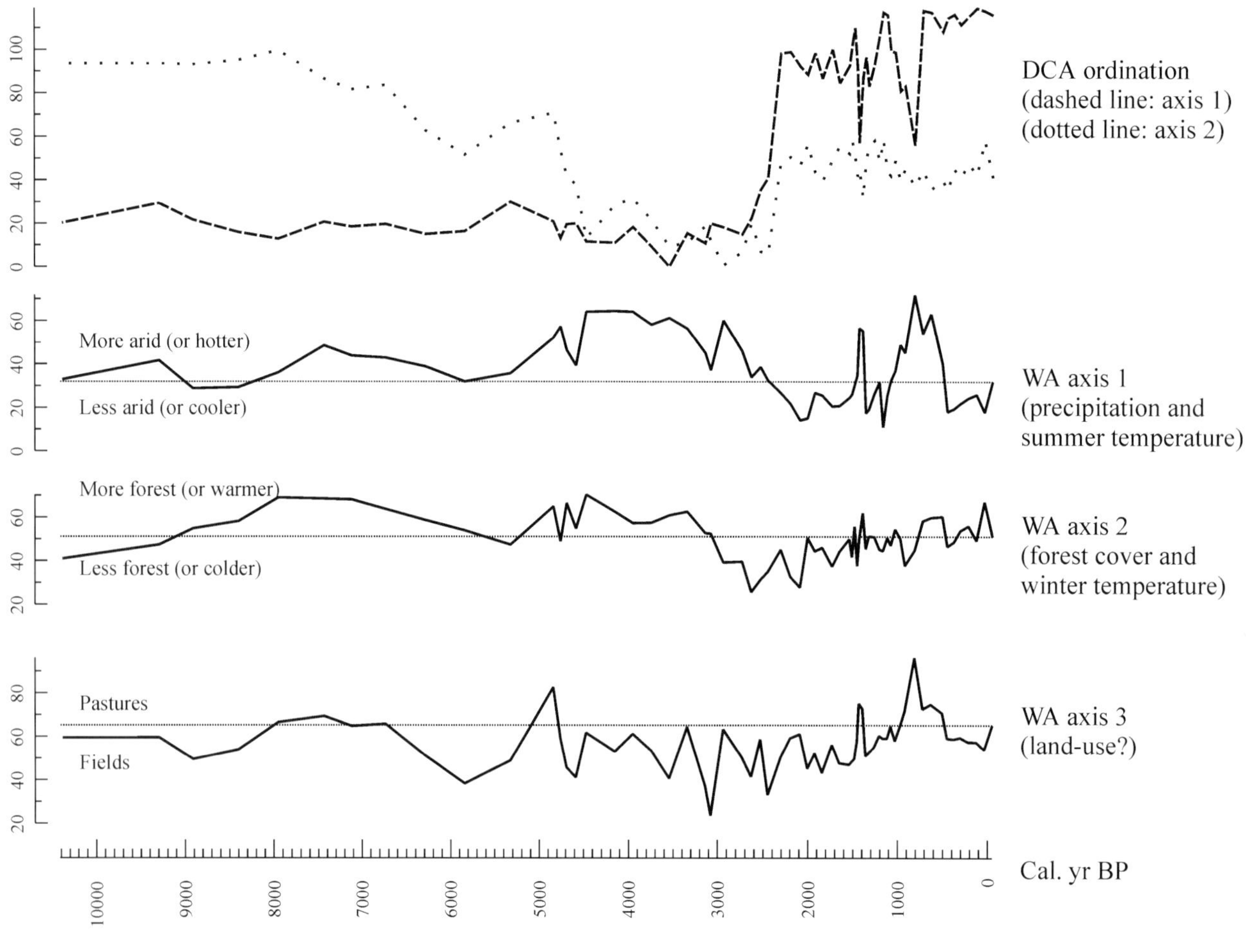

Figure 6.9. Environmental changes at Lake Imera: rainfall, forest cover and land-use gradients produced by weighted averaging. Dominant trends in the pollen data (DCA ordination) are shown at the top for comparison. Chronology based on linear interpolation of calibrated ^{14}C ages.

Comparing these gradients with the dominant trends in the Imera pollen data, it is obvious that lake level has been the primary determinant of pollen spectra at this site (**Fig. 6.9**; Appendix B). DCA axis 1 has weak correlations with the WA gradients, but appears to be strongly related to sediment parameters such as organic content and magnetic susceptibility.

The secondary trend, DCA axis 2, is not related to any of the WA gradients. It is correlated with forest taxa such as *Carpinus, Zelkova* and *Thalictrum*, and negatively correlated with indicators of open vegetation, *Silene* and *Pinus*. Quite possibly, this represents a forest-cover gradient, given the relationship between this trend and charcoal concentrations. If this assessment is correct, it means that WA axis 2 is a reflection of temperature changes only (*cf.* **Fig. 6.20**). It also means that forest cover at Imera increased gradually during the early Holocene, declined at the time of fires between 6000 and 5000 cal BP, peaked between approximately 4500 and 2400 cal BP, before declining again after repeated fires, stabilising at a level similar to the present one.

Table 6.4. Stages in vegetation development at Lake Imera.

Zone	Time (cal BP)	Inferred vegetation	TWINSPAN (Chapter Four)
I-1	10000–5000	Meadows and mixed woods	Grassland-forest border
I-2	5000–3200	Oak-dominated savanna	Grassland-forest border
I-3	3200–2400	Open coniferous woodland	Grassland-forest border
I-4	2400–1600	Treeless grasslands	Mountain grassland
I-5	1600–0	Grassland with few trees	Mountain grassland

In terms of vegetation development, these data are synthesised in **Table 6.4**. The vegetation surrounding Lake Imera went from moist subalpine meadows with a scattering of trees, through a stage in which oak was the dominant tree and the ground flora was dry grassland; oaks were replaced by conifers as the climate cooled, and these in turn gave way to treeless grasslands after a series of destructive fires. The continued impact of grazing, burning and small-scale cropping has maintained this essentially deforested landscape through until the present-day.

The Lake Bareti Record

Lake Bareti is situated only 3 km WNW of Lake Imera, separated from it by a low hill called Mt Bareti (1670 m). Lake Bareti is much larger than Lake Imera, but is still quite shallow. It lacks the distinct terraces seen at Lake Imera because it has a small, intermittent outflow to the west. Like Imera, however, the lake's water balance is determined by direct precipitation, runoff, groundwater and evaporation, as it has no permanent inflow. The lake basin is home to a pine plantation and some abandoned fishing huts. Today, the surrounding land is used mainly for grazing and haymaking.

Sedimentology and Dating

Prof. Brigitta Ammann, Willi Tanner (Institute of Plant Sciences, Bern University) and I collected the Lake Bareti core from a boat in August 2003. We used a Perspex piston corer with a barrel 130 cm long and obtained a core of that length from the centre of the lake where the water was 155 cm deep. The material was dark brown, homogenous gyttja with an increasing proportion of silt in the lower 30 cm. As the core contained no visible stratigraphic changes, no sediment description is given here.

Loss-on-ignition and moisture content provide more insightful information on the sedimentology of the Lake Bareti core (**Fig. 6.10**). Organic content peaks at around 50 percent in the middle part of the core. The upper 30 cm, and particularly the lower 30 cm, have a higher proportion of inorganic material – as much as 75 percent at a depth of 100 cm. At various points in the lake's history, either inorganic input increased through catchment erosion, or the lake's organic productivity declined.

One radiocarbon date was obtained on bulk carbon at 122–124 cm depth. This yielded an age of 900±40 ^{14}C bp (OZH-070), corresponding to calendar years between 1032–1217 AD. Since that time the Lake Bareti sediments have been accumulating at an average rate of 1.3 to 1.7 mm per annum.

Pollen Diagram

The Bareti material contained abundant, well-preserved pollen from a great diversity of plants. The pollen diagram (**Fig. 6.11**) is divided into two pollen assemblage zones:

Zone B-1. Non-arboreal pollen dominates the first zone. Amongst these, Poaceae, Chenopodiaceae and *Artemisia* are the most prevalent. Arboreal pollen occupies a subordinate position, comprising less than 35 percent of the dryland pollen sum. *Pinus* is dominant, followed by *Quercus, Fagus, Carpinus betulus* and *Ostrya*-type. Amongst aquatic pollen, Cyperaceae, *Myriophyllum* and *Equisetum* are well represented, as are *Pediastrum, Botryococcus, Spirogyra* and other fungal remains. Charcoal concentrations are low throughout the early part of the Bareti record.

Zone B-2. The second zone differs very little in pollen composition to the first. The major difference lies in the abundance of arboreal pollen, which in this zone averages 55 percent. Increasing proportions of *Plantago lanceolata, P. major* and *Urtica* in the upper 30 cm suggest some localised disturbance in the catchment in the last century. The age of this disturbance is confirmed by the appearance of *Ambrosia*-type pollen in the upper part of the record. Disturbance had a considerable impact on the wetland vegetation, as Cyperaceae, *Myriophyllum, Alisma* and some algal species increase quite dramatically in the upper part. This may relate to recent eutrophication due to construction works and the establishment of pine plantations during the Soviet Era or from the earlier peopling of the Tsalka Plateau by Pontic Greeks displaced from Turkey in the late 19th and early 20th centuries.

Numerical Analysis

Overwhelmingly, the Lake Bareti record demonstrates stability in highland vegetation over the past 800–900 years. Weighted averaging produced three trends (**Fig. 6.12**). The first of these closely resembles the ratio of arboreal pollen to non-arboreal pollen

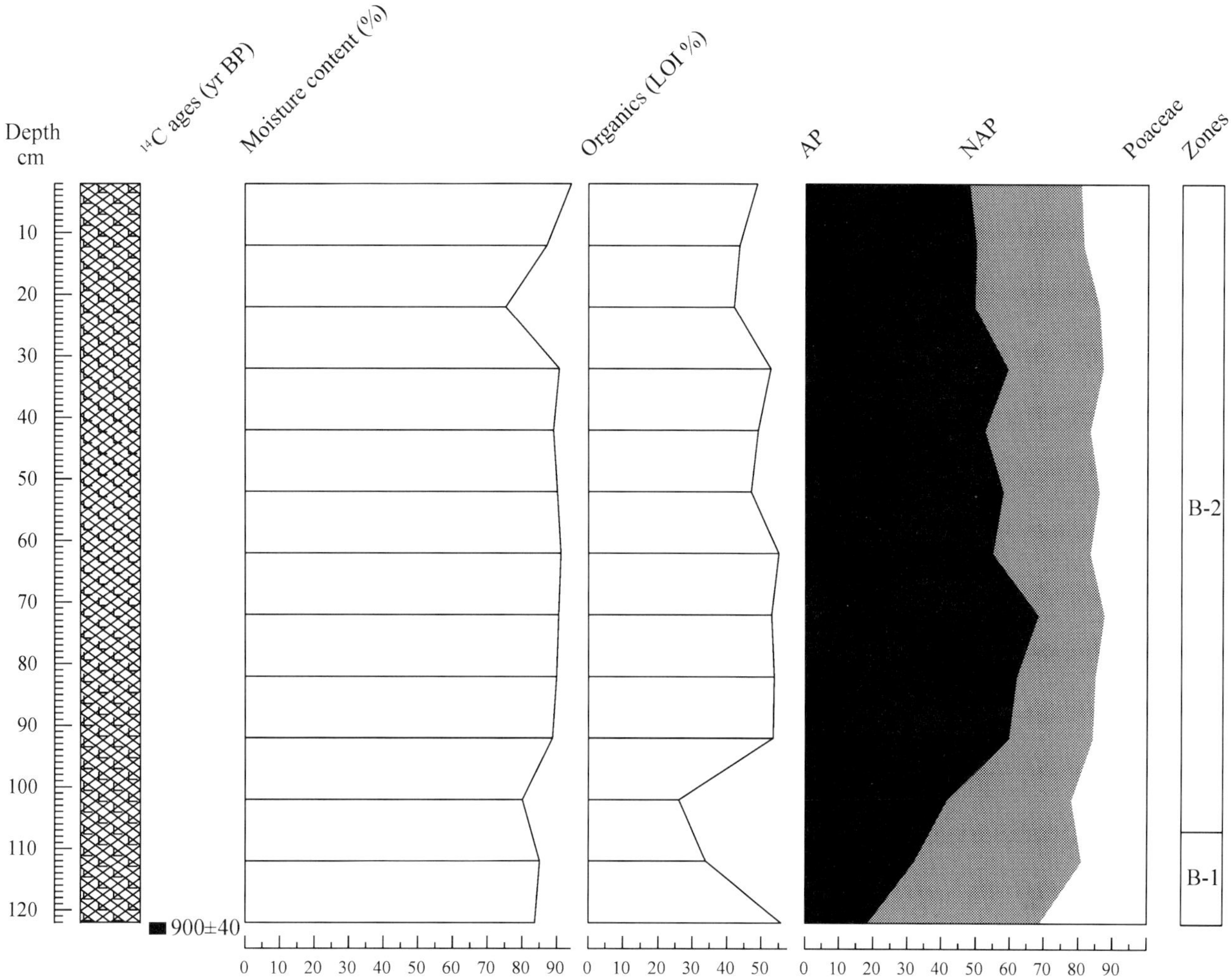

Figure 6.10. The Lake Bareti core: lithology, radiocarbon dating, moisture and organic content determinations, and pollen summary. Pollen zones at right.

and relates to moisture. As the nearby Lake Imera record has shown, this trend may relate simply to lake regression, which, by reducing pollen source area, increases the relative abundance of herbaceous pollen. At Lake Bareti this is reinforced by the loss-on-ignition, Cyperaceae and *Myriophyllum* curves, which indicate a reduced lake level in the lower and upper parts of the record. Between 900 and 700 years ago, Lake Bareti existed as a shallow wetland. It is tempting to associate this regression with the climatic anomaly often called the 'Mediaeval Warm Period'.[23]

[23] Bradley *et al.* 2003.

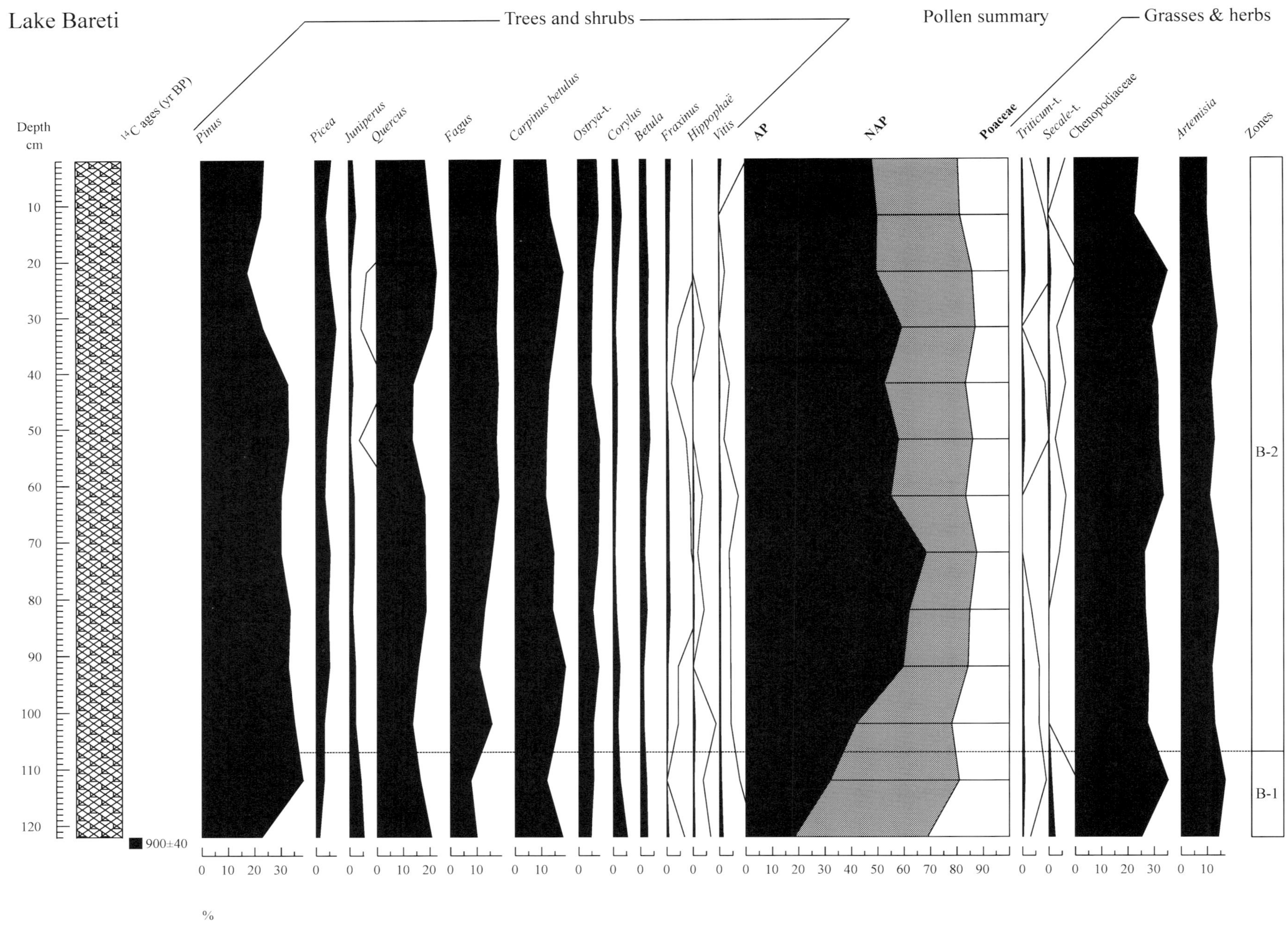
Lake Bareti
Trees and shrubs
Pollen summary
Grasses & herbs
Depth cm
^{14}C ages (yr BP)
Pinus
Picea
Juniperus
Quercus
Fagus
Carpinus betulus
Ostrya-t.
Corylus
Betula
Fraxinus
Hippophaë
Vitis
AP
NAP
Poaceae
Triticum-t.
Secale-t.
Chenopodiaceae
Artemisia
Zones
B-2
B-1
900±40
%

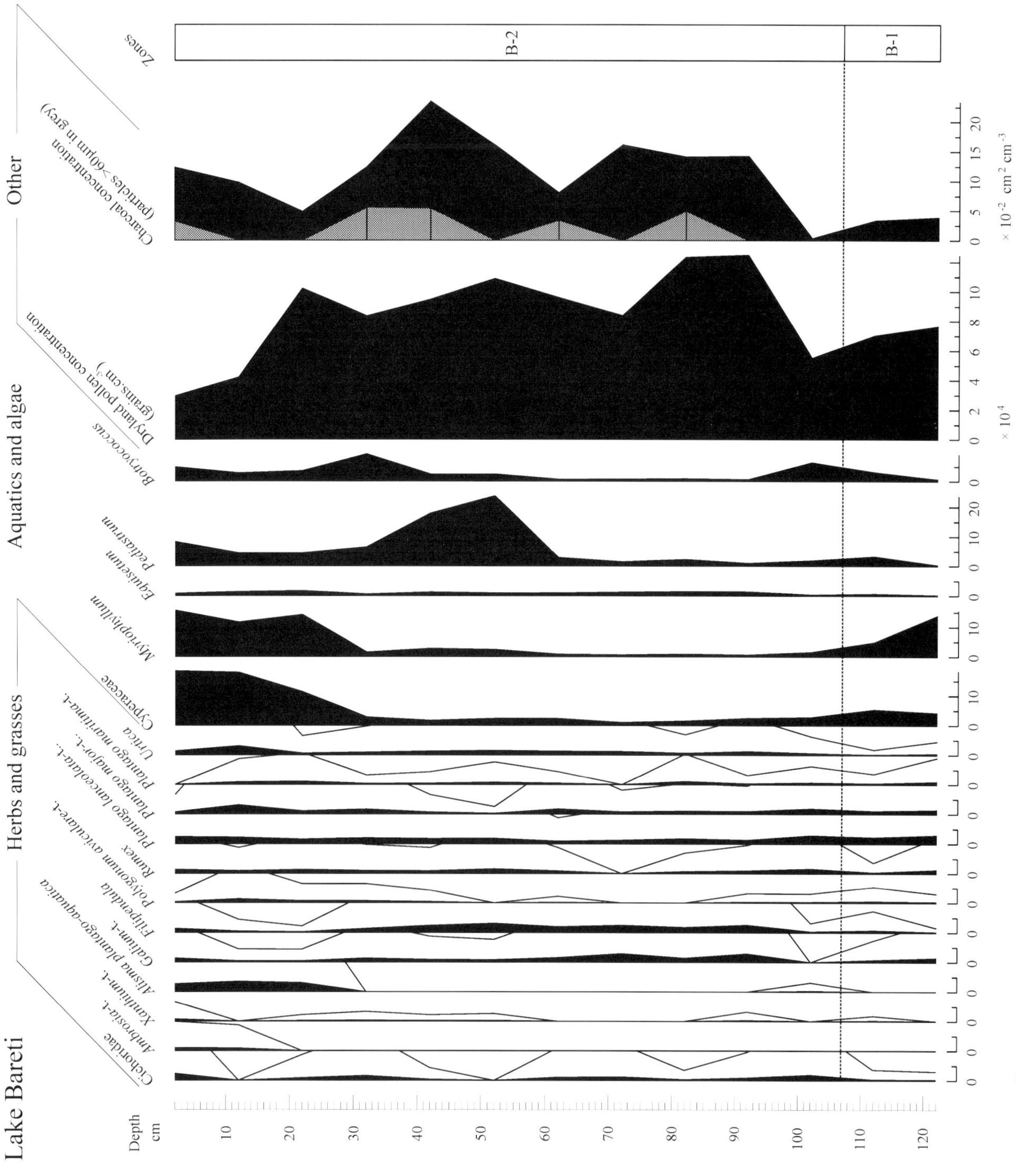

Figure 6.11. The Lake Bareti pollen diagram. Percentages calculated from separate arboreal and non-arboreal pollen sums. See Appendix E for the complete diagram.

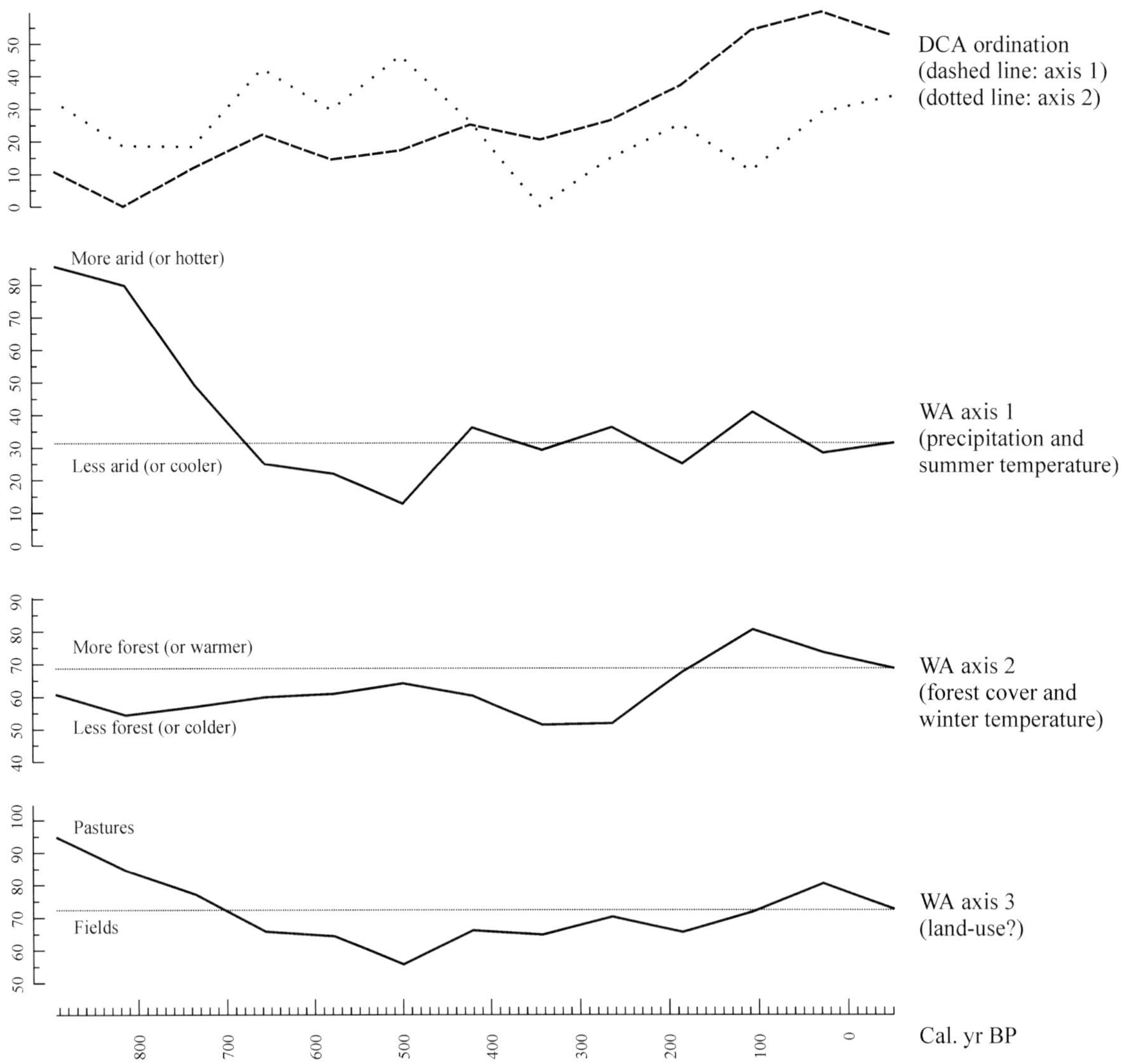

Figure 6.12. Environmental changes at Lake Bareti: rainfall, forest cover and land-use gradients produced by weighted averaging. Dominant trends in the pollen data (DCA ordination) are shown at the top for comparison.

The second trend, of forest cover and temperature, seems to suggest that some afforestation or temperature increase has occurred in recent decades. *Quercus* pollen has become more abundant in the Bareti sediments recently, whereas *Pinus* has declined somewhat. This is surprising given that pines have presumably increased in the lake catchment after being planted there following World War II. The same recent increase is seen in the Imera material and may reflect a broad-scale pattern. It is probably the dominant trend in the Bareti material, as indicated by a correlation (r^2=0.63) between the DCA ordination and WA axis 2 (**Fig. 6.12**; Appendix B). It is probable that this gradient represents temperature fluctuations, since TWINSPAN classified pollen zone B-1 as 'forest-steppe' and B-2 as 'mountain grassland, suggesting greater forest cover during the dry period of the Mediaeval.

The third trend appears to indicate that grazing was of increased importance during the late Mediaeval Period and the 20th century compared to other times. The paucity of dung fungi in the Bareti material unfortunately prevent any independent analysis of this trend, although pollen types associated with grazing do appear in greater abundance during the upper and lower parts of the record. These anthropogenic indicators are associated with lower charcoal concentrations, suggesting that land-use patterns did indeed shift.

While these trends are suggestive of changes, they were changes of a very low magnitude. For at least the last 900 years, mountain grasslands much like the present ones have existed in the landscape around Lake Bareti.

The Lake Aligol Record

Twelve kilometres west of Lake Bareti lies Lake Aligol. This small wetland is situated mid-way between the town of Santa and the Tsalka Reservoir (**Fig. 6.2**). The site is now bypassed by British Petroleum's Baku-Tbilisi-Ceyhan oil pipeline. The site's elevation is 1534 m above sea level. Its catchment has muted relief and is composed of Upper Pliocene to Quaternary lava flows (dolerite and andesite-basalt), and the lake itself is approximately 200 by 300 m in diameter. Although the site has a potential catchment area of 1 km^2, overland flow is negligible due to the porosity of the surrounding volcanic plains. Santa Township relies on springs for their water supply, suggesting a significant groundwater increment in this terrain.

The water of the lake is generally harder and higher in cations than other volcanic lakes of the region (**Table 6.1**). This is probably due to the lake being much reduced in volume, thereby concentrating chemical solutes within the remaining water body. The weathered boulders at the southern end of the lake suggest a basin of considerable age, such that dissolved salts may have been accumulating in the basin water for a significant length of time. The size of the catchment, its erosion history and groundwater may also influence the water chemistry.

Lake Aligol occupies a closed basin dammed by a basaltic outcrop (**Figs 6.14** and **6.15**). The water depth in the centre of the lake was 64 cm in early September 2003, and it is reported as having been dry during ornithological surveys carried out in March and April 2000.[24] A distinct lake terrace on the eastern side of the basin and

[24] Anon. 2002.

Figure 6.13. Photograph of Lake Aligol, viewed from the southern shore, showing the basalt outcrop that dams the lake (foreground), black soil along the oil pipeline route (right) and the village of Santa (background, at foot of hill).

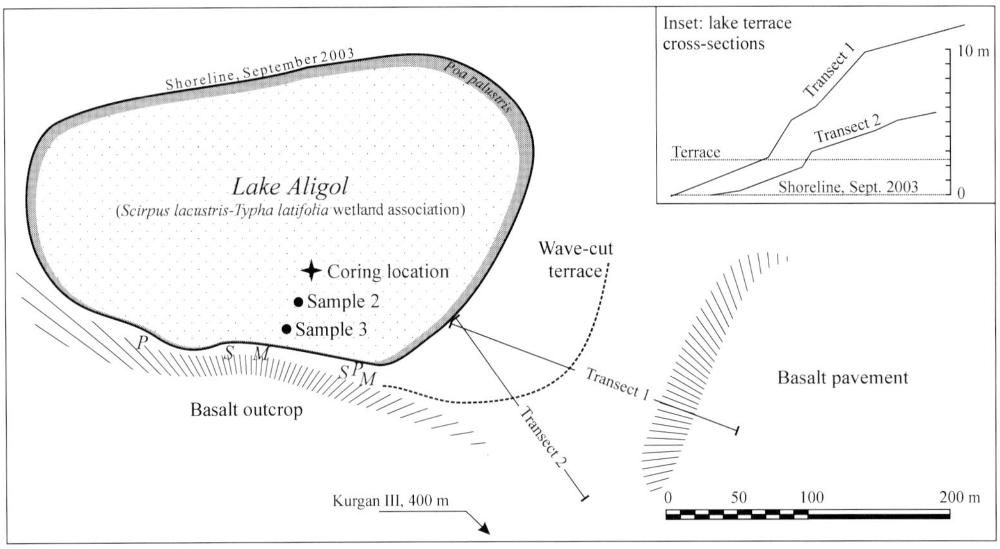

Figure 6.14. Map of the morphometry of Lake Aligol, including sampling locations and cross-sectional profiles of the lake terrace (source: ground survey, 2003). Shrubs at foot of basalt outcrop: *P*: *Populus nigra*; *S*: *Salix caprea*; *M*: *Malus orientalis*.

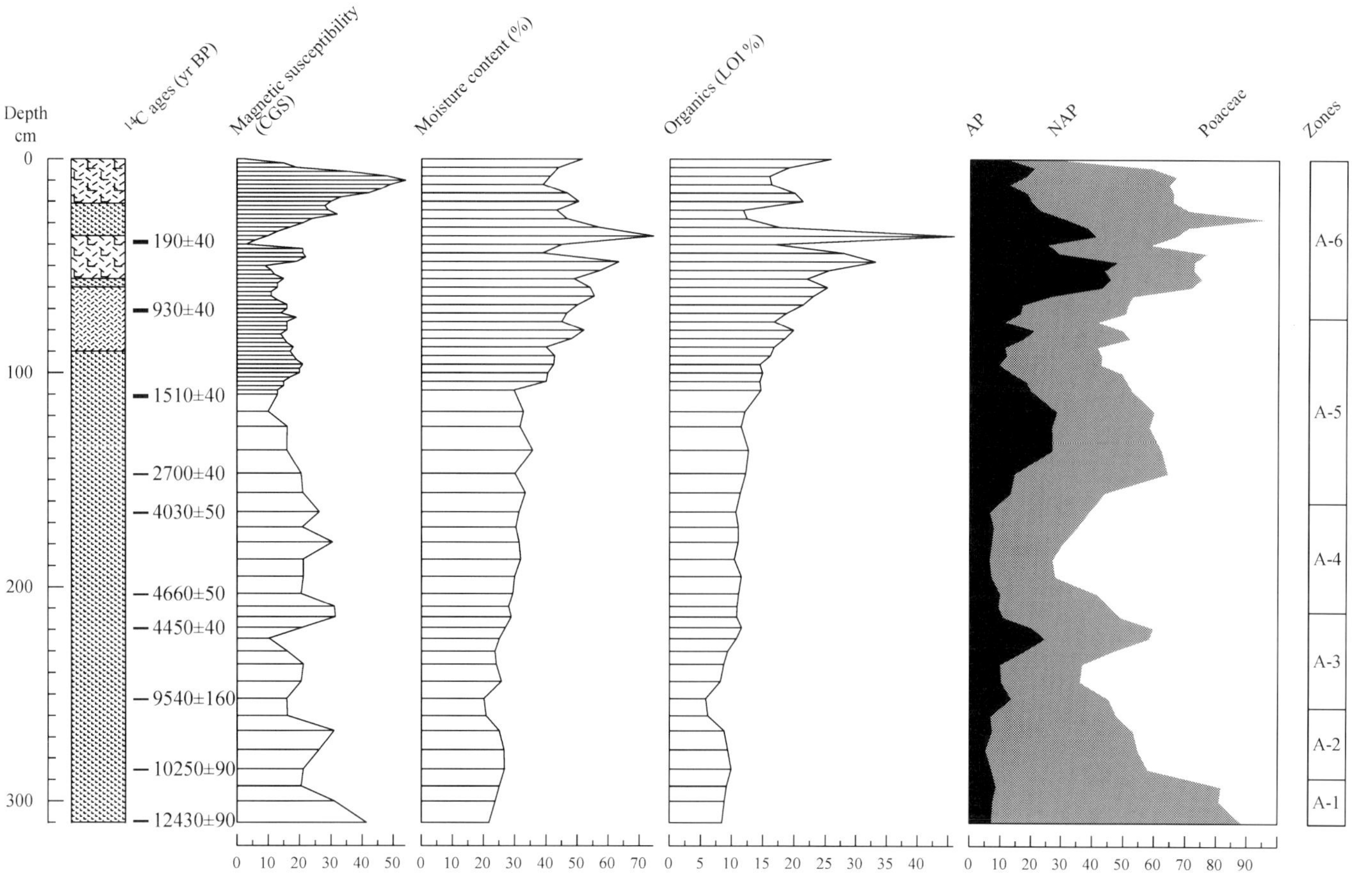

Figure 6.15. The Lake Aligol core: lithology, radiocarbon dating, magnetic susceptibility, moisture and organic content, and pollen summary diagram. Pollen zones at right.

black organic soil along the excavated pipeline route, however, provide evidence that the lake level has been significantly higher in the past (**Fig. 6.14**). The current area occupied by wetland vegetation and standing water covers six hectares.

Land use around Lake Aligol presently consists of rangeland grazing and haymaking. There are several recent pine plantations on the hills behind the village of Santa (**Fig. 6.13**), and terraced fields and gardens within the village. The presence of several Bronze Age 'kurgans' (burial mounds) in the area between the lake and the Tsalka Reservoir speaks of quite different land uses in the past.

Vegetation

The vegetation surrounding Lake Aligol is very similar to that of much of the Tsalka Plateau, *i.e.* mountain steppic grassland (Festucetea valesiacae). There are a

Table 6.5. Comparison of plant community composition and pollen spectra at Lake Aligol ('pollen 1' = lake centre sample; 'pollen 3' = near-shore sample). See **Table 4.2** for explanation and legend.

Trees and shrubs:	% cover	% pollen 1	% pollen 2	% pollen 3	Pollen type represented
Salix caprea					*Salix*
Populus nigra					*Populus*
Rosa spinosissima					
Malus orientalis					*Malus*-type
Herbaceous plants:					
Silene ruprechtii					
Pulsatilla violacea					*Ranunculus*-type
Ranunculus transcaucasicus					*Ranunculus*-type
Thalictrum foetidum					*Thalictrum*-type
Sedum oppositifolium					*Crassulaceae*
Potentilla argentea					*Potentilla*-type
Medicago polychroa					Fabaceae
Trifolium ambiguum					*Trifolium repens*-type?
Trifolium arvense					*Trifolium repens*-type
Trifolium pratense					*Trifolium pratense*-type
Lotus caucasicus					*Lotus*-type
Hypericum perforatum					
Galium verum					*Galium*-type
Echium vulgare					
Leonurus quinquelobatus					
Salvia verticillata					Lamiaceae
Thymus rariflorus					*Mentha*-type
Plantago lanceolata					*Plantago lanceolata*-type
Plantago major					*Plantago major*-type
Achillea setacea					*Anthemis*-type
Artemisia absinthium					*Artemisia*
Cichorium intybus					*Cichorium*-type
Allium albidum					
Luzula spicata					
Bromus japonicus					Poaceae
Bromopsis variegata					Poaceae
Agropyron repens					Poaceae
Melica taurica					Poaceae
Dactylis glomerata					Poaceae

Herbaceous plants:	% cover	% pollen 1	% pollen 2	% pollen 3	Pollen type represented
Festuca valesiaca					Poaceae
Festuca ovina					Poaceae
Koeleria cristata					Poaceae
Agrostis planifolia					Poaceae
Phleum phleoides					Poaceae
Phleum pratense					Poaceae
Wetland plants:					
Epilobium hirsutum					
Mentha arvensis					*Mentha*-type
Alisma plantago-aquatica					*Alisma*
Sagittaria sagittifolia					
Butomus umbellatus					
Juncus effusus					
Juncus articulatus					
Scolochloa festucea					Poaceae
Poa palustris					Poaceae
Calamagrostis neglecta					Poaceae
Typha latifolia					*Typha*
Scirpus lacustris					Cyperaceae
Carex vesicaria					Cyperaceae
Aquatic plants:					
Polygonum amphibium					*Polygonum amphibium*-type
Myriophyllum spicatum					
Lemna trisulca					*Lemna*-type
Sparganium neglectum					

number of cultivated trees in Santa village, mostly poplars, willows and fruit trees, and pine plantations nearby. A few dwarf shrubs are scattered along the southern shore of the lake, *i.e. Malus orientalis, Salix caprea, Populus nigra* and *Rosa spinosissima.* The dominant species in the grasslands are *Festuca valesiaca* (60–70 percent), *Bromopsis variegata, Agrostis planifolia* (55–65 percent) and *Phleum phleoides* (40–45 percent). A number of other grasses and herbaceous plants were found in our survey and are presented in **Table 6.5**.

The wetland vegetation is dominated by *Scirpus lacustris* (60 percent) and *Typha latifolia* (20 percent). Besides these, the wetland plant association includes the subaquatic plants listed in **Table 6.5**. Because the lake is uniformly shallow, there is no obvious zonation of wetland vegetation in relation to water depth. Aquatic species in the lake include *Polygonum amphibium, Sparganium neglectum, Myriophyllum spicatum* and *Lemna trisulca.*

The pollen spectra of surface sediments (**Table 6.5**) shows that Poaceae are dominant both in the vegetation and palynologically. This relationship is complicated, however, by the presence of Poaceae in the wetland flora. The same complication arises with *Mentha*-type pollen, produced both by plants in the wetland (*Mentha arvensis*) and those growing on the dry, rocky slope south of the lake (*Thymus rariflorus*). The shrubs that grow at the foot of this slope are very poorly represented in pollen spectra; arboreal pollen is dominated instead by long-distance transported *Pinus, Quercus, Fagus* and *Carpinus.*

Sedimentology and Dating

The Aligol core was collected on 10th September 2003 using a Russian corer and 'sticky clay' auger. The sediments were mostly compact silts, but somewhat organic in the upper metre. The auger penetrated to a depth of 310 cm in extremely hard, green-grey, clay sediments.

Table 6.6. Sediment description of the Aligol core.

Depth (cm)	Colour (Munsell)	Description
0–5	10YR2/1	Organic detritus
5–21	10YR3/2	Crumbly, tessellated silty peat with iron stains
21–36	10YR3/1	Black silt with brown bands and organic material
36–43	10YR3/2	Compact brown peat
43–48	10YR3/1	Crumbly, tessellated silty peat
48–56	10YR2/1	Dark brown peat
56–61	10YR1.7/1	Black silts with crumbly texture
61–84	10YR3/1	Compact brown peat with sand particles (iron stains 67-71 cm)
84–90	10YR2/1	Crumbly, tessellated silty peat
90–165	10YR2/2	Dark brown, compact silt
165–214	10YR4/1	Compact grey silt
214–236	10YR1.7/1	Black silt with abundant charred particles
236–252	10YR4/1	Compact grey silt
252–267	10YR6/1	Light grey silt (microscopy revealed abundant graminoid phytoliths)
267–310	10GY3/1	Extremely compact, grey-green silt

The complex lithology of the Lake Aligol deposits is described in **Table 6.6**. The upper metre of the core is characterised by dark brown sediments varying in organic and silt content. Several of the organic strata have iron stains, suggestive of redox reactions, and a crumbly, tessellated structure, typical of a dried lakebed. The composition of these sediments was difficult to determine visually – due to lake regressions, it is possible that the peat layers have taken on a more silty texture through exposure to the atmosphere and subsequent oxidation. The loss-on-ignition data therefore provide a better guide to sediment composition (**Fig. 6.15**).

Loss-on-ignition values are highest between 36 and 72 cm depth (**Fig. 6.15**), corresponding to phases of peat deposition in the Aligol basin. The upper metre of the core has higher organic content than the deeper sediments, which are mainly compact silts. Sediments with the lowest organic content were the light-grey silts between 252–267 cm. These data suggest that Lake Aligol did not exist as a permanent water body during the earlier part of the sediment record.

Magnetic susceptibility measurements (**Fig. 6.15**) show slightly elevated values between 310 and 200 cm, lower susceptibility from 200 to 30 cm, and a phase of high susceptibility in the upper 30 cm. In comparison with the loss-on-ignition curve, the magnetic susceptibility curve is negatively-correlated. Interestingly, the magnetic curve indicates enrichment of the sediments in the upper 30 cm, corresponding to a low organic phase. This suggests that catchment erosion, thought to be the source of magnetic particles in this basin, increased in more recent times. Other possible erosion events occurred at 220–200 cm (mid Holocene) and in the lower part of the record (Pleistocene-Holocene transition).

Ten samples were dated radiometrically. Pre-treatment of these involved removing the inorganic component by gravity separation (as for pollen), although no attempt was made to remove organic detritus or chitin. The results are shown in **Figure 6.15** and indicate that the Aligol record begins some 14,500 years ago, making this the longest highland pollen record from the Caucasus. Two dates are reversed – 219 and 203 cm. It is difficult to reject either of these dates. The older one, 4660±50 ^{14}C bp, is perhaps less likely to have been contaminated by younger carbon through the use of the 'sticky clay' auger, but dead carbon introduced to the sediment by an erosion event (see above) may have increased its radiocarbon age. The younger date was obtained from the charcoal layer in the Aligol core, and its age agrees much better with the corresponding charcoal layer and changes in sedimentation in the Imera core (Zones I–1 to I–2). Rapid fluctuations and possible hiatuses in sediment accumulation at Aligol mean a degree of uncertainty in deriving an age model. These uncertainties, where possible, are taken into account below.

Sediment accumulation rates in Lake Aligol have been extraordinarily slow. The late Pleistocene material accumulated at around 0.1 mm per annum, rising to three times that amount through the earliest few centuries of the Holocene. Rates then slackened (0.06 to 0.09 mm p.a.), and these lasted until the mid Holocene. Rates increased between 5500–4500 cal BP (0.4 – 1.0 mm p.a.), before decreasing once again to 0.1 mm p.a. The late Holocene has seen a steady increase in accumulation rates, especially in the last 200 years, in which sediment has accumulated at 1.7 mm per annum. Hence Lake Aligol's history of sedimentation is very similar to that of Lake Imera, with slow early-mid Holocene rates, rapidly increasing with peat formation in the last 2000 years (**Fig. 6.16**).

This shared history is reflected in the inferred lake-level history of Lake Aligol. The lake was seasonally dry from the late glacial to the mid-late Holocene, with the most severe desiccation associated with the Younger Dryas. Two thousand years ago, the lake's water balance changed and lake levels stabilised. After a regression during the Mediaeval Period, the lake became a permanent water body and in that state it remained until very recently. The strong similarity between the Imera, Aligol and other Tsalka Plateau sedimentary records[25] suggests a regional pattern of lake-level change.

Pollen Diagram

Pollen preservation in the Lake Aligol core was generally very good, with the exception of the late glacial and earliest Holocene (pre-11 k cal BP) sediments. Pollen percentages are calculated from separate arboreal and non-arboreal sums to account for the retraction in pollen source area during each lake-regressive phase. The pollen diagram (**Fig. 6.17**) is divided into six pollen assemblage zones:

Zone A-1. In the parlance of Northern European glacial stratigraphy, the Lake Aligol record begins in the Bølling-Allerød interstadial. Pollen spectra from this period are characterised by very high proportions of Chenopodiaceae, *Artemisia* and *Anthemis*-type pollen. In the arboreal group, *Juniperus* and *Ephedra* are dominant. Considered as a whole, these pollen spectra indicate conditions of extreme aridity and the existence of semidesert vegetation with scattered juniper and ephedra shrubs. Sedges grew in the wetland vegetation, along with *Butomus*, and *Pediastrum* bloomed in the water itself. Carbonised particles are constantly represented. Toward the end of this first zone, *Juniperus* and Chenopodiaceae decline, while *Artemisia*, other Asteraceae and *Ephedra* all rise to a conspicuous peak around 12,000 cal BP.

Zone A-2. The second zone represents the first millennium of the Holocene. A succession of arboreal taxa is seen – beginning with *Betula*, followed by *Corylus*, then *Quercus* and a suite of more mesophilic taxa (*Abies, Fagus, Carpinus betulus*, etc.). Poaceae rises in abundance through the zone, along with *Polygonum aviculare*, though it is difficult to interpret these because of the possibility that they grew on the dried

[25] See Margalitadze 1995.

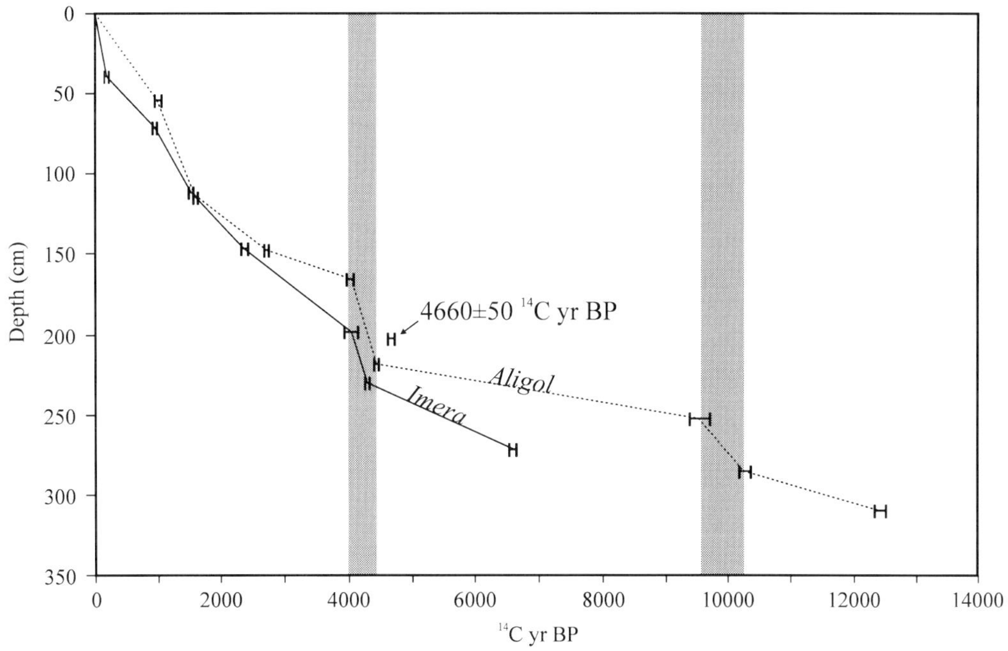

Figure 6.16. Age-depth profiles for records from Lake Imera (solid line) and Lake Aligol (dotted line). The graph shows a period of rapid sediment accumulation at both sites between 4000–4500 ^{14}C bp (4400–5100 cal BP), and a similar event associated with the beginning of the Holocene at 10,000 ^{14}C bp (11,500 cal BP).

lake-bed. They could represent the expansion of grassy steppe communities in the Aligol area. The persistence of *Artemisia* and *Ephedra* indicates that conditions remained quite cool and arid. An interesting aspect of this zone is its abundance of *Triticum*-type pollen. Rather than representing the inception of agriculture, wheat pollen here was probably produced by wild species growing amongst the wetland flora. This is an argument previously invoked by Bottema to explain high *Cerealia*-type proportions at Söğütlü,[26] near Lake Van, in an area also known for wild cereals.

Zone A-3. The third zone witnesses the decline of *Artemisia*, Chenopodiaceae, *Ephedra* and *Juniperus*, the four taxa that epitomize late glacial pollen spectra from Aligol. They are replaced by a more mesic assemblage, in which *Abies, Fagus, Carpinus betulus, Ostrya*-type, *Ulmus, Zelkova*-type and *Tilia* are well represented amongst arboreal taxa, and Poaceae, *Aster*-type, *Thalictrum,* Apiaceae, *Campanula, Dipsacus, Filipendula* and Fabaceae become important herbaceous pollen types. This pollen assemblage is very similar to Lake Imera's assemblage from the same period (Zone I-1).

[26] Bottema 1995.

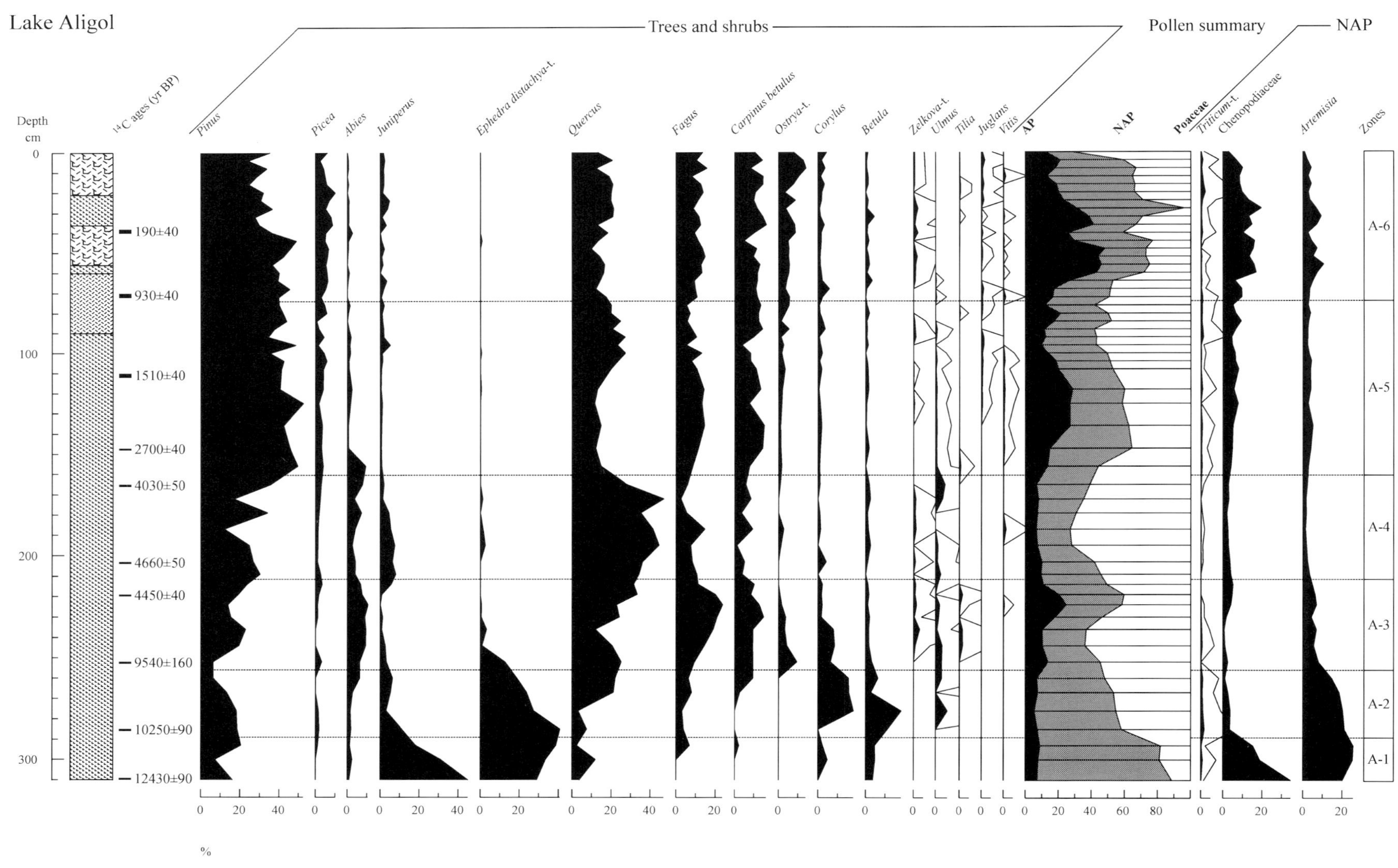
Lake Aligol
Trees and shrubs
Pollen summary
NAP
Depth cm
¹⁴C ages (yr BP)
190±40
930±40
1510±40
2700±40
4030±50
4660±50
4450±40
9540±160
10250±90
12430±90
Pinus
Picea
Abies
Juniperus
Ephedra distachya-t.
Quercus
Fagus
Carpinus betulus
Ostrya-t.
Corylus
Betula
Zelkova-t.
Ulmus
Tilia
Juglans
Vitis
AP
NAP
Poaceae
Triticum-t.
Chenopodiaceae
Artemisia
Zones
A-6
A-5
A-4
A-3
A-2
A-1
%

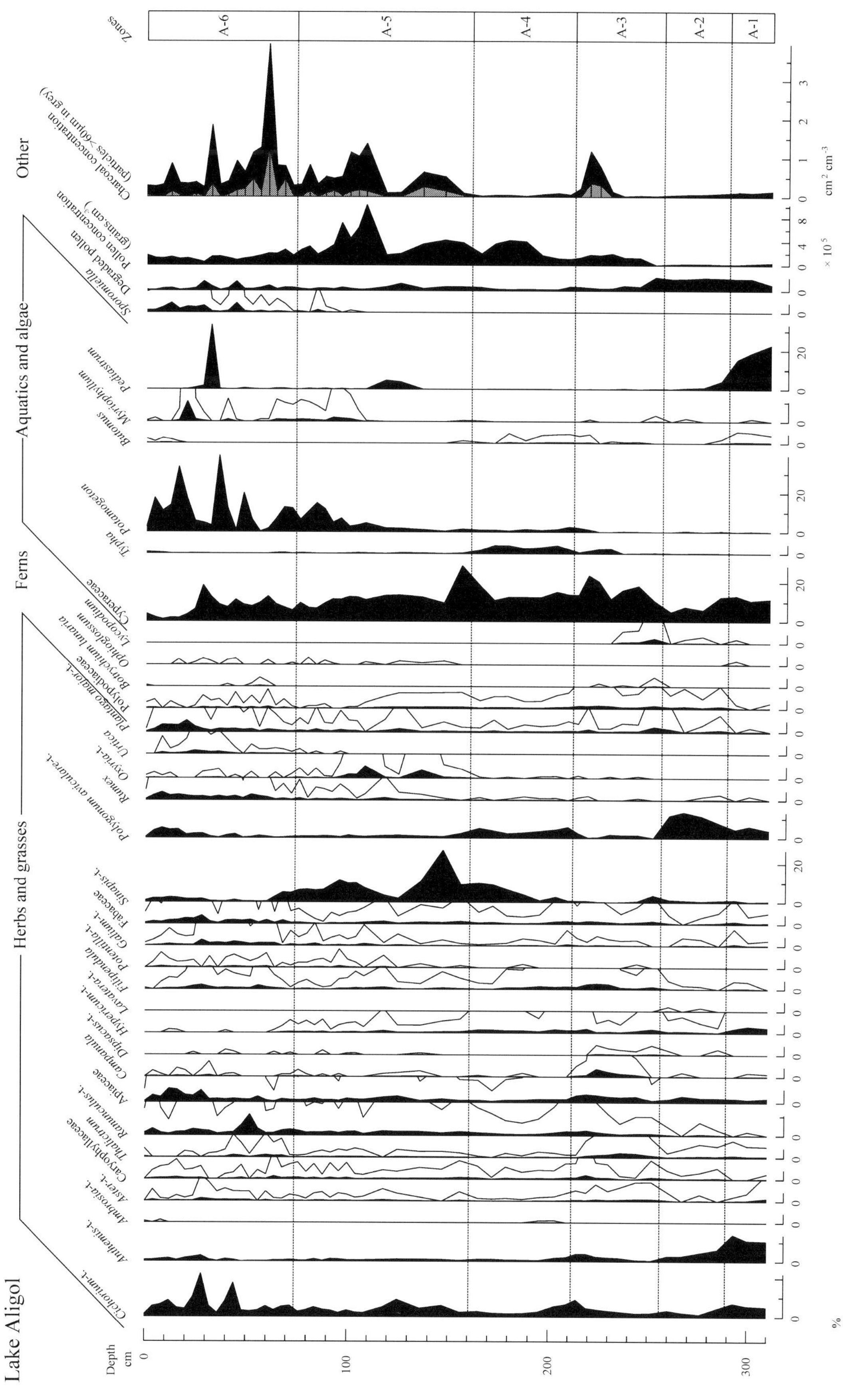

Figure 6.17. The Lake Aligol pollen diagram. Percentages calculated from separate arboreal and non-arboreal pollen sums. See Appendix E for the complete diagram.

The similarity extends to the fire record: as at Imera, charcoal concentrations peak between 7000 and 5000 years ago (based on the date from the charcoal layer).

Zone A-4. This zone matches Zone I-2 from Lake Imera, in which oak pollen achieves its highest representation among arboreal taxa. At Aligol, the oak maximum is accompanied by elevated proportions of *Juniperus*, Poaceae, *Hypericum perforatum*-type, *Polygonum aviculare*-type and Brassicaceae. These taxa suggest a steppic vegetation type with oaks and junipers. Increases in *Typha* and *Potamogeton*-type indicate that water levels in the lake had begun to increase, at least seasonally. In the upper part of this zone *Quercus* gradually declines as *Pinus* and *Abies* move in from higher altitudes.

Zone A-5. The fifth zone sees *Pinus* rise to prominence amongst arboreal taxa, initially accompanied by *Abies*, which quickly vanished because of fires in the first millennium BC. *Juglans* pollen appears around the same time. Non-arboreal taxa indicate a return to less steppic conditions, with subalpine meadow taxa such as *Dipsacus* and *Campanula* returning. Brassicaceae (*Sinapis*-type) percentages are very high throughout this zone. It has been suggested that these may reflect plants (*e.g. Cardamine* or *Rorippa* spp.) growing in the wetland vegetation.[27] This hypothesis is supported by the concomitant increase in *Oxyria*-type,[28] which represents various *Rumex* species that grow, *inter alia*, in wet or seasonally inundated places (*R. crispus, R. maritimus, R. pulcher* and *R. conglomeratus*).[29] Toward the end of Zone A-5, the habitat for these weedy species diminished, as higher *Potamogeton* and *Myriophyllum* proportions indicate increasing water levels.

Zone A-6. The final zone in the Aligol diagram is defined by increasing arboreal pollen values overall, related to the deepening of the lake. In compositional terms, however, there is little change from Zone A-5 to A-6. *Ostrya*-type and *Picea* are the only two taxa to show a substantial increase. Amongst herbs and grasses, Poaceae declines somewhat and open ground taxa such as Chenopodiaceae, *Artemisia,* Apiaceae, *Rumex, Plantago major*-type and Fabaceae increase. Spores of the dung fungus, *Sporormiella*, become quite abundant, especially in the upper 30 cm. This final phase appears to bear the distinct marks of heavy grazing, even down to the depletion of sedges (Cyperaceae) in the vegetation. Charred particles are exceedingly abundant through the lower part of the zone.

[27] J. van Leeuwen, Inst. Plant Sci., Bern University, pers. comm.
[28] Pollen-morphological nomenclature in this study generally follows Moore *et al.* 1991.
[29] Ketskhoveli 1975.

Numerical Analysis

A number of problems that relate to dating and sedimentation rates affect the Aligol record. Both Aligol and Imera are shallow lake basins and their sedimentary history is complex, as reflected in pollen influx rates from these sites (**Fig. 6.18**). Periods of increased pollen accumulation coincide, paradoxically, with declining arboreal pollen ratios. It may be that the prominent peaks in pollen accumulation at both sites between 5500–4500 and 1500–1000 cal BP are the product of differential pollen deposition brought about by changing lake levels or alterations to the density of wetland vegetation, but this remains very much an open question.

Slow sedimentation during the early Holocene means that the period from 11,000 to 5000 years ago is represented by as few as five pollen samples, approximately one for every thousand years. The same applies to the period 4500 to 3000 years ago, represented by only two samples. Obviously it is difficult to extract meaningful environmental trends over such long time spans with so few samples. Hence the trends extracted by weighted averaging must be viewed with a considerable degree of caution. No doubt future research on lake sediments from Southern Georgia will help refine the coarse picture drawn here.

The first trend (**Fig. 6.19**) represents moisture variations. It clearly shows the dry conditions of the late-glacial gradually subsiding during the early Holocene. Slightly drier conditions returned, as at Lake Imera, during the oak maximum and later, during the Mediaeval Period. The correlation (r^2=0.71) between this trend and the separate DCA of the Aligol data (**Fig. 6.19**; Appendix B) indicates that Holocene rainfall changes have exerted a strong influence on vegetation development.

The second trend differs quite sharply from the first. It has moderate values during the Bølling-Allerød interstadial. With the onset of the Younger Dryas, values plummet dramatically. If this trend is interpreted as representing forest cover, it indicates that the landscape was rendered treeless by events 12,000 years ago. If it represents temperature, it shows that winters became rapidly and considerably colder. The early Holocene birch-hazel advance led to a short-lived afforestation of the highlands at the end of this cold stage. The advance of typical woodland species (*Quercus, Fagus, Carpinus*), however, seems to have been delayed by the aridity of the early Holocene climate, and, around 5000 years ago, by a number of fires. It was not until 4500 years ago that tree cover (in the form of scattered oaks) increased. Subsequent increases can be seen during the late Holocene – around 3000 years ago and 1000 years ago, respectively. None of these, however, are sufficient to suggest that an unbroken forest canopy blanketed the Tsalka landscape. At best, forests were confined to places with suitable drainage and soil moisture, while scattered trees may have occurred more widely. A series of fires after the Mediaeval Period led to the demise of whatever trees remained near Lake Aligol in historic times.

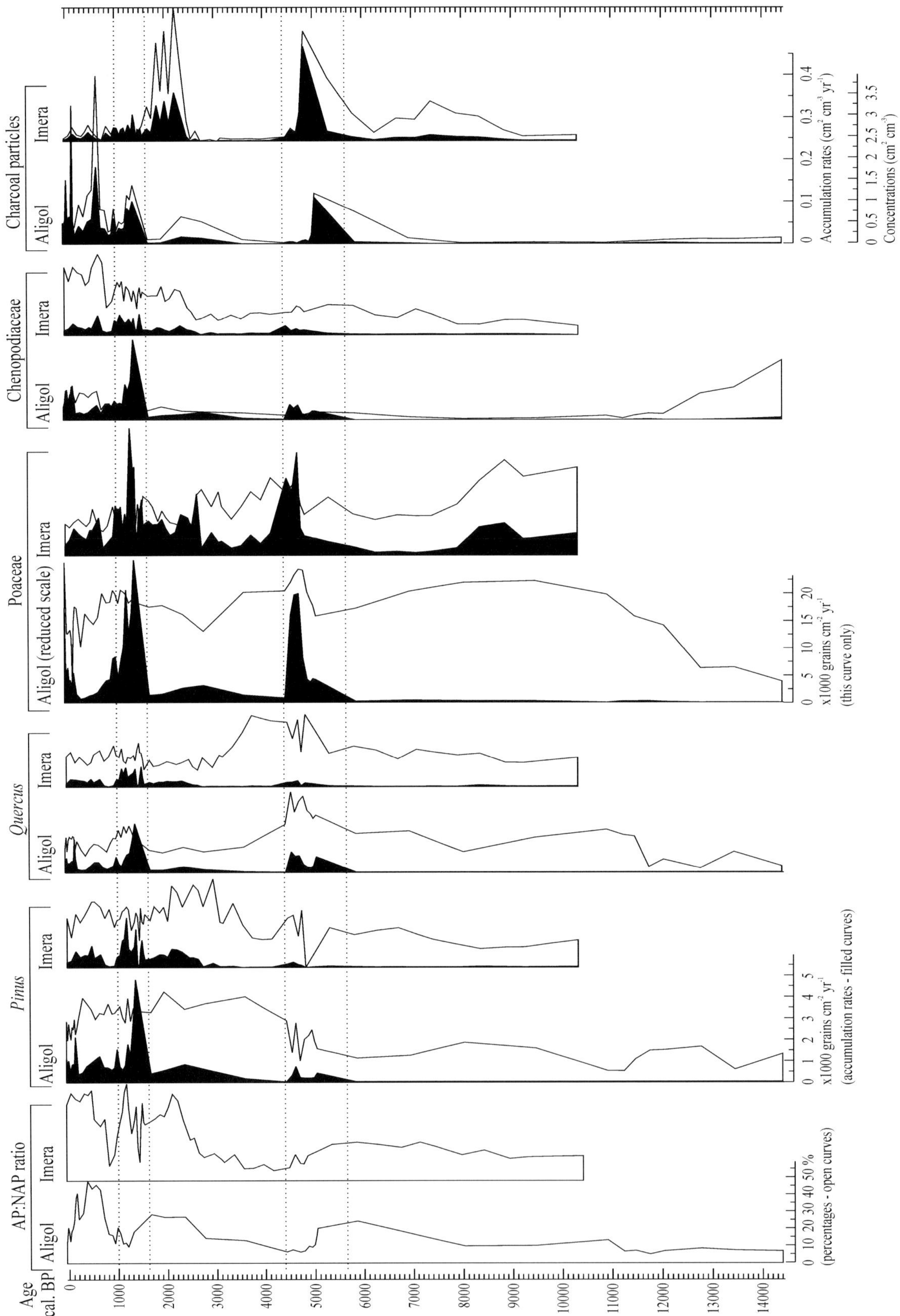

Figure 6.18. A comparison of pollen influx (shaded) and pollen percentages (open curves) from Lakes Imera and Aligol, plotted on a calendar timescale. Arboreal to non-arboreal pollen (AP:NAP) ratios are shown at left and concentrations and influx of charred particles at right.

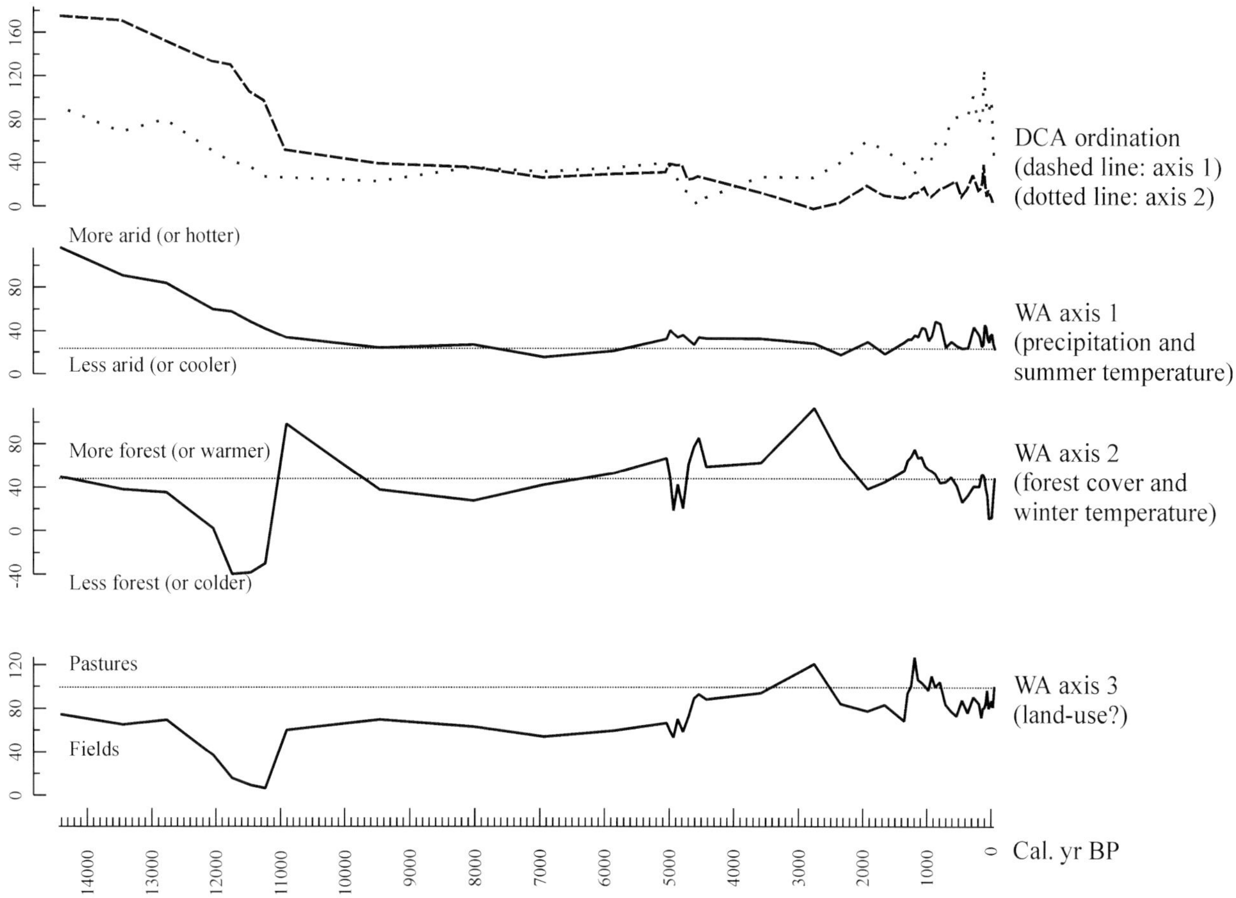

Figure 6.19. Environmental changes at Lake Aligol: rainfall, forest cover and land-use gradients produced by weighted averaging. Dominant trends in the pollen data (DCA ordination) are shown at the top for comparison.

The final trend has been interpreted elsewhere as indicating changes in land-use. A sharp decline during the Younger Dryas may reflect the presence of wild cereals in the wetland vegetation, as discussed above. If the remainder of the record is representative of grazing intensity, then it appears that grazing increased in the Aligol area 4500 years ago and continued through until approximately 2500 years ago. It resumed in the Mediaeval Period, perhaps as the climate became more amenable to pastoral activity, and has since remained an important economic practice on the Tsalka Plateau.

These trends are summarised in terms of vegetation development in **Table 6.7**. During the late glacial, semidesert vegetation with scattered juniper and ephedra shrubs prevailed near Lake Aligol. The climate at that time was hyper-arid, but seems not to have been particularly cold; in fact, summers may have been warmer than at

Table 6.7. Stages in vegetation development at Lake Aligol.

Zone	Time (cal BP)	Inferred vegetation	TWINSPAN (Chapter 4)
A-1	14.5k–12k	Glacial semidesert-savanna	*Artemisia* steppe
A-2	12k–11k	*Artemisia* steppe with birch & hazel	*Artemisia* steppe
A-3	11000–5000	Meadows and mixed woods	Grassland-forest border
A-4	5000–4000	Oak-juniper savanna	Grassland-forest border
A-5	4000–900	Open coniferous woodland	Grassland-forest border
A-6	900–0	Treeless grasslands	Grassland-forest border

present. The next stage was one of *Artemisia* steppe, with *Ephedra* shrubs and advancing birch and hazel trees. This peculiar assemblage, which has no modern analogue, may best be explained by conditions of extreme cold. In the early to mid Holocene, subalpine meadows prevailed in the Tsalka highlands, interspersed with mixed mesophytic trees and patches of beech woodland. More steppic conditions came after 5000 years ago, when oaks, junipers and grassland taxa spread, perhaps forming a kind of savanna or open woodland. This was succeeded by open coniferous woodlands and grassy subalpine meadows, until the demise of woodland species in the last millennium and the instigation of heavy grazing and burning.

Discussion

The data presented in this chapter record a remarkably dynamic set of Holocene vegetation changes in the highlands of southern Georgia. The Tsalka Plateau has hot, droughty summers and very cold, snowy winters. These climatic factors alone were sufficient to explain many of the observed variations, especially during the late Pleistocene and early-mid Holocene. Human influences are also apparent from as early as the Chalcolithic Period (5000 cal BP).

As previously noted, botanical researchers have long argued over whether the presently treeless highlands of Southern Caucasus were forested in the past. Small copses of trees in sheltered locations convinced many botanists that, "during the Holocene… the entire territory of Georgia was occupied by forests".[30]

Far fewer have been proponents of the idea that forests were never extensive in this part of the Caucasus. Margalitadze, on the basis of several pollen diagrams from the Javakheti and Tsalka Plateaux, concluded that "grassy phytocoenosae were predominant

[30] Nakhutsrishvili 1999, section 6.

in the South Georgian Highlands during the Holocene".[31] Like the pollen diagrams presented in this chapter, Margalitadze's diagrams were produced from the sediments of small lakes wetlands that appear to have experienced seasonally low water levels through the early and mid Holocene. This constraint made interpretation of the results particularly difficult,[32] since there was no way of accounting for changes in pollen source area at that time.

The modern pollen data set presented in Chapter Four identified climatic and forest-cover trends in the present-day pollen spectra of the study area thought to be independent of pollen source-area variations. The application of the taxa weights generated by that analysis allowed the past forest cover of the Tsalka Plateau to be estimated. The results confirm that grassy communities prevailed throughout the Holocene, but they also demonstrate that open woodlands did, at various times, expand into the highlands as climatic conditions became more favourable. Forests, *senso stricto*, may have grown in protected situations and on better-drained slopes, but on the open plains near Lakes Imera and Aligol, the trees that advanced during the mid Holocene formed a scattered cover. The term 'savanna', in its broadest sense, is perhaps the best structural analogue for this vegetation.

Temperate savannas often occur on flat terrain where the soils are deeply cracking clays.[33] The soils developed from igneous bedrock on the Tsalka Plateau are clay-rich; they are boggy and poorly drained in winter, desiccating and cracking in summer. Cracking soils tend to damage tree roots and favour plants with fine, fibrous root systems, especially grasses. On edaphic grounds it is therefore not surprising that grassy vegetation has dominated the Tsalka Plateau throughout the Holocene. Yet trees can and do establish on the Plateau, so soil was not the only factor limiting forest cover.

Climate Changes and Human Impacts

In attempting to disentangle the influences of climate and humans on past vegetation dynamics, it is useful to compare climatic trends drawn from the pollen data to the regional signal of Lake Van. The Van record is based on two proxies: oxygen isotopes, which in the highly evaporated waters of Lake Van primarily reflect changes in regional water balance and atmospheric circulation,[34] and the Mg:Ca ratio, which is an indicator of lake salinity in the Van basin and hence of past lake levels and effective precipitation.[35] Because Lake Van is situated in the same altitudinal zone as lakes

[31] Margalitadze 1977, p. 144.
[32] Dr N. Margalitadze, Tbilisi Botanical Institute, pers. comm.
[33] Archibold 1995.
[34] Prof. Neil Roberts, University of Plymouth, pers. comm.
[35] Lemke and Sturm 1997.

Aligol, Imera and Bareti, the magnitude and direction of regional-scale climatic events could be assumed to be the same at all four sites.

Lake Van's isotopic record is compared to environmental trends from the Tsalka Plateau in **Figure 6.20**. The most notable feature of the Van curves is the substantial peak in both lake salinity and $\delta^{18}O$ enrichment between 11,000 and 10,000 cal BP. But as Roberts has rightly pointed-out,[36] the Lake Van varve chronology places the beginning of the Holocene at 10,500 cal BP rather than 11,500 cal BP. It is more parsimonious to associate the peaks in Mg:Ca and $\delta^{18}O$ with the Younger Dryas stadial, which saw temperatures drop sharply across Europe at this time. At the Pleistocene-Holocene transition, lake levels at Van were drawn down by an estimated 260 m, an event attributed to increased aridity.[37]

When a similar pattern is sought in the Lake Aligol environmental trends, the outcome is striking. The Younger Dryas is clearly portrayed at 11,500 cal BP, but not, as might be expected, in the curve representing precipitation. The Aligol material suggests a dramatic decline in the trend representing forest cover and winter temperature. In contrast, the precipitation trend indicates that rainfall gradually increased through the terminal Pleistocene to peak 6000–7000 years ago. There is no evidence for an episode of *increased* aridity during the Younger Dryas from Lake Aligol, even though the climate then was more arid than it is presently.

A number of hypotheses could explain this apparent contradiction. It may be that the Aligol trends, because they were derived from modern analogues, are insensitive to no-analogue pollen spectra. That is, the high proportions of *Artemisia* and *Ephedra* pollen associated with the Younger Dryas event at Aligol are unreplicated in any modern plant community in the study area; their pollen types may even represent plant species no longer extant. Hence the apparent trend towards colder winters and very open landscapes could be an artefact of the method. This hypothesis could have merit, but for the extraordinarily good agreement between the Aligol and Van curves. It seems very likely, accounting for probable dating discrepancies, that both curves record the same event.

Perhaps Aligol's Younger Dryas signal is merely one of forest cover and does not reflect changing winter temperatures. Indeed, Aligol and Imera's WA axis 2 curves from the mid to late Holocene have little in common. The Imera curve follows the Van oxygen isotope record closely, whereas Lake Aligol deviates from it considerably. Dating error cannot explain this divergence, as the precipitation trends from Aligol and Imera are quite similar. It could be argued that the Imera record is therefore sensitive to regional climatic changes, while Aligol is subject to highly localised changes

[36] Roberts 2002.
[37] Wick *et al.* 2003.

in vegetation. Still, this hypothesis does not explain why the Younger Dryas is so plainly registered at Aligol.

If one accepts the Aligol trend as a true reflection of temperature, then it is possible that colder winters, not aridity, were the defining feature of the Younger Dryas in the study area. Hence continentality, rather than aridity *per se*, peaked some 12,000 years ago. How, then, can the low lake levels and enrichment of $\delta^{18}O$ at Lake Van be explained? Lemke and Sturm,[38] in deriving a model for oxygen isotopes for Lake Van, assumed an overall temperature reduction of 5 °C during the Younger Dryas. This decline was insufficient to fully explain $\delta^{18}O$ values from that period, so they factored a 20 percent reduction in humidity into the model, which accounted for the observed values and also agreed with carbonate geochemistry (Mg:Ca) results indicating lowered lake levels.[39] From this the authors concluded that the climate around Lake Van during the Younger Dryas was cold and very dry.

Most of the Van region's rainfall arrives during the cooler months, carried by westerlies from the Mediterranean and originating over the Atlantic. During the Younger Dryas, the influence of the Atlantic thermohaline circulation temporarily weakened, decreasing temperatures throughout Europe.[40] If, as the Aligol record suggests, winters in the region were much colder and more continental than now, winter precipitation would have been much reduced. The dry air of the continental winters could deny Lake Van its source of melt-water and mimic a reduction in humidity. Moreover, the switching-off of moisture-laden westerlies could increase evaporation during cold, dry winters and shift the rainfall pattern to warmer months. Atmospheric moisture and cloud cover generated by thermal convection currents, for instance, are more likely to have increased during summer, sustaining semidesert-steppe plant life but still exposing Lake Van to intense evaporation. The Younger Dryas was most certainly arid, but perhaps not more than during the previous millennium. Roberts and Wright have even suggested that the late Pleistocene climate of Western Asia was cloudier or had *greater* effective moisture due to reductions in temperature and warm season evaporation.[41]

Further support for this hypothesis may be found in the ecology of *Artemisia* and *Ephedra*, two taxa distinctly associated with the Younger Dryas at lakes Van, Aligol and others in the region.[42] Both of these xerophytes demand some moisture through the growing season.[43] If the Younger Dryas were indeed the period of greatest aridity in the last 14,000 years, it would be unlikely to favour these taxa over Chenopodiaceae.

[38] Lemke and Sturm 1997.
[39] Lemke and Sturm 1997.
[40] Roberts 1998.
[41] Roberts and Wright 1993.
[42] van Zeist and Bottema 1991.
[43] Chikov 1983; El-Moslimany 1990.

Ephedra and *Artemisia* retained an important place in the early Holocene vegetation around Lake Van long after lake levels had risen, while Chenopodiaceae died away as the Pleistocene came to a close.

The early Holocene brought increased humidity to Southern Georgia. The Aligol and Imera records indicate that the climate became gradually wetter, peaking at the same time as water levels rose to their highest stand in Lake Van.[44] Lakes Aligol and Imera, however, remained low or seasonally dry throughout this period of increased precipitation and sedimentation rates were exceedingly slow (or later desiccation destroyed evidence of previous high lake levels). Pollen data from these sites suggests a very slow expansion of mixed deciduous communities at lower altitudes and the presence of subalpine herbaceous plants locally, indicating a climate with relatively cool, humid summers. Evidence for cool conditions can also be found in the Tsavkisi pollen record from near Tbilisi (previous chapter). This alone, however, does not explain the slow pace of afforestation. Wright *et al.*[45] have invoked early Holocene aridity to explain delayed woodland expansion across the region, but climatic seasonality, human impact and migrational lags may also be considered important factors.[46]

Forest cover trends from lakes Aligol and Imera during the Holocene have little in common. The Imera trend follows the Van temperature variations closely (**Fig. 6.20**). This might be expected because Lake Imera is twice the size of Lake Aligol and its pollen record consequently has a larger, more regional, source-area. Interpreted in these terms, local and regional events can be differentiated. While woodlands expanded regionally between 9000 and 7000 cal BP, around Lake Aligol forest cover remained very restricted. Between 7000 and 5000 cal BP, woodland began to expand in the Aligol area, accompanied by ever increasing charcoal concentrations. The Lake Imera charcoal record is similar, and both suggest an increasing occurrence of fire in the vegetation. These fires were not regional but local events, clearly demonstrated by a high proportion of large charcoal particles (see Figs 6.8 and 6.19).

These fires are perhaps the first indelible mark of human impact on the vegetation history of the Caucasus region. Natural ignition can be disqualified for several reasons: Firstly, the apparent trend toward lower rainfall between 6000 and 5000 cal BP at Aligol and Imera (**Fig. 6.20**) is not seen in the Lake Van or Tsavkisi Wetland records, suggesting that the apparent drying of the Tsalka Plateau climate at this time is a function not of precipitation, but of heightened summer temperatures or seasonal shifts in rainfall. Secondly, if hotter summers led to more frequent fires, there is no reason why these fires should have ceased abruptly 5000 years ago, while summer temperatures continued to increase. Thirdly, the Lake Van charcoal record shows no

44 Wick *et al.* 2003.
45 Wright *et al.* 2003.
46 Roberts 2002.

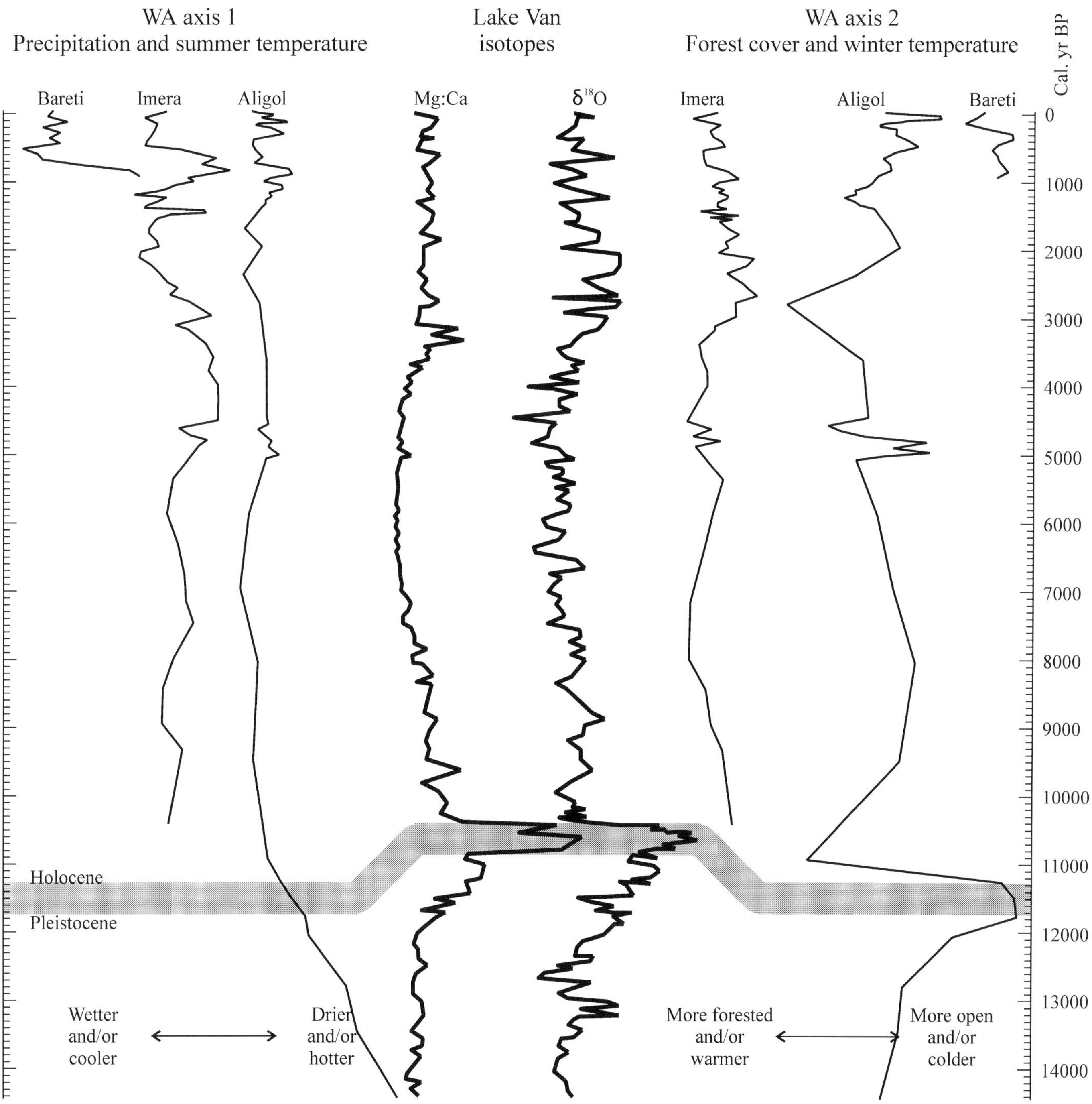

Figure 6.20. Environmental trends from the Tsalka Plateau compared to a regional climatic trend. Comparison of the Lake Van isotopic records (bold curves at centre) with the results of weighted averaging applied to pollen records from the Tsalka Plateau (all shown at the same scale). The Younger Dryas stadial is indicated by grey shading, adjusted to account for the underestimated age of the Lake Van material.

evidence of abrupt changes through this critical period, so it seems improbable that burning related to regional climatic change. Finally, archaeological evidence demonstrates that the Tsalka Plateau was already home to Chalcolithic pastoralists for whom grazing was an economic mainstay.[47] Roberts has hypothesised that human activity and aridity acted together to delay postglacial woodland expansion in Western Asia;[48] this hypothesis is examined further in Chapter Eight.

Five thousand years ago fires ceased abruptly and a scattering of oaks populated the grasslands. Whether oak trees drew down water levels in Tsalka Plateau lakes or the hot, dry summers increased evaporation, low lake levels are characteristic of this phase. Both Van and Imera indicate that temperatures began to decline after 3500 cal BP, allowing conifers to replace the more thermophilous oaks on the Tsalka Plateau. In the absence of fire, conifer woods became relatively dense, but this changed with the reinstatement of a fire regime around 2500–2300 years ago. Fire banished fir trees from the Tsalka Plateau and the vegetation became as open as it is now. High lake levels followed, implying a significant reduction in seasonal drought and continentality. And since deforestation can elevate groundwater in flat terrain, lake levels may have been influenced directly by human impact.

A firm correspondence between the Lake Van and Tsalka Plateau records through the last two thousand years is difficult to detect. The Tsalka Plateau pollen and lake level data indicate a substantial change in precipitation around one thousand years ago, coincident with the so-called Mediaeval Warm Period. Lake levels at Aligol, Imera and Bareti were much reduced (at least seasonally) through the Mediaeval and it seems that the climate briefly reverted to its mid Holocene state. Yet the Mediaeval here differs from previous climatic events in that it is registered only by the trend of precipitation and summer temperature (**Fig. 6.20**). The Mediaeval Warm Period in Europe is thought to be the result of increased anticyclonic circulation in summer and a strong westerly influence through winter.[49] Persistent westerlies would be expected to bring moist, relatively mild winters to the Tsalka Plateau. Low lake levels are thus best explained by intense summer evaporation, but this question may be better addressed using palaeohydrological methods.

Concluding Remarks

The Tsalka Plateau has yielded a record of changing vegetation, lake levels, climate and human impact spanning the past 14,000 years. The dynamism of its vegetation history is perhaps unexpected in such an unvaried landscape, but it must be remembered

[47] Kiguradze and Sagona 2003.
[48] Roberts 2002.
[49] Bradley *et al.* 2003.

that the flat terrain, harsh winters and seasonal aridity that are characteristic features of the Tsalka Plateau today are the very features that make its vegetation so sensitive to climatic changes and human impacts.

Specifically, the Tsalka Plateau's vegetation history passes from dry semideserts in the late Pleistocene, through the cold *Artemisia* steppes of the Younger Dryas, into a stage of subalpine meadows and advancing woodland that lasted until oak trees became established in open savanna vegetation 5000 years ago. As summers cooled in the late Holocene, oaks were gradually replaced by conifers; these, in turn, gave way to open grasslands in the last 2500 years.

Humans played a critical role in the development of the vegetation, manipulating its density and composition with fire. Six thousand years ago or even earlier, it seems people were using fire to forestall the advance of woodland on subalpine pastures. More recently, between 2500 and 2000 years ago, fire was used to clear the vegetation and since then frequent, low intensity fires and incessant grazing have preserved the Tsalka Plateau's rich mountain grasslands in a treeless state. The utility of small lakes is plainly revealed through their ability to trace changes of a local character that are obscure in pollen diagrams from large lakes.

In terms of past climate, the Tsalka Plateau records complement the regional climatic signal deduced from Lake Van, Eastern Anatolia. The Tsalka Plateau environmental trends indicate a hyper-arid climate 14,000 years ago led on to the dry, extremely cold winters of the Younger Dryas. The slow advance of woodland in the early Holocene is difficult to explain in climatic terms alone, but conditions were perhaps seasonally drier, cooler and cloudier than at present. By the mid Holocene, summer temperatures had increased, a trend that was reversed 3500–2500 years ago as climates cooled. A period of intensified evaporation followed 1000 years ago, before climates similar to the present established. This interpretation differs in some respects to the arguments of Wick *et al.* and Wright *et al.*,[50] particularly in relation to the Younger Dryas.

I now return to the original question posed at the beginning of this chapter: Was the Tsalka Plateau ever covered in forest during the Holocene? The evidence suggests that it was not. However, this is not to say that it was never treed. At certain, brief moments, scattered trees expanded into the mountain grasslands when climatic conditions and human activities allowed them to do so. On slopes and in gullies, pockets of forest probably existed in the past just as they do now. But the Tsalka Plateau is primarily a pastoral landscape and gives no clear indication of having been anything else for the last 11,000 years.

[50] Wick *et al.* 2003; Wright *et al.* 2003.

CHAPTER SEVEN

HOLOCENE PALAEOVEGETATION MAPS OF GEORGIA

INTRODUCTION

The foregoing chapters have dealt in detail with the vegetation history of individual sites, tracing the development of different plant communities in temporal terms. The aim of this chapter is to incorporate these records into a broader *spatial* analysis of the history of the study area's vegetation, by mapping the frequency of the most important pollen types from every location for various periods. The resulting maps provide a starting point for understanding the palaeoecology of the taxa involved, for examining changes in their distribution and abundance due to climatic, edaphic, anthropogenic and other factors, and for reconstructing the palaeovegetation at various moments in time.

The approach adopted here is very similar to Huntley and Birks' *Atlas of Past and Present Pollen Maps for Europe*.[1] The main difference is that 'isopoll' maps are not used here. This is because they can be misleading where the terrain is rugged and distribution of sites patchy. Instead, pollen frequency is represented by graduated symbols placed on the geographic location of each study site; a method used by Lennart von Post, the 'father' of palynology.[2]

The maps were constructed using a database of pollen spectra from 39 cores including 232 fossil spectra.[3] Modern pollen data were collated from 164 samples of lake and wetland sediment (**Fig. 7.1**, **Table 7.1**); moss- and soil-derived pollen spectra were carefully avoided. Modern and fossil data come from the pollen data in this study, from published pollen diagrams and from unpublished data generously provided by Jacqueline van Leeuwen, Erika Gobet and Brigitta Ammann at the Institute of Plant Sciences, Bern. While some unpublished data were available in raw counts, pollen percentages from published diagrams had to be measured by hand, introducing a minor source of error.

[1] Huntley and Birks 1983.

[2] Faegri and Iversen 1989.

[3] Unfortunately, the Dziguta pollen record (Arslanov *et al.* 2007) was published after the compilation of the database and was not included in this analysis. It is included in Connor and Kvavadze, 2008.

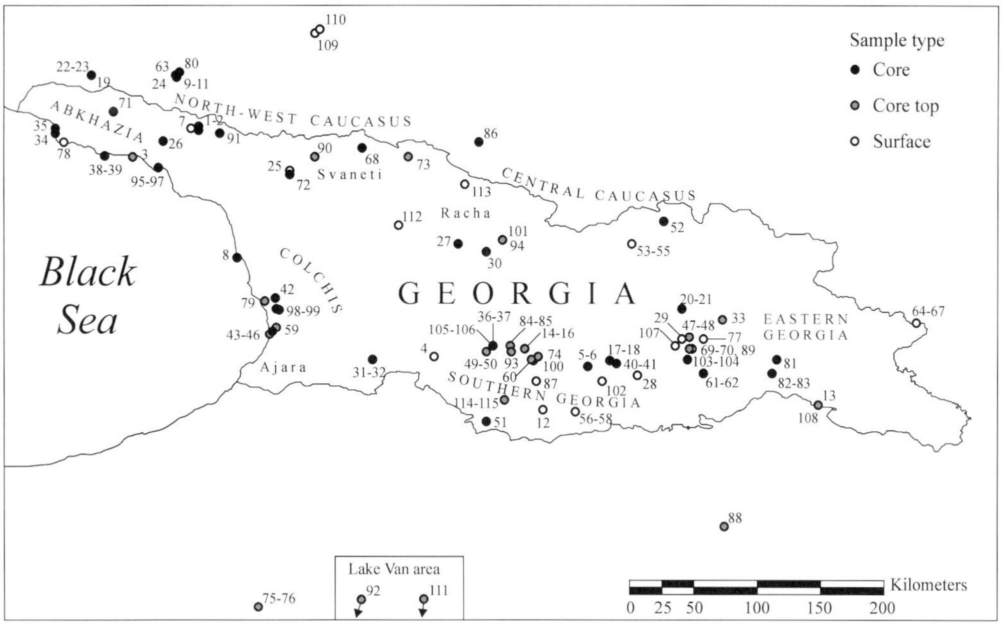

Figure 7.1. Location map of pollen sites used in mapping and analysis. Site numbers refer to Table 7.1.

Samples were selected for each 1000-year interval for the last 14,000 years. I adopted a calibrated timescale to allow the vegetation history to be compared to archaeological and historical data that do not use the radiocarbon timescale. This, of course, introduces a source of dating error, but this is probably less significant than the errors generated in assigning ages to poorly dated pollen diagrams. Dating was by linear interpolation between dated levels. Where few dates were available, I occasionally used reliable biostratigraphic markers (*i.e.* the Pleistocene-Holocene boundary, the expansion of *Juglans*) as an age control.

Radiocarbon-dated pollen diagrams from the study area are the exception rather than the rule: 42 published pollen diagrams (apart from those used in this analysis) could not be used because they lack any independent chronological control. For the dated sites combined, 127 dates were available, an average of 3.3 per core and a median of 2, which highlights the discrepancy between a few well-dated sites and many poorly-dated ones (**Table 7.1**). This comes from the generally inorganic nature of sediments in the study area and the inaccessibility of accelerator mass spectrometer (AMS) facilities. The available radiometric dates were accepted as correct representations of the age of the pollen. In some cases, ^{14}C ages are clearly erroneous, so this is taken into account in the discussion below.

Table 7.1. Location of pollen sites used in this chapter, their dating control and sources.

No.	Site, sample no.	Type	Sub-region	Lat.	Long.	Elev.	^{14}C	From	To	Source
1	Adange	Core	Abkhazia	43.28	41.28	1750	3	3000	0	Kvavadze and Rukhadze 1989
2	Adange 1	Surface	Abkhazia	43.27	41.23	2750	-	-	-	Kvavadze and Rukhadze 1989
3	Akhali Ateni 511	Core top	Abkhazia	43.07	40.82	-20	-	-	-	Kvavadze and Connor 2005
4	Akhaltsikhe 183	Surface	S Georgia	41.70	42.93	1345	-	-	-	Klopotovskaya 1973
5	Aligol	Core	S Georgia	41.63	44.02	1550	10	12000	0	Connor and Sagona 2007
6	Aligol 1-3	Surface	S Georgia	41.63	44.02	1550	-	-	-	Chapter Six
7	Amtkel	Core	Abkhazia	43.25	41.28	1830	2	2000	0	Kvavadze and Rukhadze 1989
8	Anaklia	Core	Colchis	42.38	41.55	5	5	6000	0	Serebryannyi *et al.* 1984
9	Arkhyz 1-2	Surface	Abkhazia	43.65	41.15	2582	-	-	-	Kvavadze and Efremov 1995
10	Arkhyz 3	Surface	Abkhazia	43.65	41.15	2761	-	-	-	Kvavadze and Efremov 1995
11	Arkhyz 4	Surface	Abkhazia	43.62	41.13	2884	-	-	-	Kvavadze and Efremov 1995
12	Avchalagol 1-4	Surface	S Georgia	41.33	43.70	2050	-	-	-	Connor *et al.* 2004
13	Baidos Kalo	Core top	E Georgia	41.37	45.65	680	-	-	-	Kvavadze and Stuchlik 1993
14	Bakuriani 1	Surface	S Georgia	41.77	43.47	1600	-	-	-	Dokturovskii 1936
15	Bakuriani 2	Core top	S Georgia	41.75	43.57	1700	-	-	-	Margalitadze 1995
16	Bakuriani 11	Surface	S Georgia	41.75	43.57	1650	-	-	-	Margalitadze 1995
17	Bareti 1-3	Surface	S Georgia	41.67	44.17	1630	-	-	-	Connor *et al.* 2004
18	Bareti B	Core	S Georgia	41.67	44.17	1630	1	1000	0	Chapter Six
19	Bathing Chamois	Core	NW Caucasus	43.63	40.53	2490	2	5000	0	Kvavadze and Efremov 1994
20	Bazaleti A	Core top	E Georgia	42.03	44.68	875	-	-	-	Gogichaishvili 1984b
21	Bazaleti B	Core	E Georgia	42.03	44.68	875	1	3000	0	Kvavadze and Connor 2005
22	Bezymianka 1	Surface	Abkhazia	43.63	40.53	2485	-	-	-	Kvavadze and Efremov 1994
23	Bezymianka 2	Surface	Abkhazia	43.63	40.53	2548	-	-	-	Kvavadze and Efremov 1994
24	Bluzovoe	Core top	NW Caucasus	43.63	41.13	2884	-	-	-	Kvavadze and Efremov 1996
25	Bolshoi Khodzhal	Surface	Abkhazia	42.98	41.92	2510	-	-	-	Kvavadze and Rukhadze 1989
26	Chamlukh	Core	Abkhazia	43.18	41.03	1650	1	4000	0	Yazvenko 1994
27	Cheliagele	Core	Racha	42.47	43.10	1100	25	7000	0	van Leeuwen, unpubl.
28	Cherepanov 1-4	Surface	S Georgia	41.57	44.37	1430	-	-	-	Connor *et al.* 2004
29	Chili Tba 1-3	Surface	E Georgia	41.82	44.68	1130	-	-	-	Chapter Four
30	Devis Tba	Core	Racha	42.42	43.30	1380	?	4000	300	van Leeuwen, unpubl.
31	Didajaris Tba A	Core top	Ajara	41.68	42.50	1850	-	-	-	Margalitadze 1995
32	Didajaris Tba B	Core	Ajara	41.68	42.50	1850	6	11000	0	van Leeuwen, unpubl.
33	Ertso	Core top	E Georgia	41.95	44.97	1200	-	-	-	Margalitadze 1995
34	Gagra 471	Core	Abkhazia	43.23	40.27	5	2	7000	0	Kvavadze and Rukhadze 1989
35	Gagra 607	Core	Abkhazia	43.27	40.27	0	2	6000	4000	Kvavadze and Rukhadze 1989
36	Gomnis Tba 1-2	Core	S Georgia	41.77	43.35	1850	2	12000	0	Margalitadze 1971
37	Gomnis Tba 3	Surface	S Georgia	41.77	43.35	1800	-	-	-	Margalitadze 1995
38	Gudauta 120	Core	Abkhazia	43.08	40.62	0	1	11000	0	Kvavadze and Rukhadze 1989
39	Gudauta 55	Core	Abkhazia	43.08	40.62	0	2	7000	5000	Kvavadze and Rukhadze 1989

No.	Site, sample no.	Type	Sub-region	Lat.	Long.	Elev.	^{14}C	From	To	Source
40	Imera	Core	S Georgia	41.65	44.22	1610	6	7000	0	Kvavadze and Connor 2005
41	Imera 1-4	Surface	S Georgia	41.65	44.22	1610	-	-	-	Connor *et al.* 2004
42	Imnati	Core	Colchis	42.10	41.82	2	4	7000	0	Neishtadt *et al.* 1965
43	Ispani I	Core top	Colchis	41.85	41.78	2	-	-	-	Dokturovskii 1936
44	Ispani II-A	Core top	Colchis	41.87	41.80	2	-	-	-	Dokturovskii 1936
45	Ispani II-B	Core top	Colchis	41.87	41.80	2	-	-	-	Kaffke *et al.* 2002
46	Ispani II-C	Core	Colchis	41.87	41.80	2	4	6000	0	Connor *et al.* 2007
47	Jvari	Core top	E Georgia	41.83	44.73	571	-	-	-	Chapter Five
48	Jvari 1	Surface	E Georgia	41.83	44.73	570	-	-	-	Chapter Four
49	Kakhisis Didi Chaobi	Core top	S Georgia	41.73	43.30	1780	-	-	-	Margalitadze 1995
50	Kakhisis Didi Chaobi 2	Surface	S Georgia	41.73	43.30	1800	-	-	-	Margalitadze 1995
51	Kartsakhi	Core	S Georgia	41.25	43.30	1800	2	8000	0	Margalitadze and Ramishvili 1972
52	Kazbegi	Core	Central C.	42.63	44.55	3360	2	7000	3000	Margalitadze 1995
53	Keli 1-2	Surface	Central C.	42.47	44.33	2775	-	-	-	Kvavadze 1993
54	Keli 3-4	Surface	Central C.	42.47	44.33	2900	-	-	-	Kvavadze 1993
55	Keli 5-7	Surface	Central C.	42.47	44.33	3050	-	-	-	Kvavadze 1993
56	Khachal east lake	Surface	S Georgia	41.32	43.93	2634	-	-	-	Chapter Four
57	Khachal pool 1-3	Surface	S Georgia	41.32	43.93	2640	-	-	-	Chapter Four
58	Khachal west lake	Surface	S Georgia	41.32	43.93	2633	-	-	-	Chapter Four
59	Kobuleti 39	Core top	Colchis	41.90	41.83	10	-	-	-	Kvavadze and Connor 2005
60	Ktsiis Zeda Vake	Core top	S Georgia	41.68	43.62	2090	-	-	-	Margalitadze 1995
61	Kumisi	Core	E Georgia	41.58	44.83	469	1	1000	0	Chapter Five
62	Kumisi 2-4	Surface	E Georgia	41.58	44.83	469	-	-	-	Connor *et al.* 2004
63	Kvartsevoe	Core	NW Caucasus	43.63	41.12	2726	1	6000	0	Kvavadze and Efremov, 1996
64	Lagodekhi lakes 1	Surface	E Georgia	41.93	46.35	2750	-	-	-	Kvavadze and Efremov 1995
65	Lagodekhi lakes 2-3	Surface	E Georgia	41.93	46.35	2783	-	-	-	Kvavadze and Efremov 1995
66	Lagodekhi lakes 4	Surface	E Georgia	41.93	46.35	2920	-	-	-	Kvavadze and Efremov 1995
67	Lagodekhi lakes 5-6	Surface	E Georgia	41.93	46.35	3043	-	-	-	Kvavadze and Efremov 1995
68	Lashkhrashi	Core	Svaneti	43.13	42.43	2000	1	3000	0	Margalitadze and Kimeridze 1985
69	Lisi	Core top	E Georgia	41.75	44.73	624	-	-	-	Kvavadze 1999b
70	Lisi 1-7	Surface	E Georgia	41.75	44.73	624	-	-	-	Kvavadze and Stuchlik 1991
71	Luganskoe	Core	NW Caucasus	43.38	40.68	2428	2	5000	0	Kvavadze and Efremov, 1996
72	Malyi Khodzhal	Core	Abkhazia	42.95	41.92	2030	1	1000	0	Kvavadze and Rukhadze 1989
73	Mestia Chala	Core top	Abkhazia	43.07	42.75	1500	-	-	-	Margalitadze 1995
74	Nariani	Core top	S Georgia	41.70	43.67	2050	-	-	-	Margalitadze 1995
75	Pasinler CL	Core top	E Turkey	39.98	41.70	2000	-	-	-	Connor, unpubl.
76	Pasinler SH1	Core top	E Turkey	39.98	41.70	2000	1	600	0	Connor, unpubl.
77	Patara Gldanis Tba	Surface	E Georgia	41.82	44.83	530	-	-	-	Chapter Four
78	Pitsunda 101-105	Surface	Abkhazia	43.17	40.33	5	-	-	-	Kvavadze and Rukhadze 1989
79	Poti	Core top	Colchis	42.08	41.75	5	-	-	-	Neishtadt 1957
80	Rybnoe	Core	NW Caucasus	43.65	41.15	2151	2	2000	0	Kvavadze and Efremov 1996

No.	Site, sample no.	Type	Sub-region	Lat.	Long.	Elev.	^{14}C	From	To	Source
81	Sagarejo	Core	E Georgia	41.68	45.35	550	2	14000	0	Gogichaishvili 1984b
82	Sakhare	Core	E Georgia	41.58	45.32	800	1	2000	0	Chapter Five
83	Sakhare 1-4	Surface	E Georgia	41.58	45.32	800	-	-	-	Connor *et al.* 2004
84	Sakochavi	Core top	S Georgia	41.77	43.47	1600	-	-	-	Margalitadze 1995
85	Sakochavi 7	Surface	S Georgia	41.77	43.47	1700	-	-	-	Margalitadze 1995
86	Salkanalla	Core	Central C.	43.17	43.25	2150	6	5000	0	Serebryanny and Malyasova 1981
87	Samsari 9	Surface	S Georgia	41.53	43.65	2250	-	-	-	Margalitadze 1995
88	Sevan	Core top	Armenia	40.53	44.98	1898	7	6500	0	Saiadian 1985
89	Shavi Tba 1-7	Surface	E Georgia	41.75	44.75	622	-	-	-	Kvavadze and Gabashvili 1998
90	Shavluga	Core top	Abkhazia	43.07	42.10	2050	-	-	-	Margalitadze 1995
91	Sibista	Core	Abkhazia	43.23	41.43	2160	2	2000	0	Kvavadze and Rukhadze 1989
92	Sogutlu	Core top	E Turkey	38.65	42.08	1646	3	7500	0	Bottema 1995
93	Solosveli	Core top	S Georgia	41.73	43.48	1700	-	-	-	Margalitadze 1995
94	Soseva	Core top	Racha	42.50	43.42	1800	-	-	-	Margalitadze 1995
95	Sukhumi 36	Core	Abkhazia	43.00	41.00	0	1	20000	0	Kvavadze and Rukhadze 1989
96	Sukhumi 721	Core	Abkhazia	43.00	41.00	0	2	10000	0	Kvavadze and Rukhadze 1989
97	Sukhumi 723	Core	Abkhazia	43.00	41.00	0	4	10000	0	Kvavadze and Rukhadze 1989
98	Supsa	Core	Colchis	42.02	41.85	8	4	8000	0	Kvavadze 1982
99	Supsa 2	Core	Colchis	42.03	41.83	10	1	2000	0	Kvavadze and Connor 2005
100	Tabatsquri 3-5	Surface	S Georgia	41.67	43.63	2100	-	-	-	Margalitadze 1995
101	Tbata	Core top	Racha	42.50	43.42	1800	-	-	-	Margalitadze 1995
102	Trialeti 2-4	Surface	S Georgia	41.53	44.12	1520	-	-	-	Connor *et al.* 2004
103	Tsavkisi	Core	E Georgia	41.68	44.72	1100	9	7000	0	Chapter Five
104	Tsavkisi 1-4	Surface	E Georgia	41.68	44.72	1100	-	-	-	Connor *et al.* 2004
105	Tseros Tbis	Core	S Georgia	41.77	43.35	1920	2	6000	0	Margalitadze 1967
106	Tseros Tbis 4	Surface	S Georgia	41.77	43.35	1700	-	-	-	Margalitadze 1995
107	Tsodoreti 1-6	Surface	E Georgia	41.77	44.63	940	-	-	-	Kvavadze and Stuchlik 1991
108	Udabno Gareji 1	Surface	E Georgia	41.37	45.65	600	-	-	-	Kvavadze and Stuchlik 1993
109	Urukh Basin 109	Surface	N Caucasus	43.92	42.10	1020	-	-	-	Klopotovskaya 1973
110	Urukh Basin 111	Surface	N Caucasus	43.95	42.13	870	-	-	-	Klopotovskaya 1973
111	Van Lake	Core top	E Turkey	38.42	42.67	1648	v	12000	0	Wick *et al.* 2003
112	Western Georgia 21	Surface	Colchis	42.60	42.68	520	-	-	-	Klopotovskaya 1973
113	Western Georgia 7	Surface	Svaneti	42.88	43.15	1520	-	-	-	Klopotovskaya 1973
114	Zresi	Core top	S Georgia	41.40	43.43	1700	-	-	-	Margalitadze 1995
115	Zresi 6	Surface	S Georgia	41.40	43.43	1700	-	-	-	Margalitadze 1995

Key to abbreviations: Central C. – Central Caucasus; Lat. – latitude (decimal degrees); Long. – longitude (decimal degrees); ^{14}C – number of radiocarbon dates (v – varve ages); From – basal age in ^{14}C bp; To – most recent sample age (0 = modern).

As for sample selection, the sample nearest to each 1000-year interval was read from the diagram. The sample was rejected if it appeared anomalous or where high Asteraceae percentages suggested pollen degradation. The next closest sample was then used. Samples were not 'smoothed' with the surrounding samples because this was often impractical. A drawback of this approach, apart from dating uncertainty, is that rare taxa are often missed.

The pollen sum was calculated from the total of tree and shrub pollen reported in each diagram. The reason for doing this was to be able to compare pollen diagrams calculated from a dryland pollen sum with those calculated from an arboreal sum only. Trace proportions marked with a '+' sign were assigned a value of 0.5 percent. Proportions were then recalculated to a total of 100 percent. Grass and herb (non-arboreal) pollen percentages were calculated individually from a sum including trees, shrubs and the non-arboreal pollen type of interest.

Finally, the data were mapped using geographic co-ordinates based on published site descriptions. Maps were plotted using ArcGIS software and a Georgia base map kindly provided by Dr Lawrence Crissman, director of the ACASIAN database, Griffith University.[4]

This chapter is organised in three sections: the first one deals with the regional history of individual pollen taxa through the last 14,000 years, the second part is a reconstruction of Georgia's vegetation history, while the third part is an attempt to derive quantitative climatic estimates for the past 12,000 years.

Such a task has not previously been attempted in the Caucasus region. This is in spite of a long history of diverse palynological investigation there. Continental-scale reconstructions, such as those provided by Peterson[5] and Tarasov *et al.*,[6] are too coarse to give a detailed impression of vegetation change in the Caucasus. Kvavadze and Rukhadze[7] and Kvavadze *et al.*[8] have reconstructed the Holocene vegetation and climate of Abkhazia, but a general overview of Georgia's vegetation history, from the arid steppes to the coastal swamps, is notably lacking. The Holocene history of individual pollen taxa is poorly known, with the exception of *Zelkova*.[9] The aim of this chapter, then, is to incorporate the pollen data presented in previous chapters into a broader regional context.

4 Crissman 1995; ESRI 2005.
5 Peterson 1983.
6 Tarasov *et al.* 1998.
7 Kvavadze and Rukhadze 1989.
8 Kvavadze *et al.* 1992.
9 Kvavadze and Connor 2005.

Holocene Changes in the Distribution and Abundance of Major Pollen Taxa

The distribution and abundance of pollen taxa reflect, on a regional scale, spatial variations in the composition of plant communities. Within this context, the past behaviour of each individual pollen type tells us much about temporal changes in plant species' altitudinal range, population expansions and contractions, migrations and extinctions.[10] With a clear understanding of the ecology of the species represented, we gain a clearer insight into the factors (climatic, edaphic, anthropogenic, etc.) that influenced the distribution and abundance of plants in the past.

Each of Georgia's main pollen types is discussed in some detail, with reference to the ecology of its parent plants, its pollen representation, and the history of the taxon in the study area over the last 14,000 years. Rather than present this exhaustive ecological and palaeoecological information here, I have placed it in the appendix and give a synopsis below.

In the broadest of terms, the history of each pollen type can be grouped into one of five temporal categories: (A) those that peaked during the late Pleistocene, (B) those that were most abundant at the beginning of the Holocene, (C) those that rose to prominence in the early-mid Holocene, (D) taxa that were most prevalent during the mid Holocene, and finally, (E) those that have increased in the last few millennia (**Table 7.2**). Although no taxon fits perfectly into this scheme, it is a convenient way of discussing temporal changes in the Georgian vegetal landscape through the Holocene.

Table 7.2. Pollen types that share a common history, grouped according to the time of first maximum abundance in Georgia over the last 14,000 years.

Group	Maximum abundance	Pollen types
A	Late Pleistocene	*Ephedra, Juniperus, Artemisia*, Chenopodiaceae
B	Early Holocene	*Betula, Corylus, Salix, Rumex*
C	Early-mid Holocene	*Abies, Carpinus betulus, Fagus, Tilia, Ulmus/Zelkova*
D	Mid Holocene	*Castanea, Quercus, Hedera, Ilex*
E	Late Holocene	*Juglans, Olea, Ostrya-type, Picea, Pinus, Pterocarya, Plantago*

Group A — Late Pleistocene

Group A includes shrubs and herbaceous plants that are characteristic of semidesert-steppe vegetation today. *Ephedra* and *Juniperus* are xerophytic conifers, most abundant in the arid scrub vegetation of Eastern Georgia, although *Juniperus* is much more

[10] Huntley and Birks 1983; van der Knaap *et al.* 2005.

widespread and some species occur in forests and alpine vegetation. *Artemisia* and Chenopodiaceae are herbaceous plants that are well adapted to aridity and are prevalent in lowland semidesert-steppes, as well as being somewhat weedy.

All of these taxa peaked during the late Pleistocene, when the climate was presumably much more arid than it is today. If pollen data are any guide, there appears to be a relationship between *Juniperus* and Chenopodiaceae on the one hand, and *Ephedra* and *Artemisia* on the other. *Juniperus* and Chenopodiaceae declined during the Younger Dryas, while *Ephedra* and *Artemisia* expanded greatly.[11] All four taxa were much reduced through the early Holocene. The mid Holocene triggered an expansion in *Juniperus* and Chenopodiaceae and a decline in *Ephedra* and *Artemisia.* Chenopodiaceae and *Artemisia* have expanded during the late Holocene in some places (**Fig.** 7.2).

The history of these four taxa suggests that desert-steppe vegetation was distributed in the continental parts of the Caucasus during the late Pleistocene. The association between *Ephedra* and *Artemisia* is perhaps most easily explained by increased air humidity and reduced temperature, compared to the period in which *Juniperus* and Chenopodiaceae were dominant. This hypothesis could also explain why *Juniperus* and Chenopodiaceae increased through the mid Holocene, when the climate may have been relatively warm and dry.[12] The late Holocene expansion of Chenopodiaceae and *Artemisia* is probably the result of widespread localised disturbance. It is less clear what impact humans have had on the two gymnosperms: *Juniperus* is killed by fire and promoted by grazing, while *Ephedra* appears to be the opposite.

Group B — Early Holocene

Group B includes several small trees, shrubs and a herb that share the common characteristic of being light-demanding; they are overshadowed by larger growing deciduous trees and are often found at forest edges. *Betula* and *Corylus* are small trees and shrubs that are a minor component in most mixed forests. Some *Betula* species are dominant trees at the upper tree line and *Corylus* are most prevalent in the understorey of deciduous forests. *Salix* are moisture-loving trees and shrubs, often found along watercourses. *Rumex* are grassland herbs regarded as good indicators of open vegetation and grazing.[13] *Salix* and *Rumex* both have poor pollen production, so their history is somewhat patchy.

All four taxa were present during the late Pleistocene in Georgia. *Betula*, it seems, was even present at relatively high altitudes, allowing it to expand rapidly in the mountains once the Younger Dryas came to an end (**Fig.** 7.3). *Corylus* and *Salix* followed

[11] Carrión *et al.* 2003 observed a similar pattern in Spain.
[12] The so-called 'xerothermic' maximum: Gogichaishvili 1984b.
[13] Behre 1986; Huntley and Birks 1983.

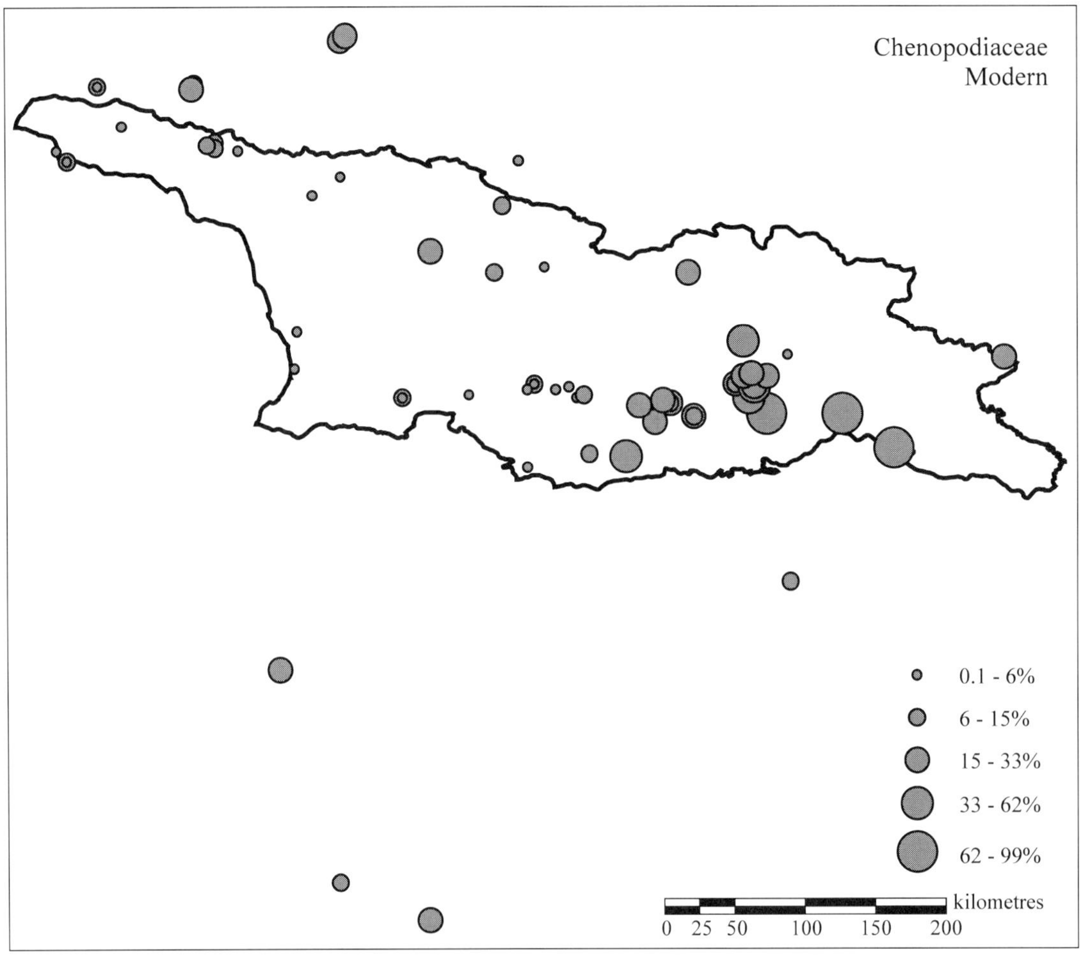

Figure 7.2 (a). The modern distribution of Chenopodiaceae pollen in the study area.

soon after, competing with *Betula* for light and space in early Holocene woods. *Rumex*, which was present in the Pleistocene flora, also expanded considerably in the treeless parts of the study area. The four Group B taxa declined during the early-mid Holocene as more competitive deciduous trees moved in to supplant them (see Group C, below). In general, all four remained relatively scarce until the late Holocene, when they experienced a pronounced resurgence.

The woods that spread over the mountain slopes during the earliest millennia of the Holocene were unlike anything that is seen today. These woods must have been quite open in structure and short in stature. As in northern Europe, their composition was determined by the speed at which *Betula, Corylus* and *Salix* colonise new areas. The major difference in the Caucasus was that these pioneer taxa only dominated at higher elevations, since the lowlands were already home to forest refugia. The late Holocene rise of *Betula, Corylus, Salix* and *Rumex* is almost certainly the result of widespread forest disturbance at lower and middle elevations.

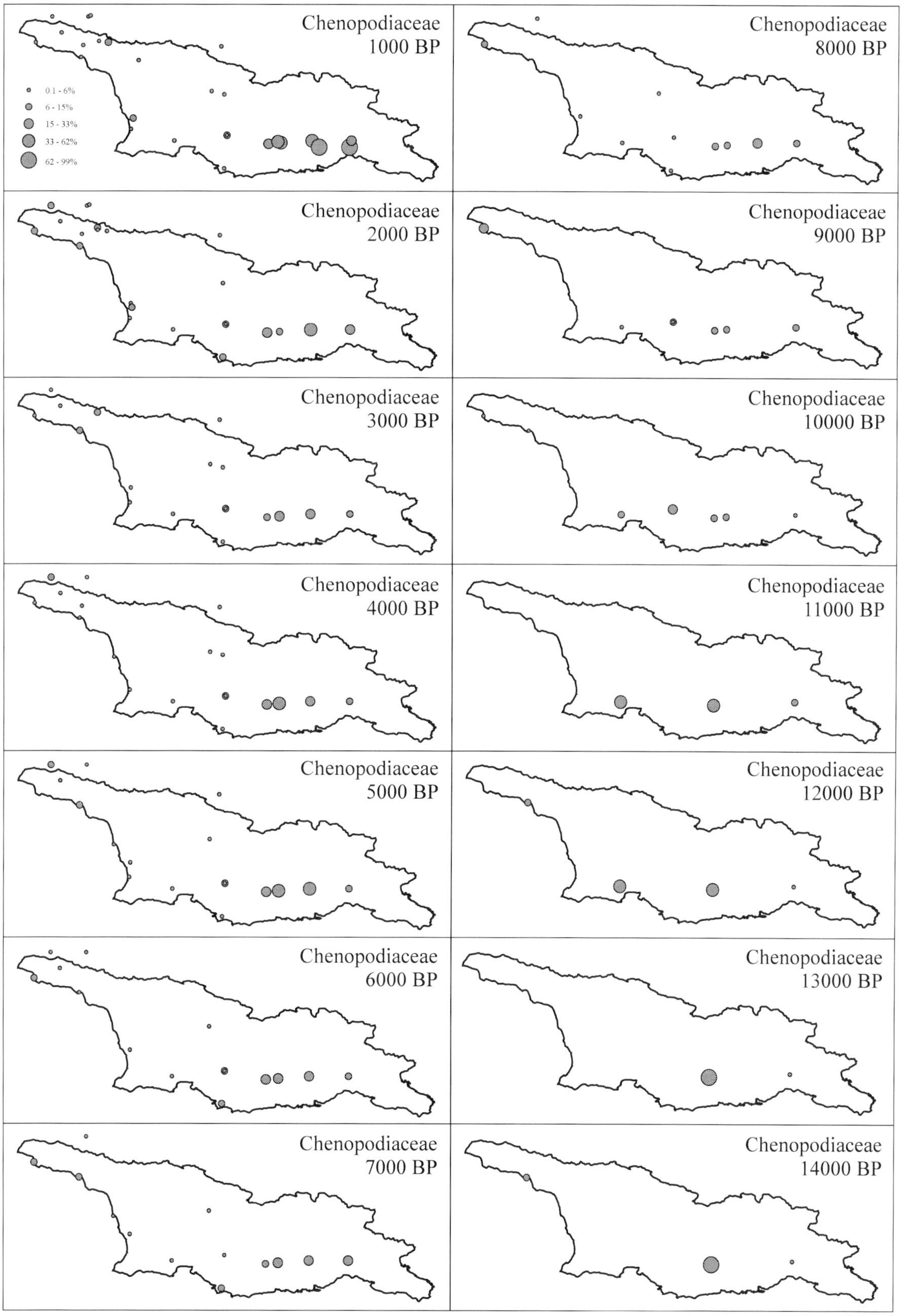

Figure 7.2 (b). Changes in Chenopodiaceae (Group A) pollen abundance in Georgia, 14,000–1,000 cal BP.

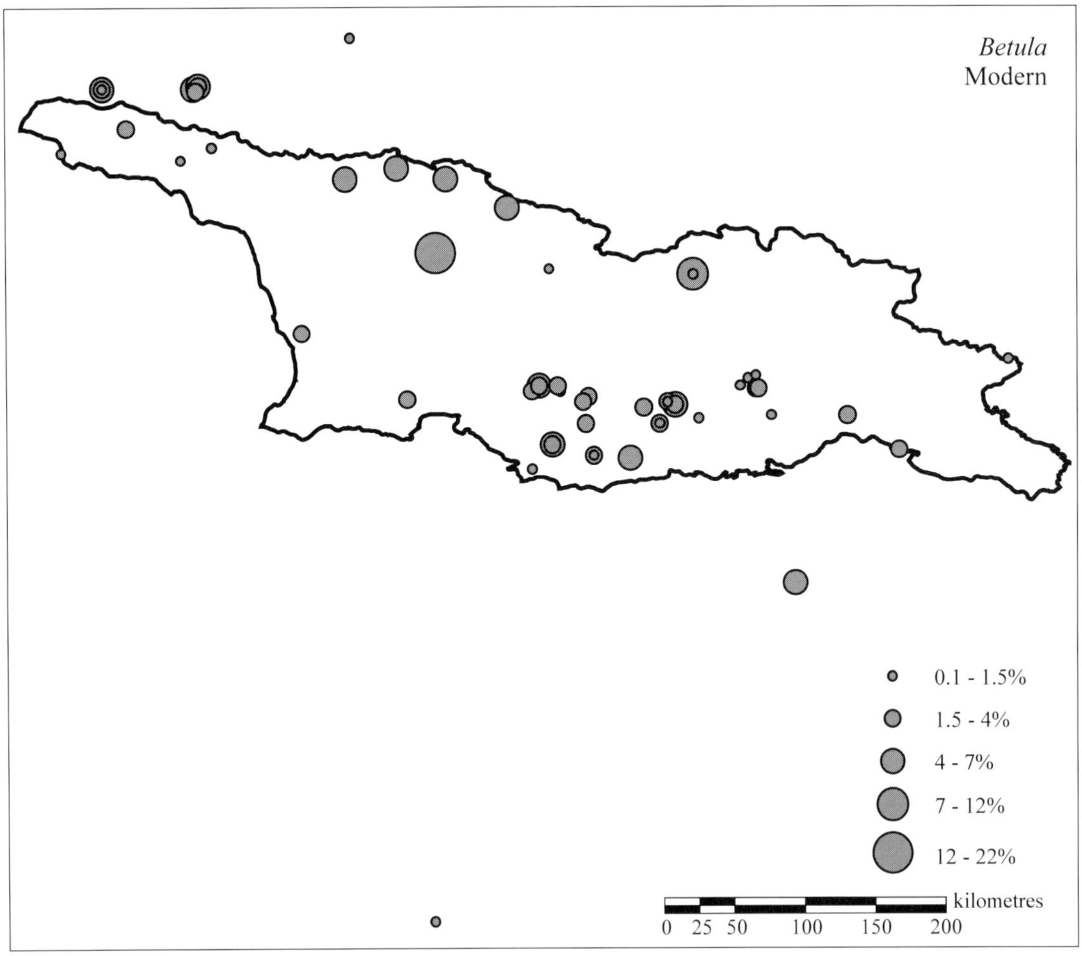

Figure 7.3 (a). The modern distribution of *Betula* pollen in the study area.

Group C — Early-mid Holocene

The pollen types that fall into Group C all represent trees of mixed deciduous, mesophytic forests. *Abies nordmanniana*, although a conifer, is a common component of deciduous forests. *Carpinus betulus* and *Tilia* are widespread and usually sub-dominant in Caucasian forests, while *Fagus orientalis* is the most important forest dominant in the South Caucasus region. *Ulmus* and *Zelkova* species have diverse ecological tolerances and are discussed separately in the appendix. All of these Group C taxa grow together in mixed forests, especially in the more humid parts of Georgia; they avoid the droughty, continental areas of Eastern Georgia.

Group C taxa were present in Pleistocene refugia in the Black Sea lowlands. In these forests, *Abies* seems to have been a major contributor. Through the early Holocene

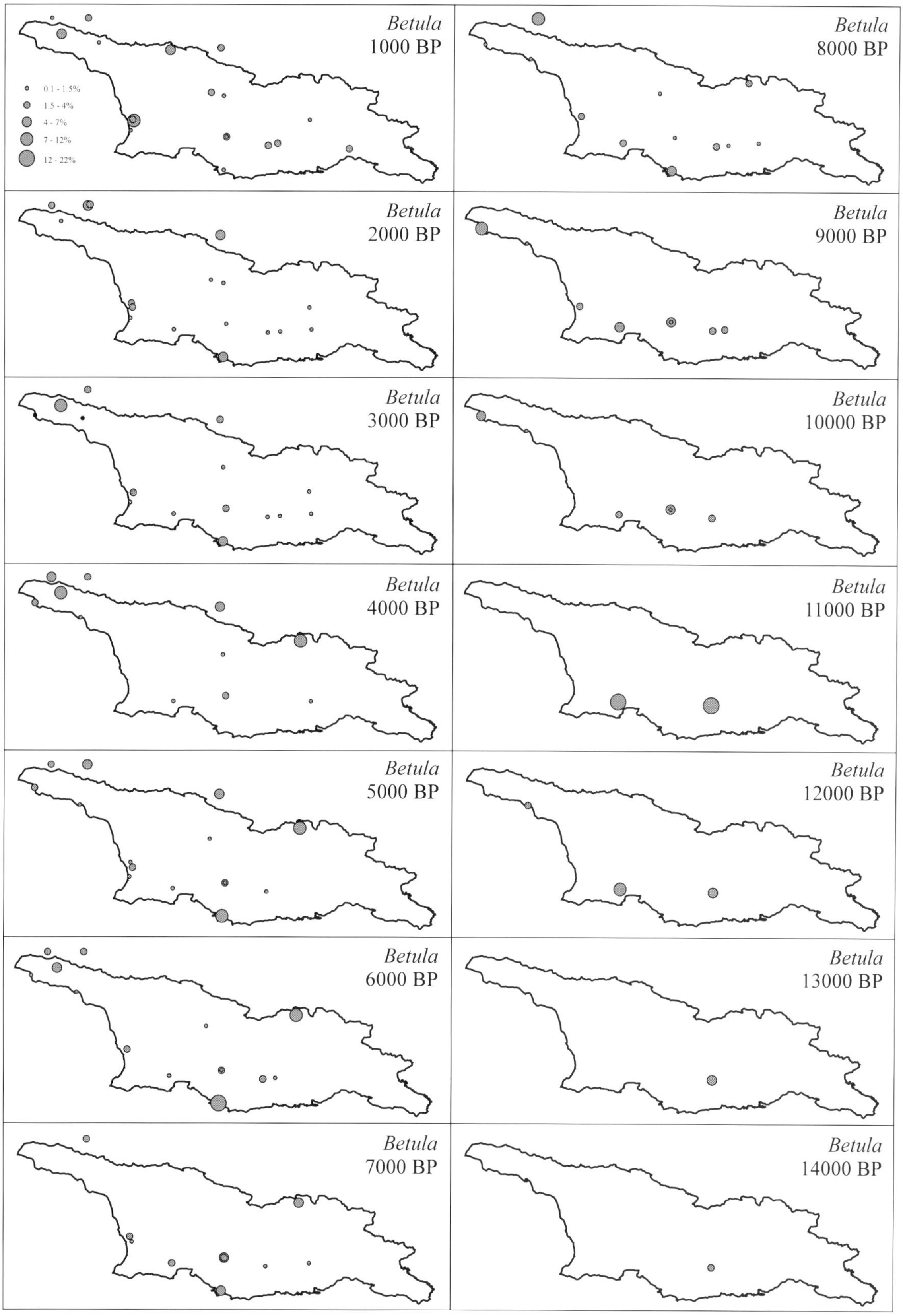

Figure 7.3 **(b).** Changes in *Betula* (Group B) pollen abundance in Georgia, 14,000–1,000 cal BP.

Abies, Fagus, Carpinus betulus and *Tilia* expanded slowly out of the lowlands to establish in the middle and upper forest belts. *Fagus orientalis* was the slowest to migrate, such that it became widespread only 7000 years ago (**Fig.** 7.4). The mid Holocene was unfavourable to these taxa in the semi-arid parts of Georgia, and elsewhere *Fagus orientalis* reached such a level of dominance that *Carpinus betulus* and *Tilia* declined. During the late Holocene, *Carpinus betulus* and *Fagus orientalis* have increased in some places. The distribution of *Abies,* on the other hand, has become much more restricted.

Group C taxa expanded to replace the pioneer taxa of Group B during the early-mid Holocene. Their expansion was mitigated by possible early Holocene aridity and, for *Fagus,* the inability to disperse seed over large areas. In some places in Southern Georgia, it seems that fire was at least partly responsible for displacing Group C taxa during the mid Holocene, at a time when the climate may have also become less amenable to these taxa. The widespread late Holocene decline of *Abies* can be attributed to the

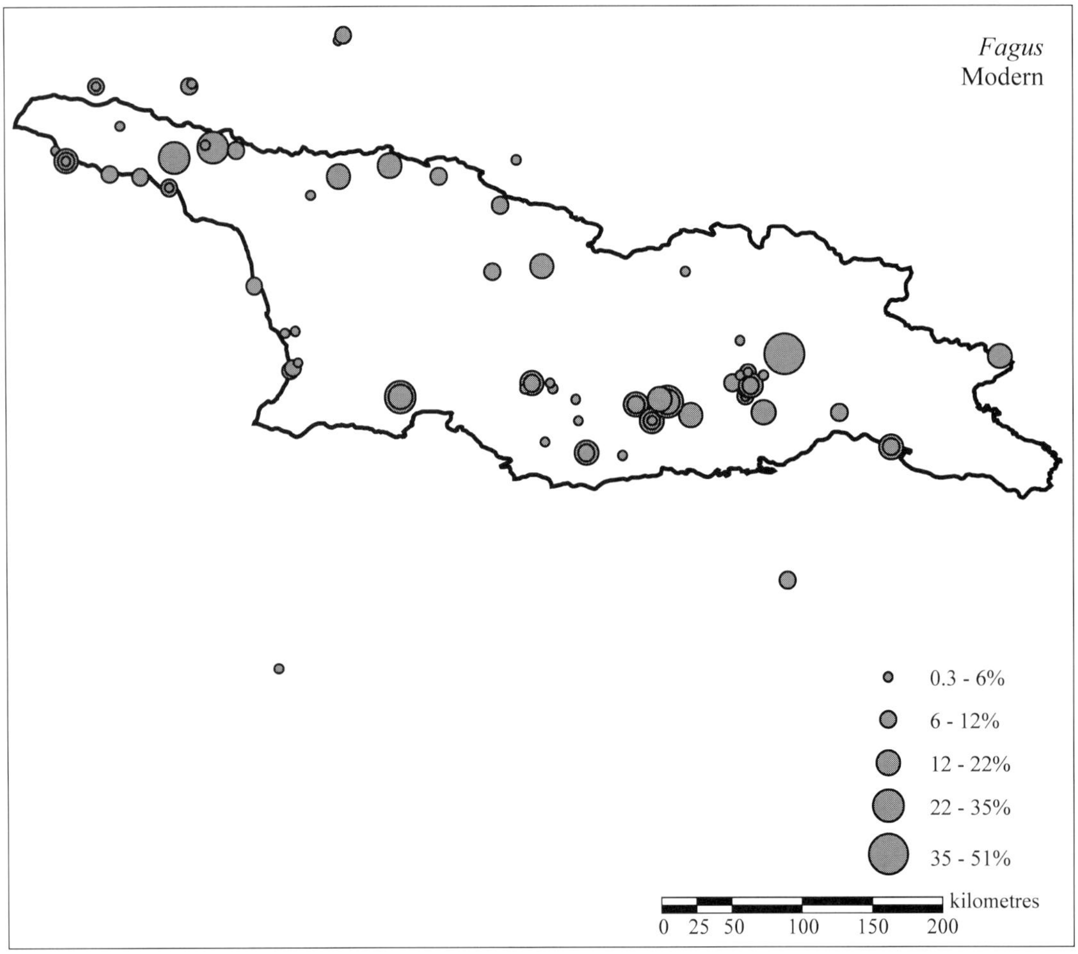

Figure 7.4 (**a**). The modern distribution of *Fagus* pollen in the study area.

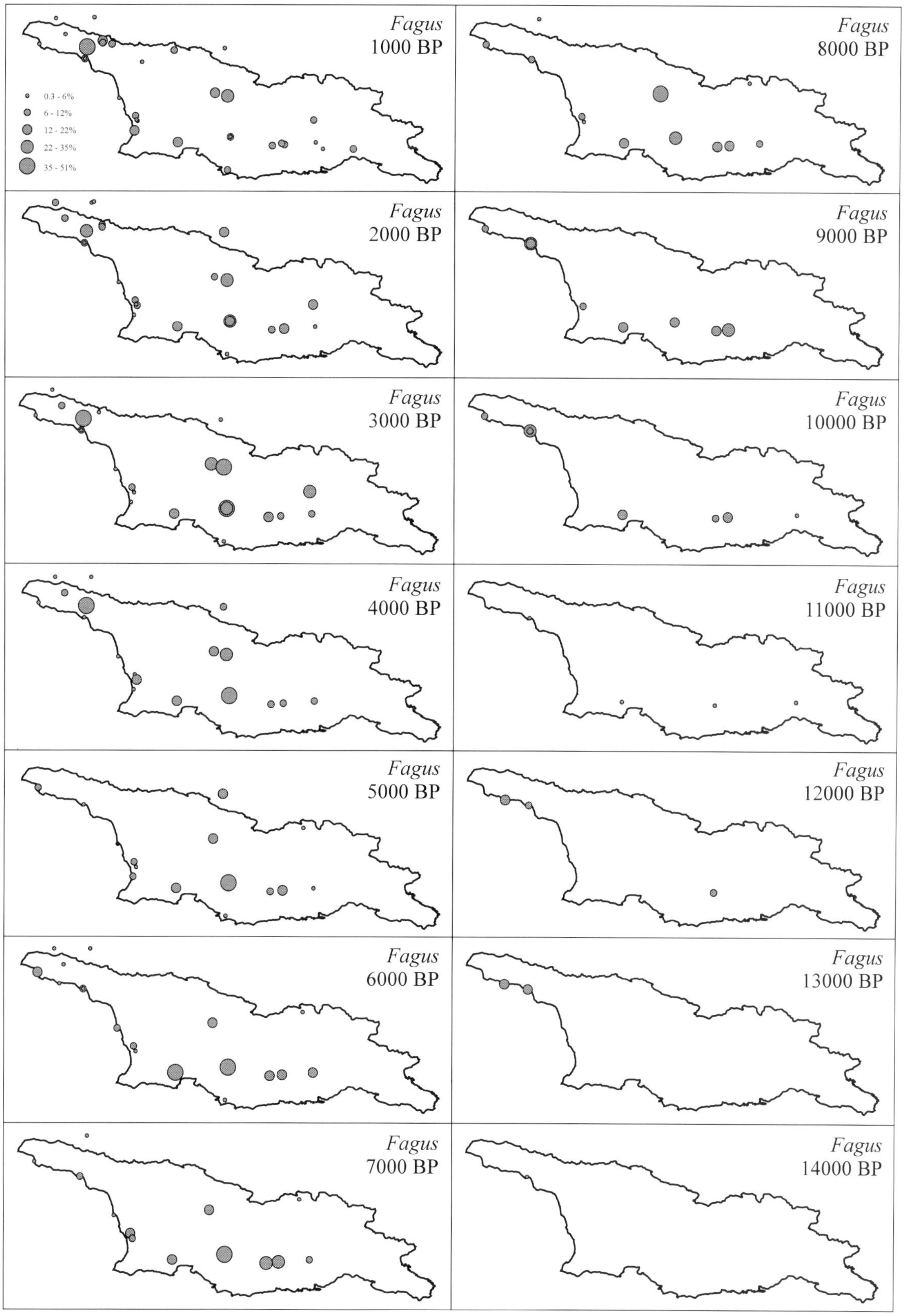

Figure 7.4 (b). Changes in *Fagus* (Group C) pollen abundance in Georgia, 14,000–1,000 cal BP.

increasing role of fire in forest vegetation. On the other hand, the abandonment of open land in both Eastern and Western Georgia has allowed *Fagus* and *Carpinus betulus* to increase their spread in the lower forest belt.

Group D — Mid Holocene

Those taxa that peaked during the mid Holocene are thermophiles, lending support to the idea that the mid Holocene was a time of increased temperatures (at least seasonally) across the Caucasus region. *Castanea sativa* is a tree with a fragmented distribution in the South Caucasus, growing mostly on warm, south-facing slopes in the foothills around the Colchis Lowland (Western Georgia) and the Alazani Valley (Eastern Georgia). *Quercus* species are much more widely distributed, occurring in middle elevation forests across the region, but dominating woodland vegetation where summers are hot and droughty. Highland oak (*Quercus macranthera*) is a subalpine species common in Southern Georgia and Armenia. *Hedera* and *Ilex* are evergreens confined to low-altitude Colchic forests and occur mostly in the west of the country.

Castanea and *Quercus* were present at low altitudes during the late Pleistocene – *Castanea* in the west and *Quercus* in the east. Both taxa expanded during the early-mid Holocene, but it was only during the mid Holocene that they became dominant in several pollen records (**Fig.** 7.5). *Hedera* and *Ilex* likewise displayed a maximum distribution in Colchis and Abkhazia during the mid Holocene. All four taxa declined between 4500 and 3000 years ago and have not regained their former importance in the vegetation in recent times.

Changing climates have exerted a profound influence on the histories of *Castanea, Quercus, Hedera* and *Ilex*. Milder winters would be expected to benefit Group D taxa over others. In Southern Georgia, the onset of milder temperatures in the mid Holocene, accompanied by slightly drier conditions, may have encouraged *Quercus* to dominate. Warmer temperatures in Colchis may have assisted the spread of *Castanea, Ilex* and *Hedera* there. Colder winters through the late Holocene reduced Group D taxa considerably, but in many circumstances this factor alone is insufficient to explain the decline of *Quercus* and *Castanea*. These trees are very much intertwined with human history, having long been exploited by humans for fuel, timber, food and fodder. Past human activity is thought to have reduced their populations to a significant degree, since they occupy soils favoured for agriculture.[14] On parts of the coastal plain, *Castanea* was set back by paludification during the late Holocene.[15]

[14] Dolukhanov 1966.
[15] Connor *et al.* 2007.

Group E — Late Holocene

This last group includes an assortment of taxa that achieved maximum abundances in the late Holocene. *Juglans* and *Pterocarya* are large thermophilous trees of the Juglandaceae family found in lowland gallery forests and, in the case of *Juglans*, more commonly in cultivation. *Picea* and *Pinus* are cold-tolerant coniferous trees, distributed in the upper forest belt; *Picea* is less drought-tolerant than *Pinus* and is hence associated with Western Georgia. *Ostrya*-type pollen represents *Carpinus orientalis*, a tree or shrub found in the forest understorey, but most abundant in scrub and shibliak communities. *Olea europaea* is the common olive, native to the circum-Mediterranean region, and *Plantago* represents grassland herbs that occur widely in open vegetation, especially pastures.

Although these seven taxa peaked during the late Holocene, some of them were also locally abundant during the late Pleistocene and early Holocene. *Juglans*, *Pterocarya*, *Picea* and *Pinus* were present in lowland refugia of Western and Eastern Georgia during the late Pleistocene. *Juglans* and *Pterocarya* were probably dominant trees on some floodplains in Eastern Georgia, but declined during the early Holocene to the point of local extinction. *Picea* and *Pinus* were likewise much reduced. *Ostrya*-type increased with the early Holocene advance of forest trees, but faded rapidly into the background. Around 4000 years ago, *Picea, Pinus* and *Ostrya*-type began to expand (**Fig.** 7.6). Two thousand years ago they were joined by *Juglans, Pterocarya* and *Olea europaea* and all these taxa increased at an accelerated pace more recently. If the pollen records are indicative of plant abundance, then *Picea, Pinus, Pterocarya* and *Carpinus orientalis* are more common now than at any time during the entire Holocene.

Several hypotheses could help explain these patterns. For *Picea* and *Pinus,* cooler winter temperatures during the late Holocene may have assisted their late expansion through the mountain forests of Georgia. This hypothesis has some merit, but fails to explain why thermophiles like *Pterocarya, Juglans, Olea europaea* and *Ostrya*-type expanded at the same time at lower altitudes. For *Juglans* and *Olea europaea*, the most straightforward explanation is that people planted them. Most archaeobotanical finds of *Juglans* post-date the Hellenistic Period,[16] suggesting that walnuts were introduced to cultivation via contact with Greece and Rome. It is probable that olives were similarly introduced.

Human activities during the late Holocene could have also contributed to the rise of *Carpinus orientalis*, *Pterocarya* and *Plantago*, as well as *Picea* and *Pinus*, through the abandonment of pastures and fields (favouring *Pterocarya* and *Carpinus orientalis*), deforestation (favouring *Plantago* and *Picea*) and forest fire (favouring *Pinus*). Although a climatic change seems the most probable reason for the late Holocene expansion of

[16] Lisitsyna and Prishchepenko 1977; Rusishvili 1990.

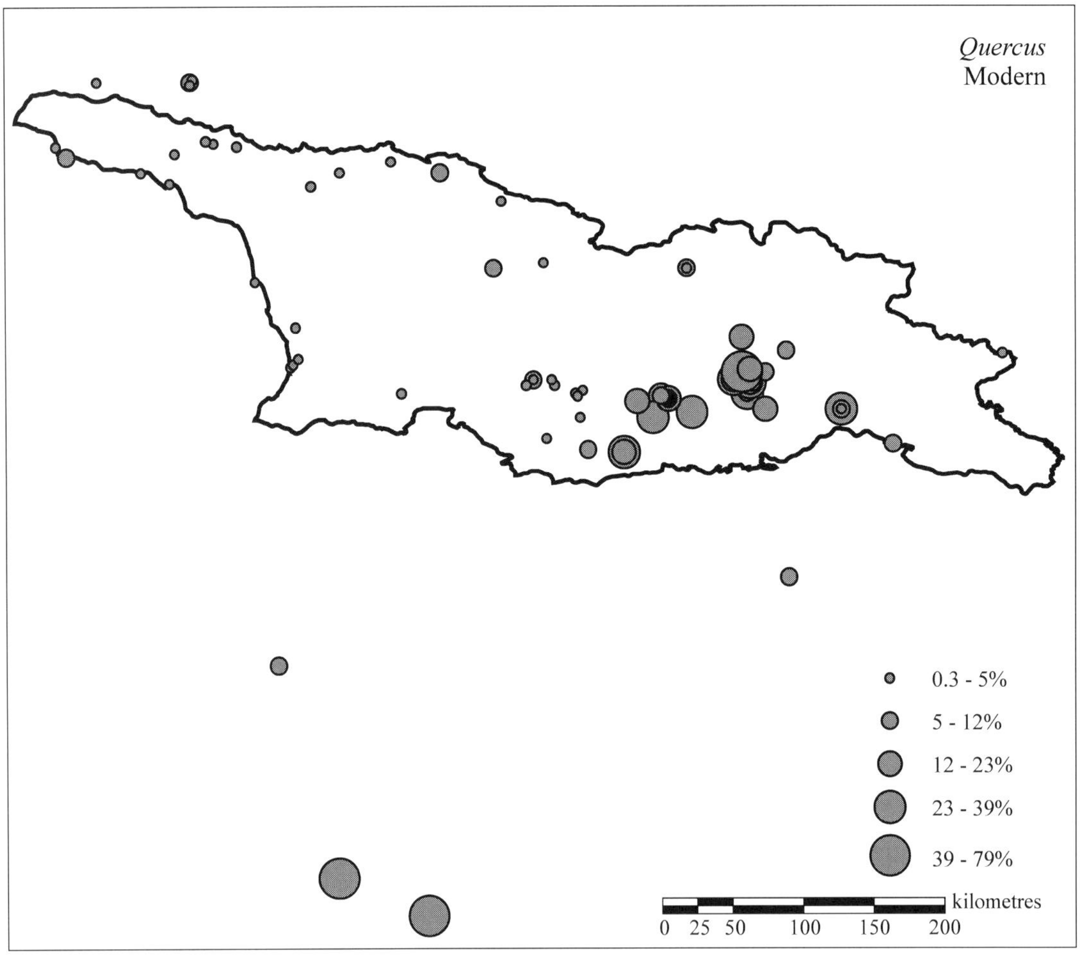

Figure 7.5 (a). The modern distribution of *Quercus* pollen in the study area.

Picea and *Pinus*, they and other taxa have also benefited from recent changes in human economy and land-use.

Other taxa, including *Acer, Alnus, Frangula, Pistacia, Rhododendron,* Rosaceae, *Vitis,* Poaceae and *Cerealia*-type, have histories that do not easily fit into one of the five categories above. They are discussed in detail in the appendix.

Each of the categories provides an insight into the climatic and ecological conditions prevailing at various stages through the past 14,000 years. During the late Pleistocene, tree lines were depressed by cold winters and aridity; forests were confined to the Black Sea lowlands in Western Georgia and the floodplains of Eastern Georgia. Increasing moisture during the early Holocene led to the rapid colonisation of the mountains by pioneer species such as *Betula* and *Corylus*, but these were quickly supplanted by *Fagus, Carpinus* and other mesophiles. The mid Holocene appears to have

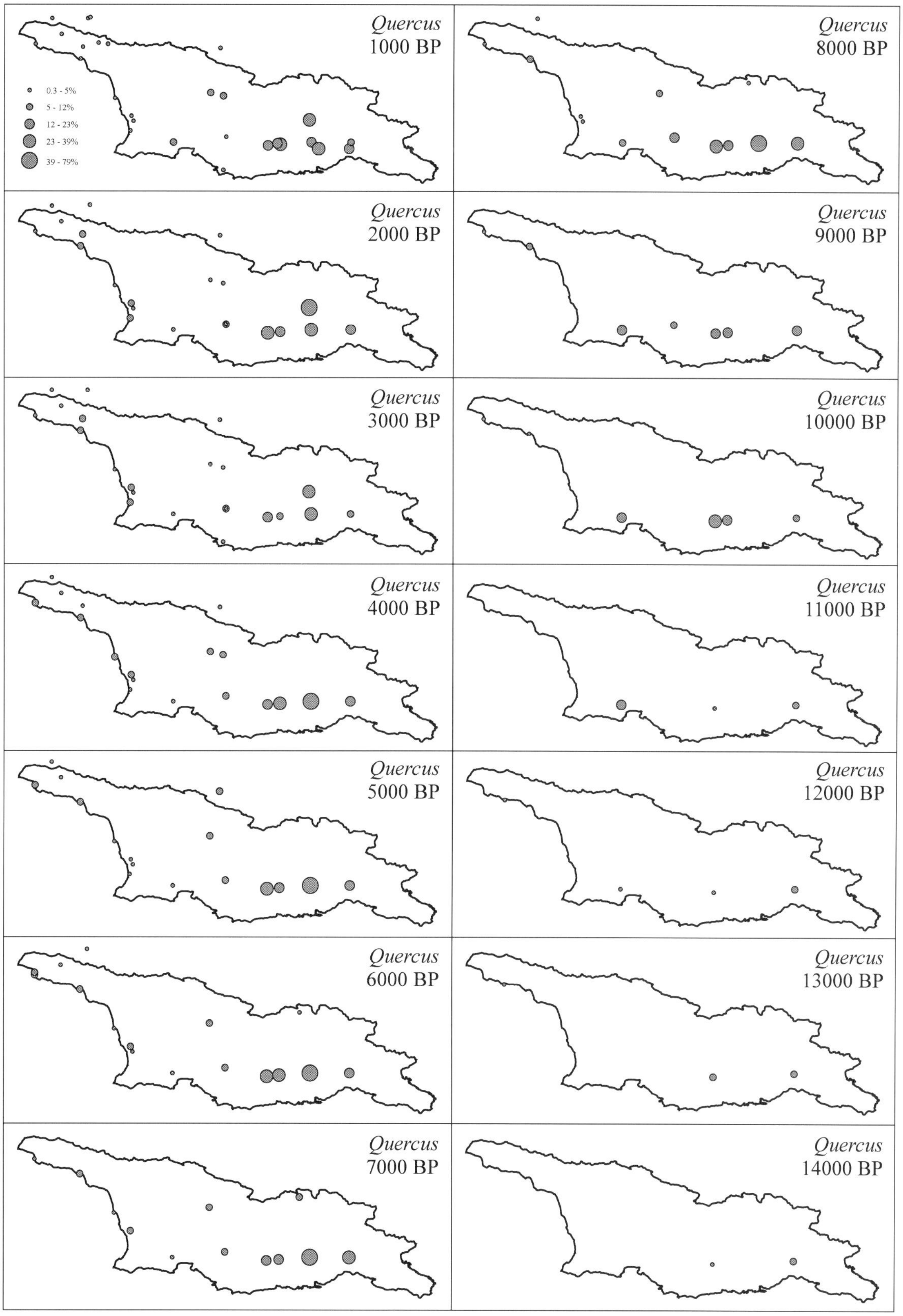

Figure 7.5 (b). Changes in *Quercus* (Group D) pollen abundance in Georgia, 14,000–1,000 cal BP.

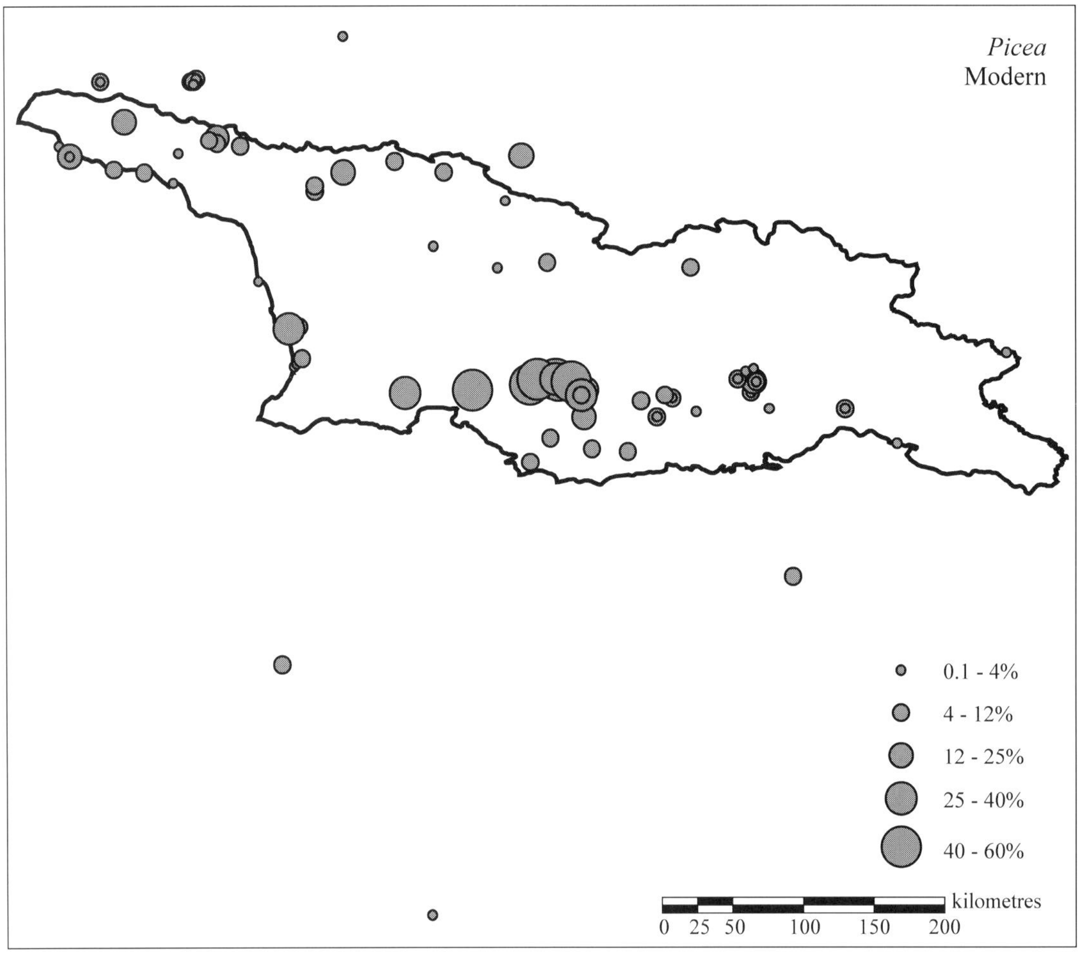

Figure 7.6 (a). The modern distribution of *Picea* pollen in the study area.

been somewhat warmer – *Quercus* expanded at all altitudes, and *Castanea* increased in the lowlands. Conditions changed again in the late Holocene, as these thermophiles retreated and cold-tolerant conifers widened their distribution. Many mid-late Holocene changes, however, bear the obvious imprint of human activity.

Palaeovegetation Maps for Georgia — 14,000 cal BP to the Present

The histories of the individual taxa discussed above have considerable ecological and geographical value. They allow us to trace changes in altitudinal or spatial distribution, population expansions and contractions, migrations and extinctions. But plants do not exist in isolation. Their movements may be important to the composition of plant communities, but may have little impact on the structure of the vegetation. Hence, in this section, an attempt is made to reconstruct the past vegetation of the study area using the available pollen data.

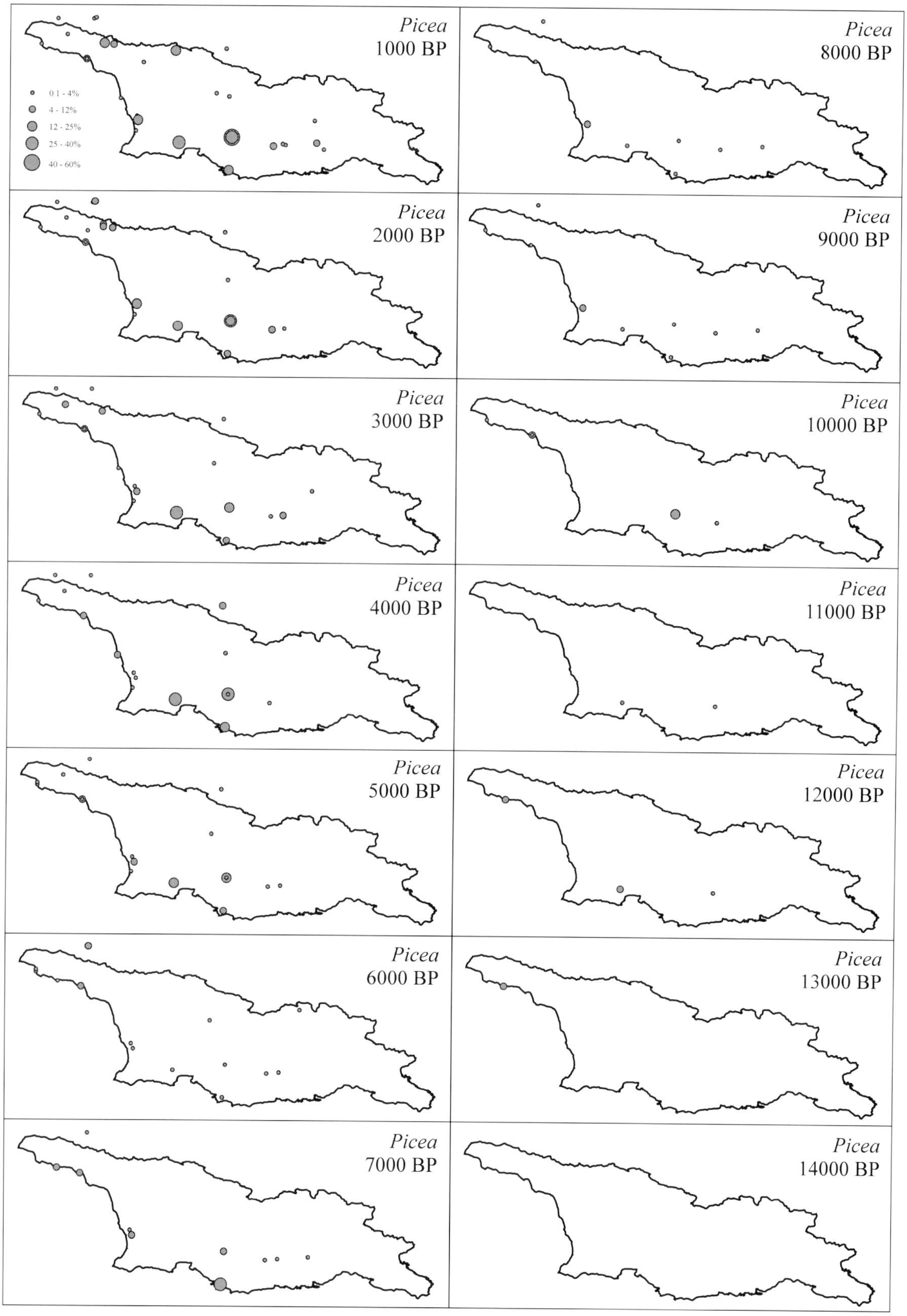

Figure 7.6 (b). Changes in *Picea* (Group E) pollen abundance in Georgia, 14,000–1,000 cal BP.

Method

There is no one, incontrovertible method for the reconstruction of past vegetation. Most attempts are relatively subjective, relying on a knowledge of the present vegetation, how it relates to present soils and hydrology, comparing the present vegetation to the pollen signature it produces, and then applying all this knowledge to the interpretation of palaeoecological, archaeological or other historical data to give an approximation of how the landscape may have looked in the past.[17] The spatial scale of interest, which may range from the environs of an archaeological site to a whole continent, largely determines how this past vegetation is then mapped.[18]

In recent decades palynologists have made a concerted effort to *quantify* their observations. Up until the 1960s, most analysts relied on "qualitative, intuitive interpretation of pollen diagrams", paying little attention to the need to link modern plant communities to pollen spectra.[19] Since then a great deal of work has examined the relative pollen productivity of various trees and other plants,[20] providing a numerical basis for interpreting pollen spectra in terms of the composition of the vegetation that produced it. The problem of pollen source area has so far frustrated attempts to accurately reconstruct vegetation mosaic from pollen data, but considerable progress has been made both in understanding pollen source area and in modelling landscape openness.[21] Because of the great complexity of interactions between different pollen types within the one pollen sum, a common approach has been to analyse surface samples and compare these semi-quantitatively with fossil pollen assemblages.[22]

Statistical methods have also been applied to match fossil spectra with modern analogues in the vegetation,[23] thus removing the bias associated with subjectively comparing modern and fossil pollen spectra. This method runs into difficulty where no modern analogue has been found for the particular pollen spectrum in question and is an especial problem in heavily human-impacted landscapes.[24]

An alternative approach involves linking individual pollen types to the plants that produce them and the environments in which these plants are found. The most ambitious example of this is the BIOME project, which relates each pollen taxon to possible plant functional types and thence to biomes, assigning each fossil pollen spectrum to a biome via a matrix of probabilities.[25] This method circumvents the problem

[17] Malmer and Regnell 1986.
[18] Malmer and Regnell 1986; Prentice *et al.* 2000.
[19] Wright 1967.
[20] Andersen 1970; Andersen 1973; Bradshaw 1981; Bunting 2003 and references therein.
[21] Davis 2000; Gaillard and Berglund 1998; Prentice 1985; Sugita 1993; Sugita *et al.* 1999.
[22] Birks and Birks 1980, pp. 237–255.
[23] Anderson *et al.* 1989; Overpeck *et al.* 1985.
[24] Birks 1973; Webb 1973; Webb 1980.
[25] Prentice *et al.* 1996.

of finding modern analogues for pollen spectra because it relies on plant functional types and not pollen taxa *per se*. The method has been applied to produce palaeovegetation maps of Europe, the Former Soviet Union, China, North America, Africa and most of the Northern Hemisphere.[26]

At the continental scale, the BIOME reconstructions are convincing representations of the present day vegetation. It is more difficult to assess their applicability at the sub-regional scale: a certain "noisiness" is produced in mountainous areas and along biome boundaries.[27] It comes as no surprise that some of the reconstructed biomes for the Caucasus, with its complex relief and varied vegetation, do not accord to the present vegetation,[28] although this may have as much to do with the quality of the original dataset as it does with the errors inherent in the model.

Pollen data from the preceding chapters will shortly be incorporated into a BIOME model for the Eastern Mediterranean-Black Sea-Caspian Sea region,[29] but until this is completed, I have adopted the same approach used by Huntley to construct palaeovegetation maps for Europe.[30] This method uses TWINSPAN to classify pollen data into ecologically meaningful and similar groups, and mapping to trace the movements of these groups through time. It has an advantage over some other multivariate methods in that it does not assume that fossil pollen spectra will have a modern analogue. It merely identifies the prevailing trend in the data and classifies pollen spectra according to similarities in their composition. The indicator species and thresholds produced by the analysis have ecological and biogeographical meaning and can be used subsequently to classify other pollen spectra.

The dataset described at the beginning of this chapter was reduced to include only those pollen diagrams where both arboreal and non-arboreal pollen were reported. This is because of the importance of taxa like Poaceae in mountain grasslands and Chenopodiaceae and *Artemisia* in semideserts and steppes. An attempt to use arboreal pollen alone failed because of the prevalence of long-distance-transported *Pinus* and *Alnus* pollen in treeless environments. I excluded *Cerealia*-type pollen from the analysis because few palynologists in the study area have counted or reported it. The Sagarejo and Bazaleti records[31] were also excluded because they skewed the result considerably; they appear to be outliers perhaps because of fluvial pollen transport or redeposition. TWINSPAN pseudo-species cut levels were set to 1, 2, 4, 10, 20, 30,

[26] Prentice *et al.* 1996; Prentice *et al.* 2000; Tarasov *et al.* 2000; Tarasov *et al.* 1998; Williams *et al.* 1998; Yu *et al.* 1998.

[27] Prentice *et al.* 1996, p. 191.

[28] Tarasov *et al.* 1998.

[29] Eastern Mediterranean-Black Sea-Caspian Corridor Biomes, the palynology working group of IGCP project 521, www.bridge.bris.ac.uk/projects/EMBSECBIO.

[30] Huntley 1988.

[31] Gogichaishvili 1990.

40, 60 and 80 percent to ensure that major taxa were selected as indicators and to reduce the influence of chance occurrences.

The data are presented on a base map of Georgian bioclimatic zones. This map was produced from a digital elevation model and climatic data using the IsoCluster algorithm in ArcGIS software.[32] This yielded sixteen groups based on similarities in elevation, temperature and rainfall (**Fig.** 7.7). These groups were then extrapolated to the entire landscape of Georgia through a maximum likelihood classification. The resulting map (**Fig.** 7.8) shares broad similarities with Ketskhoveli's superbly detailed *Map of Reconstructed Vegetation Cover*,[33] but is not intended to replace it. Instead it provides a useful bioclimatic backdrop for examining vegetation changes at individual sites.

Table 7.3. TWINSPAN classification of pollen taxa (eigenvalue = 0.71)

Euxinian and Euro-Siberian provinces		**Irano-Turanian province**	
Abies	*Hippophaë*	*Acer*	*Rhamnus* and
Alnus	*Ilex*	*Ephedra*	*Paliurus*
Betula	*Picea*	*Fraxinus*	Rosaceae (arboreal)
Carpinus betulus	*Pinus*	*Juglans*	*Salix*
Castanea	*Pterocarya*	*Juniper*	*Vitis*
Corylus	*Rhododendron*	*Ostrya* and	*Artemisia*
Fagus	*Tilia*	*Carpinus orientalis*	Chenopodiaceae
Frangula	*Ulmus* and *Zelkova*	*Pistacia*	*Plantago*
Hedera		*Quercus*	Poaceae
			Rumex

Classification

The major TWINSPAN division in the data set, explaining 23 percent of variance, separates samples high in *Alnus* from those with more Chenopodiaceae, Poaceae, *Quercus* and *Ostrya*-type (**Fig.** 7.9). In biogeographical terms this divides the moist Euxinian and Euro-Siberian botanical provinces from the more arid Irano-Turanian province.[34] The pollen taxa best associated with these provinces are given in **Table** 7.3. The Euxinian group is split from the Euro-Siberian by its greater representation of *Carpinus betulus* and *Ulmus-Zelkova*. The Euro-Siberian group is characterised by comparatively high proportions of *Picea* and *Pinus* and is generally associated with higher altitudes.

[32] ESRI 2005.
[33] AN GSSR 1964; Ketskhoveli 1959.
[34] Zohary 1973.

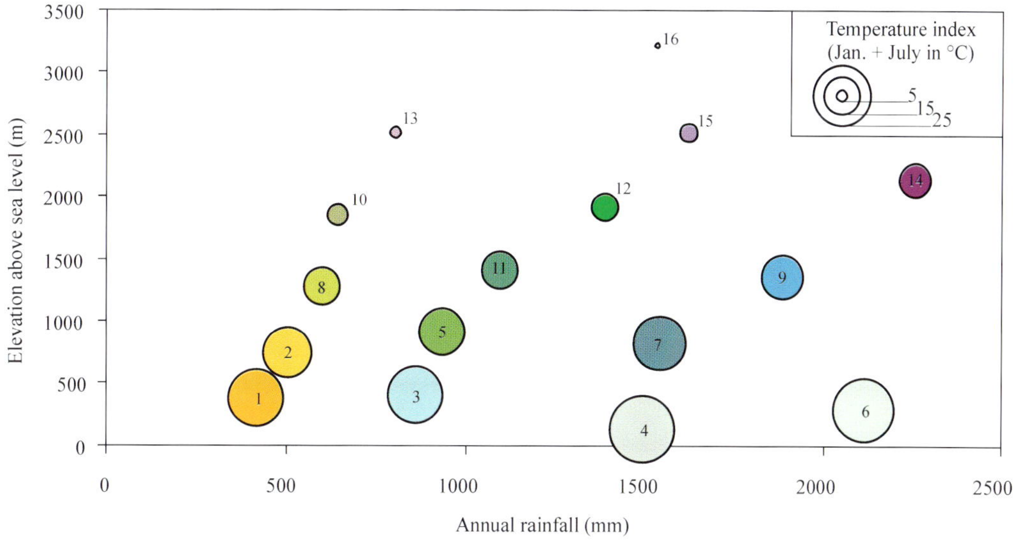

Figure 7.7. Average rainfall, elevation and temperature values for Georgian bioclimatic zones shown in Figure 7.8.

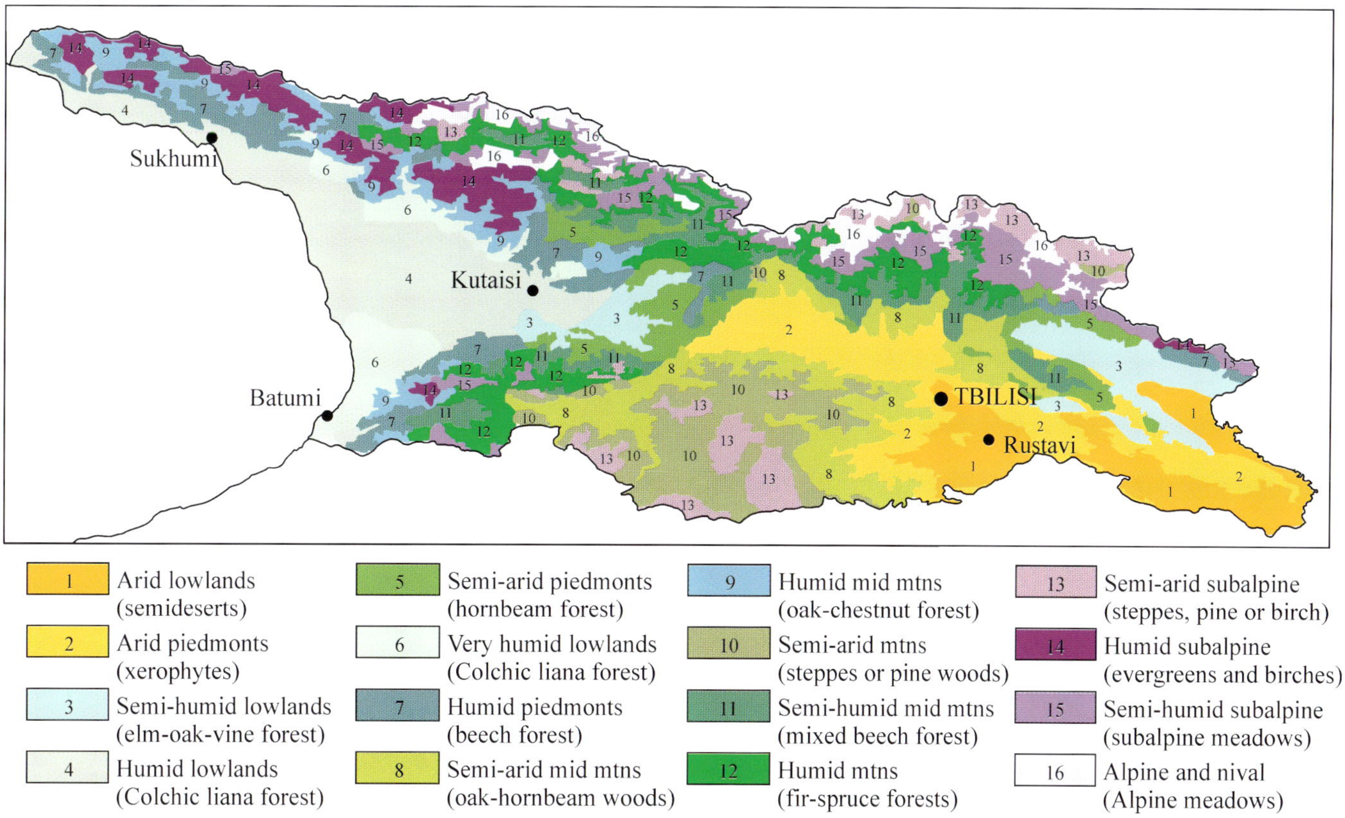

Figure 7.8. Bioclimatic zones of Georgia. Major vegetation units corresponding to each zone are given in parentheses.

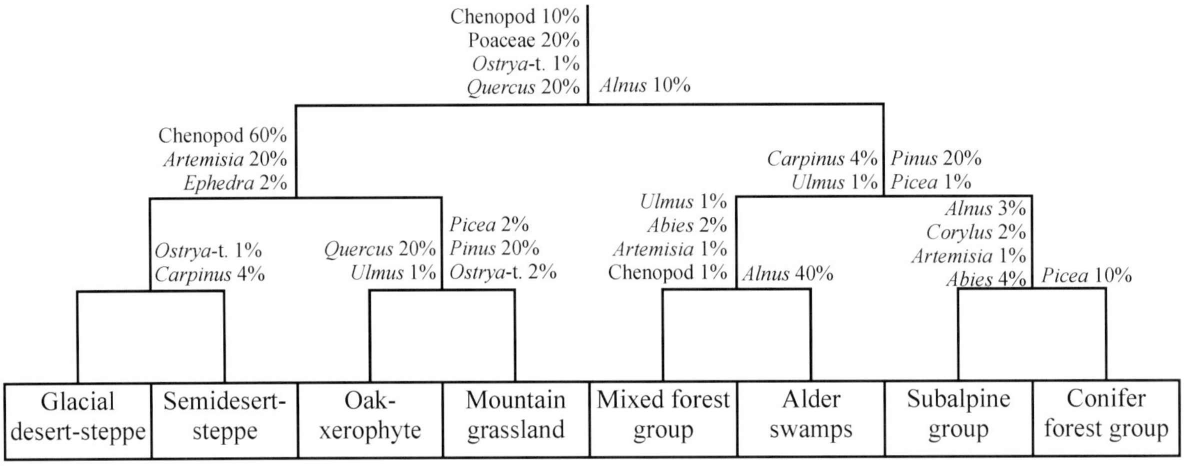

Figure 7.9. TWINSPAN classification dendrogram for the Georgian pollen dataset.

The Irano-Turanian group is divided on the basis of *Ephedra, Artemisia* and Chenopodiaceae proportions. The group with more of these taxa is a steppe and semi-desert group, while the other group has analogues in oak forests, xerophyte scrub and mountain grasslands.

Eight sub-groups have been chosen for mapping because these most simply and accurately reflect broad patterns in the study area's present-day vegetation. The groups are:

- Glacial desert-steppe (no modern analogue)
- Semidesert-steppe
- Oak-xerophyte group (oak forest, open woods and xerophytic scrub)
- Mountain grassland
- Mixed forest group (beech, hornbeam, elm, fir and chestnut forests)
- Alder swamp-forest
- Subalpine vegetation
- Coniferous forest group (forests and open woods of pine and spruce)

The present-day map of these groups (**Fig. 7.10**), despite some gaps, clearly portrays the major vegetation types in the study area. Alder swamps are located on the Black Sea coast, and above these is a zone of mixed deciduous forest, with alpine vegetation on the Main Caucasus and Anticaucasus Ranges. The coniferous forests of the Akhaltsikhe-Bakuriani area are clearly delineated, and east of these we pass over the treeless mountain grasslands of the South Georgian Uplands, through the narrow oak

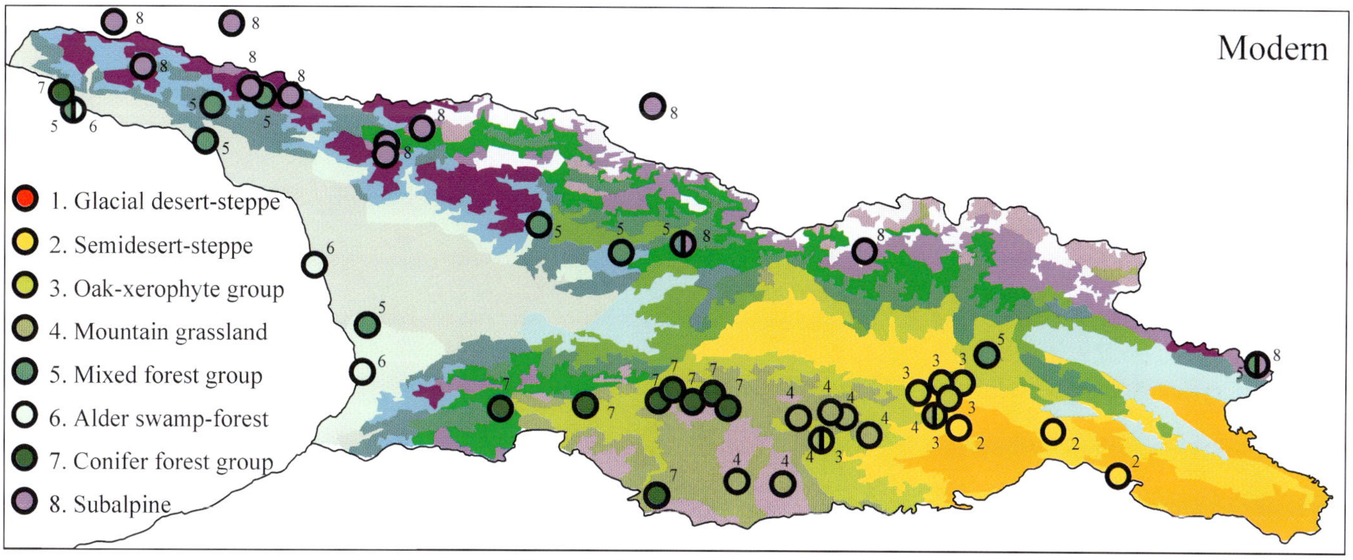

Figure 7.10. Present-day reconstructed vegetation at pollen sites in Georgia.

forest belt, along the xerophytic lower tree line into the semidesert-steppe region of Eastern Georgia.

Around 9 percent of the samples have a pollen signal that is not entirely consistent with their present-day vegetation, and these are listed in **Table 7.4**. In general, the misclassifications are easily explained by long-distance pollen transport in mountain areas, by the site having a larger pollen source area than accounted for in the vegetation description provided by the authors, the mixing of pollen from different altitudinal vegetation belts, or by sampling surface sediments from lake edges rather than the centre. The misclassifications tend to push the pollen-based vegetation units upwards because of pollen transport to higher elevations.[35]

The palaeovegetation maps that follow are limited by the paucity of sites in the late Pleistocene, early Holocene and the arid parts of the study area, as well as problems with dating where there are few absolute dates available. Not all vegetation types will be captured in the samples examined, and the successional transitions between some of them will be missed because of the 1000-year time interval chosen. Nevertheless, I believe they provide a good overview of the general stages of late Quaternary palaeovegetation development in the study area.

[35] Prentice *et al.* 1996.

Table 7.4. Misclassified pollen-vegetation relationships.

Site and sample no.	Source	Pollen-based vegetation	Observed vegetation
Lagodekhi 6	KE	3. Oak-xerophyte group	7. Subalpine
Cherepanov 2-3	C	4. Mountain grassland	3/2. Oak forest, mountain grassland
Imnati core top	N	5. Mixed forest group	6/5. Alder swamps, mixed forests
Lisi 2	KS	5. Mixed forest group	4/2. Mountain grassland, steppe
Lagodekhi 1, 2, 4, 5	KE	5. Mixed forest group	7. Subalpine
Urukh 109, 111	K	6. Alder swamps	5/8. Mixed and conifer forests
Akhaltsikhe 183	K	8. Conifer forest group	4. Mountain grassland
Ktsiis Zeda Vake	M	8. Conifer forest group	4. Mountain grassland
Nariani	M	8. Conifer forest group	4. Mountain grassland
Kartsakhi	M	8. Conifer forest group	4. Mountain grassland

Sources: C = Connor *et al.*, 2004; K = Klopotovskaia, 1973; KE = Kvavadze and Efremov, 1995; KS = Kvavadze and Stuchlik, 1991; M = Margalitadze, 1995; N = Neishtadt *et al.*, 1965.

Palaeovegetation Development from 14,000 cal BP to the Present

During the late Pleistocene, from 14,000 to 11,500 cal BP, mixed deciduous forests grew in the Black Sea lowlands, providing refuge to thermophilous trees in *Abies*-dominated vegetation (**Fig.** 7.11). At the same time in Southern Georgia the landscape was one of hyper-arid semidesert-steppe, a community that apparently extended far to the west, into what is now a 'humid mountain zone' where *Picea* forests grow today.

In the early Holocene the mixed deciduous forests in Abkhazia changed to alder forests, perhaps as *Abies* and other trees migrated upslope and the lowlands were paludified by rising watertables fed by glacial meltwater. Glacial desert-steppe vegetation persisted into the early Holocene at several sites in Southern Georgia, suggesting that the climate remained arid until after 10,000 cal BP, when desert-steppes gave way to grasslands and alpine meadows.

Nine thousand years ago, steppic vegetation had vanished from the study sites. A grassy oak-xerophyte community populated the Tsalka Plateau. Further west, on the moist slopes of the Trialeti Range, subalpine meadows gave way to mixed deciduous forests, while in Abkhazia and Colchis, alder forests and mixed forests were probably widespread.

At 8000 cal BP we can begin to see the differentiation of the vegetation on geographical lines – the mixed deciduous forests of Western Georgia and the oak woodlands of the East, both hemmed in by subalpine vegetation on the surrounding high mountains. This situation changed surprisingly little during the next few thousand years. Apart from the appearance of alder swamps at various locations in Colchis, the vegetation of the mountains remained stable from 8000 cal BP until after 5000 cal BP.

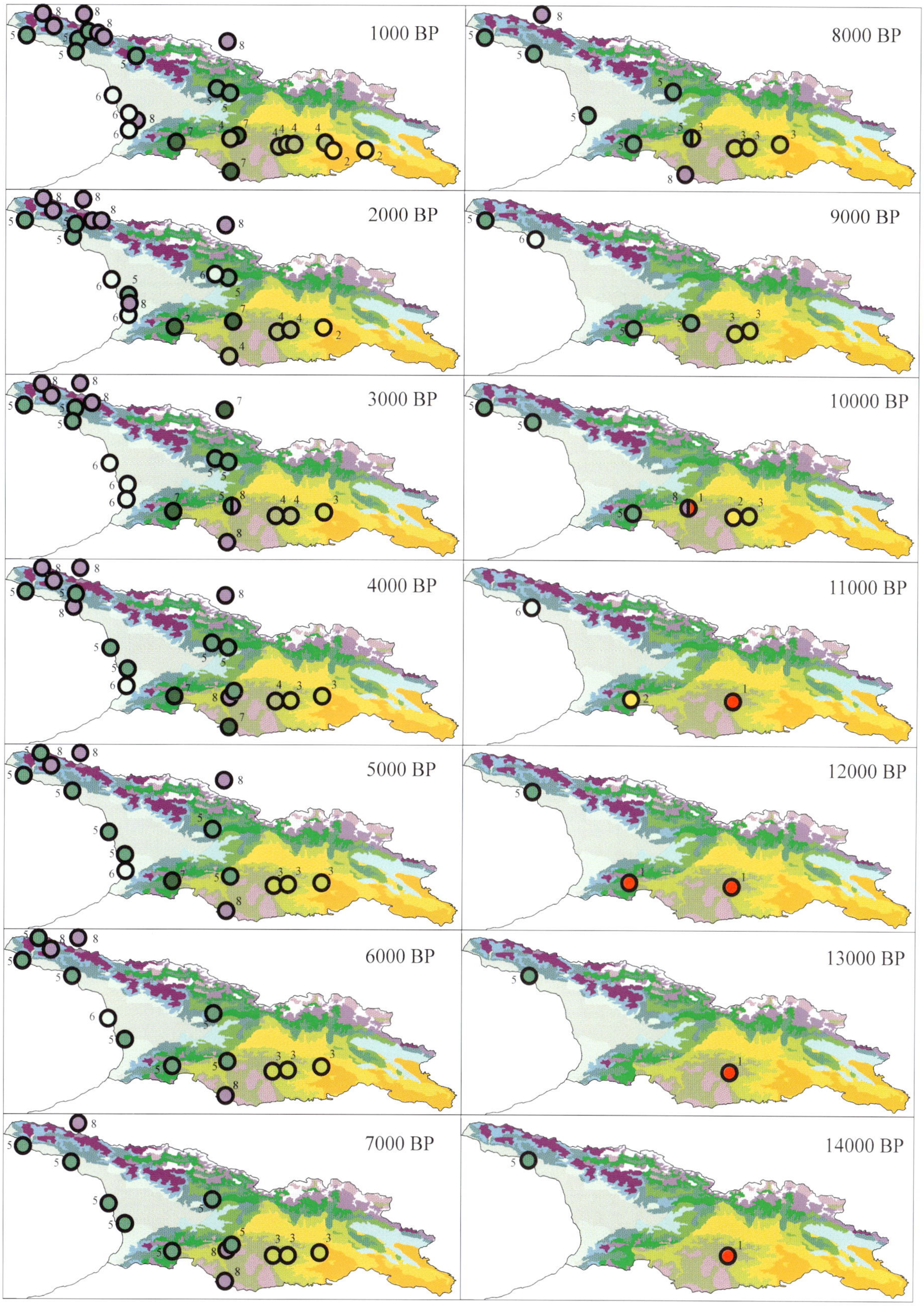

Figure 7.11. Palaeovegetation maps for Georgia – 14,000 to 1,000 cal years before present. Symbols are the same as those used in Figure 7.10.

At 5000 cal BP, coniferous forest appears on the maps for the first time, in the mountains of Ajara. By 4000 cal BP, this community had spread eastwards at the same time as mountain grassland vegetation appeared in Southern Georgia and areas of subalpine vegetation had expanded to encompass some sites formerly in mixed deciduous forest.

A thousand years later the area of *Alnus* swamps along the Black Sea coast had grown considerably as sea levels rose. Mountain grasslands expanded on the Tsalka Plateau, and subalpine vegetation in the Bakuriani and Javakheti areas. Coniferous forest appeared briefly at Salkanalla in the North Caucasus.

Two thousand years ago the landscape of the entire study area had changed dramatically. No longer did it resemble the simple divisions of the early Holocene. Oak-xerophyte vegetation disappears from the map, replaced by steppe vegetation at lower altitudes and mountain grasslands in elevated places. Coniferous forest invaded the Bakuriani area, formerly home to mixed forests and subalpine meadow vegetation. 'Subalpine' vegetation appears at one site on the sea coast, probably as a result of deforestation or fluvial pollen transport. Only in Abkhazia does the vegetation remain the same.

By 1000 cal BP, mountain grasslands had enveloped the highlands of Southern Georgia, grading into steppe vegetation in the east and open pine forest in the Javakheti region. The Colchis coastline was completely paludified and mixed forests confined to the piedmonts and mountainsides. The last 1000 years saw the expansion of alder swamps and subalpine vegetation in Abkhazia, the re-invasion of coniferous forests in the Bakuriani area, and the re-establishment of oak-hornbeam forest near Tbilisi.

Several points of interest emerge from this analysis. Firstly, the desert-steppes of the late Pleistocene lingered on in the continental parts of the study area for at least a thousand years after the Holocene began. This suggests a slow change from hyper-arid to more humid conditions. Secondly, deciduous trees were dominant in forests almost everywhere during the early-mid Holocene, with coniferous forest appearing only 5000 years ago in Ajara and subsequently spreading through Southern Georgia. It seems that the spread of conifers was stalled by conditions during the early-mid Holocene and accelerated during the late Holocene. Thirdly, the late Holocene appears to have been a time of rapid and fundamental reorganisation in Georgia's vegetation. For example, the Tsavkisi site (Chapter Five), classified as oak-xerophyte vegetation for most of its 8000-year history, changed to steppe 2000 years ago and mountain grassland 1000 years ago, before reverting to oak-xerophyte in recent years. A drawback of this classification is that it cannot adequately describe the *density* of past vegetation units, such that the oak-xerophyte group includes oak forest as well as arid scrub and highland savanna. Such nuances are better accentuated by the numerical analyses in previous chapters.

In the following section, the same dataset is used to provide some insight into the reasons for Georgia's changing palaeovegetation. By estimating past climatic parameters, the relative roles of climate, human impact and other factors in explaining vegetation change can be more objectively evaluated.

Holocene Climatic Change in Georgia – a Modelling Experiment

A significant branch of recent palaeoenvironmental research has aimed to reconstruct quantitative estimates of past climate from palaeoecological data.[36] In all of these studies, the modern analogue is the cornerstone of the reconstruction.

But it must be admitted that in spite of its attractiveness, the modern analogue technique has many inherent problems. Reconstructing climatic parameters carries the assumption that modern pollen spectra are in equilibrium with the current climate and that fossil pollen spectra were in equilibrium with past climate.[37] The modern analogue technique is also unable to adequately address problems of palaeoecological 'convergence' and 'divergence',[38] where different vegetation units may produce similar pollen assemblages due to pollen representation and taxonomic factors (convergence) or when the same vegetation unit produces pollen assemblages that differ due to source-area and fossilisation factors (divergence). And the problem of 'no-modern-analogue' pollen spectra is well documented.[39]

In areas where vegetation dynamics are well understood, some authors have avoided the no-analogue problem by using forest simulation models. By comparing simulated palaeovegetation with fossil pollen data, these models can be 'tuned' to accurately estimate past climatic parameters.[40]

Vegetation dynamics, in a modelling sense, are poorly understood in the Caucasus region, so modern analogues provide the most appropriate basis for palaeoclimatic reconstruction. Despite its constraints and assumptions, the climatic trends that can be produced from pollen data are very useful for hypothesis testing. Pollen-derived precipitation and temperature trends can be compared to similar reconstructions derived from other proxies, such as chironomids, tree rings, geochemistry and stable isotopes. In the future, comparisons such as these will bring about a more refined understanding of the palaeoclimate and help to both inform and improve reconstructions based on pollen data.

[36] Cheddadi *et al.* 1998; Guiot 1990; Magny *et al.* 2001; Peyron *et al.* 2000; Prebble and Shulmeister 2002; Sepp and Birks 2001; Solovieva *et al.* 2005.
[37] Birks and Birks 1980; Jackson and Williams 2004.
[38] Jackson and Williams 2004.
[39] Anderson *et al.* 1989; Jackson and Williams 2004; Overpeck *et al.* 1985.
[40] Hall and McGlone 2001.

Method

For the purposes of estimating past climatic parameters, I have adopted an approach that combines ordination scores and elevation data in a model based on non-parametric multiplicative regression.[41] Rather than being a direct method, which links pollen spectra to modern analogues via a distance metric, this method attempts to circumvent problems of no-analogue, divergent and convergent spectra by linking past spectra to *gradients produced by modern analogues*. It can therefore be conceptualised as an indirect method.

To begin with, the data set described in this chapter was reduced to a few relevant trends using detrended correspondence analysis (DCA – see Chapter Three). This provided ordination scores for each fossil and modern sample. The result is shown in **Figure 7.12**. Next, correlations between the modern sample scores and environmental variables were sought. Precipitation was positively correlated with DCA axis 1 ($r^2 = 0.55$); July and January temperature were positively correlated with DCA axis 2 ($r^2 = 0.10$ and 0.08) and negatively correlated with DCA axis 3 ($r^2 = 0.10$). Temperature reconstructions, therefore, must be viewed with much more scepticism than rainfall estimates.

The next stage of the method involved the habitat-modelling computer program Hyperniche.[42] This program builds non-linear models of species abundance based on predictors. In this case, the predictor variables were input as modern DCA scores, while the response was the modern climate dataset.[43] The model form selected was Local Mean Non-Parametric Multiple Regression (LM-NPMR) with a Gaussian weighting function.

Hyperniche searches for models that best fit the relationship between the ordination scores and climatic variables. Selection of the 'best' model involved a trade-off between goodness-of-fit and parsimony: often a model with a lower xR^2 (cross-validated R^2) value and broader tolerance gave a better prediction from patchy data. The selected models (**Fig. 7.13**, **Table 7.5**) were then used to estimate palaeoclimatic parameters for fossil pollen data based on site elevation and ordination scores. Data were mapped in ArcGIS,[44] and points were interpolated by kriging based on the exponential semivariogram model, which seems to be the most appropriate model for pollen dispersion.[45]

[41] McCune and Mefford 2004.
[42] McCune and Mefford 2004.
[43] AN GSSR 1964.
[44] ESRI 2005.
[45] Davis 2000; Johnston 1998.

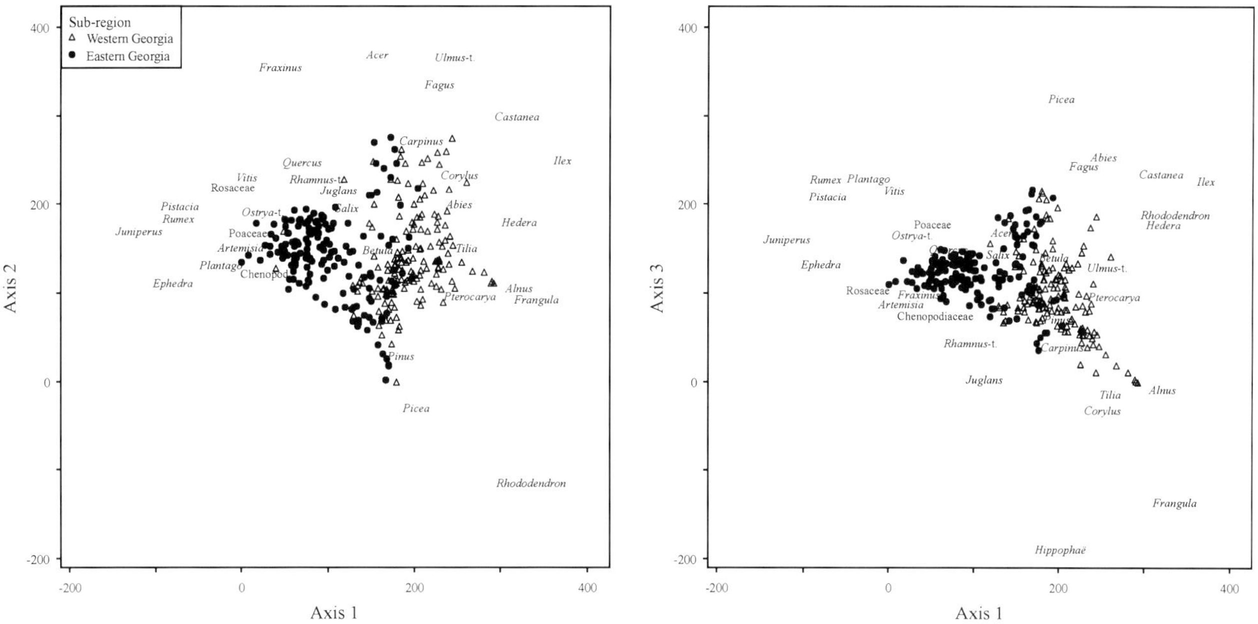

Figure 7.12. DCA ordination of the Georgian pollen dataset. Result for axes 1 and 2 (left) and axes 1 and 3 (right).

Table 7.5. Predictors used in the selected models, tolerances of each, and correlation coefficients for the three models. ANS: Average Neighbourhood Size (a measure of model flexibility).

Model	Predictors (and tolerances)	ANS	Fit (xR^2)
Annual precipitation	Axis 1 (36.2), Axis 2 (68.6) and Elevation (458)	16.5	0.81
July temperature	Axis 3 (29.6) and Elevation (305)	18.8	0.72
January temperature	Axis 1 (60.3), Axis 2 (68.6) and Elevation (305)	16.3	0.68

Results

Scatterplots of observed climatic parameters versus predicted values are shown in **Figure** 7.14. Predicted rainfall values are quite close to the rainfall data obtained from the Georgian Atlas.[46] Most overestimates relate to high *Picea* percentages in the Akhaltsikhe Basin of Southern Georgia. Underestimates occur at sites where Chenopodiaceae are disproportionately common due to long-distance pollen transport (in

[46] AN GSSR 1964. Data in the Georgian Atlas are modelled from a discontinuous network of meteorological stations and may not be entirely accurate representations of climate at a given point. Despite this constraint, these data are still the best available for Georgia as a whole.

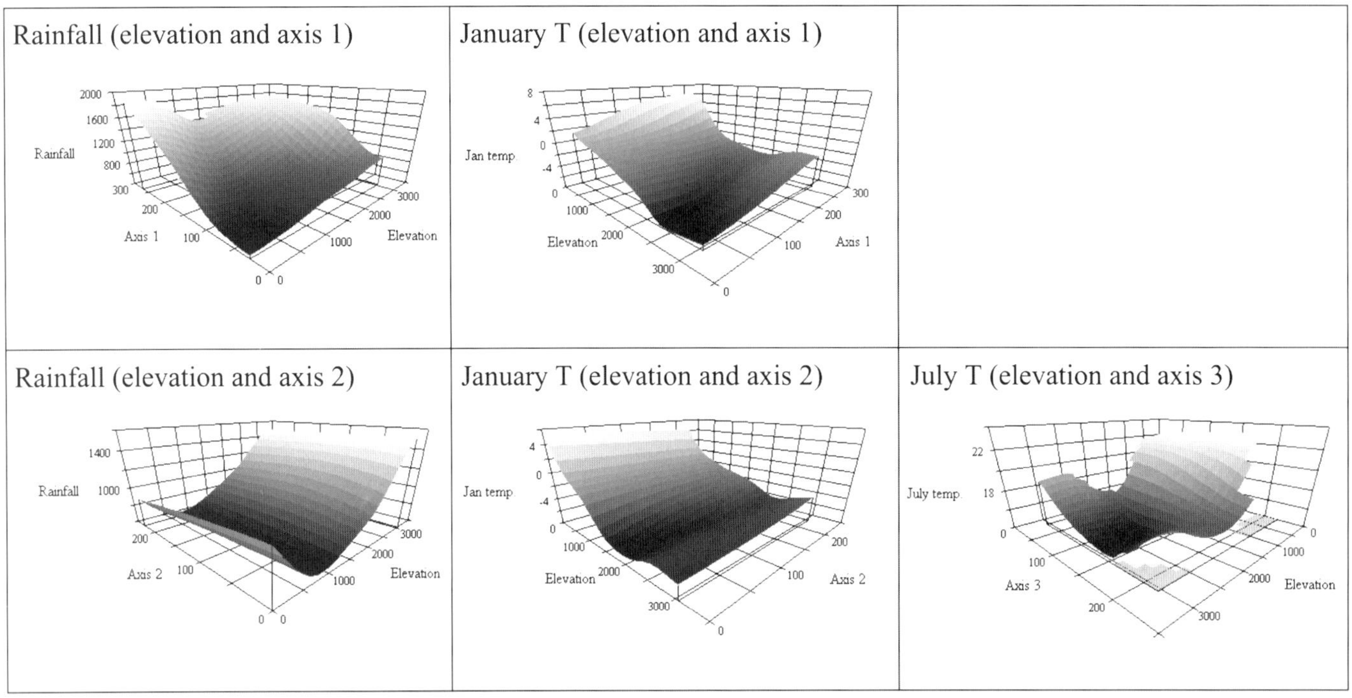

Figure 7.13. NPMR models used to derive palaeoclimatic estimates from DCA scores and elevation. The graphs are three-dimensional response surfaces, showing modelled values (on the vertical axis) compared to predictors (diagonal axes).

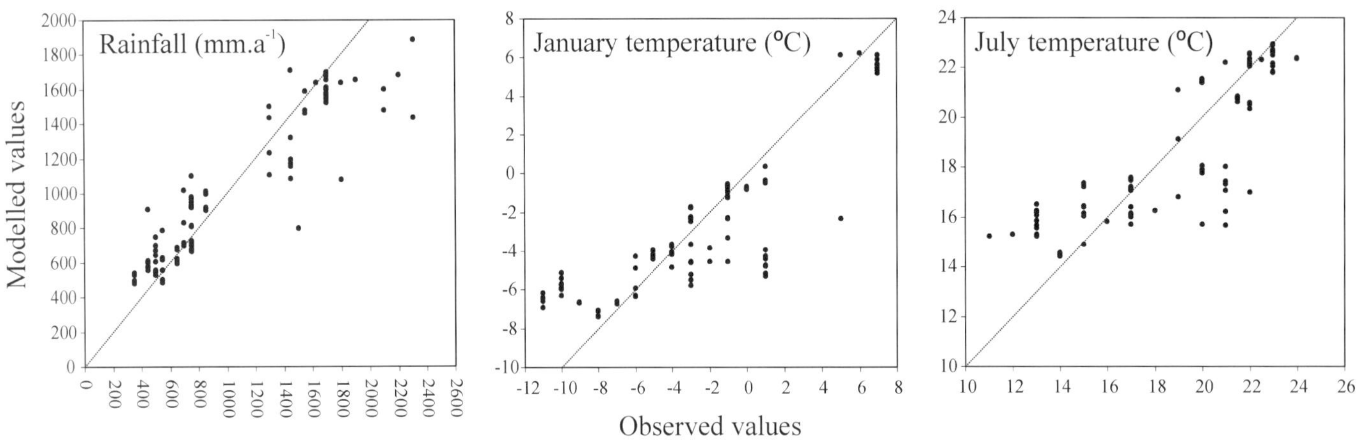

Figure 7.14. Scatterplots of modelled versus observed climatic parameters for modern data. Dotted line on each graph indicates a one-to-one relationship.

alpine sites) and local disturbance (in forested sites). Temperature predictions are overestimated in some treeless sites because of pollen transport from lower altitudes; temperatures are underestimated in places where conifer pollen is dominant. Even though the correlation coefficients produced by this method are equivalent to those in studies that employ the 'best modern analogue' technique,[47] temperature reconstructions are not reported here because of the weak relationship between ordination scores and temperature parameters. Much of the variance explained by the temperature models must be simply due to the inclusion of elevation as a predictor, which allows the model to mimic altitudinal lapse-rates. The stronger relationship between rainfall and ordination scores (r^2=0.55) suggests that past rainfall trends may be reconstructed with greater confidence and, for this reason, only those reconstructions are discussed below.[48]

Figure 7.15 shows reconstructed precipitation patterns for the last 12,000 years. The reconstructed pattern for the present-day, accounting for the patchy distribution of sites, quite closely resembles the actual pattern (**Fig.** 2.7). A certain loss of detail can be ascribed to interpolation between data points, which does not take altitudinal variations into account. The high rainfall zone of Western Georgia, and especially the mountain slopes of Ajara and the Main Caucasus Range, is clearly portrayed. The arid lowlands of Eastern Georgia are also accurately delineated in this reconstruction. Somewhat less certainty surrounds the gradation between the arid and humid zones, as there are no published investigations of lake or wetland pollen sites for the Shida Kartli and Imereti regions.

If we turn now to the reconstructed history of rainfall for the Holocene, 12,000 years ago arid conditions prevailed across Southern Georgia with somewhat moister climates in the Black Sea refugium. Late-glacial rainfall values are higher than expected and this is probably due to the Local Mean estimator in the model.[49] Aridity waned through the early Holocene, with a pattern resembling the modern one established between 9000 and 8000 cal BP. The lead up to the mid Holocene saw a partial contraction of the arid zone, with higher rainfall penetrating further eastward than it does today. By 5000 cal BP, however, the distinction between the humid west and arid east had sharpened.

Over the next few thousand years, the rainfall pattern in Western Georgia appears to have moved from the south to the north, producing drier conditions in Ajara 3000 years ago. Increased precipitation is predicted for 2000 cal BP, with higher rainfall

[47] *Cf.* Cheddadi *et al.* 1998; Guiot 1990; Magny *et al.* 2001, *etc.*

[48] Experimental results of temperature reconstructions can be found in Connor 2006.

[49] The Local Mean estimator tends to compress values at the extreme ends of the predictor variables' range, such that high and low DCA scores (*e.g.* late Pleistocene) tend to be pushed closer to those in the centre. When the analysis was re-run using a Local Linear estimator instead, the lack of analogues for the late Pleistocene resulted in negative rainfall estimations in some cases and lower xR^2 values.

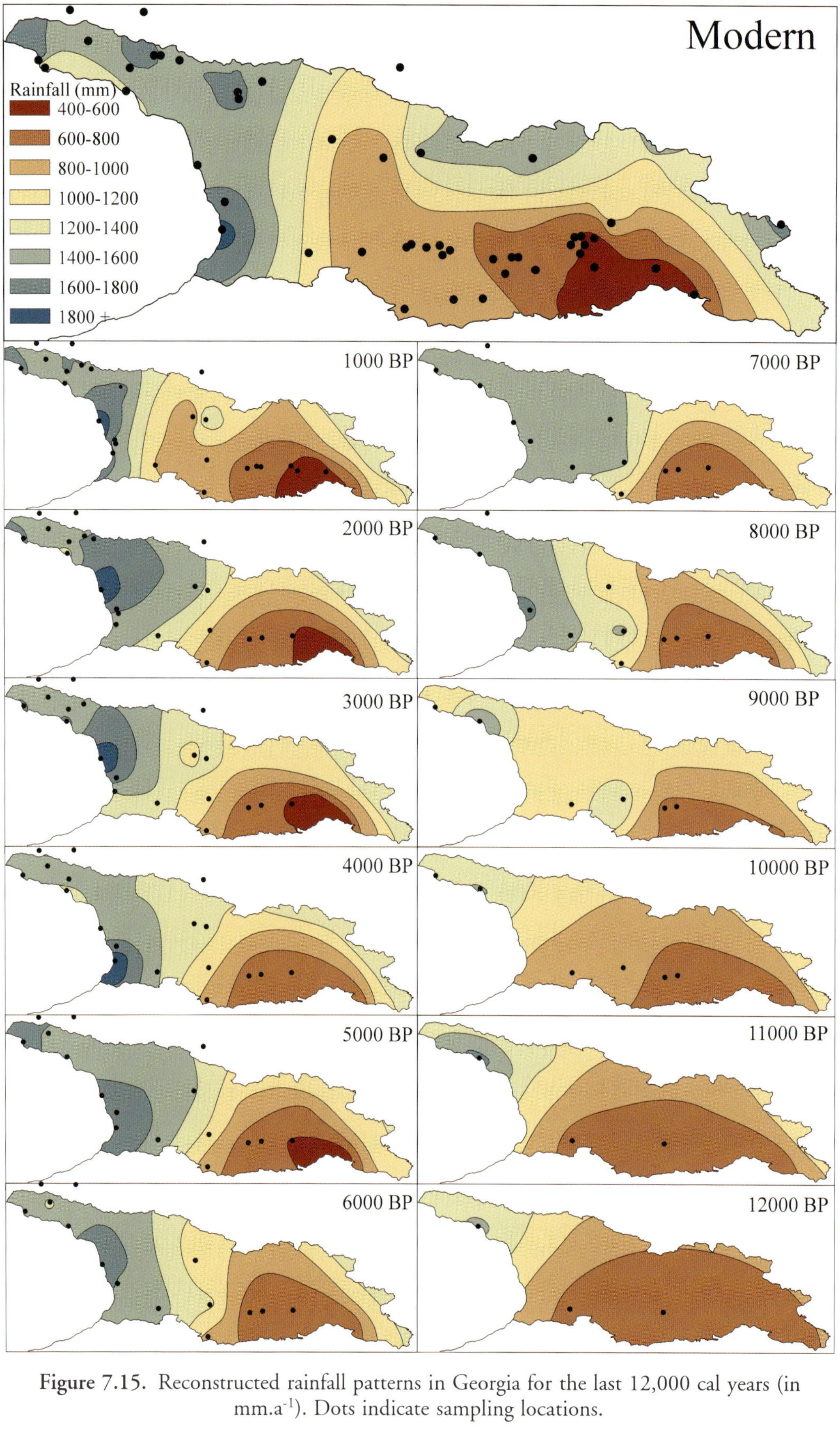

Figure 7.15. Reconstructed rainfall patterns in Georgia for the last 12,000 cal years (in $mm.a^{-1}$). Dots indicate sampling locations.

throughout Western Georgia. The Mediaeval Period saw a contraction of the humid zone and a consequent expansion of the drier zone westward, producing much steeper precipitation gradients in Western Georgia. A similar pattern is reconstructed for the present-day, with a return to high rainfall values in Ajara.

Local Trends

Maps of reconstructed climate are useful for tracing major spatial shifts over time, but give only a coarse indication of change at individual sites. By applying the same models to pollen data from previous chapters, palaeoclimatic trends can be reconstructed for the Tbilisi area, Tsalka Plateau and Black Sea coast. This has already been attempted for the Tbilisi and Tsalka sites in Chapters Five and Six, although the trends produced by weighted averaging were difficult to interpret because temperature, rainfall and forest cover were all interrelated. And yet these trends were produced from complete pollen assemblages and with reference to good modern analogues in Southern Georgia; as such they provide a semi-independent proxy for assessing the palaeoclimatic models put forward in this section.

Figure 7.16 shows reconstructed rainfall trends for four sites with long Holocene records – Aligol, Imera, Tsavkisi and Ispani. The trends for the two Tsalka Plateau sites, Aligol and Imera, are very close, indicating aridity through the late Pleistocene, a gradual increase in rainfall through the early Holocene, peaking 6000 years ago, before declining 5000–3500 years ago, then rising again until 2000 years ago and fluctuating

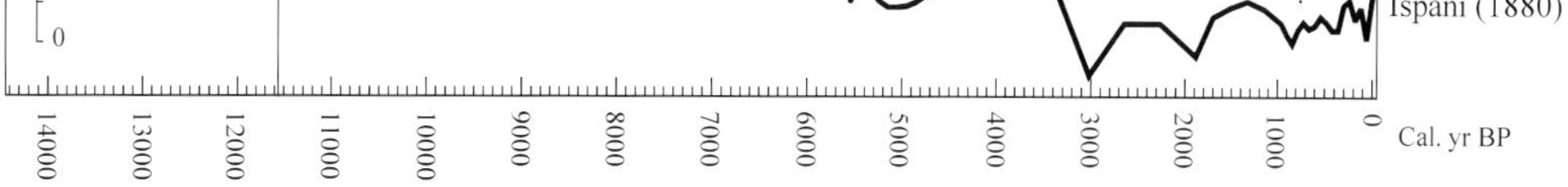

Figure 7.16. Modelled Holocene palaeoclimatic (rainfall) trends based on pollen data from Southern Georgia, presented in order of altitude. Figures in parentheses beside site names are present-day reconstructed values. Vertical scales for each climatic parameter are given at left (vertical scale on the Ispani rainfall curve has been reduced by ten times). The dotted lines are Weighted Averaging environmental trends from Chapters Five and Six included for comparison.

subsequently. Interestingly, the Tsavkisi trend follows this one not at all. Rather than decreased rainfall 5000–3500 years ago, it exhibits increased values, and declines thereafter. In this respect the Tsavkisi rainfall trend is closer to the trend reconstructed for Ispani, on the Black Sea coast, and to the Lake Van oxygen isotope record.

A degree of uncertainty surrounds these reconstructed trends and their interpretation. Late Pleistocene estimations are particularly difficult to accept, although in relative terms they probably reflect the *direction* of change. The modelling method presented here is experimental and to my knowledge has not been used previously in the palynological literature; better results may have come from applying the method to a simpler region, where palynological difficulties associated with mountainous and semi-arid environments are minimised and the climatic dataset is more reliable. And because these reconstructions are all based on the same pollen dataset, there is a danger of constructing circular arguments in support of past climatic changes. It is very difficult to disentangle the late Holocene conifer ingression, for instance, from possible human impacts and the effects of progressive soil impoverishment.

I merely offer these reconstructions as a possible model for Holocene climatic change in Georgia. As such, they are testable hypotheses, and I hope that in future alternative sources of palaeoclimatic data from the Caucasus will be brought to both inform and improve them.

Conclusions

The foregoing chapter attempts to chart a new course through the Holocene vegetation history of Georgia, from the late Pleistocene's upland deserts and lowland refugia, via the early-mid Holocene's advancing forests and expansion of thermophiles in the mid Holocene, through the rapid vegetation changes of the mid-late Holocene to the varied, anthropogenic landscapes of today.

The maps confirm the widespread aridity and high seasonality of the late Pleistocene, the more moderate, humid climate of the early-mid Holocene, followed by warmer mid Holocene conditions and more variable, generally cooler climates during the late Holocene. Climate change has thus exerted a profound influence on the course of vegetation development over the past 14,000 years.

But the present-day complexity of the study area's vegetation must be explained, in part, by the long and varied impact of humans. Through burning, grazing, cropping and felling, humans have disrupted the successional pathways of the beech forests of Western Georgia and the oak woods of Eastern Georgia, giving rise to a number of plant communities that seem to have been absent during the early and mid Holocene.

In the following chapter, the comparative roles of climate change and human impacts on vegetation development in Southern Georgia will be evaluated in the light of independent palaeoclimatic and archaeological data.

CHAPTER EIGHT

HUMAN IMPACT ON THE VEGETATION OF SOUTHERN GEORGIA

> During the Holocene, human activity has become one of the major influences on the vegetation cover. Its role has increased unremittingly, especially in recent times. Extensive areas are now covered in cultural vegetation. The most fertile forest and steppe soils were developed the earliest, particularly under oak trees and in oak-hornbeam woods. Virgin steppe soils have vanished. And, with the irrigation of ever-expanding areas of Eastern Transcaucasia, semideserts formerly used as winter pastures are increasingly displaced by agriculture. Forest areas have been reduced, and their structure and composition have changed.[1]

Dolukhanov was one of the few botanists to acknowledge the long and varied impact of human activities on Caucasian vegetation. His view stands in stark contrast to the majority, for whom human impact is conceptualised as a very recent phenomenon.[2]

And yet evidence from around the world suggests that the antiquity and extent of past human impacts may have been severely underestimated by ecologists and historians alike. Grove and Rackham[3] have eloquently argued that the Mediterranean landscape is not the result of recent and catastrophic human impact, but the diverse and colourful product of a long symbiosis between vegetation, humans, livestock and fire. Andrič and Willis[4] analysed pollen data from different floristic regions in Slovenia, showing that the early Holocene vegetation was remarkably uniform; only when human impact intensified did the vegetation of different areas diverge to create today's rich and distinct landscape mosaic. In the semi-arid rangelands of Western Asia, Roberts[5] has suggested that humans had a hand in delaying early Holocene forest expansion by burning the vegetation, thus preserving open landscapes since prehistoric times. Essentially the same argument has been invoked to explain the persistence of fire-promoted moorland vegetation in southwest Tasmania, where the climate and

[1] Dolukhanov 1966, p. 228.

[2] Badenkov *et al.* 1990; Gulisashvili 1964; Ketskhoveli 1959; Nakhutsrishvili 1999; Sakhokia 1961; Sosnovskii 1915; Tumajanov 1971; Walter 1974.

[3] Grove and Rackham 2001.

[4] Andric and Willis 2003.

[5] Roberts 2002.

soils are capable of supporting rainforest of much greater biomass.[6] On the Australian continent, biomass burning has a relationship with human history that began as early as 140,000 years ago.[7] In Indonesia, humans may have been burning the vegetation some 200,000 years ago.[8]

On a larger scale, the global fire simulations by Bond *et al.*[9] have demonstrated that many of the world's ecosystems would cease to exist in the absence of fire. And Ruddiman[10] has put forward the controversial suggestion that biomass burning beginning with early Eurasian agriculture slowly added carbon to the atmosphere, causing the global climate to diverge from the path expected by insolation effects and palaeoclimatic analogy. The 'greenhouse effect' thus began 8000 years ago. Following pandemics of bubonic plague, he argues, the afforestation of abandoned farmland sequestered enough carbon that temperature variations such as the Little Ice Age, to a surprising extent, can be explained in anthropogenic terms.[11]

Fire looms large in all of these studies. In many parts of the world, fire histories have been instrumental in understanding past changes in the vegetation, and particularly in identifying phases of human impact.[12] In Europe, fire histories are yet to be fully embraced in palaeoenvironmental research. This seems almost incomprehensible, since as early as 1941 Johannes Iversen used charcoal analysis to demonstrate that Neolithic people cleared European forests with fire.[13] This is the well-known *landnám* phase that often followed the 'elm decline'. With some notable exceptions, few European researchers since Iversen's day have paid much attention to the ecological role of fire.[14]

Unfortunately, this has led to the widespread perception that fire is bad, destroys habitats and turns forest into desert.[15] The neglect of fire history has also severely impeded our knowledge of past human impacts in Western Asia. Most pollen diagrams from this region lack charcoal records,[16] leaving open the question of how, when and if early agriculturalists cleared the land for the planting of crops. The Caucasus is no exception – there are no published fire histories from this region of great botanical and archaeological significance – yet fire is an important agro-pastoral tool in the Caucasus today.

[6] Fletcher and Thomas 2007.
[7] Kershaw 1994; Moss and Kershaw 2000; Singh and Geissler 1985.
[8] Wang *et al.* 1999.
[9] Bond *et al.* 2005.
[10] Ruddiman 2003.
[11] Ruddiman 2003.
[12] *E.g.* Carrión and van Geel 1999; Dodson and Mooney 2002; Edwards 1988; Naveh 1990; Patterson *et al.* 1987; Tinner *et al.* 1998; Willis *et al.* 1998; Yasuda *et al.* 2000.
[13] Iversen 1941.
[14] Tinner *et al.* 2005.
[15] Grove and Rackham 2001.
[16] See van Zeist and Bottema 1991.

One of the earliest European botanists to visit Georgia, Joseph de Tournefort, was given to remark that, in the highlands of Southern Georgia, "all manner of Trees have been long banish'd..."[17] This chapter draws heavily upon the archaeological and historical record of this ancient region to place the vegetation and fire histories of Southern Georgia in context, and in so doing, answer the question of how, when and if its trees were banished. This history is not without its problems: like any history it is selective in its scope; there are many uncertainties in relation to chronology and dating; good early Holocene data are lacking, while historical data for the late Holocene are often too detailed for fair comparison with the palaeoenvironmental record. Nonetheless, in the prophetic words of the archaeologist Charles Burney, "seeing through a glass darkly is surely better than not seeing at all".[18] In that spirit I present the following discussion, which is arranged in archaeological periods for the sake of convenience and because the relationship between environmental change and cultural change is often more suggestive than might at first appear.

Upper Palaeolithic (*c.* 30,000–8,500 BC)

The Palaeolithic takes in the great sweep of time from the appearance of hominids (**Fig. 8.1**) in the South Caucasus some 1.8 million years ago to the earliest centuries of the Holocene.[19] Perhaps the most significant cultural event through this time was the arrival of modern humans in the Caucasus region some 32,000 years ago and the displacement of Neanderthal populations.[20] These Upper Palaeolithic hunters established themselves in caves in the foothills and lowlands of Western Georgia, where the climate must have been more amenable to human habitation than further east where, as yet, Upper Palaeolithic finds are lacking.[21]

Various authors have suggested that, at the last glacial maximum, snow lines were depressed by 800–1000 m along the Main Caucasus Range,[22] valley glaciers descended to an altitude of 1000–1500 m above sea level in the west (compared to 2300–2700 m today), and to 1800–2400 m (2500–3200 m today) in the more arid east.[23] Tree lines in Western Georgia were perhaps depressed to 600–800 m above sea level,[24] pushing forest vegetation into the foothills and lowlands. Several Upper Palaeolithic sites in Western Georgia were thus located near the upper tree line; a fact reflected in

[17] Tournefort 1718, p. 227.
[18] Burney and Lang 1971. Paraphrased from St. Paul's first letter to the Corinthians, 13: 12.
[19] Gabunia and Vekua 1995.
[20] Tushabramishvili *et al.* 2002.
[21] Kiguradze and Menabde 2004; Vekua and Lordkipanidze 1998.
[22] Tsereteli and Maisuradze 1976.
[23] Janelidze *et al.* 2000; Serebryanny 1984; Tsagareli 1964. Very few glacial features in the Caucasus have been dated, so their exact age is unknown.
[24] Kvavadze *et al.* 1992.

Figure 8.1. Facial reconstructions of Georgia's earliest hominids, the so-called 'Dmanisi man' (left) and 'Dmanisi woman', photographed at the State Museum of Georgia in 2001.

faunal assemblages comprising both forest and steppe taxa: wild goat, bison, Caucasian tur, cave bear, wild boar, hamster, deer, auroch, mouflon, horse and others.[25] Other Upper Palaeolithic faunal assemblages are dominated by just one species, for example *Capra caucasica*, comprising 85 percent of remains at Ortvale Klde.[26] Modern humans seem to have been more selective in their diet than the Neanderthal populations they replaced,[27] and some caves were perhaps occupied only seasonally for the hunting of specific highland prey.[28] This brought humans into frequent conflict with other cave-dwelling carnivores – cave bears, cave hyenas and wolverines – all of which became extinct in the Caucasus during the late Pleistocene and early Holocene.[29] Many large herbivores, notably aurochs, bison, roe deer and moose, also became much less abundant in Georgia at the Pleistocene-Holocene transition.[30] Palaeolithic humans in the Caucasus probably exerted a strong and highly selective impact on the fauna, using the foothill cave sites to hunt in the open landscapes inhabited by large mammals.

[25] Bendukidze 1979; Tsereteli *et al.* 1982; Vekua and Lordkipanidze 1998.
[26] Tushabramishvili *et al.* 2002; Vekua and Lordkipanidze 1998.
[27] Tushabramishvili and Vekua 1977.
[28] Vekua and Lordkipanidze 1998.
[29] Bendukidze 1979; Dolukhanov 1980.
[30] Bendukidze 1979.

It would not be surprising, however, if the majority of Upper Palaeolithic human populations were clustered around the Black Sea coast. At its lowest stand, the sea level stood between 90 and 155 metres below its present level and the Black Sea was in fact a huge freshwater lake.[31] In the Georgian sector of the Black Sea basin, the shelf drops off steeply, so the Pleistocene coastline was only a few kilometres distant from the present one in most areas, making seasonal hunting excursions to the mountains far from impossible. The coastline of this huge lake no doubt had a milder climate and more abundant plant and animal resources than elsewhere, and hence may have provided a focus for human habitation in the late Pleistocene.

Around 14,500 years ago, a global cycle of increased solar radiation began in the northern latitudes, triggering a northward movement in the inter-tropical convergence zone and higher rainfall in the Sahara.[32] In Georgia, forests began to move upward and outward in Abkhazia,[33] arid juniper savanna spread through the highlands of Southern Georgia and Armenia,[34] and pine woodlands grew on the piedmonts of Eastern Georgia.[35] This process was curtailed by the Younger Dryas, which pushed vegetation zones downslope and saw plants like *Ephedra* and *Artemisia* expand in desert-steppe vegetation. Some authors have suggested that the main characteristic of the climate of Western Asia at this time was a switch to conditions of hyper-aridity rather than of cold.[36] Increasing aridity associated with the Younger Dryas is not evident in the palaeoclimatic trends reconstructed in Chapters Six and Seven, so some explanation is required.

The Younger Dryas was caused by a temporary interruption to the Gulf Stream, which carries warm water from the equatorial regions of the Atlantic Ocean northward to Europe. If the salinity of the Atlantic Ocean is diluted, or its connection to the Southern Ocean is cut off, the Gulf Stream loses its ability to transport heat to the North Atlantic, and Europe is plunged into glacial conditions.[37] As conditions in the North Atlantic currently have a strong influence on the winter climate of Western Asia,[38] it follows that an interruption to the thermohaline circulation should have a considerable seasonal impact in the Caucasus.

I would argue that, instead of hyper-aridity during the Younger Dryas, the continental interiors of the Caucasus and adjacent regions experienced a climate of extreme seasonality. Deprived of the moderating influence of the North Atlantic, winters must

[31] Ballard *et al.* 2000; Chepalyga 1984; Tvalchrelidze *et al.* 2004.
[32] Burroughs 2005.
[33] Kvavadze and Rukhadze 1989.
[34] Grichuk and Airapetian 1976. See also Chapter Six.
[35] Gogichaishvili 1984b.
[36] Lemke and Sturm 1997; Wick *et al.* 2003; Wright *et al.* 2003.
[37] Broecker 1994; Broecker 1997; Knorr and Lohmann 2003.
[38] Cullen *et al.* 2002.

have been very cold and dry. Summer conditions, however, were perhaps governed by the northward shift of the intertropical convergence zone and intensified monsoonal influences or thermal convection. Summers were thus hotter and perhaps more humid than now; a conclusion also arrived at independently by Roberts and Wright.[39]

The extraordinary seasonality and unpredictability of the late-glacial climate has been invoked to explain why agriculture was not adopted in Western Asia during the Pleistocene.[40] Yet Hillman *et al.*[41] have argued that these were precisely the conditions that induced people into cereal cultivation, providing evidence of probable rye cultivation at Abu Hureyra around 13,000 cal BP. In theory, conditions of high seasonality would necessitate the collection of foodstuffs that could survive long periods of storage.[42] Climatic conditions at the terminal Pleistocene allowed the area of wild cereals to expand in the Fertile Crescent,[43] and the considerable amounts of cereal pollen in records from Southern Georgia might suggest that this was a phenomenon of significant geographic extent.[44] The Caucasus is known as a centre of crop plant diversity[45] and it is possible that these wild cereals grew amongst the wetland fringing vegetation[46] where moist soils provided a kind of refuge from aridity. Did Upper Palaeolithic humans encounter these wild cereal stands in their peregrinations across the highlands in search of game? While there is no direct evidence from the Caucasus that these wild plant resources were exploited during the late Pleistocene,[47] in the pre-agricultural context, such evidence is likely to be very difficult to find.[48]

When the Younger Dryas came to an end, the renewed strength of the Gulf Stream brought milder winters to the Caucasus region and vegetation belts responded with steady upward migration (**Fig. 8.2**). Although rainfall increased at the beginning of the Holocene, the climate remained somewhat dry for the first few millennia, a phenomenon that Wright *et al.*[49] linked to high summer temperatures and drought produced by an orbital cycle of increased summer insolation. The widespread, early Holocene expansion of *Pistacia* in Western Asia has logically been interpreted as an indication of early Holocene dry climates with mild winters.[50] As in the Younger Dryas, the prevalence of *Ephedra*, *Artemisia* and Poaceae during the early Holocene

[39] Roberts and Wright 1993.
[40] Richerson *et al.* 2001.
[41] Hillman *et al.* 2001.
[42] Blumler 1996.
[43] Hillman 1996.
[44] That is, records from Lake Aligol (Chapter Six), Gomnis Lake (E. Gobet, unpubl.) and Didajaris Lake (J. van Leeuwen, unpubl.).
[45] Vavilov 1992.
[46] *Cf.* Bottema 1995.
[47] Lisitsyna and Prishchepenko 1977.
[48] Harris 1984.
[49] Wright *et al.* 2003.
[50] Rossignol-Strick 1995; Stevens *et al.* 2001; van Zeist and Bottema 1991; Wick *et al.* 2003.

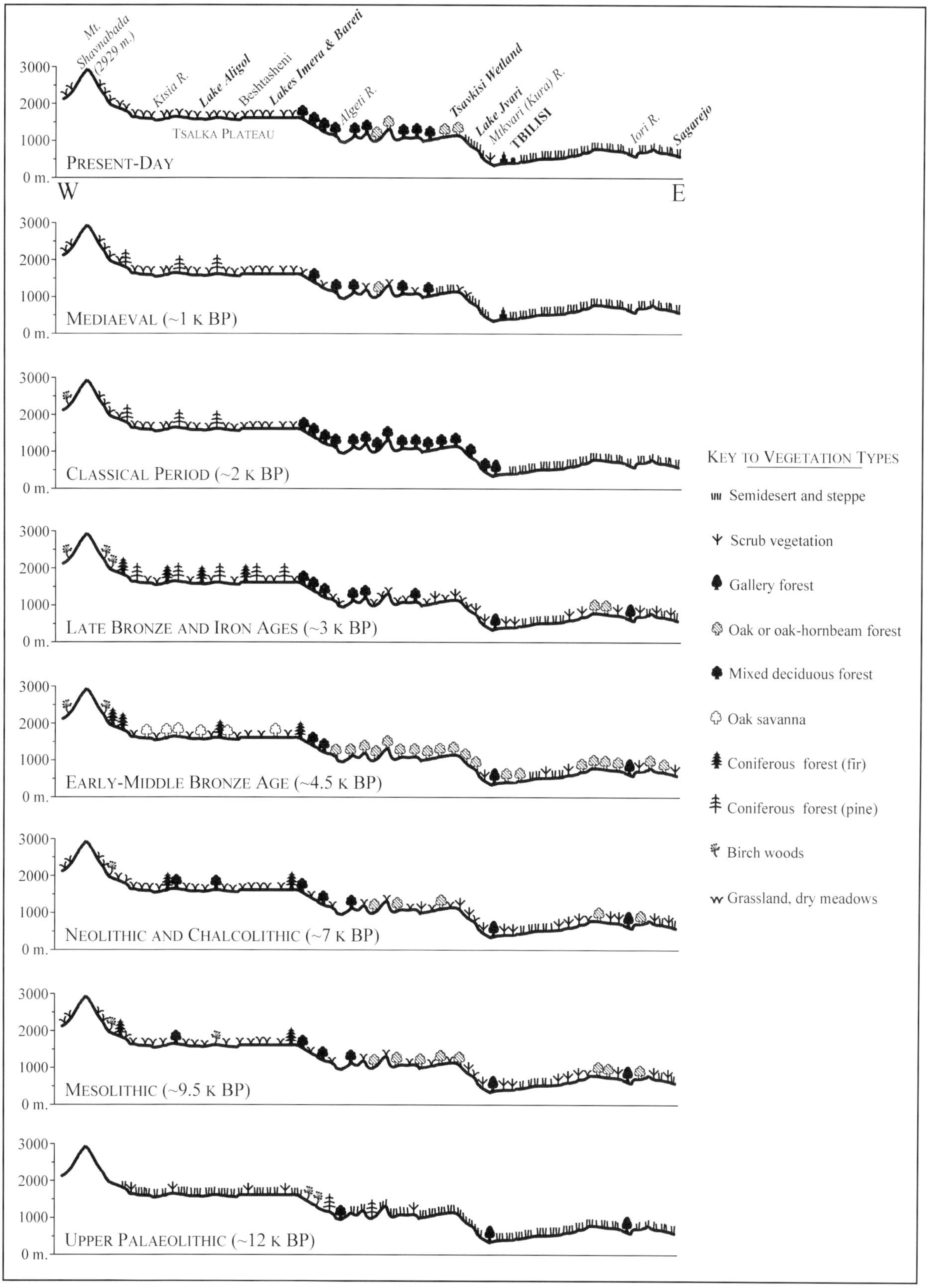

Figure 8.2. A generalised scheme of Holocene vegetation changes in Southern Georgia for selected archaeological periods. The location of the altitudinal transect is shown in Figure 8.3. Pollen sites and other features are shown on the uppermost transect (present-day).

may mitigate against this hypothesis, since these three taxa are not especially favoured by summer drought, though they can tolerate dry winters.[51] Roberts and colleagues noted the apparent contradiction between the supposed droughty summers of the early Holocene on the one hand, and the presence of mesic tree species and high lake levels on the other.[52] Palaeoclimatic trends from Southern Georgia also suggest that the summer drought hypothesis may need revision.

A possible alternative explanation is that the peak in Northern Hemisphere insolation during the early Holocene produced high summer humidity across Western Asia through thermal convection. Relative to the present climate, the concentration of annual rainfall in the warmer months would have been conducive to the development of steppic grassland and savanna vegetation in the continental interiors of Western Asia. *Pistacia*, which "tolerates protracted winter drought, had little competition from the more drought-stressed oak"[53] and thus became abundant. High lake levels could be maintained by low evaporation rates during summer, even if the winter rainfall input were considerably reduced. For reasons yet unclear, rainfall in the Caucasus and Western Asia increased quite slowly through the first millennia of the Holocene.[54] At lakes Van and Aligol, this led to the slow demise of *Ephedra* in the vegetation and the expansion of grasses, while at Didajaris Lake in the humid mountains of Ajara, woodlands of *Fraxinus* and *Quercus* began to expand.[55]

In summary, the Upper Palaeolithic period was one of climatic extremes. Summers were hot; winters were cold and especially arid. Humans adapted to this seasonality by congregating in Western Georgia and exploiting resources over large areas, especially in the treeless uplands. Although a clear impact of humans on the vegetation during the Upper Palaeolithic is difficult to discern, their impact on the fauna was much more clearly manifested. And compared to Western Europe at the same time, the Caucasus and much of Western Asia were home to considerable human populations; populations whose impact on the vegetation in the future would be considerable.

MESOLITHIC (*c.* 8500–7000 BC)

The period of the Caucasian Mesolithic was one of rapid environmental reorganisation throughout Western Asia. In Georgia, Mesolithic sites are not confined only to the Colchis lowlands and foothills, but are distributed through the highlands of Western Georgia, Southern Georgia, Armenia and Azerbaijan, and this is thought to represent a considerable population expansion.[56]

[51] Chikov 1983; El-Moslimany 1990; Rossignol-Strick 1995.
[52] Roberts *et al.* 2001.
[53] Stevens *et al.* 2001, p. 753.
[54] Lemke and Sturm 1997.
[55] J. van Leeuwen, unpublished data.
[56] Kiguradze and Menabde 2004.

Like the Upper Palaeolithic that preceded it, Mesolithic sites are concentrated in Western Georgia, but often at considerably higher altitudes (*e.g.* Tsona, 2100 m; Kudaro, 1600 m). Mountain caves that were abandoned during the coldest phases of the late Pleistocene were reoccupied during the Mesolithic.[57] Western Georgian Mesolithic sites are characterised by a suite of microliths and bone tools, and faunal remains indicative of forested surrounds.[58] Evidently the climate in Western Georgia had changed to the point where forests had expanded up the mountain slopes from their Colchian refugia, but not to the point where their present altitudinal limit was achieved.[59] Afforestation occurred rapidly in Western Georgia and may explain the increasing role of foraging, fishing and gathering in the Mesolithic economy at the same time as hunting declined in importance at several sites.[60]

A second group of Mesolithic sites occurs in Southern Georgia. These are mostly open settlements and are centred on the broader Tsalka Plateau (**Fig. 8.3**).[61] All of these sites are nowadays situated in mountain steppe landscapes. They are differentiated from their Western Georgian counterparts in the almost exclusive use of obsidian in the manufacture of stone tools. Faunal remains here also differ from those in Western Georgia; the Tsalka sites have yielded assemblages that indicate the prevalence of open vegetation – wild horse, onager, mouflon, aurochs, maral, Caucasian tur, gazelle and wild boar.[62] Pollen data from the Tsalka Plateau indicate that birch and hazel pioneer woodlands had begun to encroach on the uplands during the Mesolithic. In general, however, these environments remained relatively open, allowing Mesolithic people in Southern Georgia to continue hunting the same animals that had sustained their predecessors through the Upper Palaeolithic. According to Bendukidze,[63] human impact on Georgia's fauna increased sharply during the Mesolithic and intensified through subsequent periods.

During the late Mesolithic the waning influence of Pleistocene cold and aridity allowed Georgia's vegetation to anticipate its present form in several respects (**Fig. 7.11**). While the Colchian landscape remained much the same as ever, mesic forests of elm spread in the mountains of Ajara,[64] mixed woods of *Corylus, Carpinus, Ulmus* and *Fraxinus* established on the western flanks of the Trialeti Range,[65] and the same taxa became important around the Tsalka Plateau slopes and margins. In Eastern Georgia the early Holocene saw the disappearance of Pleistocene-relict populations of

[57] Dzhanelidze 1980.
[58] Bendukidze 1979; Kushnareva 1997.
[59] Kvavadze and Rukhadze 1989; Margalitadze 1995; Margalitadze 1998; Tumajanov 1971.
[60] Kushnareva 1997.
[61] Kikodze and Koridze 1978.
[62] Bendukidze 1979.
[63] Bendukidze 1979.
[64] J. van Leeuwen, unpublished data.
[65] Margalitadze 1995.

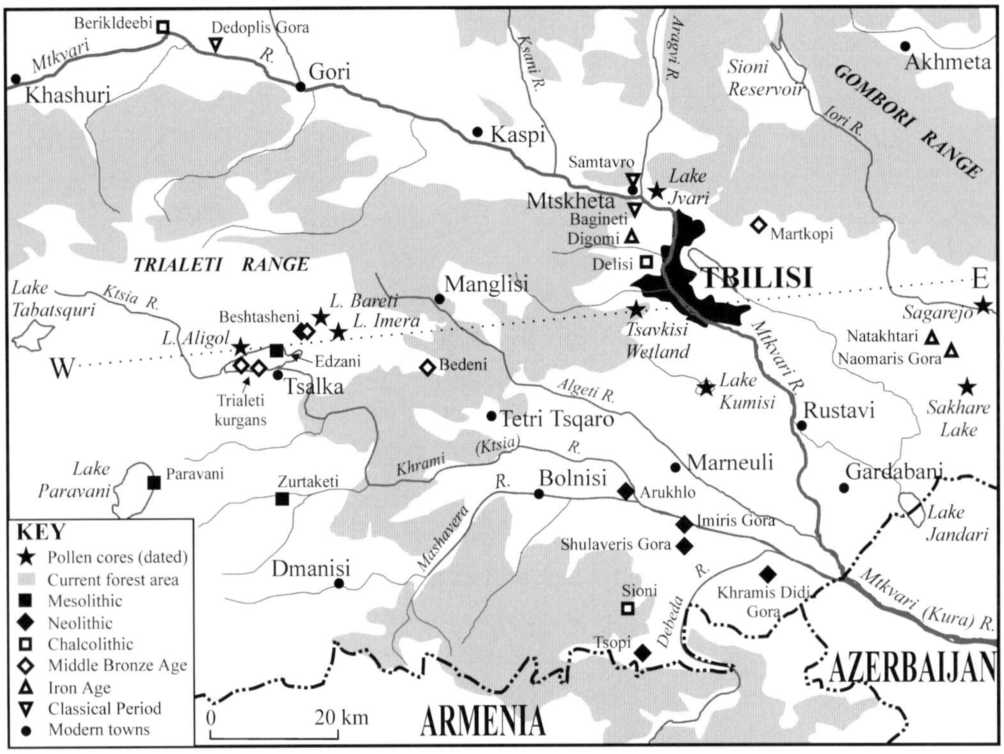

Figure 8.3. Map of Southern Georgia showing palynological sites and some of the archaeological sites mentioned in the text. Forest areas, modern settlements and the W-E transect in Figure 8.2 are also marked.

Pterocarya and *Juglans* from the Iori River gallery forests.[66] These thermophiles probably became extinct because of localised changes in floodplain sedimentation and geomorphology associated with deglaciation of the Main Caucasus Range.[67] If palaeoclimatic trends from Chapter Seven and Lake Van are interpreted literally, temperatures and precipitation fluctuated and rose slowly during the Mesolithic, producing conditions that might ordinarily favour *Pterocarya* and *Juglans*.

Toward the end of the Mesolithic, the appearance of obsidian tools in Western Georgia gives certain proof of active trade connections between communities living in lowland Colchis and those in Southern Georgia, where the main obsidian sources are located.[68] The emergence of this obsidian-based lithic industry in the eighth millennium BC and stronger interconnections with peoples south of the Caucasus is regarded as an important precursor to the Neolithic in Southern Georgia.[69]

[66] Gogichaishvili 1984b.
[67] See Appendix A.
[68] Kiguradze and Menabde 2004; Kikodze and Koridze 1978.
[69] Kiguradze and Menabde 2004.

NEOLITHIC (c.7000–4800 BC)

In the Caucasus, it is supposed that the earliest, aceramic Neolithic settlements appeared in the Black Sea lowlands of Western Georgia.[70] Changes in material culture, economic subsistence and settlement types around 7000 BC point to a sedentary existence more in keeping with the Neolithic than the preceding Mesolithic.[71] Neolithic Colchians eschewed the dankness of caves and instead established open-air settlements, often in close proximity to rivers.[72] Usually these sites have only very thin layers of cultural accumulation, so organic remains have rarely been preserved. In a few places, soil discolouration left by post-holes suggest they built rectangular, wooden or wattle-and-daub structures.[73] Archaeologists have distinguished two aceramic Neolithic traditions in Western Georgia on the basis of stone tool assemblages: the Anaseuli tradition, developed in the Colchis lowlands with a reliance on flint and obsidian blades, and the Paluri tradition, a stone flake industry occurring in the highland zone.[74]

Our picture of faunal exploitation in Western Georgia at this time is very limited and subject to considerable dating uncertainty. However, the appearance of domestic pig, sheep, cow and dog remains alongside red deer, roe deer and wild boar fossils in early Neolithic assemblages from Darkveti provide a glimpse of agricultural beginnings.[75] Pig remains, in particular, suggest sedentary human populations and rich forest resources of beech mast and acorns.[76] At other sites, wild herbivores and fish remained the predominant catch of aceramic Neolithic humans.[77] Palynological evidence from both lacustrine and archaeological contexts attests to the widespread occurrence of forest vegetation and expanding role of oak, hornbeam, elm, chestnut and *Zelkova* in Western Georgia during the Neolithic.[78] This occurred as the last vestiges of Pleistocene aridity faded away to produce a rainfall pattern of more or less modern character (**Fig.** 7.15).

Interestingly, the establishment of near-modern climatic conditions in Western Georgia coincides with changes in the level of the Black Sea. Evidence for an abrupt flooding of the Black Sea with saline Mediterranean waters around 6000 BC (7150 ^{14}C bp) comes from drowned strandlines, a change from freshwater to marine molluscs and other proxies.[79] As well as impacting on the regional climate, it has been

[70] Chataigner 1995; Gogitidze 1978; Kiguradze and Menabde 2004; Nebieridze 1972.
[71] Kiguradze and Menabde 2004.
[72] Chataigner 1995.
[73] Kiguradze and Menabde 2004.
[74] Grigolia 1977; Kiguradze and Menabde 2004.
[75] Bendukidze 1979; Nebieridze 1978.
[76] Kushnareva 1997.
[77] Dolukhanov 1980; Kushnareva 1997.
[78] Kalandadze *et al.* 1977; Kvavadze and Connor 2005; Kvavadze and Dzheiranashvili 1989; Kvavadze and Rukhadze 1989; Neishtadt *et al.* 1965.
[79] Ballard *et al.* 2000; Ryan *et al.* 1997; Uchupi and Ross 2000.

suggested that the rapid submergence of 100,000 km^2 of lowland landscapes may have contributed to the migration of Neolithic communities into adjacent regions.[80]

It may be more than coincidence that archaeological sites in Western Georgia see the first manifestations of the ceramic Neolithic around 6000 BC. These appeared initially along the northern coast of Colchis; spreading subsequently to encompass the entire Colchis lowland.[81] With them came more obvious signs of sophisticated foraging, gathering and perhaps cultivation; suggested by occasional finds of grindstones, hoes and other implements.[82] These communities produced a limited range of coarse ceramic jars, red or brown in colour and sometimes ornamented with simple, geometric patterns.[83] Although macrofossil evidence is exceedingly scanty, Caucasian archaeologists believe that the early agriculturalists of Colchis grew or gathered millet and rye[84] in preference to the traditional 'founder' crops of Western Asia – wheat, barley, flax and legumes.[85]

Leaving aside the Neolithic of Western Georgia, let us now turn our attention eastward. Until relatively recently it was assumed that Eastern Georgia became home to Neolithic communities somewhat later than Western Georgia, during the sixth millennium BC, when the so-called Shulaveri agricultural settlements began to appear on the alluvial soils of the Kura lowlands.[86] A recent re-examination of lithic assemblages suggests that the aceramic Neolithic is indeed represented in Southern Georgia: at Dmanisi, Lake Paravani and on the Tsalka Plateau at Beshtasheni.[87] All of these open-air sites are located in the present-day mountain grassland zone at altitudes up to 2100 m. While little is known about them apart from their obsidian-based lithic industry in the Paluri tradition, they bridge the chronological gap between the Mesolithic sites and Shulaveri settlements. Regrettably, no palaeozoological or palaeobotanical material is preserved at these sites.

This, it would seem, is all that can be said of the aceramic Neolithic in Southern Georgia. But some startling evidence for occupation and human activity in this area comes from an unexpected source – the pollen records of Lake Imera and Tsavkisi Wetland. Sediments from these sites contain increasing amounts of microscopic charcoal from the early Neolithic onward, suggesting frequent and repeated burning. Fires in the landscape of Southern Georgia apparently peaked around 6000 BC, a date coincident with the emergence of the late Neolithic in Georgia (**Fig. 8.4**). Any suggestion that these were natural fires can be quickly dismissed by examining the pollen

80 Ryan *et al.* 1997.
81 Gogitidze 1978; Nebieridze 1972.
82 Kushnareva 1997.
83 Kiguradze and Menabde 2004.
84 Kiguradze and Menabde 2004; Kikvidze 1975.
85 Zohary 1996.
86 Chataigner 1995; Kushnareva 1997; Munchaev 1975.
87 Kiguradze and Menabde 2004.

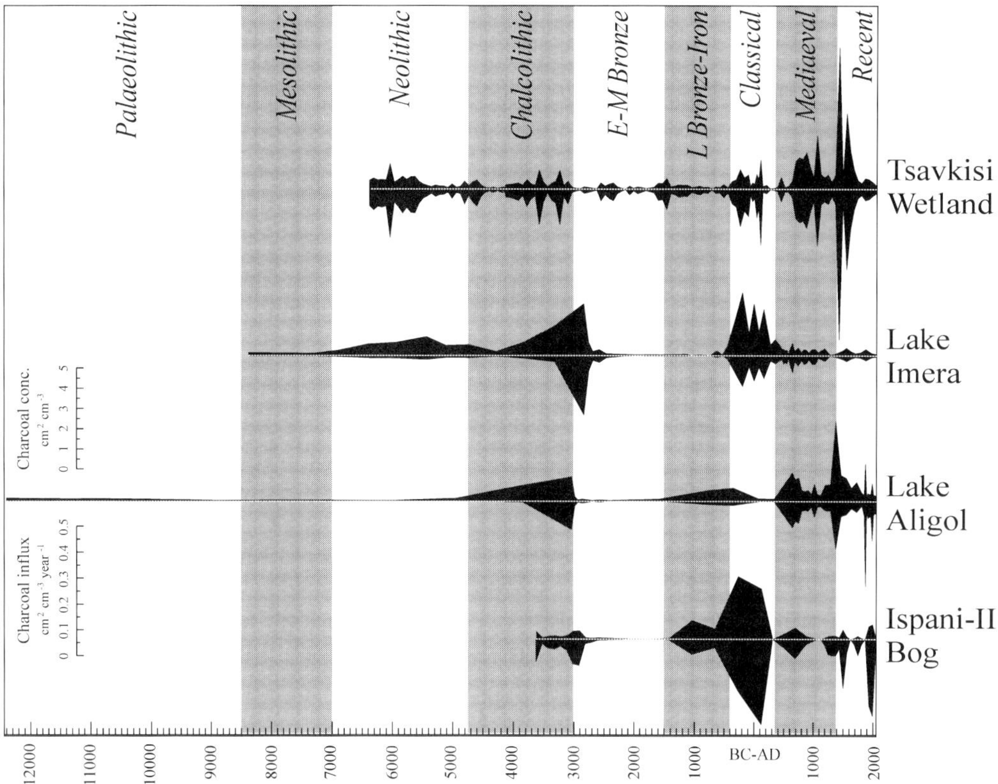

Figure 8.4. Fire records from Southern Georgia in relation to archaeological periods. Charcoal concentrations are shown above the line and influx below. The graphs indicate a Neolithic peak in burning, followed by a clear late Chalcolithic increase and a Bronze Age decline. Burning increased during the Classical Era and subsequent periods.

records themselves – this increase in vegetation burning corresponds to a very pronounced expansion of mesic forest taxa across the whole of Georgia (while the local vegetation of Imera and Tsavkisi remained conspicuously open) and a steady increase in precipitation throughout the region.[88]

Roberts[89] has recently and controversially put forward the suggestion that Neolithic humans deliberately used fire to delay the expansion of oak woodland in Western Asia. "In concert with a range of natural agencies," he argues, Neolithic and Chalcolithic populations acted "to retard the establishment and spread of woodland at an ecotone where moisture stress would always have limited plant growth".[90] Thus in the semi-arid rangelands of Western Asia, the ecological disjunction between increased precipitation

[88] Chapter Seven; see also Wick *et al.* 2003.
[89] Roberts 2002.
[90] Roberts 2002, p. 1009.

and stalled afforestation may be explained by humans manipulating the vegetation with fire. Here, in the semi-arid uplands of Southern Georgia, so ideally placed to test this assertion, is convincing evidence in support of Roberts' hypothesis.

The Neolithic desire to preserve open landscapes speaks volumes about the economy of that time. Even though we have no data on the subsistence strategies of the aceramic Neolithic people of the South Georgian Uplands, one might assume that they were primarily concerned either with hunting or livestock breeding. The great abundance of dung fungal spores in zone T-1 of the Tsavkisi Wetland pollen diagram (see **Fig. 5.19**) suggests that large herbivores were as common in that landscape during the Neolithic as at any time more recently. Whether wild or domestic, these animals provided a rich resource, and one that could be maintained with judicious use of fire. In this regard some things have changed little in Southern Georgia: herders today intentionally burn large tracts of the mountain plateaux each October to promote fresh grass growth in the following spring.[91] And, like their Neolithic counterparts, today's transhumant herders leave little trace of their seasonal camps when they return to the lowlands with their flocks and herds for the winter.

At the beginning of the sixth millennium BC, ceramic Neolithic communities began to settle in the lowland valleys of the Kura and Araxes: in southeastern Georgia, northeastern Azerbaijan and northwestern Armenia. Unlike the situation in Western Georgia, the people of the Shulaveri culture established clusters of farmsteads that were intensely settled over long periods. The multi-layered accumulation of mud bricks and other cultural debris at these sites produced conspicuous mounds that even today rise above the otherwise flat, alluvial plains.[92] The settlement types of the Shulaveri culture were much more closely linked to Anatolia than Colchis, which shared many features with the North Caucasian Neolithic.[93] Key late-Neolithic sites in Eastern Georgia include Shulaveris Gora, Imiris Gora and Khramis Didi Gora, which is the largest Neolithic site in the South Caucasus. Radiocarbon dated to the period from 5800 to 4600 BC, these sites are characterised by closely spaced agglomerations of dome-roofed, mud-brick roundhouses. Ceramics from the earliest levels of the Shulaveri settlements are notable for their rarity and crude manufacture.[94] Obsidian and bone implements, in contrast, are distinguished by their number, diversity and quality (**Fig. 8.5**).

The most celebrated aspect of the Shulaveri settlements is their clear evidence of agriculture. Several species of wheat, hulled and naked barley, rye, oats, millet, lentils, peas and beans have been identified in macrofossil assemblages from these sites.[95] It

[91] Connor *et al.* 2004.
[92] Connor and Sagona 2007; Kiguradze 1986; Kiguradze and Menabde 2004.
[93] Kushnareva 1997.
[94] Kiguradze and Menabde 2004.
[95] Lisitsyna and Prishchepenko 1977; Rusishvili 1990; Yanushevich and Rusishvili 1984.

Figure 8.5. Some stone tools from early agricultural sites in Southern Georgia. The one on the right is from Khramis Didi Gora, the middle one from Gadachrili Gora and the black obsidian blade at left is from Imiris Gora.

has often been claimed that these finds, particularly in relation to wheat, represent the independent development of agriculture in the Caucasus.[96] The appearance of domestic grape pips at Shulaveris Gora, Khramis Didi Gora and Shomu Tepe appears to support Georgia's claim to be the home of wine-making and viticulture; but, as with wheat, this remains an open question. There is some merit in the idea that the Shulaveri settlements were originally sited to take advantage of wild cereal stands, and in such circumstances local domestication would be far from impossible.[97] Evidence reported from some aceramic Neolithic sites in the Armenian highlands suggests that wild cereals and wild fauna were exploited prior to and during the adoption of domestic varieties.[98] Against independent agricultural development in the Caucasus are the obvious material links between the Shulaveri and other contemporaneous cultures of Western Asia,[99] as well as the fact that the crops gathered by the Shulaveri communities were essentially the same as those planted further south.[100] Until more archaeological work

[96] Kushnareva 1997; Lisitsyna 1984; Lisitsyna and Prishchepenko 1977.
[97] Hillman and Davies 1990.
[98] Mellaart 1975, p. 201.
[99] Glumac and Anthony 1992.
[100] Zohary and Hopf 1993.

is undertaken with a greater attention to radiocarbon dating, this issue will remain unresolved. Suffice it to say that the early agricultural settlements of the Shulaveri type have a strong local character no doubt influenced by the environment in which they developed.

The surroundings of the Shulaveri settlements are relatively well studied. Wild plant remains from Arukhlo and Imiris Gora are entirely those of herbs and forbs still common in the steppe vegetation – species of *Polygonum, Lithospermum, Sideritis, Trifolium, Papaver, Chenopodium, Amaranthus, Rumex* and various grasses.[101] Pollen analysis of cultural deposits and irrigation ditches at these two sites showed Chenopodiaceae, Asteraceae, *Artemisia*, Lamiaceae and Poaceae to be the most common pollen taxa, with tree pollen all but absent.[102] Faunal assemblages from the Shulaveri settlements contain abundant remains of domestic cattle, sheep, pigs and goats.[103] Far fewer and more diverse were the remains of wild fauna: red deer, wild sheep, onagers, aurochs, wild horses, gazelles, wild boar and to a lesser extent birds, small mammals and tortoises.[104] These three, independent lines of palaeoenvironmental evidence point to the persistence of open vegetation in the Kura River lowlands during the Neolithic.

The exceptionally rich pollen record from Tsavkisi Wetland exhibits some noteworthy changes around the time of the Georgian late Neolithic. Of particular interest is the substantial peak in *Triticum*-type (wheat) pollen coincident with a decline in grazing and fire (see **Figs 5.19** and **9.3**). This might suggest that cropping had extended well into the foothills, pushing the more established practice of herding to higher altitudes. Why would this be the case? Perhaps intense demand for cereals in the Shulaveri communities made cropping on the rugged foothills worthwhile. Or, as the orbital influence of higher summer insolation began to fade during the mid Holocene, summer drought became more common in Eastern Georgia's lowlands and necessitated cultivation in more elevated places. Evidence for artificial irrigation is also known from this period.[105] Whatever the answer, it is clear that socio-economic, political or environmental events at the end of the Neolithic prompted a move away from a sedentary, agrarian lifestyle to one where herding assumed a greater role.

Chalcolithic (*c.* 4800–3000 BC)

It has been argued that a general cultural continuity stretched from the Caucasian Neolithic until the end of the Chalcolithic period around 3000 BC.[106] During this period a socio-economic structure was gradually established that served highland

[101] Lisitsyna and Prishchepenko 1977.
[102] Dzhanelidze 1980; Gogichaishvili 1984a; Gogichaishvili 1990.
[103] Dzhanelidze 1980; Vekua 1984.
[104] Bendukidze 1979; Vekua 1984.
[105] Mardaleishvili and Janelidze 1984.
[106] Chataigner 1995.

communities of the subsequent Early Bronze Age. The Chalcolithic, or Eneolithic as it is also called, in the South Caucasus includes the key site of Sioni, as well as Delisi, Berikldeebi, Alikemek Tepesi, Leila Tepe and Tekhut.[107] A series of recurring cultural traits and patterns define the Chalcolithic of the eastern South Caucasus, known collectively as the Sioni culture.[108] This itself can be classified into two phases: an early Chalcolithic phase stretching from 4800 to 4000 BC, and a middle to late Chalcolithic phase covering the 4th millennium (*i.e.* 4000–3000 BC). It was the later of these two phases that gave rise to the ubiquitous Kura-Araxes culture, aspects of which begin to manifest more clearly from 3600–3500 BC.[109]

Compared to the clustered distribution of Neolithic settlements, Chalcolithic sites are scattered more widely and at various altitudes. The two periods also differ architecturally. The multi-layered mound sites of the Neolithic gave way to Chalcolithic sites that were generally flat, pock-marked with bell-shaped and cylindrical pits, and poor in architectural remains – mostly small wattle-and-daub dwellings.[110] Chalcolithic ceramics are characterised by a relative sameness of form and finish. Mostly jars and bowls, they were built by hand, usually from clay tempered with inclusions of grit, mica, grog or chaff. Chalcolithic ceramics from Georgia usually have either a pale finish (Sioni pottery) or are burnished to a lustrous black.[111] It is the wide distribution of this red-black pottery that gives certain indication of strong cultural and socio-economic interactions among the highland communities of the South Caucasus and Eastern Anatolia through the Chalcolithic, the same interactions that saw Georgian obsidian traded throughout the region.[112]

Complementary palaeoecological analyses have shown that wheat, barley and grapes were probably the most commonly grown crops.[113] In comparison to the Neolithic, however, the role of grain agriculture is thought to have become subsidiary to livestock breeding, particularly the transhumant herding of sheep and goats.[114] Chalcolithic communities were by no means nomadic, but may have moved back and forth between lowland and highland pastures with the changing seasons.[115] While expanding trade connections across the highlands may have encouraged the herding of livestock (a commodity that walks to market), it could also be argued that changing environmental conditions contributed to the shift toward a pastoral economy.

[107] Kiguradze and Sagona 2003.
[108] Menabde *et al.* 1978.
[109] Kiguradze and Sagona 2003.
[110] Kiguradze and Sagona 2003.
[111] Kiguradze and Sagona 2003.
[112] Kiguradze and Sagona 2003; Palumbi 2003.
[113] Lisitsyna and Prishchepenko 1977.
[114] Dzhanelidze 1980; Piotrovskii 1955.
[115] Kiguradze and Sagona 2003.

Numerous palaeoenvironmental analyses show that the region experienced a mid Holocene peak in precipitation between 4000 and 2000 BC.[116] Annual rainfall increased and temperatures rose during the Chalcolithic, leading to an upward migration of altitudinal vegetation belts perhaps hundreds of metres above their present level.[117] Around Lake Van in Eastern Anatolia, oak woodland (forest-steppe) reached its maximum extent, corresponding to a peak in lake levels.[118] Higher rainfall may have rendered lowland floodplains uninhabitable or unpredictable,[119] leading to a shift away from cereal-based agriculture.

In addition, orbitally driven changes in insolation led to the desiccation of the Sahara and the reorganisation of regional weather systems around this time.[120] The same probably had a significant impact on the seasonality of rainfall in semi-arid Southern Georgia. Even though annual rainfall was increasing through the Chalcolithic, much of this increase came from enhanced westerly airstreams originating in the North Atlantic.[121] In consequence, winters became milder and wetter, while summers became hotter and drier (see Chapter Seven). These conditions would have enhanced drought susceptibility and crop failure in the steppic lowlands. In such circumstances, the highland zone may have provided a more stable base for cropping and animal husbandry.

The extension of forest vegetation and upward migration of tree lines clearly posed an economic challenge to highland communities as arable land was gradually consumed by woodland. Chalcolithic farmers responded to this afforestation in the same way as their Neolithic predecessors had – with fire. Charcoal records from across the Tsalka Plateau to the hills of Tbilisi clearly demonstrate the degree to which humans were deliberately maintaining open landscapes to sustain their pastoral economy (**Fig. 8.4**). It should be emphasised that these fires were not indiscriminate, nor were they universal. There is no suggestion that Chalcolithic farmers destroyed established areas of forest. Their interest must have been in sustaining an agro-pastoral system of ancient inheritance, and this was achieved by using fire (and the livestock themselves) to prevent afforestation. Toward the end of the Chalcolithic, the effort expended in maintaining open pastures seems to have intensified. Then, as the period came to a close around 3000 BC, the fires that had burned with ever-increasing frequency and intensity through the preceding millennium were abruptly snuffed out.

[116] Gogichaishvili 1984b; Margalitadze 1995; Margalitadze 1998; Roberts and Wright 1993; van Zeist and Bottema 1991; Wick *et al.* 2003.

[117] Kvavadze and Connor 2005; Kvavadze and Rukhadze 1989; Margalitadze 1998.

[118] Lemke and Sturm 1997; Wick *et al.* 2003.

[119] *Cf.* Oguchi and Oguchi 1998.

[120] Goodfriend 1991; Ritchie *et al.* 1985.

[121] Roberts and Wright 1993.

Archaeologists speak of a regional 'collapse' around 3000 BC that saw the abandonment of some sites and loss of complexity at others.[122] This seems to have been more a political or economic event than a cultural one, for the same traits that emerge during the latter part of the Chalcolithic are carried through uninterrupted to the Early Bronze Age. It seems that a significant reorganisation of the region's political and economic sphere occurred around 3000 BC. In places, pastures that were so assiduously maintained during the Chalcolithic were overrun by trees during the Bronze Age. Fire was banished from the landscape and even grazing seems to have declined. All of these changes relate not to any abrupt climatic event, but to a technological one – the arrival of Bronze Age metallurgy.

Early and Middle Bronze Age (*c.* 3000–1500 BC)

The Early and Middle Bronze Ages have been the subject of great archaeological interest over the past few decades, giving rise to the definition of a series of 'cultures' known, in chronological order, as Kura-Araxes, Martkopi, Bedeni and Trialeti.[123] By 3000 BC, the communities of farmers and stock-breeders that emerged during the late Chalcolithic were fully established across the highlands of the South Caucasus and eastern Anatolia. The distribution of these sites in Georgia in many ways resembles present-day settlement patterns, with a significant emphasis on highland environments.

Despite some clear regional traits, the Kura-Araxes cultural assemblage is overwhelmingly homogenous, suggesting a high degree of social cohesion across the mountain rangelands.[124] This cultural assemblage in Georgia included a diversity of architectural styles, ranging from wooden-framed, free-standing rectangular houses with wattle-and-daub walls to more typically Near Eastern, mud-brick dwellings of rectilinear or circular plan. Within these varied forms, the internal use of space was remarkably uniform, each with a centrally placed terracotta hearth set into the floor. The universal occurrence of red-black, burnished pottery of this time is considered the hallmark of the Kura-Araxes culture.[125] Metallurgy also developed at this time, attested by bronze artefacts such as axes, spearheads, earrings, dress-pins and other ornaments.

Around 2500 BC, the communities of Southern Georgia began to adopt elaborate burial customs that contrasted starkly with the simple inhumations favoured by the Kura-Araxes culture of the Early Bronze Age.[126] Monumental barrow tombs

[122] Kiguradze and Sagona 2003.
[123] Burney and Lang 1971; Glumac and Anthony 1992; Kuftin 1941; Kushnareva 1997; Marro and Hauptmann 2000; Munchaev 1975; Rubinson 1977; Sagona 1984; Sagona 2004; Sagona 2004; Smith and Rubinson 2003; Zhorzhikashvili and Gogadze 1974.
[124] Sagona 1984.
[125] Burney and Lang 1971.
[126] Sagona 2004.

('kurgans') were constructed for the burial of important persons, suggesting fundamental changes in social structure. Burial chambers were either dug into the ground and covered with planks or built entirely of oaken logs.[127] In each case the chamber was covered with a tumulus of stone and earth rising up to 15 m high, making these mounds prominent features in the otherwise flat landscapes of Martkopi, Bedeni and Tsalka. The burials themselves were accompanied by symbols of power and wealth: vessels of precious metal, jewellery, pottery, weaponry, furs, fabrics and wheeled vehicles.[128] The most elaborate examples of this mode of burial are the later Trialeti kurgans from the highland Tsalka Plateau. Amongst the opulent funerary offerings was an embossed silver goblet, with depictions of humans and animals in a style thought to reflect the Tsalka Plateau's elevated standing in sphere of trade and exchange stretching as far as Assyria and the eastern Mediterranean.[129]

While the mortuary architecture and customs of the Middle Bronze Age are well known, the same cannot be said of the economy of the period.[130] This is largely attributable to the apparent absence of settlements, though it could be argued that this 'absence' is in fact a chronological misinterpretation.[131] The perceived lack of settlements has promoted the idea that Middle Bronze Age communities were transhumant rather than sedentary.[132] Although this is possibly a regional phenomenon, recent evidence from Sos Höyük in eastern Anatolia suggests, conversely, that this period may have seen increased sedentism; evidence that includes more permanent architecture and a high proportion of cattle and pigs in faunal assemblages.[133]

More than previous periods, the Early and Middle Bronze Age were characterised by increased precipitation and temperatures across the length and breadth of Georgia. Warm and seasonally dry conditions in the highlands were accompanied by equally warm but wetter conditions at lower altitudes. These varied conditions promoted the expansion of oak savanna on the Tsalka Plateau and dense oak-ash forest above Tbilisi. The *Quercus* maximum in Southern Georgia, however, comes at least 1000 years later than its equivalents at Lake Van in eastern Anatolia and Lake Zeribar in Iran.[134] This considerable migrational lag cannot be adequately explained in terms of the palaeoclimate. Instead it was human activity that delayed oak's expansion in Southern Georgia, namely widespread burning and grazing during the previous Chalcolithic period.

[127] Gobejishvili 1980; Sagona 2004.
[128] Sagona 2004.
[129] Rubinson 2003; Sagona 2004.
[130] Kushnareva 1997.
[131] Sagona 2004.
[132] Kushnareva 1997; Puturidze 2003.
[133] Howell-Meurs 2001; Sagona 2004; Sagona and Sagona 2000.
[134] van Zeist and Bottema 1991; Wick *et al.* 2003.

The absence of vegetation fires during the Bronze Age (**Fig. 8.4**) argues in favour of a new method of vegetation management. Forest cover trends from the Tsalka Plateau (Chapter Six) indicate that tree cover did not blanket the highlands, but fluctuated considerably during the Early Bronze Age, when the development of metal tools allowed woodland to be maintained, moulded and utilised without the need for landscape-scale fires. Axe in hand, a Bronze Age farmer was able to shape the vegetation in a much more selective way than his Chalcolithic ancestors could do with fire. Any doubt that these people could have had such a wide-ranging impact should be allayed by remembering that they had the labour wherewithal to build the kurgans, the largest of which were artificial mounds of 21,000 cubic metres of stone and earth, contained oak-lined burial chambers up to 175 square metres in area, and had paved ritual roads several hundred metres long leading westward to the chamber entrance.[135]

A valuable insight into prehistoric wood exploitation comes from a pioneering macrofossil analysis of the Beshtasheni fortress on the Tsalka Plateau.[136] The site contained 45 percent oak and 30 percent fir macrofossils during the Middle Bronze Age, a pattern wholly consistent with the pollen evidence. Chalcolithic assemblages, in contrast, were predominantly (60 percent) of alder, a fire-enhanced streamside tree. A similar shift from exploitation of riverine taxa during the Chalcolithic to woodland taxa during the Bronze Age is recorded at Sos Höyük.[137] In both these studies, the authors concluded that deforestation of riversides forced Bronze Age folk to exploit woodlands farther from their settlements.

I regard this deforestation hypothesis as only partly applicable to Southern Georgia for several reasons. One is the growing complexity and diversification of Middle Bronze Age economies,[138] which saw greater specialisation in crafts such as metallurgy, carpentry and pottery. I find it difficult to accept that a society that could build the kurgans and had sturdy axes to hand would wittingly cut poor quality wood like alder in preference to oak. Secondly, the emergence of metallurgy itself must have had a profound impact on wood use. Smelting in the prehistoric context required large amounts of wood charcoal, and for this purpose trees like *Carpinus* were probably sought after.[139] It may be no coincidence that *Carpinus* pollen all but disappears from the Tsalka Plateau records during the Bronze Age. Oak, on the other hand, was more valued for construction, as its almost exclusive use in building kurgan burial chambers attests.[140] Thirdly, pollen evidence clearly shows that oaks were relatively abundant in

[135] Gobejishvili 1978; Kushnareva 1997; Narimanishvili 2004; Puturidze 2003.

[136] Yatsenko-Khmelevskii and Kandelaki 1941.

[137] Longford 2005. At Sos Höyük, Salicaceae dominate Chalcolithic assemblages (76 percent). Middle Bronze Age levels contained less Salicaceae (18 percent) and more *Pinus* (42 percent) and *Quercus* (24 percent).

[138] Puturidze 2003.

[139] Gardner 2002; Rackham 2001.

[140] Burney and Lang 1971; Gobejishvili 1980.

Bronze Age landscapes, so their exploitation seems entirely natural. The prior exploitation of alders and willows may relate simply to their relative abundance in the largely treeless highland environment of the Chalcolithic, their proximity to riverside settlements and their ease of collection in the absence of metal tools.

In conclusion, the Early to Middle Bronze Age saw the emergence of a new relationship between society and environment. This occurred as populations became more sedentary, economic activities diversified, farming intensified, and metallurgical innovation changed the way vegetation density was managed. The use of timber and joinery in the kurgans and their wheeled vehicles reveals a level of sophistication in this craft hitherto unprecedented. As for the kurgans themselves, one could indeed ask whether they were powerful statements of land tenure placed ostentatiously in the partially wooded plateau landscape.[141]

Late Bronze and Iron Ages (*c.* 1500–400 BC)

The Late Bronze and Iron Ages, more than any other period, were afflicted by climatic changes that had a profound impact on the distribution and economy of prehistoric settlements. Pollen and isotopic records from across the region indicate a progressive and lasting decrease in both temperature and precipitation from the beginning of the second millennium BC.[142] Coniferous trees on the Tsalka Plateau responded to this climate change positively, displacing the oaks that had been so widespread during the Early and Middle Bronze Age. On the hills above Tbilisi, oak-ash forest gave way to open vegetation. Even in southern Colchis the climate became less humid. On the Main Caucasus Range, glaciers reactivated around 1000 BC.[143]

Climate change has often been invoked as a factor in the destabilisation of socioeconomic structures.[144] Perhaps the most celebrated example is the collapse of the Akkadian Empire around 2200 BC, an event some have linked to a sudden and sustained aridification event in Mesopotamia.[145] In Southern Georgia, there are signs that the climatic instability of the second millennium BC brought about changes in the human sphere. Not least amongst these is the widespread abandonment of the highlands that had been the focus of so much activity during the Early and Middle Bronze Age.[146]

Settlements of the Late Bronze and Iron Ages in Southern Georgia were instead concentrated in the lowland valleys and piedmonts, where the climatic conditions

[141] Connor and Sagona 2007.
[142] Chapter Seven; Lemke and Sturm 1997; Roberts *et al.* 2001; Wick *et al.* 2003.
[143] Serebryanny 1984; Serebryanny and Orlov 1982.
[144] Burroughs 2005; Diamond 2004.
[145] Cullen *et al.* 2000; Dalfes *et al.* 1997.
[146] Burney and Lang 1971.

may have been more amenable to farming.[147] Compared to the Middle Bronze Age, when the scarcity of weapons is thought to signify relatively peaceable political relations,[148] the Late Bronze and Iron Ages saw a dramatic increase in weaponry manufacture – swords, daggers, arrowheads, axes, maces and the like.[149]

Gogichaishvili,[150] in reviewing the available pollen data from the lowlands of Eastern Georgia, concluded that human impact on the vegetation intensified dramatically during the Late Bronze and Iron Ages. Her conclusion is reaffirmed by pollen data from Tsavkisi Wetland showing the destruction of oak-ash forests and their replacement with open steppe-phrygana vegetation. Livestock grazing on the foothills around Tbilisi must have been exceedingly intense during the Late Bronze and Iron Ages, as the extraordinary abundance of dung fungal spores and phrygana pollen demonstrates (see **Fig. 5.19**). The demise of wild fauna in Eastern Georgia, especially onagers, aurochs, beavers, wild horses, moose, Caucasian tur and wild goats, was to a large degree attributable to intensified human impacts during the mid to late Holocene.[151] Human impact on the Tsalka Plateau at this time is less obvious because populations had largely abandoned the highlands. Even so, high proportions of *Triticum*-type pollen in the Imera record may indicate that Iron Age highlanders had resorted to growing wheat in shallow lake basins where a source of water could be assured. Around Lake Aligol it seems that the lack of human activity allowed pine and fir trees to consolidate into a forest – the densest vegetation in all its Holocene history (see **Fig. 6.19**). Yet this afforestation was all too brief, cut short by destructive fires that had long been in abeyance on the Tsalka Plateau. In their wake, fire-sensitive fir trees were eradicated from the landscape.

The Late Bronze-Iron Age phase of intensive human impact in Georgia corresponds to changes occurring around the same time in southwestern Turkey. A good deal of palaeoenvironmental research has been devoted to unearthing the causes of impact events known collectively as the 'Beyşehir Occupation Phase'.[152] Because the beginning of the phase corresponded more or less with the eruption of Santorini around 1600 BC,[153] some questioned whether the deposition of volcanic ash provided the impetus for the forest clearance and cultivation episodes recorded in pollen diagrams. Although tephra deposition temporarily affected lake productivity, the Beyşehir Occupation Phase did not begin in many places until centuries later, pointing to political

[147] Furtwängler *et al.* 1998; Kvavadze 1999c; Kvavadze and Todria 1992; Pitskhelauri 1979; Pizchelauri 1984.

[148] Burney and Lang 1971; Lang 1966, p. 46.

[149] Abramishvili 2003.

[150] Gogichaishvili 1990.

[151] Bendukidze 1979.

[152] Bottema and Woldring 1990; Eastwood *et al.* 1998; Eastwood *et al.* 1999; Roberts 1990; Roberts *et al.* 1997; Vermoere *et al.* 2002.

[153] Guichard *et al.* 1993.

instability as a probable trigger for deforestation and cultivation.[154] At the same time, deforestation in eastern Anatolia is indicated by increased sedimentation in the Euphrates,[155] the almost wholesale loss of woodland fauna in vertebrate assemblages at Sos Höyük,[156] and a decreasing proportion of woodland taxa in macrobotanical data from the same site.[157] This clearly was a very widespread pattern of human impact.

Around 1200 BC, the Hittite Empire of Anatolia crumbled under the repeated depredations of the 'Sea Peoples', ushering in the so-called Anatolian Dark Age and unleashing a cascade of political reorganisations that reverberated across the Near East.[158] It has been argued that colder, drier climates across Europe and the Mediterranean during the second millennium BC led to crop failure and famine in the Sea Peoples' homelands, driving them out *en masse* in search of new territory.[159] Southern Georgia was by no means immune from the political ramifications of this regional event.[160] Nor was it immune from subsequent invasion of the Scythians, who laid waste to much of Eastern Georgia in the seventh century BC; nor indeed the imperial reach of the Achaemenid Empire of Persia in the fifth century BC.[161] Climatic change compounded political instability by forcing populations to congregate in the lowlands and foothills, where such population pressures seem to have taken a substantial toll on the vegetation and fauna. As in more recent times, people of the highlands responded to the destabilisation of their political and economic structures by once again taking up the ancient practice of herding.

Contacts with Greece and Rome (400 BC–330 AD)

So much has been written on historical-era Colchis and Iberia that it is not possible to review here the extent of their labyrinthine interactions with the Greeks, Romans, Achaemenids, Urartians, Armenians, Medes, Pontians, Albanians, Cimmerians, Scythians and other neighbouring peoples during the 1st millennium BC.[162] Suffice to say that these cultures had both direct and indirect influences on the culture of Georgia of the time, through migrations not only of artistic ideas, but of ethnic groups as well.[163] The date of 400 BC is a convenient starting point as it coincides with Xenophon's march through the highlands of Anatolia and the southwestern Caucasus, the

[154] Eastwood *et al.* 1998; Roberts *et al.* 1997.
[155] Kuzucuoglu 2003.
[156] Howell-Meurs 2001.
[157] Longford 2005; Newton 2004.
[158] Lang 1966.
[159] Burroughs 2005.
[160] Lang 1966.
[161] Tsetskhladze 2003.
[162] Braund 1994; Burney and Lang 1971; Lang 1966; Salia 1983.
[163] Knauß 2005; Tsetskhladze 2003.

weakening of Persian control in Eastern Georgia, and, most importantly, the rise of the Iberian Kingdom in that region.[164] Southern Georgia at this time was a frontier zone between the Armenians and Iberians, lying within the territory of the latter.

The date of 400 BC is also significant in the palaeoenvironmental record, for it was around this time that fir trees were burnt out from the surroundings of Lake Imera.[165] This initial fire was followed by a series of even more intense blazes through the following centuries that obliterated woodland from that sector of the Tsalka Plateau altogether. Although it is difficult to tie these fires to particular historical events, the impact of war should not be overlooked – especially the imperial expeditions of Xenophon, Alexander and Pompey. It was in 66 BC that Pompey invaded Georgia with his army, on a campaign ostensibly to rid the Roman Empire of the troublesome Eupator of Pontus, Mithridates VI. One contemporary source recorded that the Georgians put up fearsome resistance to the Roman legionaries, who "drove the barbarians into a dense forest. These people are terrible forest fighters, hiding in the woods and darting out unexpectedly. Pompey surrounded this forest with his army, *set it on fire*, and pursued the fugitives when they ran out".[166] Of course, it cannot be claimed that the Lake Imera fires were the result of Pompey's campaign, the movements of which are largely unknown,[167] but the almost regional increase in fires through this turbulent period (**Fig. 8.4**) forces me to consider that fire may have been as much an instrument of war as it was a tool of farming. War's indirect impact, far from the battlefields, was probably the same – by promoting transhumance and pastoralism, war also promoted the torching of forests on a large scale.

Somewhat cool climates persisted in Southern Georgia throughout the 1st millennium BC. In the highlands, precipitation rose gradually during this period, but in the lowlands the opposite trend is observed.[168] Judging from dung spores preserved in the Tsavkisi and Jvari sediments, this period was one of intense grazing on the piedmonts about Tbilisi. At the very end of the first millennium BC, it appears that the climate again became warmer and wetter across Georgia. At once the pastures surrounding Tsavkisi Wetland were abandoned and enveloped by dense regrowth forests of hornbeam. Was it this warmer, wetter climate that brought about such a fundamental shift in vegetation and land-use? The answer, it seems, is no; for the Lake Jvari record shows convincingly that grazing and burning continued in some piedmont areas unabated, maintaining the openness of the vegetation. Climatic factors, at best, played a

[164] Lang 1966.

[165] See Chapter Six. Although no charcoal record is yet available for the Gomnis Lake pollen diagram, fir and beech declined rapidly there (Margalitadze 1971; Gobet unpublished data) around the same time as the Lake Imera fires, suggesting widespread human impact in Southern Georgia.

[166] Appian 1913, 15: 103.

[167] Braund 1994, p. 161 ff.

[168] Chapter Seven.

minor role. Because the wholesale transformation of the Tsavkisi landscape coincides directly with the appearance of walnut and olive pollen, there can be little doubt that socio-economic and cultural factors were at play.

By the first century AD, the Kingdom of Iberia had assimilated many of the trappings of Graeco-Roman life. The geographer Strabo (ca. 64 BC – 25 AD) gives a description of the Iberians that reveals much about their agricultural economy, settlements, cultural references, and their political alliances with the mountain tribes:

> The greater part of Iberia is so well built up in respect to cities and farmsteads that their roofs are tiled, and their houses as well as their market places and other public buildings are constructed with architectural skill... The plain of the Iberians is inhabited by people who are rather inclined to farming and to peace, and they dress after both the Armenian and Median fashion; but the major, warlike portion occupy the mountainous territory, living like the Scythians and the Sarmatians, of whom they are both neighbours and kinsmen; however they engage also in farming.[169]

Archaeological excavations have since validated much of Strabo's description.[170] The substantial, two-storied villa at Dedoplis Gora is but one example. Its buildings were constructed of mud-brick with large wooden beams supporting a tiled roof. Auxiliary structures contained querns and ovens for baking bread, pithoi for wine storage, various farming implements (ploughs, hoes, etc.), an armoury store, as well as abundant and skilfully crafted ceramics, glassware and ornaments. Macrofossil remains included wheat, millet, peaches, cherries, grapes and nuts.[171]

Around the ancient Iberian capital of Mtskheta, the strongholds of Harmozice[172] and Seusamora mentioned by Strabo have also been excavated, revealing not only substantial fortifications, but also a Roman bath-house and large civic buildings (**Fig. 8.6**). Extraordinarily rich and abundant finds from the nearby necropolis of Samtavro indicate a population strongly influenced by surrounding cultures and with a high degree of social stratification.[173]

In my opinion, the rapid expansion of forest and orchards near Tsavkisi Wetland reflects a changing social landscape. The ancient Iberians had shown their readiness to adopt elements of Greek and Roman culture, much as they had assimilated Persian influences earlier. Evidently, some Iberian farmers responded to this by abandoning pastoralism and growing imported fruit trees like walnuts and olives. Compared to pastures, walnut groves carry much stronger overtones of sedentism, land ownership

[169] Strabo 1928, 11: 3: 1-3. Scythians and Sarmatians were North Caucasian tribes that Strabo called the "nomads of the north": Strabo 1928, 11: 3: 5.

[170] Braund 1994, p. 205 ff.; Lordkipanidze 1994, p. 158 ff.

[171] Gagoshidze and Tsotselia 1991.

[172] Evidently a corruption of the Georgian *Armaz-tsikhe*, meaning 'fortress of Armazi'. Armazi was the Georgian equivalent of Ahura-Mazda, the fire god of Zoroastrian religion, who was revered above all other gods in the pantheon of ancient Iberia: Burney and Lang 1971.

[173] Apakidze *et al.* 1958; Braund 1994; Lordkipanidze 1994.

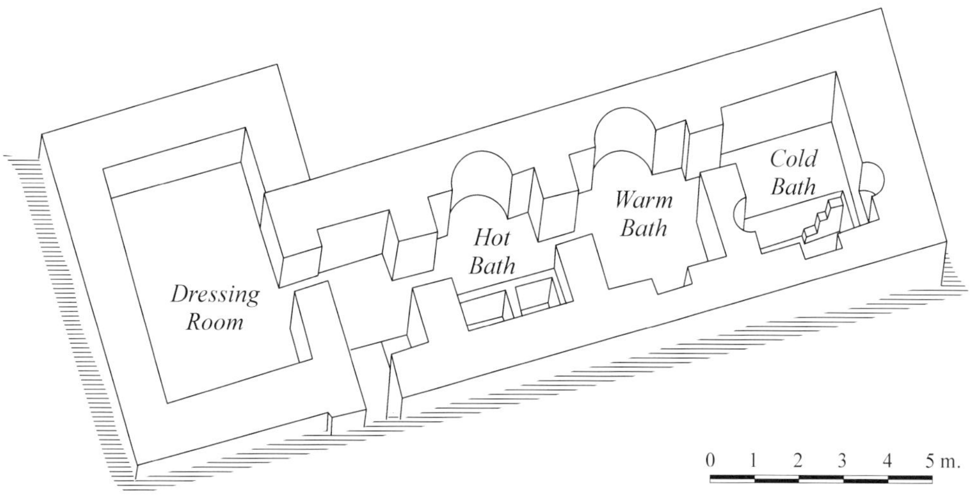

Figure 8.6. Axonometric projection of the remains of the bath complex at Bagineti (Harmozice), 1st century AD (source: Apakidze *et al.*, 1958, p. 176).

and prosperity, and it was these three characteristics that seem to have been the hallmarks of the Iberian Kingdom in the first century AD.

From Christianity to the Middle Ages (330 AD–1400 AD)

The adoption of Christianity as the state religion of Iberia makes a convenient starting point for the next phase of its history, for this decision by the then Iberian king, Mirian, had far-reaching effects on Georgia's culture, literature, language, political alliances and the structure of its feudal monarchy.[174] It also marks the beginning of a long and complex stretch of history in which Georgia weakened as a political entity, falling under Persian suzerainty from 378 AD and having its monarchy dissolved during the 6th century. Persian domination was followed by the capture of Tbilisi by the Greeks in 627 AD and then Arabs in 655 AD.

The Georgians and their fellow Christians in Armenia, despite their oppression, succeeded in preserving their Church, language and sense of identity through years of foreign domination. During the 8th century, Arab forces swept through the whole of Georgia, leaving such a wake of destruction that a contemporary source records: "Now at that time the lands of Kartli, Armenia, and Ran had been devastated, and there were no dwellings nor food at all for men or beasts."[175] A more evocative description, notable for its attitudes toward fires and forests, is given of the aftermath of the Turkish invasion of 1080 AD:

[174] Lang 1966
[175] Thomson 1996, p. 245.

> They consumed the land and massacred everyone wherever he might escape – to forests, cliffs, caves, or holes in the ground. In those times there was neither sowing nor harvest. *The land was ruined and turned into forest*; in place of men, beasts and animals of the field made their dwelling there… *Fire, an unaccustomed adversary*, consumed all dwellings.[176]

In geopolitical terms, Georgia remained a frontier and vassal state for most of the first millennium AD. Then, during the reign of King David the Builder toward the end of the 11th century, Georgia rose to prominence as a political force, annexing parts of Armenia and regaining Tbilisi from the Muslim emirs who had controlled it for centuries past.[177] What many consider to be Georgia's 'Golden Age' followed during the reign of Queen Tamar (1184–1213 AD). This period saw Georgia's sphere of influence grow to encompass almost the entire Caucasus region: from the North Caucasian tribes in the north, the Black Sea and Trabzon Empire in the west, the Caspian Sea in the east, to what are today northwestern Iran, Armenia and eastern Turkey in the south.[178] The beginning of the 13th century was the zenith of Georgian language, arts, culture and religion.[179]

It is not my aim here to analyse the innumerable environmental changes that may or may not have influenced and been influenced by Georgia's turbulent socio-economic situation during the Mediaeval. The data presented in this study provide insufficient detail to deal with this question except in the most general terms. This aside, several points are of interest. The first concerns the charcoal record (**Fig. 8.4**), which indicates that burning was more common in the vegetation from *ca* 330–1000 AD compared to the period *ca* 1000–1400 AD. A second point relates very much to this first; that is, the decline in burning during Georgia's 'Golden Age' corresponds to an apparent reduction in rainfall at all altitudes (compare **Figs 8.16** and **9.3**). This suggests that it was not the flammability of the vegetation that contributed most to its ignition, but the prevailing political climate. As I have previously argued, episodes of political stability are characterised by stability in agro-pastoral systems, whereas political instability favours transhumant pastoralism over sedentary agriculture. Landscape fires in Georgia today are lit most commonly by herders, rather than by those who sow grain, cultivate the vine or tend their orchards.

Much has been written about the Mediaeval climate of northwestern Europe, which is thought to have been somewhat warmer than the previous and subsequent centuries.[180] In Southern Georgia, the Mediaeval climate was certainly different to its

[176] Thomson 1996, pp.310-11, my italics.
[177] Burney and Lang 1971.
[178] AN GSSR 1964.
[179] Lordkipanidze 1994; Salia 1983. It was around this time that the famous Georgian epic poem, *Vepkhistqaosani* (*The Knight in the Panther's Skin*), is thought to have been penned by Shota Rustaveli (Rustaveli 1966).
[180] *E.g.* Bradley *et al.* 2003; Broecker 2001; Hughes and Diaz 1994; Lamb 1965.

precedent, but the description 'Mediaeval Warm Period' is not entirely fitting here. While the Georgian Mediaeval seems to have seen slightly warmer winter temperatures on the Tsalka Plateau, the main characteristic of the climate was increased seasonal aridity, witnessed particularly in low lake levels at Imera and Bareti. At least for the highlands of Southern Georgia, the climate reverted to a state that more closely resembled its Early to Middle Bronze Age character (*cf.* **Fig. 7.16**). It could be asked whether the same climatic conditions that saw the Trialeti Kurgan culture flourish in the highlands during the Bronze Age helped to foster Georgia's Mediaeval prosperity in the important highland centres of Tsalka, Akhalkalaki, Timogvi and Vardzia.

The irruption of the Mongol hordes from Central Asia sounded the death knell for Georgia's Golden Age.[181] Between 1235 and 1240 AD the Georgian feudal lords each and separately came to terms with the Mongol leaders, thus ending the unification won by David the Builder over a century before. Not until 1334 AD did the Georgian population reunite under King Giorgi the Brilliant (1314–1346 AD). Again, this period of autonomy was short-lived: in the year 1386 Tamerlane came to Georgia. According to chroniclers of the time, the devastation his armies inflicted during this and subsequent campaigns was absolute: they massacred the population, burnt their cities, churches and villages, laid waste to vineyards, fields and orchards.[182] At the dawn of the 15th century, after decades of brutal war and a catastrophic outbreak of the Black Death, Georgia was left defeated, depopulated and deforested.

Recent History (1400 AD – Present)

Following Tamerlane's campaigns, Georgia was divided amongst a number of its feudal lords and became the subject of prolonged Turkish and Persian territorial disputes over the next few centuries. Toward the end of the 16th century Russia also entered into this conflict; eventually Georgia sided with their Christian neighbours to the north against the Muslim Persians and Turks. Thus Georgia was subsumed by the Russian Empire during the 19th century and its successor, the Soviet Union, during the twentieth.[183]

This more recent history played out in a climate somewhat cooler and wetter than the preceding Mediaeval. Permanent snow in the highlands of Southern Georgia described by the 18th century geographer Vakhushti could be interpreted as evidence for the Little Ice Age.[184] Human impact on the vegetation intensified and especially so in the forests along lowland rivercourses, on the Main Caucasus Range and in Colchis.[185]

181 Allen 1929; Lang 1966.
182 Salia 1983.
183 Salia 1983.
184 Dzhanelidze 1980.
185 Badenkov *et al.* 1990; Dzhanelidze 1980; Ketskhoveli 1959.

Deforestation in these areas must have been sufficiently serious to warrant Georgian rulers employing forest guards and enacting forest protection laws from the beginning of the 15th century.[186] At least in Southern Georgia, intense fires ensured that the land did not become "ruined and turned into forest" in Tamerlane's wake (**Fig. 8.4**).

Perhaps in consequence of deforestation near settlements, tree pollarding and coppicing were practised widely at this time.[187] Poplars, willows, elms, ash and *Zelkova* trees were commonly pollarded on short rotation (**Fig. 8.7**), oaks were often shredded to produce construction timbers, and other trees and shrubs were coppiced for fencing, brushwood and firewood.[188] Such activities could well have their genesis in the ancient past, especially in Southern Georgia where wood was always scarce due to the dry climate and long history of burning and grazing.

The transformation of lowland landscapes during the Soviet period impacted heavily on their biota.[189] In Western Georgia, swamps were drained and turned over to intensive mechanised agriculture; in the arid lowland steppes of Eastern Georgia, irrigation converted vast winter pastures into collectivised agricultural fields.

Yet in the hills and highlands of Southern Georgia, the vegetation remained more or less in its prehistoric state, bitten down by livestock, felled and fired by villagers, so that when Joseph de Tournefort arrived there at the behest of Louis XIV, he observed a landscape little different to the present one – "full of Grass; but all manner of Trees have been long banish'd out of it".[190]

In spite of all Georgia's political vicissitudes, had he visited centuries or even millennia before, he may well have made the same observation.

[186] Ketskhoveli 1959, p. 345.
[187] Ketskhoveli 1959, pp. 81 ff.
[188] Ketskhoveli 1957, pp. 263 ff.
[189] Badenkov *et al.* 1990; Bendukidze 1979; Dolukhanov 1966.
[190] Tournefort 1718, p. 227.

Figure 8.7. Dr Eliso Kvavadze standing beside an ancient pollarded *Zelkova carpinifolia* tree in the Babaneuri Reservation, Eastern Georgia. The diameter of this tree's trunk is 1.22 m (breast height). Most of Babaneuri's old, large *Zelkova* trees have been pollarded in the past.

CHAPTER NINE

CONCLUSIONS

This study has revealed the complex history landscapes and vegetation in Southern Georgia through the last 14,000 years. The steppic and xerophytic plant communities that currently prevail in parts of Southern Georgia are not recent developments, but have existed in one form or another for the entire Holocene period. Paradoxically, when forests expanded throughout the Caucasus region during the early-mid Holocene, the landscapes of Southern Georgia remained stable, open and relatively treeless. This was in spite of a clear trend towards higher rainfall and temperatures.

I argue that the ecological void between the expected vegetation and the climate of that period was filled by Neolithic and Chalcolithic human populations. Through burning and grazing, these people preserved and maintained the open landscapes upon which their agro-pastoral economy depended. Not until the dawn of the Bronze Age, some 5000 years ago, did the emergence of metallurgy foster a new relationship with the vegetation. The axe allowed highland populations to mould and utilise the vegetation in a highly selective and deliberate fashion. Some parts of Southern Georgia responded to the extinguishment of landscape fires with rapid afforestation, while areas close to population centres remained conspicuously open. The increasing complexity of the Bronze Age economy thus manifested in the increasing complexity of the vegetation.

The turbulent history of Southern Georgia in historical times brought dramatic changes to its plant communities. Moments of stability brought stability to agricultural systems and allowed the expansion of orchards, forests and fields; political instability on the other hand forced many to adopt transhumance and saw the widespread expansion of open vegetation and burning. To a significant degree, the heterogenous landscapes and various phytogeographical provinces of Georgia in the present-day can be directly attributed to the differential effects of human activity in time and in space, superimposed on a land of great bioclimatic diversity.

I hope that these conclusions will stimulate discussion on the causes of vegetation change in neighbouring regions and lead to a careful re-evaluation of the pollen evidence. In particular, this study has shown that traditional interpretations of pollen data are likely to produce misleading results in a semi-arid environment. Changing lake levels have a strong bearing on the pollen source area, such that low lake levels are representative of local vegetation and high lake levels of more regional vegetation. This relationship is most vividly expressed in the ratio of tree pollen to non-tree pollen.

While this ratio is commonly regarded as an indication of forest cover in humid regions, in the shallow, steppe lakes of Southern Georgia it is a much more reliable indicator of lake levels.

Other considerations include the interpretation of grass pollen, which is produced by both steppic and wetland vegetation, and the difficulties in detecting xerophyte scrub vegetation using pollen data. This work has highlighted the need for future research to include analysis of microscopic charcoal and fungal remains as an adjunct to radiocarbon-dated pollen data. Statistical methods, in particular, show great promise in aiding the interpretation of the often-cryptic pollen diagrams from semi-arid and mountainous environments. Even though trends may not be immediately apparent, changes in temperature and precipitation, forest cover and land-use can be effectively isolated from background noise using modern analogues.

Indeed, the Late Pleistocene seems to have been a time of intense aridity and seasonality, with a cold phase centred on the Younger Dryas. Moisture increased during the early Holocene, albeit slowly, so it was not until after 9000 years ago that rainfall came to resemble its present-day aspect. The mid Holocene seems to have been a time of increased temperatures and was followed by somewhat cooler and more variable conditions during the late Holocene. While climate has had a strong bearing on the course of Holocene vegetation change in Georgia, within the bounds of Southern Georgia human impact has always played the leading role.

Archaeologists have long concerned themselves with questions of the emergence of agriculture and subsequent changes in economic strategies. In Western Asia, this question has traditionally been examined through artefact, macrobotanical and faunal assemblages in archaeological sites. Perhaps we need to broaden our view of human interactions with the environment, and consider the ancient cultural landscape as our predecessors' most significant and enduring legacy.

BIBLIOGRAPHY

AABY, B.

1988 "The cultural landscape as reflected in percentage and influx pollen diagrams from two Danish ombrotrophic mires," in *The Cultural Landscape: past, present and future,* edited by H. H. Birks, H. J. B. Birks, P. E. Kaland and D. Moe, pp. 209–228. Cambridge: Cambridge University Press.

AABY, B. and DIGERFELDT, G.

1986 "Sampling techniques for lakes and bogs," in *Handbook of Holocene Palaeoecology and Palaeohydrology,* edited by B. E. Berglund, pp. 181–194. Chichester: John Wiley and Sons.

ABRAMISHVILI, R.

2003 "Towards dating the remains of the Late Bronze Age and of the period of wide adoption of iron, discovered at the Samtavro burial ground," in *Problems of Caucasian Bronze-Iron Age Archaeology (Dziebani, supplement 10),* pp. 12–26. Tbilisi: Centre for Archaeological Studies.

ALLEN, W. E. D.

1929 "The March-lands of Georgia." *Geographical Journal* 74: 135–156.

1942 "The Caucasian borderland." *Geographical Journal* 99: 225–237.

AMMANN, B. and WICK, L.

1993 "Analysis of fossil stomata of conifers as indicators of the alpine tree line fluctuations during the Holocene," in *Oscillations of the Alpine and Polar Tree Limits in the Holocene,* edited by B. Frenzel, pp. 175–185. Stuttgart: Gustav Fischer Verlag.

AN GSSR

1964 *Atlas Gruzinskoi SSR.* Tbilisi: GUGK.

ANDERSEN, S. T.

1970 "The relative pollen productivity and pollen representation of North European trees, and correction factors for tree pollen spectra, determined by surface pollen analyses from forests." *Danmarks Geologiske Undersøgelse* 96: 1–99.

1973 "The differential pollen productivity of trees and its significance for the interpretation of a pollen diagram from a forested region," in *Quaternary Plant Ecology,* edited by H. J. B. Birks and R. G. West, pp. 108–115. Oxford: Blackwell Scientific.

ANDERSON, P. M., BARTLEIN, P. J., BRUBAKER, L. B., GAJEWSKI, K. and RITCHIE, J. C.

1989 "Modern analogues of late-Quaternary pollen spectra from the western interior of North America." *Journal of Biogeography* 16: 573–596.

ANDRIC, M. and WILLIS, K. J.

2003 "The phytogeographical regions of Slovenia: a consequence of natural environmental variation or prehistoric human activity?." *Journal of Ecology* 91: 807–821.

ANON.

2002 *BTC Project Environmental and Social Impact Assessment, 8: Environmental Baseline.* Tbilisi: British Petroleum.

APAKIDZE, A. M., GOBEDZHISHVILI, G. F., KALANDADZE, A. N. and LOMTATIDZE, G. A.

1958 *Mtskheta I: Armazis–Khevi.* Tbilisi: Izdatel'stvo AN Gruzinskoi SSR.

APPIAN
1913 *Romans History: Mithradic Wars* (translated by H. White). London: Heinemann.

ARABULI, G. J.
2003 "Reliktovyi aridnyi mozhzhevel'nikovyi les b okrestnostiakh Chachuskogo zakaznika (Vostochnaia Gruziia, Iorskoe plato)," in *Konservatsiia Aridnykh i Semiaridnykh Ekosistem v Zakavkaz'e Sbornik Nauchnykh Trudov,* edited by O. Abdaladze and I. K. Baridze, pp. 12–15. Tbilisi: NACRES.

ARCHIBOLD, O. W.
1995 *Ecology of World Vegetation.* London: Chapman and Hall.

ARNOL'DI, L. V.
1933 "Limnologicheskii ocherk ozer Akhalkalakskogo Plato," in *Dzhavakhetiia,* pp. 159–206. Tiflis: Zak. GIZ.

ARSLANOV, K. A., DOLUKHANOV, P. M. and GEI, N. A.
2007 "Climate, Black Sea levels and human settlements in Caucasus Littoral 50,000–9000 BP." *Quaternary International* 167–168: 121–127.

AUSTAD, I.
1988 "Tree pollarding in western Norway," in *The Cultural Landscape: Past, Present and Future,* edited by H. H. Birks, H. J. B. Birks, P. E. Kaland and D. Moe, pp. 11–29. Cambridge: Cambridge University Press.

AVAKOV, G. S.
1982 "Estestvennoe vosstanovlenie lesov v okrestnostiakh Tbilisi." *Priroda* 11: 43–49.

BADENKOV, Y. P., BORUNOV, A. K., MANDYCH, A. F., ROMASHKEVICH, A. I. and TARGULIAN, V. O.
1990 "Caucasia," in *The Earth as transformed by Human Actions: global and regional changes in the biosphere over the past 300 years,* edited by B. L. Turner, II, pp. 513–531. Cambridge: Cambridge University Press.

BALLARD, R. D., COLEMAN, D. F. and ROSENBERG, G. D.
2000 "Further evidence of abrupt Holocene drowning of the Black Sea shelf." *Marine Geology* 170: 253–261.

BEHRE, K.-E.
1986 "The interpretation of anthropogenic indicators in pollen diagrams." *Pollen et Spores* 23: 225–245.
1990 "Some reflections on anthropogenic indicators and the record of prehistoric occupation phases in pollen diagrams from the Near East," in *Man's role in the shaping of the Eastern Mediterranean Landscape,* edited by S. Bottema, G. Entjes-Nieborg and W. van Zeist, pp. 219–230. Rotterdam: A.A. Balkema.

BENDUKIDZE, O. G.
1979 *Golotsenovaia Fauna Pozvonochnykh Gruzii.* Tbilisi: Metsniereba.

BENGTSSON, L. and ENELL, M.
1986 "Chemical analysis," in *Handbook of Holocene Palaeoecology and Palaeohydrology,* edited by B. E. Berglund, pp. 423–451. Chichester: John Wiley and Sons.

BENNETT, K. D.
1996 "Determination of the number of zones in a biostratigraphical sequence." *New Phytologist* 132: 155–170.
2002 *Psimpoll 4.10 (Computer Programme).* Uppsala: Department of Earth Sciences, Uppsala Universitet.

BERG, L. S.
1950 *Natural Regions of the U.S.S.R.* New York: Macmillan.

BEUG, H.-J.
1967 "Contribitions to the postglacial vegetational history of northern Turkey," in *Quaternary Paleoecology,* edited by E. J. Cushing and H. E. Wright, Jr., pp. 349–356. London: Yale University Press.
2004 *Leitfaden der Pollenbestimmung für Mitteleuropa und angrenzende Gebeite.* München: Verlag Dr Friedrich Pfeil.

BIRKS, H. H. and BIRKS, H. J. B.
2000 "Future uses of pollen analysis must include plant macrofossils." *Journal of Biogeography* 27: 31–35.

BIRKS, H. J. B.
1973 "Modern pollen rain studies in some arctic and alpine environments," in *Quaternary Plant Ecology,* edited by H. J. B. Birks and R. G. West, pp. 142–169. Oxford: Blackwell Scientific.
1995 "Quantitative palaeoenvironmental reconstructions," in *Statistical Modelling of Quaternary Science Data,* edited by D. Maddy and J. S. Brew, pp. 161–254. Cambridge: Quaternary Research Association.

BIRKS, H. J. B. and BIRKS, H. H.
1980 *Quaternary Palaeoecology.* London: Edward Arnold.

BIRKS, H. J. B. and GORDON, A. D.
1985 *Numerical methods in Quaternary pollen analysis.* London: Academic Press.

BIRKS, H. J. B., JONES, V. J. and ROSE, N. L.
2004 "Recent environmental change and atmospheric contamination on Svalbard as recorded in lake sediments – synthesis and general conclusions." *Journal of Paleolimnology* 31: 531–546.

BLACKFORD, J. J.
2000 "Charcoal fragments in surface samples following a fire and the implications for interpretation of subfossil charcoal data." *Palaeogeography, Palaeoclimatology, Palaeoecology* 164: 33–42.

BLUMLER, M. A.
1996 "Ecology, evolutionary theory and agricultural origins," in *The Origins and Spread of Agriculture and Pastoralism in Eurasia,* edited by D. R. Harris, pp. 25–50. Washington D.C.: Smithsonian Institution Press.

BOBROV, A. E., KUPRIANOVA, L. A., LITVINTSEVA, M. V. and TARASEVICH, V. F.
1983 *Spory i Pyl'tsa.* Leningrad: Nauka.

BOCK, J. H., JOLLS, C. L. and LEWIS, A. C.
1995 "The effects of grazing on alpine vegetation: a comparison of the central Caucasus, Republic of Georgia, with the Colorado Rocky Mountains, U.S.A." *Arctic and Alpine Research* 27: 130–136.

BOND, W. J., WOODWARD, F. I. and MIDGLEY, G. F.
2005 "The global distribution of ecosystems in a world without fire." *New Phytologist* 165: 525–538.

BORTENSCHLAGER, S.
1987 "Este pollenanalytische Untersuchungen bei Kazbegi im Zentralkaukasus." *Sitzunberichte – Österreichische Akademie der Wissenschaften, Math.-Natur. Klasse* 196: 11–20.

BOTCH, M. S. and MASING, V. V.
1983 "Mire ecosystems in the U.S.S.R.," in *Ecosystems of the World 4B – Mires: Swamp, Bog, Fen and Moor – Regional Studies,* edited by A. J. P. Gore, pp. 95–152. Amsterdam: Elsevier Scientific.

BOTTEMA, S.
1974 Late Quaternary Vegetation History of Northwestern Greece. Groningen: V. R. B. Offsetdrukkerij.
1975 "The interpretation of pollen spectra from prehistoric settlements (with special attention to Liguliflorae)." *Palaeohistoria* 17: 17–35.
1986 "A late Quaternary pollen diagram from Lake Urmia (northwestern Iran)." *Review of Palaeobotany and Palynology* 47: 241–261.
1992 "Prehistoric cereal gathering and farming in the Near East: the pollen evidence." *Review of Palaeobotany and Palynology* 73: 21–33.
1995 "Holocene vegetation of the Van area: palynological and chronological evidence from Sögütlü, Turkey." *Vegetation History and Archaeobotany* 4: 187–193.

BOTTEMA, S. and WOLDRING, H.
1984 "Late Quaternary vegetation and climate of southwestern Turkey, part II." *Palaeohistoria* 26: 123–149.
1990 "Anthropogenic indicators in the pollen record of the Eastern Mediterranean," in *Man's role in the shaping of the Eastern Mediterranean Landscape,* edited by S. Bottema, G. Entjes-Nieborg and W. van Zeist, pp. 231–264. Rotterdam: A. A. Balkema.

BRADLEY, R. S., HUGHES, M. K. and DIAZ, H. F.
2003 "Climate in Medieval time." *Science* 302: 404–405.

BRADSHAW, R. H. W.
1981 "Modern pollen-representation factors for woods in south-east England." *Journal of Ecology* 69: 45–70.

BRAIDWOOD, R. J. and HOWE, B.
1960 "Investigations in Iraqi Kurdistan." *Studies in Ancient Oriental Civilisation* 31. Chicago: University of Chicago.

BRAUND, D.
1994 *Georgia in Antiquity*. Oxford: Clarendon Press.

BROECKER, W. S.
1994 "Massive iceberg discharges as triggers for global climate change." *Nature* 372: 421–424.
1997 "Thermohaline circulation, the Achilles heel of our climate system: will man-made CO_2 upset the current balance?." *Science* 178: 1582–1588.
2001 "Was the Medieval Warm Period global?." *Science* 291: 1497–1499.

BROWICZ, K.
1982 (and subsequent years) *Chorology of Trees and Shrubs in South-West Asia and Adjacent Regions*, 6 volumes. Warszawa: Polish Scientific Publishers.

BROWN, T. A., NELSON, D. E., MATHEWES, R. W., VOGEL, J. S. and SOUTHON, J. R.
1989 "Radiocarbon dating of pollen by accelerator mass spectrometry." *Quaternary Research* 32: 205–212.

BUITENHUIS, H.
1990 "Archaeozoological aspects of late Holocene economy and environment in the Near East," in *Man's Role in the Shaping of the Eastern Mediterranean Landscape,* edited by S. Bottema, G. Entjes-Nieborg and W. van Zeist, pp. 195–205. Rotterdam: A.A. Balkema.

BUNTING, M. J.
2002 "Detecting woodland remnants in cultural landscapes: modern pollen deposition around small woodlands in northwest Scotland." *The Holocene* 12: 291–301.
2003 "Pollen-vegetation relationships in non–arboreal moorland taxa." *Review of Palaeobotany and Palynology* 125: 285–298.

BURNEY, C. and LANG, D. M.
1971 *The Peoples of the Hills: Ancient Ararat and Caucasus*. London: Weidenfeld and Nicolson.

BURROUGHS, W. J.
2005 *Climate Change in Prehistory*. Cambridge: Cambridge University Press.

CALCOTE, R.
1995 "Pollen source area and pollen productivity: evidence from forest hollows." *Journal of Ecology* 83: 591–602.

CARRIÓN, J. S., SÁNCHEZ-GÓMEZ, P., MOTA, J. F., YLL, R. and CHAÍN, C.
2003 "Holocene vegetation dynamics, fire and grazing in the Sierra de Gádor, southern Spain." *The Holocene* 13: 839–849.

CARRIÓN, J. S. and VAN GEEL, B.
1999 "Fine-resolution Upper Weichselian and Holocene palynological record from Navarrés (Valencia, Spain) and a discussion about factors of Mediterranean forest succession." *Review of Palaeobotany and Palynology* 106: 209–236.

CAVALLI-SFORZA, L. L.
1996 "The spread of agriculture and nomadic pastoralism: insights from genetics, linguistics and archaeology," in *The Origins and Spread of Agriculture and Pastoralism in Eurasia,* edited by D. R. Harris, pp. 51–69. Washington D.C.: Smithsonian Institution Press.

CAVALLI-SFORZA, L. L., PIAZZA, A. AND MENOZZI, P.
1994 *The History and Geography of Human Genes*. Princeton: Princeton University Press.

CHATAIGNER, C.
1995 *La Transcaucasie au Néolithique et au Chalcolithique*. Oxford: B.A.R. International Series 624.

CHEDDADI, R., LAMB, H. F., GUIOT, J. and VAN DER KAARS, S.
1998 "Holocene climatic change in Morocco: a quantitative reconstruction from pollen data." *Climate Dynamics* 14: 883–890.

CHEPALYGA, A. L.
1984 "Inland sea basins," in *Late Quaternary Environments of the Soviet Union,* edited by A. A. Velichko, pp. 229–247. Minneapolis: University of Minnesota Press.

CHESTER, P. I. and RAINE, J. I.
2001 "Pollen and spore keys for Quaternary deposits in the northern Pindos Mountains, Greece." *Grana* 40: 299–387.

CHIKOV, P. S., (ed.)
1983 *Atlas Arealov i Resursov Lekarstvennykh Rastenii SSSR*. Moskva: GUGK.

CHOCHIEVA, K. and MAMATSASHVILI, N.
1996 "Spore-pollen complexes of Chaudian deposits bearing the decidua of Colchis Redwood Forest." *Bulletin of the Georgian Academy of Sciences* 154.

CLARK, K. R.
1993 "Non-parametric multivariate analyses of changes in community structure." *Australian Journal of Ecology* 18: 117–143.

CLARK, R. L.
1982 "Point count estimation of charcoal in pollen preparations and thin sections of sediments." *Pollen et Spores* 24: 523–535.

CONNOR, S. E.,
2003 "Postcard from Tbilisi." *Quaternary Australasia* 21: 24–25.
2006 *A Promethean Legacy: Late Quaternary Vegetation History of Southern Georgia, Caucasus*. Unpublished PhD thesis. Melbourne: University of Melbourne.

CONNOR, S.E. and KVAVADZE, E.V.
2008 "Modelling late Quaternary changes in plant distribution, vegetation and climate using pollen data from Georgia, Caucasus." *Journal of Biogeography* 36: 529–545.

CONNOR, S. E. and SAGONA, A.
2007 "Environment and society in the late prehistory of southern Georgia, Caucasus," in *Les Cultures du Caucase (VIe-IIIe millénaires avant notre ère): leurs relations avec le Proche-Orient,* edited by B. Lyonnet, pp. 21–36. Paris: CNRS Editions.

CONNOR, S. E., THOMAS, I. and KVAVADZE, E. V.
2007 "A 5600-year history of changing vegetation, sea levels and human impacts from the Black Sea coast of Georgia." *The Holocene* 17: 25–36.

CONNOR, S. E., THOMAS, I., KVAVADZE, E. V., ARABULI, G. J., AVAKOV, G. S. and SAGONA, A.
2004 "A survey of modern pollen and vegetation along an altitudinal transect in southern Georgia, Caucasus region." *Review of Palaeobotany and Palynology* 129: 229–250.

CRISSMAN, L.
1995 *Georgia Administrative Dataset 1989 Census.* Griffith: Australian Centre for Asian Spatial Information and Analysis, Griffith University.

CULLEN, H. M., DEMENOCAL, P. B., HEMMING, S., HEMMING, G., BROWN, F. H., GUILDERSON, T. and SIROCKO, F.
2000 "Climate change and the collapse of the Akkadian empire: evidence from the deep sea." *Geology* 28: 379–382.

CULLEN, H. M., KAPLAN, A., ARKIN, P. A. and DEMENOCAL, P. B.
2002 "Impact of the North Atlantic Oscillation on Middle Eastern climate and streamflow." *Climatic Change* 55: 315–338.

CZEREPANOV, S. K.
1995 *Vascular Plants of Russia and Adjacent States (the former USSR).* Cambridge: Cambridge University Press.

DALFES, H. N., KUKLA, G. and WEISS, H., EDS.
1997 *Third Millennium BC Climate Change and Old World Collapse.* Berlin: Springer.

DARGIE, T. C. D.
1986 "Species richness and distortion in reciprocal averaging and detrended correspondence analysis." *Vegetatio* 65: 95–98.

DAVID, F.
1997 "Holocene tree limit history in the northern French Alps: stomata and pollen evidence." *Review of Palaeobotany and Palynology* 97: 227–237.
2001 "Modern pollen precipitation from an elevational transect in central Jordan and its relationship to vegetation." *Journal of Biogeography* 28: 1195–1210.

DAVIS, A. M.
1980 "Modern pollen spectra from the tundra-boreal forest transition in northern Newfoundland, Canada." *Boreas* 9: 89–100.

DAVIS, B. A. S., BREWER, S., STEVENSON, A. C., GUIOT, J. and data contributors
2003 "The temperature of Europe during the Holocene reconstructed from pollen data." *Quaternary Science Reviews* 22: 1701–1716.

DAVIS, M. B.
2000 "Palynology after Y2K – understanding the source area of pollen in sediments." *Annual Review of Earth and Planetary Science* 28: 1–18.

DAVIS, O. K. and SHAFER, D. S.
2006 "*Sporormiella* fungal spores, a palynological means of detecting herbivore density." *Palaeogeography, Palaeoclimatology, Palaeoecology* 237: 40–50.

DAVIS, P. H., (ed.)
1982 *Flora of Turkey and the East Aegean Islands, volume seven.* Edinburgh: Edinburgh University Press.

DEARING, J. A. and FOSTER, I. D. L.
1986 "Lake sediments and palaeohydrological studies," in *Handbook of Holocene Palaeoecology and Palaeohydrology,* edited by B. E. Berglund, pp. 67–90. Chichester: John Wiley and Sons.

DENK, T., FROTZLER, N. and DAVITASHVILI, N.
2001 "Vegetational patterns and distribution of relict taxa in humid temperate forests and wetlands of Georgia (Transcaucasia)." *Biological Journal of the Linnean Society* 72: 287–332.

DIAMOND, J.
2004 *Collapse: How Societies Choose to Fail or Succeed.* New York: Viking.

DODSON, J. R. and MOONEY, S. D.
2002 "An assessment of historic human impact on south-eastern Australian environmental systems, using late Holocene rates of environmental change." *Australian Journal of Botany* 50: 455–464.

DOKTUROVSKII, V. S.
1931 "O torfianikakh Zakavkaz'ia." *Torfianoe Delo* 2: 29–32.
1936 "Materialy po izucheniiu torfianikov Zakavkaz'ia." *Pochvovedenie* 2: 183–202.

DOLUKHANOV, A. G.
1966 "Rastitel'nostii Pokrov," in *Kavkaz,* edited by N. V. Dumitrashko, pp. 223–251. Moscow: Nauka.
1978 "The timberline and the subalpine belt in the Caucasus Mountains, USSR." *Arctic and Alpine Research* 10: 409–422.
1989 *Lesnaia rastitel'nost' Gruzii, I.* Tbilisi: Metsniereba.
1992 *Lesnaia rastitel'nost' Gruzii, II.* Tbilisi (unpublished manuscript cited in Nakhutsrishvili, 1999).

DOLUKHANOV, P. M.
1980 "Paleogeography and prehistoric settlement in Caucasus and in Central Asia during the Pleistocene and Holocene." *Annali dell'Instituto Orientale, Napoli* 40: 49–87.

DZHANELIDZE, C. P.
1980 *Paleogeografiia Gruzii v Golotsene.* Tbilisi: Metsniereba.
1984 "K rekonstruktsii prirodnykh uslovii epokhi sushchestvovaniia rannezemledel'cheskoi kul'tury vostochnoi Gruzii," in *Chelovek i okruzhaiushchaia ego sreda: materialy po arkheologii Gruzii i Kavkaza,* edited by T. N. Chubinishvili, pp. 7–11. Tbilisi: Metsniereba.

EASTWOOD, W. J., ROBERTS, N. and LAMB, H. F.
1998 "Palaeoecological and archaeological evidence for human occupance in southwest Turkey: the Beyşehir occupation phase." *Anatolian Studies* 48: 69–86.

EASTWOOD, W. J., ROBERTS, N., LAMB, H. F. and TIBBY, J. C.
1999 "Holocene environmental change in southwest Turkey: a palaeoecological record of lake and catchment-related changes." *Quaternary Science Reviews* 18: 671–695.

EDWARDS, K. J.
1988 "The hunter-gatherer/agricultural transition and the pollen record in the British Isles," in *The Cultural Landscape: Past, Present and Future,* edited by H. H. Birks, H. J. B. Birks, P. E. Kaland and D. Moe, pp. 255–266. Cambridge: Cambridge University Press.

EJRNÆS, R.
2000 "Can we trust gradients extracted by Detrended Correspondence Analysis?." *Journal of Vegetation Science* 11: 565–572.

EJRNÆS, R. and BRUUN, H. H.
2000 "Gradient analysis of dry grassland vegetation in Denmark." *Journal of Vegetation Science* 11: 573–584.

ELLIOT, M. B.
1999 "Modern pollen-vegetation relationships in Northland, New Zealand." *New Zealand Journal of Botany* 37: 131–148.

EL-MOSLIMANY, A. P.
1990 "Ecological significance of common nonarboreal pollen: examples from drylands of the Middle East." *Review of Palaeobotany and Palynology* 64: 343–350.

ESRI.
2005 *ArcMap 9.1 (Computer Programme).* Redlands, California: ESRI Inc.

FAEGRI, K. and IVERSEN, J.
1989 *Textbook of Pollen Analysis (IV Edition).* Caldwell, New Jersey: Blackburn Press.

FALL, P. L.
1992 "Spatial patterns of atmospheric pollen dispersal in the Colorado Rocky Mountains, USA." *Review of Palaeobotany and Palynology* 74: 293–313.

FLANNERY, K. V., Ed.
1976 *The Early Mesoamerican Village.* New York: Academic Press.

FLETCHER, M.-S. AND THOMAS, I.
2007 "Holocene vegetation and climate change from near Lake Pedder, south-west Tasmania, Australia." *Journal of Biogeography* 34: 665–677.

FREI, T.
1997 "Pollen distribution at high elevation in Switzerland: evidence for medium range transport." *Grana* 36: 34–38.

FURTWÄNGLER, V. A. E., KNAUSS, F., MOTZENBÄCKER, I., GAGOSHIDZE, J. and KVAVADZE, E. V.
1998 "Archäologische Expedition in Kachetien 1997: Ausgrabungen in Širaki." *Eurasia Antiqua* Band 4: 309–364.

GABUNIA, L. and VEKUA, A.
1995 "A Plio-Pleistocene hominid from Dmanisi, East Georgia, Caucasus." *Nature* 373: 509–512.

GAGNIDZE, R., (ed.)
1985 *Sakartvelos Flora X.* Tbilisi: Metsniereba.

GAGNIDZE, R.
2005 *Vascular Plants of Georgia: a Nomenclatural Checklist.* Tbilisi: Universali.

GAGNIDZE, R., GVINASHVILI, T., SHETEKAURI, S. and MARGALITADZE, N. A.
2002 "Endemic genera of the Caucasian flora." *Feddes Repertorium* 113: 616–630.

GAGNIDZE, R. and MARGALITADZE, N. A.
1998 "Holocene history of the vegetation of the high-mountain regions of the botanico-geographical provinces of Georgia," in *Plant Life in High Mountains: Proceedings of Kazbegi IV International Symposium, Kazbegi, 1996,* edited by G. S. Nakhutsrishvili and O. Abdaladze, pp. 35–38. Tbilisi: Institute of Botany.

GAGOSHIDZE, J. and TSOTSELIA, M.
1991 "Arameultsartseriani pirpitebi dedoplis goridan," in *Amierkavkasiis Istoriis Problemebi,* edited by G. Maisuradze, pp. 47–78. Tbilisi: Metsniereba.

GAILLARD, M.-J. and BERGLUND, B. E. (eds)
1998 *Quantification of Land Surfaces Cleared of Forests during the Holocene – Modern Pollen/Vegetation/Landscape Relationships as an Aid to the Interpretation of Fossil Pollen Data.* Stuttgart: Gustav Fischer Verlag.

GAJEWSKI, K., LÉZINE, A.-M., VINCENS, A., DELESTAN, A., SAWADA, M. and the African Pollen Database
2002 "Modern climate-vegetation-pollen relations in Africa and adjacent areas." *Quaternary Science Reviews* 21: 1611–1631.
GARDNER, A. R.
2002 "Neolithic to Copper Age woodland impacts in northeast Hungary? Evidence from the pollen and sediment chemistry records." *The Holocene* 12: 541–553.
GARDNER, A. R. and WILLIS, K. J.
1999 "Prehistoric farming and the postglacial expansion of beech and hornbeam: a comment on Küster." *The Holocene* 9: 119–121.
GAUCH, H. G.
1982 *Multivariate Analysis in Community Ecology*. Cambridge: Cambridge University Press.
GEDYE, S. J., JONES, R. T., TINNER, W., AMMANN, B. and OLDFIELD, F.
2000 "The use of mineral magnetism in the reconstruction of fire history: a case study from Lago di Origlio, Swiss Alps." *Palaeogeography, Palaeoclimatology, Palaeoecology* 164: 101–110.
GERSAMIA, T.
1984 *Old Tbilisi*. Tbilisi: Sabchota Sakartvelo.
GIRLING, M. A. and GREIG, J.
1985 "A first fossil record for *Scolytus scolytus* (F.) (elm bark beetle): its occurrence in elm decline deposits from London and their implications for Neolithic elm decline." *Journal of Archaeological Science* 12: 347–351.
GLUMAC, P. and ANTHONY, D.
1992 "Culture and environment in the prehistoric Caucasus: the Neolithic through the Bronze Age," in *Chronologies in Old World Archaeology,* edited by R. W. Ehrich, pp. 196–206. Chicago: University of Chicago Press.
GOBEJISHVILI, G.
1978 *Tetri Tsqaros Nasoplari*. Tbilisi: Metsniereba.
1980 *Kultura Bedenskikh kurganov*. Tbilisi: Metsniereba.
GODWIN, H.
1975 *The History of the British Flora*. Cambridge: Cambridge University Press.
GOGICHAISHVILI, L. K.
1984a "Rastitel'nost' nizmennostei i predgorii Kvemo-Kartli v Golotsene," in *Chelovek i okruzhaiushchaia ego sreda: materialy po arkheologii Gruzii i Kavkaza,* edited by T. N. Chubinishvili, pp. 11–15. Tbilisi: Metsniereba.
1984b "Vegetational and climatic history of the western part of the Kura River basin," in *Paleoclimates, Paleoenvironments, and Human Communities in the Eastern Mediterranean Region in Later Prehistory,* edited by J. L. Bintliff and W. van Zeist, pp. 325–341.
1990 "Vegetation and human occupation in the lowlands and foothills of eastern Georgia in the middle Holocene," in *Man's Role in the Shaping of the Eastern Mediterranean Landscape,* edited by S. Bottema, G. Entjes-Nieborg and W. van Zeist, pp. 265–269. Rotterdam: A. A. Balkema.
GOGICHAISHVILI, L. K., KVACHAKIDZE, R. K., KIKAVA, G. S. and SHARASHIDZE, R. B.
1977 "Golotsenovaia istoriia lesnoi rastitel'nosti verkhov'aikh reki Beloi Aragvi," in *Palinologicheskie issledovania v Gruzii,* pp. 114–123. Tbilisi: Metsniereba.
GOGICHAISHVILI, L. K., RAMISHVILI, M. N. and SAQVARELIDZE, M. D.
1975 *The Morphology of the Pollen of Some Plants Causing Allergic Diseases*. Tbilisi: Metsniereba.
GOGITIDZE, S.
1978 *Samkhret-aghmosavlet Shavizghvrispiretis Neolituri Kultura*. Tbilisi: Metsniereba.

GOODFRIEND, G. A.
1991 "Holocene trends in ^{18}O in land snail shells from the Negev Desert and their implications for changes in rainfall source areas." *Quaternary Research* 35: 417–426.

GRICHUK, V. P.
1984 "Late Pleistocene vegetation history," in *Late Quaternary Environments of the Soviet Union,* edited by A. A. Velichko, pp. 155–178. Minneapolis: University of Minnesota Press.

GRICHUK, V. P. and AIRAPETIAN, T. A.
1976 "Razrez pleistotsenovykh kontinental'nykh otlozhenii Gamzachinmanskoi kotloviny." *Izvestiia A. N. Arm. SSR* 6: 31–39.

GRIGOLIA, G.
1977 *Tsentraluri Kolkhetis Neoliti Paluri.* Tbilisi: Metsniereba.

GROSSGEIM, A. A.
1945 *Flora Kavkaza, vol. III.* Baku: Izdatel'stvo AN AzSSR.
1948 *Rastitel'nostyi Pokrov Kavkaza.* Moscow: Izdatel'stvo Moskovskovo Obshchestva ispytatelei prirody.

GROVE, A. T. and RACKHAM, O.
2001 *The Nature of Mediterranean Europe: an Ecological History.* New Haven: Yale University Press.

GUDZHABIDZE, G. E., Ed.
2003 *Geologicheskaia Karta Gruzii.* Tbilisi: Kartografiia.

GUICHARD, F., CAREY, S., ARTHUR, M. A., SIGURDSSON, H. and ARNOLD, M.
1993 "Tephra from the Minoan eruption of Santorini in sediments of the Black Sea." *Nature* 363: 610–612.

GUIOT, J.
1990 "Methodology of the last climatic cycle reconstruction in France from pollen data." *Palaeogeography, Palaeoclimatology, Palaeoecology* 80: 49–69.

GULISASHVILI, V. Z.
1949 "Zakonomernoct' rasprostraneniia sosny obyknovennoi *Pinus hamata* Sosn. (*P. silvestris V. hamata*) Stev. v Zakavkaz'e." *Vestnik Tbilisskogo Botanicheskie sada AN GSSR* 58.
1964 Prirodnye zony i estestvenno-istoricheskie oblasti Kavkaza. Moscow: Nauka.

HABERLE, S. G. and BENNETT, K. D.
2001 "Modern pollen rain and lake mud-water interface geochemistry along environmental gradients in southern Chile." *Review of Palaeobotany and Palynology* 117: 93–107.

HALL, G. M. J. and MCGLONE, M. S.
2001 "Forest reconstruction and past climatic estimates for a deforested region of south-eastern New Zealand." *Landscape Ecology* 16: 501–521.

HARRIS, D. R.
1984 "Ethnohistorical evidence for the exploitation of wild grasses and forbs: its scope and archaeological implications," in *Plants and Ancient Man: Studies in Palaeoethnobotany,* edited by W. van Zeist and W. A. Casparie, pp. 63–69. Rotterdam: A.A. Balkema.

HART, D. M.
1988 "A safe method for the extraction of plant opal from sediments." *Search* 19: 293–294.

HAVINGA, A. J.
1964 "Investigation into the differential corrosion susceptibility of pollen and spores." *Pollen et Spores* 6: 621–635.

1967 "Palynology and pollen preservation." *Review of Palaeobotany and Palynology* 2: 81–98.

HEJNÝ, S.
1960 Ökologische Charakteristic der Wasser- und Sumpfpflanzen in den Slowakischen Tiefebenen. Bratislava: Slowakischen Akademie der Wissenschaften.

HICKS, S.
2001 "The use of annual arboreal pollen deposition values for delimiting tree-lines in the landscape and exploring models of pollen dispersal." *Review of Palaeobotany and Palynology* 117: 1–29.

HILL, M. O.
1979 *DECORANA: a FORTRAN Program for Detrended Correspondence Analysis and Reciprocal Averaging (Computer Programme).* Ithaca, New York: Cornell University.

HILL, M. O. and GAUCH, H. G.
1980 "Detrended correspondence analysis: an improved ordination technique." *Vegetatio* 42: 47–58.

HILLMAN, G.
1996 "Late Pleistocene changes in wild plant-foods available to hunter-gatherers of the northern Fertile Crescent: possible preludes to cultivation," in *The Origins and Spread of Agriculture and Pastoralism in Eurasia,* edited by D. R. Harris, pp. 159–203. Washington D.C.: Smithsonian Institution Press.

HILLMAN, G. and DAVIES, M. S.
1990 "Domestication rates in wild-type wheats and barley under primitive cultivation." *Biological Journal of the Linnean Society* 39: 39–78.

HILLMAN, G., HEDGES, R., MOORE, A., COLLEDGE, S. and PETTITT, P.
2001 "New evidence of Lateglacial cereal cultivation at Abu Hureyra on the Euphrates." *The Holocene* 11: 383–393.

HOROWITZ, A.
1992 *Palynology of Arid Lands.* Amsterdam: Elsevier.

HOWELL-MEURS, S.
2001 *Early Bronze and Iron Age Animal Exploitation in Northeastern Anatolia.* Oxford: British Archaeological Reports.

HUA, Q., JACOBSEN, G. E., ZOPPI, U., LAWSON, E. M., WILLIAMS, A. A., SMITH, A. M. and MCGANN, M. J.
2001 "Progress in radiocarbon target preparation at the ANTARES AMS centre." *Radiocarbon* 43: 275–282.

HUGHES, M. K. and DIAZ, H. F.
1994 "Was there a 'Medieval Warm Period', and if so, where and when?." *Climatic Change* 26: 109–142.

HUNTLEY, B.
1988 "Europe," in *Vegetation History,* edited by B. Huntley and T. Webb, III, pp. 341–383. Dordrecht: Kluwer Academic Publishers.

HUNTLEY, B. and BIRKS, H. J. B.
1983 *An Atlas of Past and Present Pollen Maps for Europe: 0 – 13 000 years ago.* Cambridge: Cambridge University Press.

INNES, J. B., BLACKFORD, J. J. and SIMMONS, I. G.
2004 "Testing the integrity of fine spatial resolution palaeoecological records: microcharcoal data from near-duplicate peat profiles from the North York Moors, UK." *Palaeogeography, Palaeoclimatology, Palaeoecology* 214: 295–307.

ISLEBE, G. A. and HOOGHIEMSTRA, H.
1995 "Recent pollen spectra of highland Guatemala." *Journal of Biogeography* 22: 1091–1099.

IVERSEN, J.
1941 "Landnám i Danmarks Stenalder." *Danmarks Geologiske Undersøgelse. II Række* 66: 1–68.

JACKSON, S. T. and WILLIAMS, J. W.
2004 "Modern analogs in Quaternary paleoecology: here today, gone yesterday, gone tomorrow?." *Annual Review of Earth and Planetary Science* 32: 495–537.

JACOBSON, G. L., JR. and BRADSHAW, R. H. W.
1981 "The selection of sites for palaeovegetational studies." *Quaternary Research* 16: 80–96.

JANELIDZE, A.
1948 "Tbilisis midamoebis da gare k'akhetis geologiis zogi sak'itkhi." *Trudy Tbilissk. Gosudarstv. Universiteta* 34: 299–309.
1955 "K'umisis t'apobis genezisi." *Akad. Nauk GSSR Trudy Geologicheskogo Instituta* 8: 89–96.

JANELIDZE, C. P., TATASHIDZE, Z., SEPERTELADZE, Z. and MARUASHVILI, L. I.
2000 "Sakartvelos buneba akhlo da shoreul geologiur tsarsulshi," in *Sakartvelos Geograpia I,* edited by Z. Tatashidze, Tbilisi: Metsniereba.

JANSSEN, C. R.
1966 "Recent pollen spectra from the deciduous and coniferous-deciduous forests of northeastern Minnesota: a study in pollen dispersal." *Ecology* 47: 804–825.
1986 "The use of local pollen indicators and of the contrast between regional and local pollen values in the assessment of the human impact on vegetation," in *Anthropogenic Indicators in Pollen Diagrams,* edited by K.-E. Behre, pp. 203–208. Rotterdam: A. A. Balkema.

JIKIA, S.
1941 *Gurjistanis Vilaetis Didi Davtari.* Tbilisi: Metsniereba.

JOHNSTON, C. A.
1998 *Geographic Information Systems in Ecology.* Oxford: Blackwell.

JOWSEY, P. C.
1966 "An improved peat sampler." *New Phytologist* 65: 245–248.

KAFFKE, A., MATCHUTADZE, I., COUWENBERG, J. and JOOSTEN, H.
2002 "Early 20th century Russian peat scientists as possible vectors for the establishment of *Calluna vulgaris* in Georgian *Sphagnum* bogs." *Suo* 53: 61–66.

KALANDADZE, A. N., KALANDADZE, K. S., VEKUA, A. K. and MAMATSASHVILI, N. S.
1977 "Ekologicheskie usloviia pozdnego pleistotsena-golotsena v predgornoi Kolkhide po raskopkam Beloi peshery," in *Paleoekologiia Drevnego Cheloveka,* pp. 172–180. Moskva: Nauka.

KAPLIN, P. A. and SELIVANOV, A. O.
2004 "Lateglacial and Holocene sea level changes in semi-enclosed seas of North Eurasia: examples from the contrasting Black and White Seas." *Palaeogeography, Palaeoclimatology, Palaeoecology* 209: 19–36.

KENT, M. and COKER, P.
1992 *Vegetation Description and Analysis.* London: Belhaven Press.

KERSHAW, A. P.
1994 "Pleistocene vegetation of the humid tropics of northeastern Queensland, Australia." *Palaeogeography, Palaeoclimatology, Palaeoecology* 109: 399–412.

KETSKHOVELI, N., (ed.)
1952 *Sakartvelos Flora VIII (Series 1).*Tbilisi: Metsniereba.

KETSKHOVELI, N.
1957 *K'ult'urul Mtsenareta Zonebi Sakartveloshi.* Tbilisi: Sakartvelos SSR Metsnierebata Ak'ademiis Gamomtsemloba.
1959 *Sakartvelos Mtsenareuli Sapari.* Tbilisi: SSSR Metsnierebata Akademiis Gamomtsemloba.
KETSKHOVELI, N. (ed.)
1971 *Sakartvelos Flora I.* Tbilisi: Metsniereba.
1973 *Sakartvelos Flora II.* Tbilisi: Metsniereba.
1975 *Sakartvelos Flora III.* Tbilisi: Metsniereba.
1980 *Sakartvelos Flora VI.* Tbilisi: Metsniereba.
1983 *Sakartvelos Flora VIII.* Tbilisi: Metsniereba.
1984 *Sakartvelos Flora IX.* Tbilisi: Metsniereba.
KHOTINSKIY, N. A.
1984 "Holocene vegetation history," in *Late Quaternary Environments of the Soviet Union,* edited by A. A. Velichko, pp. 179–200. Minneapolis: University of Minnesota Press.
KIGURADZE, T.
1986 *Neolithische Siedlungen von Kvemo-Kartli, Georgien.* München: Verlag C.H. Beck.
KIGURADZE, T. and MENABDE, M.
2004 "The Neolithic of Georgia," in *A View from the Highlands: Archaeological Studies in Honour of Charles Burney,* edited by A. Sagona, pp. 345–398. Louvain: Peeters.
KIGURADZE, T. and SAGONA, A.
2003 "On the origins of the Kura-Araxes cultural complex," in *Archaeology in the Borderlands: Investigations in Caucasia and Beyond,* edited by A. T. Smith and K. S. Rubinson, pp. 38–94. Los Angeles: Cotsen Institute of Archaeology Press.
KIKODZE, Z. and KORIDZE, E.
1978 "Paravnis dazverviti ekspeditsiis mier 1977 tsels chatarebul samushota mokle angarishi." *Sakartvelos Sakhelmtsipo Muzeumis Arkeologiuri Ekspeditsiebi* 6: 19–26.
KIKVIDZE, I. A.
1975 *Zemledeliye i zemledelcheskiy kult v Gruzii (Avtoreferat).* Tbilisi: Metsniereba.
KILIAN, M. R., VAN DER PLICHT, J., VAN GEEL, B. and GOSLAR, T.
2002 "Problematic ^{14}C-AMS dates of pollen concentrates from Lake Gosciaz (Poland)." *Quaternary International* 88: 21–26.
KLOPOTOVSKAYA, N. B.
1973 *Osnovnye zakonomernosti formirovaniia sporovo-pyl'tsevykh spektrov v gornykh raionakh Kavkaza.* Tbilisi: Metsniereba.
KNAUSS, F.
2005 "Zur Eisenzeitlischen wohnarchitektur Ostgeorgiens." *Ancient Near Eastern Studies* 42: 187–210.
KNORR, G. and LOHMANN, G.
2003 "Southern Ocean origin for the resumption of Atlantic thermohaline circulation during deglaciation." *Nature* 424: 532–536.
KOMÁREK, J. and JANKOVSKÁ, V.
2001 *Review of the Green Algal Genus Pediastrum; Implication for Pollen-analytical Research.* Berlin: J. Cramer.
KORELYAKOVA, I. L. and GORBIK, V. P.
1989 "Factori, opredelyajusshie zarastanie vodokhranilisch, i osnovnie zakonomernosti etogo processa," in *Rastitel'nost'i i bakterial'noe nasalenie Dnepra i ego*

vodokhranilisch, edited by N. V. Kondrat'eva, pp. 20–32. Kiev: Izdatel'stvo Naukova Dumka.

KRUCHININ, A. F., Ed.
1973 *Atlas Lesov SSSR.* Moskow: Glavnoe Upravlenie Geodezii i Kartografii (GUGK).

KUFTIN, B. A.
1941 *Arkeologicheskiye raskopki v Trialeti.* Tbilisi: Metsniereba.

KUPRIANOVA, L. A. and ALESHINA, L. A.
1972 *Pyl'tsa i Spory Rastenii.* Leningrad: Nauka.
1978 *Pyl'tsa Dvudol'nykh Rastenii.* Leningrad: Nauka.

KUSHNAREVA, K. K.
1997 *The Southern Caucasus in Prehistory: stages of cultural and socioeconomic development from the eighth to the second millennium B.C.* Philadelphia: The University Museum, University of Pennsylvania.

KUZNETSOV, N. I.
1909 "Printsipy deleniia Kavkaza na botaniko-geograficheskie provintsii." *Zap. Akad. Nauk po fiz.-mat. otd.* VIII.

KUZUCUOGLU, C.
2003 "Environmental changes in southern (Antalya) and south-eastern (Euphrates Valley) Turkey, at the end of the 2nd millennium BC and beginning of the 1st millennium BC," in *Identifying Changes: The Transition from Bronze Age to Iron Ages in Anatolia and its Neighbouring Regions,* edited by B. Fischer, H. Genz, E. Jean and K. Koroglu, pp. 271–281. Istanbul: Turk Eskicag Bilimleri Enstitusu.

KVAVADZE, E. V.
1978 "Novaia skhema razvitiia rastitel'nogo pokrova Kolkhidskoi nizmennosti v golotsene." *Doklady Akademii Nauk SSSR* 241: 170–173.
1982 "Novye dannye po stratigrafii i paleogeografii Golotsena Kolkhidskoi nizmennosti," in *Chetvertichnaia Sistema Gruzii,* pp. 123–130. Tbilisi: Metsniereba.
1993 "On the interpretation of subfossil spore-pollen spectra in the mountains." *Acta Palaeobotanica* 33: 347–360..
1999a "The first results of the pollen monitoring programme in the Caucasus mountains (Georgia)." *Acta Palaeobotanica* 39: 171–177.
1999b "Golotsenovye kolebaniia urovnia Ozera Lisi i izmenenia polozhenii nizhnei granitsy lesa." *P'aleobioloiis p'roblemebi* 1: 75–87.
1999c "The result of palynological studies of sediments from the cultivated layers of the late Bronze and early Iron ages in the steppe regions of Georgia." *Acta Palaeobotanica Supplement* 2: 555–559.

KVAVADZE, E. V., BUKREEVA, G. F. and RUKHADZE, L. P.
1992 *Komp'iuternaia tekhnologia rekonstruktsii paleogeograficheskikh rekonstruktsii v gorakh (na primere golotsena Abkhazii).* Tbilisi: Metsniereba.

KVAVADZE, E. V. and CONNOR, S. E.
2005 "*Zelkova carpinifolia* (Pallas) K. Koch in Holocene sediments of Georgia — an indicator of climatic optima." *Review of Palaeobotany and Palynology* 133: 69–89.

KVAVADZE, E. V. and DZHEIRANASHVILI, V. G.
1989 "Palinologicheskaia kharakteristika Golotsenovykh otlozhenii shel'fa Chernogo Moria v raione Sukhumi." *Bulletin of the Academy of Sciences of the Georgian SSR* 135: 201–204.

KVAVADZE, E. V. and EFREMOV, Y. V.
1994 "Palynological studies of Holocene lake sediments in the headwaters of the river Bezymianka (west Caucasus)." *Acta Palaeobotanica* 34: 205–214.

1995 "Peculiarities of recent pollen spectra of lake sediments in the Caucasus." *Acta Palaeobotanica* 35: 57–72.

1996 "Palynological studies of lake and lake-swamp sediments of the Holocene in the high mountains of Arkhyz (western Caucasus)." *Acta Palaeobotanica* 36: 107–119.

KVAVADZE, E. V. and GABASHVILI, I.

1998 "Reflection of anthropogenic activity in pollen spectra of bottom sediments of Lake Shavi (East Georgia)." *Bulletin of the Georgian Academy of Sciences* 157: 140–143.

KVAVADZE, E. V. and RUKHADZE, L. P.

1989 *Rastitel'nosti i Klimat Golotsena Abkhazii*. Tbilisi: Metsniereba.

KVAVADZE, E. V., SHATBERASHVILI, Z., AMIRANASHVILI, J. and ARABULI, G. J.

2004 "Palynological investigations of two burial mounds of the Middle Bronze Age of Tkemlara (southern Georgia)." *Acta Palaeobotanica* 44: 267–279.

KVAVADZE, E. V. and STUCHLIK, L.

1990 "Subrecent spore-pollen spectra and their relation to recent vegetation belts in Abkhazia (North-Western Georgia, USSR)." *Acta Palaeobotanica* 30: 227–257.

1991 "Correlation of subfossil pollen spectra with recent vegetation of the eastern border of the Trialeti Range (the Tbilisi environs)." *Acta Palaeobotanica* 31: 273–288.

1993 "Subfossil pollen spectra of the steppe regions of East Georgia." *Acta Palaeobotanica* 33: 365–376.

KVAVADZE, E. V. and TODRIA, V. T.

1992 *Ekologicheskie usloviia cheloveka epokhi pozdnei Bronzy i rannego Zheleza Udabno Garedzhi po palinologicheskim dannym (Vostochnaia Gruziia)*. Tbilisi: Metsniereba.

LAMB, H. H.

1965 "The early Medieval warm epoch and its sequel." *Palaeogeography, Palaeoclimatology, Palaeoecology* 1: 13–37.

LANG, D. M.

1966 *The Georgians*. New York: Praeger.

LEMKE, G. and STURM, M.

1997 "δ^{18}O and trace element measurements as proxy for the reconstruction of climate changes at Lake Van (Turkey): preliminary results," in *Third Millennium BC Climate Change and Old World Collapse,* edited by H. N. Dalfes, G. Kukla and H. Weiss, pp. 653–678. Berlin: Springer.

LISITSYNA, G. N.

1984 "The Caucasus — a centre of ancient farming in Eurasia," in *Plants and Ancient Man: Studies in Palaeoethnobotany,* edited by W. van Zeist and W. A. Casparie, pp. 285–292. Rotterdam: A. A. Balkema.

LISITSYNA, G. N. and PRISHCHEPENKO, L. V.

1977 *Paleo-etnobotanicheskie nakhodki Kavkaza i Blizhnego Vostoka*. Moskva: Nauka.

LONGFORD, C.

2005 *The Archaeobotany of Sos Höyük, Turkey.* Unpublished Honours thesis. Melbourne: University of Melbourne.

LORDKIPANIDZE, M.

1994 *Essays on Georgian History*. Tbilisi: Metsniereba.

LORDKIPANIDZE, O. D.

1994 "Recent discoveries in the field of classical archaeology in Georgia." *Ancient Civilizations* 1: 127–168.

LYDOLPH, P. E.

1977 *Climates of the Soviet Union*. Amsterdam: Elsevier.

MAGNY, M., GUIOT, J. and SCHOELLAMMER, P.
2001 "Quantitative reconstruction of Younger Dryas to Mid-Holocene paleoclimates at Le Locle, Swiss Jura, using pollen and lake-level data." *Quaternary Research* 56: 170–180.

MALMER, N. and REGNELL, G.
1986 "Mapping present and past vegetation," in *Handbook of Holocene Palaeoecology and Palaeohydrology,* edited by B. E. Berglund, pp. 203–218. Chichester: John Wiley and Sons.

MANGERUD, J., ANDERSEN, S. T., BERGLUND, B. E. and DONNER, J. J.
1974 "Quaternary stratigraphy of Norden, a proposal for terminology and classification." *Boreas* 3: 109–128.

MANGERUD, J., BIRKS, H. J. B. and JAGER, K. D.
1982 "Chronostratigraphical subdivisions of the Holocene: a review." *Striae* 16: 1–6.

MARDALEISHVILI, R. K. and JANELIDZE, C. P.
1984 "Sravitel'nyi analiz diagnosticheskikh pokazatelei sovremennykh i Golotsenovykh pogrebennykh pochv Marneulskoi Pavniny," in *Chelovek i okruzhaiushchaia ego sreda: materialy po arkheologii Gruzii i Kavkaza,* edited by T. N. Chubinishvili, pp. 89–91.

MARGALITADZE, N. A.
1967 "Istoriia lesov Dabadzvel'skogo plato v golotsene po dannym sporovo-pyl'tsevogo analiza." *Bulletin of the Academy of Sciences of the Georgian SSR, Biological Series* 45.
1971 "The history of forests of the north-western part of the Trialeti Range in Holocene according to the pollen analysis." *Journal of Palynology (India)* 7: 69–75.
1977 "Istoriia rastitel'nosti Dzhavakhetskogo Nagor'ia i Tsalkinskogo Plato v Golotsene," in *Palinologicheskie issledovania v Gruzii,* pp. 124–147. Tbilisi: Metsniereba.
1995 *Istoriia golotsenovoi rastitel'nosti Gruzii.* Tbilisi: Metsniereba.
1998 "Results of the palynological study of the high mountain Holocene deposits in Georgia," in *Plant Life in High Mountains: Proceedings of Kazbegi IV International Symposium, Kazbegi, 1996,* edited by G. S. Nakhutsrishvili and O. Abdaladze, pp. 138-140. Tbilisi: Institute of Botany.

MARGALITADZE, N. A. and KIMERIDZE, K. R.
1985 "Istoriia golotsenovoi rastitel'nosti Verkhnei Svaneti," in *Flora i rastitel'nost' Svaneti.* Tbilisi: Georgian Academy of Sciences Botanical Institute.

MARGALITADZE, N. A. and RAMISHVILI, M. N.
1972 "Rezul'taty palinologicheskikh issledovanii golotsenovykh otlozhenii bolot Kartsakhi i Suldy (Malyi Kavkaz)." *Bulletin of the Academy of Sciences of the Georgian SSR* 66: 397–400.

MARKGRAF, V.
1980 "Pollen dispersal in a mountain area." *Grana* 19: 127–146.

MARRO, C. and HAUPTMANN, H., (EDS)
2000 *Chronologies des pays du Caucase et de l'Euphrate aux IVe–IIIe millenaires.* Paris: De Boccard.

MARTÍN SÁNCHEZ, M., LÓPEZ GARCÍA, P. and LÓPEZ SÁEZ, J. A.
1999 "Aproximación palinológica a la evolución de la vegetación del Caúcaso y del nivel del Mar Caspio durante el Holoceno reciente." *Temas Geológico-Mineros ITGE* 26: 269–273.

MARUASHVILI, L. I.
1956 *Vakhushti Bagrationi: ego Predshestvenniki i Sovremenniki.* Moskva: Geografgiz.
1971 "Iuzhno-gruzinskoe vulkanicheskoe nagor'e," in *Geomorfologiia Gruzii,* edited by L. I. Maruashvili, pp. 346–384. Tbilisi: Metsniereba.
MATCHARASHVILI, I., ARABULI, G. J., DARCHIASHVILI, G. and GORGADZE, G.
2004 *Javakheti Wetlands: Biodiversity and Conservation.* Tbilisi: NACRES.
MCCUNE, B. and MEFFORD, M. J.
1999 *PC-Ord: Multivariate Analysis of Ecological Data (Computer Programme).* Gleneden Beach, Oregon: MjM Software Design.
2004 *HyperNiche: Nonparametric Multiplicative Habitat Modeling (Computer Programme).* Gleneden Beach, Oregon: MjM Software Design.
MELLAART, J.
1975 *The Neolithic of the Near East.* London: Thames and Hudson.
MENABDE, M., KIGURADZE, T. and KIKODZE, Z.
1978 "Kvemo kartlis arkeologiuri ekspeditsiis 1976–1977 tslebis mushaobis shedegebi." *Sakartvelos Sakhelmtsipo Muzeumis Arkeologiuri Ekspeditsiebi* VI.
MENABDE, V. J.
1961 "Cultivated flora of Georgia," in *Botanical Excursions over Georgia,* edited by M. F. Sakhokia, pp. 69–76. Tbilisi: Academy of Sciences of the Georgian SSR.
MILLER, N. F.
1991 "The Near East," in *Progress in Old World Palaeoethnobotany,* edited by W. van Zeist, K. Wasylikowa and K.-E. Behre, pp. 133–160. Rotterdam: Balkema.
MINCHIN, P. R.
1987 "An evaluation of the relative robustness of techniques for ecological ordination." *Vegetatio* 69: 89–107.
MOORE, P. D.
1986 "Hydrological changes in mires," in *Handbook of Holocene Palaeoecology and Palaeohydrology,* edited by B. E. Berglund, pp. 91–107. Chichester: John Wiley and Sons.
MOORE, P. D. and EVANS, A.
1991 "The development of valley mires in south east England." *Aquilo Serie Botanica* 30: 25–34.
MOORE, P. D. and STEVENSON, A. C.
1982 "Pollen studies in dry environments," in *Desertification and development: dry land ecology in social perspective,* edited by B. Spooner and H. S. Mani, pp. 249–267. London: Academic Press.
MOORE, P. D., WEBB, J. A. and COLLINSON, M. E.
1991 *Pollen Analysis.* Oxford: Blackwell Science Publishers.
MOSS, P. T. and KERSHAW, A. P.
2000 "The last glacial cycle from the humid tropics of northeastern Australia: comparison of a marine and terrestrial record." *Palaeogeography, Palaeoclimatology, Palaeoecology* 155: 155–176.
MUNCHAEV, R. M.
1975 *Kavkaz na zare bronzovogo veka.* Moscow: Nauka.
NACRES
2005 *Biodiversity Strategy and Action Plan – Georgia.* Tbilisi: Noah's Ark Center for the Recovery of Endangered Species.
NAKAGAWA, T., GARFI, G., REILLE, M. and VERLAQUE, R.
1998 "Pollen morphology of *Zelkova sicula* (Ulmaceae), a recently discovered relic species of the European Tertiary flora: description, chromosonal relevance and

palaeobotanical significance." *Review of Palaeobotany and Palynology* 100: 27–37.

NAKHUTSRISHVILI, G. S.
1999 "The vegetation of Georgia (Caucasus)." *Braun-Blanquetia* 15: 1–68.

NARIMANISHVILI, G.
2004 "Ritual roads at Trialeti barrows." *Journal of Georgian Archaeology* 1: 120–133.

NAVEH, Z.
1990 "Ancient man's impact on the Mediterranean landscape in Israel – ecological and evolutionary perspectives," in *Man's Role in the Shaping of the Eastern Mediterranean Landscape,* edited by S. Bottema, G. Entjes-Nieborg and W. van Zeist, pp. 43–50. Rotterdam: A. A. Balkema.

NEBIERIDZE, L.
1972 *Dasavlet Amierkavkasiis Neoliti.* Tbilisi: Metsniereba.
1978 *Darkvetis Mravalpeniani Ekhi.* Tbilisi: Metsniereba.

NEISHTADT, M. I.
1957 *Istoriia lesov i paleogeografiia SSSR v Golotsene.* Moscow: Izdat. AN SSSR.

NEISHTADT, M. I., KHOTINSKIY, N. A., DEVIRTS, A. L. and MARKOVA, N. G.
1965 "Imnatskoe boloto (Gruzinskaia SSR)," in *Paleogeografiia i khronologiia verkhnego pleistotsena i golotsena po dannym radiouglerodnogo metoda,* pp. 105–112. Moscow: Nauka.

NEUSTADT, M. I.
1971 "Über die Lage der Untergrenze des Holozäns nach den Vorstellungen der Sowjetischen Wissenschaftler." *Geologiska Föreningen i Stockholm Förhandlingar* 93: 103–116.

NEWTON, J.
2004 *Life in the Mountains of Anatolia: a Study of a Bronze Age Settlement in its Environmental Context.* Unpublished PhD thesis. Melbourne: University of Melbourne.

NIKLEWSKI, J. and VAN ZEIST, W.
1970 "A late Quaternary pollen diagram from northwestern Syria." *Acta Botanica Neerlandica* 19: 737–754.

NOIRFALISE, A. and DJAZIREI, M. H.
1965 "Contribution a la phytogéographie de la forêt Caspienne." *Bulletin de la Société royale de Botanique de Belgique* 98: 197–214.

OGUCHI, T. and OGUCHI, C. T.
1998 "Mid-Holocene floods of the Syrian Euphrates inferred from 'tell' sediments," in *Palaeohydrology and Environmental Change,* edited by G. Benito, V. R. Baker and K. J. Gregory, pp. 307–315. New York: John Wiley and Sons.

OKSANEN, J. and MINCHIN, P.
1997 "Instability of ordination results under changes in input data order: explanations and remedies." *Journal of Vegetation Science* 8: 447–454.

OLSSON, I. U.
1986 "Radiometric dating," in *Handbook of Holocene Palaeoecology and Palaeohydrology,* edited by B. E. Berglund, pp. 273–312. Chichester: John Wiley and Sons.

ORCHARD, A. E.
1985 "Myriophyllum (Haloragaceae) in Australasia. II: The Australian species." *Brunonia* 8: 173–291.

OVERPECK, J. T., WEBB, T., III and PRENTICE, I. C.
1985 "Quantative interpretation of fossil pollen spectra: dissimilarity coefficients and the method of modern analogs." *Quaternary Research* 23: 87–108.

PAEZ, M. M., SCHÄBITZ, F. and STUTZ, S.
2001 "Modern pollen–vegetation and isopoll maps in southern Argentina." *Journal of Biogeography* 28: 997–1021.
PAKHOMOV, M. M. and KLOPOTOVSKAYA, N. B.
1980 *Palinologiia kainozoia Srednei Azii i Kavkaza.* Dushanbe: Izdatel'stvo Donish.
PALUMBI, G.
2003 "Red-black pottery: Eastern Anatolian and Transcaucasian relationships around the mid-fourth millennium BC." *Ancient Near Eastern Studies* 40: 80–134.
PAPAVA, V., Ed.
2000 *World in Figures.*Tbilisi: Siakhle.
PATTERSON, W. A., EDWARDS, K. J. and MAGUIRE, D. J.
1987 "Microscopic charcoal as a fossil indicator of fire." *Quaternary Science Reviews* 6: 3–23.
PEET, R. K., KNOX, R. G., CASE, J. S. and ALLEN, R. B.
1988 "Putting things in order: the advantages of Detrended Correspondence Analysis." *American Naturalist* 131: 924–934.
PEGLAR, S. and BIRKS, H. J. B.
1993 "The mid-Holocene *Ulmus* fall at Diss Mere, south-east England: disease and human impact?." *Vegetational History and Archaeobotany* 2: 61–68.
PETERSON, G. M.
1983 "Recent pollen spectra and zonal vegetation in the western USSR." *Quaternary Science Reviews* 2: 281–321.
PEYRON, O., JOLLY, D., BONNEFILLE, R., VINCENS, A. and GUIOT, J.
2000 "Climate of East Africa 6000 ^{14}C yr B.P. as inferred from pollen data." *Quaternary Research* 54: 90–101.
PIOTROVSKII, B. B.
1955 "Razvitie skotovodstva v drevneishem Zakavkaz'e." *Sovetskaia Arkheologiia* 23: 5–15.
PISARIC, M. F. J., SZEICZ, J. M., KARST, T. and SMOL, J. P.
2000 "Comparison of pollen and stomates as indicators of alpine treelines in northwestern Canadian lake sediments." *Canadian Journal of Botany* 78: 1180–1186.
PITSKHELAURI, K. N.
1979 *Vostochnaia Gruziia v Kontse Bronzovogo Veka.* Tbilisi: Metsniereba.
PIZCHELAURI, K.
1984 *Jungbronzezeitliche bis ältereisenzeitliche Heiligtümer in Ost-Georgien.* München: Verlag C.H. Beck.
POLUNIN, O.
1969 *Flowers of Europe: a Field Guide.* London: Oxford University Press.
POTT, R.
1992 "The impact of Early Neolithic agriculture on the vegetation of Northwestern Germany," in *Evaluation of Land Surfaces Cleared from Forests by Prehistoric Man in Early Neolithic Times and the Time of Migrating Germanic Tribes*, edited by B. Frenzel, B. Gläser and L. Reisch, pp. 57–72. Stuttgart: Fischer.
PREBBLE, M. and SHULMEISTER, J.
2002 "An analysis of phytolith assemblages for the quantitative reconstruction of late Quaternary environments on the Lower Taieri Plain, Otago, South Island, New Zealand: II. palaeoenvironmental reconstruction." *Journal of Paleolimnology* 27: 415–427.
PRENTICE, I. C.
1980 "Multidimensional scaling as a research tool in Quaternary Palynology: a review of theory and methods." *Review of Palaeobotany and Palynology* 31: 71–104.

1985 "Pollen representation, source area, and basin size: toward a unified theory of pollen analysis." *Quaternary Research* 23: 76–86.

1986 "Multivariate methods for data analysis," in *Handbook of Holocene Palaeoecology and Palaeohydrology,* edited by B. E. Berglund, pp. 775–797. Chichester: John Wiley and Sons.

PRENTICE, I. C., GUIOT, J., HUNTLEY, B., JOLLY, D. and CHEDDADI, R.

1996 "Reconstructing biomes from palaeoecological data: a general method and its application to European pollen data at 0 and 6 ka." *Climate Dynamics* 12: 185–194.

PRENTICE, I. C., JOLLY, D. and BIOME 6000 PARTICIPANTS

2000 "Mid-Holocene and glacial-maximum vegetation geography of the northern continents and Africa." *Journal of Biogeography* 27: 507–519.

PRICE, M. D. R. and MOORE, P. D.

1984 "Pollen dispersion in the hills of Wales: a pollen shed hypothesis." *Pollen et Spores* 26: 127–136.

PUTURIDZE, M.

2003 "Social and economic shifts in the South Caucasian Middle Bronze Age," in *Archaeology in the Borderlands: Investigations in Caucasia and Beyond,* edited by A. T. Smith and K. S. Rubinson, pp. 111–127. Los Angeles: Cotsen Institute of Archaeology.

RACKHAM, O.

1988 "Trees and woodland in a crowded landscape — the cultural landscape of the British Isles," in *The Cultural Landscape: Past, Present and Future,* edited by H. H. Birks, H. J. B. Birks, P. E. Kaland and D. Moe, pp. 53–77. Cambridge: Cambridge University Press.

1990 "The greening of Myrtos," in *Man's Role in the Shaping of the Eastern Mediterranean Landscape,* edited by S. Bottema, G. Entjes-Nieborg and W. van Zeist, pp. 341–348. Rotterdam: A. A. Balkema.

2001 *Trees and Woodland in the British Landscape: The Complete History of Britain's Trees, Woods and Hedgerows.* London: Phoenix Press.

RENFREW, C.

1992 "World languages and human dispersals: a minimalist view," in *Transition to Modernity: Essays on Power, Wealth and Belief,* edited by J. A. Hall and I. C. Jarvie, pp. 11–68. Cambridge: Cambridge University Press..

1996 "Language families and the spread of farming," in *The Origins and Spread of Agriculture and Pastoralism in Eurasia,* edited by D. R. Harris, pp. 70–92. Washington D.C.: Smithsonian Institution Press.

RHODES, A. N.

1998 "A method for the preparation and quantification of microscopic charcoal from terrestrial and lacustrine sediment cores." *The Holocene* 8: 113–117.

RICHERSON, P. J., BOYD, R. and BETTINGER, R. L.

2001 "Was agriculture impossible during the Pleistocene but mandatory during the Holocene? A climate change hypothesis." *American Antiquity* 66: 387–411.

RITCHIE, J. C., EYLES, C. H. and HAYNES, C. V.

1985 "Sediment and pollen evidence for an early to mid Holocene humid period in the eastern Sahara." *Nature* 314: 352–355.

ROBERTS, N.

1990 "Human-induced landscape change in south and southwest Turkey during the later Holocene," in *Man's role in the shaping of the Eastern Mediterranean Landscape,* edited by S. Bottema, G. Entjes-Nieborg and W. van Zeist, pp. 53–67. Rotterdam: A. A. Balkema.

1998 *The Holocene: an environmental history*. Oxford: Blackwell.

2002 "Did prehistoric landscape management retard the post-glacial spread of woodland in Southwest Asia?." *Antiquity* 76: 1002–1010.

ROBERTS, N., EASTWOOD, W. J., LAMB, H. F. and TIBBY, J. C.

1997 "The age and causes of mid-late Holocene environmental change in southwest Turkey," in *Third Millennium BC Climate Change and Old World Collapse,* edited by H. N. Dalfes, G. Kukla and H. Weiss, pp. 409–429. Berlin: Springer.

ROBERTS, N., REED, J. M., LENG, M. J., KUZUCUOGLU, C., FONTUGNE, M., BERTAUX, J., WOLDRING, H., BOTTEMA, S., BLACK, S., HUNT, E. and KARABIYIKOGLU, M.

2001 "The tempo of Holocene climatic change in the eastern Mediterranean region: new high-resolution crater-lake sediment data from central Turkey." *The Holocene* 11: 721–736.

ROBERTS, N. and WRIGHT, H. E., JR.

1993 "Vegetational, lake-level, and climatic history of the Near East and Southwest Asia," in *Global Climates since the Last Glacial Maximum,* edited by H. E. Wright, Jr. *et al.*, pp. 194–220. Minneapolis: University of Minnesota Press.

RÖHRIG, E.

1991 "Deciduous forests of the Near East," in *Temperate Deciduous Forests,* edited by E. Röhrig and B. Ulrich, pp. 527–537. Amsterdam: Elsevier.

ROSSIGNOL-STRICK, M.

1995 "Sea-land correlation of pollen records in the eastern Mediterranean for the glacial-interglacial transition: biostratigraphy versus radiometric time-scale." *Quaternary Science Reviews* 14: 893–915.

RUBINSON, K. S.

1977 "The chronology of the Middle Bronze Age kurgans at Trialeti," in *Mountains and Lowlands: Essays in the Archaeology of Greater Mesopotamia,* edited by L. D. Levine and T. C. Young, pp. 235–249. Malibu (Calif.): Undena Publications.

2003 "Silver vessels and cylinder sealings: precious reflections of economic exchange in the early second millennium BC," in *Archaeology in the Borderlands: Investigations in Caucasia and Beyond,* edited by A. T. Smith and K. S. Rubinson, pp. 128–143. Los Angeles: Cotsen Institute of Archaeology, University of California.

RUDDIMAN, W. F.

2003 "The anthropogenic greenhouse era began thousands of years ago." *Climatic Change* 61: 261–293.

RUMMERY, T. A.

1983 "The use of magnetic measurements in interpreting the fire histories of lake drainage basins." *Hydrobiologia* 103: 53–58.

RUSISHVILI, N. S.

1990 *Kulturnye Rasteniia na Rannikh Poseleniiakh Gruzii po Paleoetnobotanicheskim Issledovaniiam (avtoreferat).* Tbilisi: AN GSSR.

RUSTAVELI, S.

1966 *The Man in the Panther's Skin (translated by M.S. Wardrop).* Tbilisi: Literatura da Khelovneba.

RYAN, W. B. F., PITMAN, W. C., III, MAJOR, C. O., SHIMKUS, K., MOSKALENKO, V., JONES, G. A., DIMITROV, P., GORÜR, N., SAKINÇ, M. and YÜCE, H.

1997 "An abrupt drowning of the Black Sea shelf." *Marine Geology* 138: 119–126.

SAARNISTO, M.

1986 "Annually laminated lake sediments," in *Handbook of Holocene Palaeoecology and Palaeohydrology,* edited by B. E. Berglund, pp. 343–369. Chichester: John Wiley and Sons.

SAGONA, A.
1984 *The Caucasian Region in the Early Bronze Age*. Oxford: B.A.R. International Series 214.
2004 "Social boundaries and ritual landscapes in late prehistoric Trans-Caucasus and highland Anatolia," in *A View from the Highlands: Archaeological Studies in Honour of Charles Burney,* edited by A. Sagona, pp. 475–538. Louvain: Peeters.

SAGONA, A., (ed.)
2004 *A View from the Highlands: Archaeological Studies in Honour of Charles Burney.* Louvain: Peeters.

SAGONA, A. and SAGONA, C.
2000 "Excavations at Sos Höyük, 1998 to 2000: fifth preliminary report." *Ancient Near Eastern Studies* 37: 56–127.

SAIADIAN, Y. V.
1985 "Sevan – prirodnyi "klimatograf" golotsena," in *Voprosy Geologii Golotsena,* pp. 61–67. Yerevan: AN Armianskoi SSR.

SAKHOKIA, M. F.
1958 "Botanicheskoe opisanie okrestnostei gor. Tbilisi i po marshrutu gor. Tbilisi – plato Shiraki," in *Botanicheskie ekskursii po Gruzii,* edited by M. F. Sakhokia. Tbilisi: Georgian Academy of Sciences.
1961 "Botanical description of the environs of Tbilisi and of the route from Tbilisi to the Shiraki Plateau," in *Botanical Excursions over Georgia,* edited by M. F. Sakhokia, pp. 7–25. Tbilisi: Academy of Sciences of the Georgian SSR.

SALIA, K.
1983 *History of the Georgian Nation.* Paris.

SCHWEIZER, G.
1975 *Untersuchungen zur Physiographie von Ostanatolien und Nordwestiran: Geomorphologische, klima- und hydrogeographische Studien im Vansee- und Rezaiyehsee-Gebiet.* Tübingen: University of Tübingen.

SEDDON, J. M., SANTUCCI, F., REEVE, N. and HEWITT, G. M.
2002 "Caucasus Mountains divide postulated postglacial colonization routes in the white-breasted hedgehog, *Erinaceus concolor.*" *Journal of Evolutionary Biology* 15: 463–467.

SEPPÄ, H. and BENNETT, K. D.
2003 "Quaternary pollen analysis: recent progress in palaeoecology and palaeoclimatology." *Progress in Physical Geography* 27: 548–579.

SEPPÄ, H. and BIRKS, H. J. B.
2001 "July mean temperature and annual precipitation trends during the Holocene in the Fennoscandian tree-line area: pollen-based climate reconstructions." *The Holocene* 11: 527–539.

SEREBRYANNY, L. R.
1984 "Mountain glaciation in the USSR in the late Pleistocene and Holocene," in *Late Quaternary Environments of the Soviet Union,* edited by A. A. Velichko, pp. 45–54. Minneapolis: University of Minnesota Press.

SEREBRYANNY, L. R. and MALYASOVA, E.
1981 "Holocene palaeogeography of Central Caucasus: a sketch." *Striae* 14: 88–92.

SEREBRYANNY, L. R. and ORLOV, A. V.
1982 "Genesis of marginal moraines in the Caucasus." *Boreas* 11: 279–289.
1993 "Debris in tills and moraines as a source of glaciological and palaeoenvironmental information: methodology and applications in the Caucasus." *The Holocene* 3: 63–69.

SEREBRYANNYI, L. R., MARGALITADZE, N. A., JANELIDZE, C. P., GEI, N. A., ILVES, E. O., SKOBEEVA, E. I., VINIEVSKAYA, E. M. and ORLOV, A. V.
1984 "Pozdne- i poslelednikovaia istoriia fitolandshaftov Kolkhidy." *Proceedings of the Academy of Sciences of the Georgian SSR, Biological Series* 10.

SHATILOVA, I. I.
1984 *Istoriia razvitiia pozdnepliotsenovoi rastitel'nosti Zapadnoi Gruzii.* Tbilisi: Metsniereba.

SHATILOVA, I. I. and MCHEDLISHVILI, N. S.
1980 *Palinologicheskie kompleksy Chaudinskikh otlozhenii Zapadnoi Gruzii i ikh stratigraficheskoe znachenie.* Tbilisi: Metsniereba.

SHATILOVA, I. I. and RAMISHVILI, M. N.
1990 *Materialy po istorii flory i rastitel'nosti Gruzii.* Tbilisi: Metsniereba.

SHATILOVA, I. I., RUKHADZE, L. P. and MCHEDLISHVILI, N.
2003 "The results of palynological study of Chaudian (Early Pleistocene) deposits of Western Georgia." *Bulletin of the Georgian Academy of Sciences* 167: 340–343.

SHATILOVA, I. I., RUKHADZE, L. P., MCHEDLISHVILI, N. and MAKHARADZE, N.
2002 "The main stages of the development of vegetation and climate of Western Georgia during the Gurian (Euopleistocene) time by the pollen records." *Bulletin of the Georgian Academy of Sciences* 166: 624–627.

SHATILOVA, I. I. and STUCHLIK, L.
1996 "The genus *Pteris* L. in the Neogene deposits of western Georgia." *Acta Palaeobotanica* 36: 5–77.

SIEGENTHALER, U. and EICHER, U.
1986 "Stable oxygen and carbon isotope analyses," in *Handbook of Holocene Palaeoecology and Palaeohydrology,* edited by B. E. Berglund, pp. 407–422. Chichester: John Wiley and Sons.

SINGH, G. and GEISSLER, E. A.
1985 "Late Cainozoic history of vegetation, fire, lake levels and climate, at Lake George, New South Wales, Australia." *Philosophical Transactions of the Royal Society of London* B 311: 379–447.

SMITH, A. G.
1965 "Problems of inertia and threshold related to post-Glacial habitat changes." *Proceedings of the Royal Society of London* B 161: 331–342.

SMITH, A. T. and RUBINSON, K. S., EDS.
2003 *Archaeology in the Borderlands: Investigations in Caucasia and Beyond.* Los Angeles: Cotsen Institute Press.

SOLOVIEVA, N., TARASOV, P. E. and MACDONALD, G. M.
2005 "Quantitative reconstruction of Holocene climate from the Chuna lake pollen record, Kola Peninsula, northwest Russia." *The Holocene* 15: 141–148.

SOSNOVSKII, D. I.
1915 "Protsessy ischeznoveniia lesov v blizhaishikh okrestnostiax Tiflisa." *Izvestiia Kavkazkago Otdela Imperatorskago Russkago Geograficheskago Obshchestva* 23: 37–51.
1933 "K voprosy o floristicheskom kharaktere Dzhavakhetii," in *Dzhavakhetiia: Materialy po izucheniiu prirodnykh resursov Akhalkalakskovo Nagor'ia,* pp. 229–235. Tiflis: Zak. GIZ.

STEVENS, L. R., ITO, E., SCHWALB, A. and WRIGHT, H. E., JR.
2006 "Timing of atmospheric precipitation in the Zagros Mountains inferred from a multi-proxy record from Lake Mirabad, Iran." *Quaternary Research* 66: 494–500.

STEVENS, L. R., WRIGHT, H. E., JR. and ITO, E.
2001 "Proposed changes in seasonality of climate during the Lateglacial and Holocene at Lake Zeribar, Iran." *The Holocene* 11: 747–755.

STEVENSON, A. C. and HARRISON, R. J.
1992 "Ancient forests in Spain: a model for land-use and dry forest management in south-west Spain from 4000 BC to 1900 AD." *Proceedings of the Prehistoric Society* 58: 227–247.

STRABO
1928 *Geography (translated by H. L. Jones)*. Cambridge Massachusetts: Harvard University Press.

STUCHLIK, L. and KVAVADZE, E. V.
1993 "Spore-pollen spectra of surface samples from *Zelkova* forests on the Babaneuri reservation (east Georgia)." *Acta Palaeobotanica* 33: 357–364.
1995 "On the problem of actuopalynology in the Carpathians and Caucasus." *Acta Palaeobotanica* 35: 73–83.
1998 "Subfossil pollen spectra of flood-plain forest of *Pterocarya pterocarpa* in the Alazani Valley (East Georgia)." *Acta Palaeobotanica* 38: 217–222.

STUIVER, M. and REIMER, P. J.
1993 "Extended ^{14}C database and revised CALIB radiocarbon calibration program." *Radiocarbon* 35: 215–230.

SUGITA, S.
1993 "A model of pollen source area for an entire lake surface." *Quaternary Research* 39: 239–244.
1994 "Pollen representation of vegetation in Quaternary sediments: theory and method in patchy vegetation." *Journal of Ecology* 82: 881–897.

SUGITA, S., GAILLARD, M.-J. and BROSTRÖM, A.
1999 "Landscape openness and pollen records: a simulation approach." *The Holocene* 9: 409–421.

SVANIDZE, G. G. and PAPINASHVILI, L. K., (eds.)
1992 *Klimat Tbilisi*. St. Petersburg: Gidrometeoizdat.

TARASOV, P. E., VOLKOVA, V. S., WEBB, T., III, GUIOT, J., ANDREEV, A. A., BEZUSKO, L. G., BEZUSKO, T. V., BYKOVA, G. V., DOROFEYUK, N. I., KVAVADZE, E. V., OPISOVA, I. M., PANOVA, N. K. and SEVASTYANOV, D. V.
2000 "Last glacial maximum biomes reconstructed from pollen and plant macrofossil data from northern Eurasia." *Journal of Biogeography* 27: 609–620.

TARASOV, P. E., WEBB, T., III, ANDREEV, A. A., AFANAS'EVA, N. B., BEREZINA, N. A., BEZUSKO, L. G., BLYAKHARCHUK, T. A., BOLIKHOVSKAYA, N. S., CHEDDADI, R., CHERNAVSKAYA, M. M., CHERNOVA, G. M., DOROFEYUK, N. I., DIRKSON, V. G., ELINA, G. A., FILIMONOVA, L. V., GLEBOV, F. Z., GUIOT, J., GUNOVA, V. S., HARRISON, S. P., JOLLY, D., KHOMUTOVA, V. I., KVAVADZE, E. V., OSIPOVA, I. M., PANOVA, N. K., PRENTICE, I. C., SAARSE, L., SEVASTYANOV, D. V., VOLKOVA, V. S. and ZERNITSKAYA, V. P.
1998 "Present-day and mid-Holocene biomes reconstructed from pollen and plant macrofossil data from the former Soviet Union and Mongolia." *Journal of Biogeography* 25: 1029–1053.

TATASHIDZE, Z., (ed.)
2000 *Sakartvelos Geograpia*. Tbilisi: Metsniereba.

TAUBER, H.
1965 "Differential pollen dispersion and the interpretation of pollen diagrams." *Danmarks Geologiske Undersøgelse. II Række* 89.
1967 "Investigations of the mode of pollen transfer in forested areas." *Review of Palaeobotany and Palynology* 3: 277–286.

TAUSCH, R. J., CHARLET, D. A., WEIXELMAN, D. A. and ZAMUDIO, D. C.
1995 "Patterns of ordination and classification instability resulting from changes in input order." *Journal of Vegetation Science* 6: 897–902.
TER BRAAK, C. J. F.
1995 "Ordination," in *Data Analysis in Community and Landscape Ecology,* edited by R. H. G. Jongman, C. J. F. ter Braak and O. F. R. van Tongeren, pp. 91–173. Cambridge: Cambridge University Press.
TER BRAAK, C. J. F. and PRENTICE, I. C.
1988 "A theory of Gradient Analysis." *Advances in Ecological Research* 18: 271–317.
THOMAS, I.
1999 "The recent history of rangelands in Western Asia," in *People and Rangelands: Proceedings of the VI International Rangelands Congress, Townsville*, edited by D. Eldridge and D. Freudenberger, pp. 1048–1052. Aitkenvale (Queensland).
THOMPSON, R., BATTARBEE, R. W., O'SULLIVAN, P. E. and OLDFIELD, F.
1975 "Magnetic susceptibility of lake sediments." *Limnology and Oceanography* 20: 687–698.
THOMSON, R. W.
1996 *Rewriting Caucasian History: the Medieval Armenian Adaptation of the Georgian Chronicles*. Oxford: Clarendon Press.
TINNER, W., CONEDERA, M., AMMANN, B., GÄGGELER, H. W., GEDYE, S., JONES, R. and SÄGESSER, B.
1998 "Pollen and charcoal in lake sediments compared with historically documented fires in southern Switzerland since AD 1920." *The Holocene* 8: 31–42.
TINNER, W., CONEDERA, M., AMMANN, B. and LOTTER, A. F.
2005 "Fire ecology north and south of the Alps since the last ice age." *The Holocene* 15: 1214–1226.
TINNER, W., CONEDERA, M., GOBET, E., HUBSCHMID, P., WEHRLI, M. and AMMANN, B.
2000 "A palaeoecological attempt to classify fire sensitivity of trees in the southern Alps." *The Holocene* 10: 565–574.
TINNER, W. and HU, F. S.
2003 "Size parameters, size-class distribution and area-number relationship of microscopic charcoal: relevance for fire reconstruction." *The Holocene* 13: 499–505.
TINSLEY, H. M. and SMITH, R. T.
1974 "Surface pollen studies across a woodland/heath transition and their application to the interpretation of pollen diagrams." *New Phytologist* 73: 547–565.
TOMESCU, A. M. F.
2000 "Evaluation of Holocene pollen records from the Romanian Plain." *Review of Palaeobotany and Palynology* 109: 219–233.
TOURNEFORT, J. P., DE
1718 *A Voyage into the Levant: perform'd by command of the late French King; containing the ancient and modern state of the Islands of the Archipelago; as also of Constantinople, the coasts of the Black Sea, Armenia, Georgia, the frontiers of Persia, and Asia Minor*. London: D. Browne, A. Bell, J. Darby, A. Bettesworth, J. Pemberton, C. Rivington, J. Hooke, R. Cruttenden, T. Cox, J. Battley and E. Symon.
TROELS-SMITH, J.
1955 "Karakterisering af løse jordarter." *Danmarks Geologiske Undersøgelse* 4: 1–73.
TSAGARELI, A. L.
1964 "Chetvertichnaia sistema," in *Geologiia SSSR: Grunzinskaia SSR,* edited by P. D. Gamkrelidze, pp. 332–352. Moskva: Nedra.
TSERETELI, D. V.
1971 "Iveriiskaiia chasti mezhgor'ia," in *Geomorfologiia Gruzii,* edited by L. I. Maruashvili, pp. 282–309. Tbilisi: Metsniereba.

TSERETELI, D. V. and MAISURADZE, G. M.
1976 "Osnovnye cherty paleogeografii Gruzii v verkhnem pleistotsene." *Mezhdunarodnaia Geografiia* 76: 354–357.
TSERETELI, L. D., KLOPOTOVSKAYA, N. B. and KURENKOVA, E. I.
1982 "Mnogosloinyi pamiatnik Apiancha (Abkhazia)," in *Chetvertichnaia Sistema Gruzii*, pp. 198–212. Tbilisi: Metsniereba.
TSETSKHLADZE, G. R.
2003 "The culture of ancient Georgia in the first millennium BC and Greater Anatolia: diffusion or migration?," in *Archaeology in the Borderlands: Investigations in Caucasia and Beyond,* edited by A. T. Smith and K. S. Rubinson, pp. 229–245. Los Angeles: Cotsen Institute of Archaeology.
TUMAJANOV, I. I.
1971 "Changes of the Great Caucasus forest vegetation during the Pleistocene and Holocene," in *Plant Life of South-West Asia,* edited by P. H. Davis, P. C. Harper and I. C. Hedge, pp. 73–87. Edinburgh: Botanical Society of Edinburgh.
TUMAJANOV, I. I. and GOGICHAISHVILI, L. K.
1969 "Osnobnye cherty poslekhvalynskoi istorii lesnoi rastitel'nosti Iorskoi nizmennosti (Vostochnaia Gruziia)," in *Golotsene: k VIII kongressu INQUA, Parizh 1969,* edited by M. I. Neustadt, pp. 183–194. Moscow: Nauka.
TURNER, S. D. and BROWN, A. G.
2004 "*Vitis* pollen dispersal in and from organic vineyards, I: pollen trap and soil pollen data." *Review of Palaeobotany and Palynology* 129: 117–132.
TUSHABRAMISHVILI, D. M. and VEKUA, A. K.
1977 "Paleolit v Gruzii," in *Palinologicheskie issledovania v Gruzii,* pp. 178–197. Tbilisi: Metsniereba.
TUSHABRAMISHVILI, N., ADLER, D. S., BAR-YOSEF, O. and BELFER-COHEN, A.
2002 "Current middle & upper Palaeolithic research in the southern Caucasus." *Antiquity* 76: 927–928.
TUTIN, T. G., HEYWOOD, V. H., BURGES, N. A., MOORE, D. M., VALENTINE, D. H., WALTERS, S. M. and WEBB, D. A., (eds.)
1980 *Flora Europaea, Volume 5*. Cambridge: Cambridge University Press.
TVALCHRELIDZE, M., LEBANIDZE, Z. and JAOSHVILI, G.
2004 "Eustatics of the Black Sea and sedimentation peculiarities during the last 20 000 years (Georgian sector of the Black Sea)." *Proceedings of the Geological Institute, Georgian Academy of Sciences, new series* 119: 656–670.
UCHUPI, E. and ROSS, D. A.
2000 "Early Holocene marine flooding of the Black Sea." *Quaternary Research* 54: 68–71.
UKLEBA, D.
1989 "Tbilisis da misi midamoebis buneba," in *Tbilisi: Ekonomiur-Geograpiuli Gamokvleva,* edited by E. Jaoshvili, pp. 22–62. Tbilisi: Sabchota Sakartvelo.
URUSHADZE, T., (ed.)
1999 *Sakartvelos Niadagebis Ruka*. Tbilisi: Kartograpia.
VAKHUSHTI BAGRATIONI
1983 "Agts'era sameposa sakartvelosa," in *Kartuli P'roza,* edited by N. Abesadze, pp. 381–454. Tbilisi: Sabch'ota Sakartvelo.
VAN DER KNAAP, W. O., VAN LEEUWEN, J. F. N., FINSINGER, W., GOBET, E., PINI, R., SCHWEIZER, A., VALSECCHI, V. and AMMANN, B.
2005 "Migration and population expansion of *Abies, Fagus, Picea,* and *Quercus* since 15000 years in and across the Alps, based on pollen-percentage threshold values." *Quaternary Science Reviews* 24: 645–680.

VAN GEEL, B.
1978 "A palaeoecological study of Holocene peat bog sections in Germany and the Netherlands, based in the analysis of pollen, spores and macro-and microscopic remains of fungi, algae, cormophytes and animals." *Review of Palaeobotany and Palynology* 25: 1–120.
VAN GEEL, B. and APTROOT, A.
2006 "Fossil ascomycetes in Quaternary deposits." *Nova Hedwigia* 82: 313–329.
VAN GEEL, B., BOHNCKE, S. J. P. and DEE, H.
1981 "A palaeoecological study of an Upper Late Glacial and Holocene sequence from "De Borchert," the Netherlands." *Review of Palaeobotany and Palynology* 31: 367–448.
VAN GEEL, B., BUURMAN, J., BRINKKEMPER, O., SCHELVIS, J., APTROOT, A., VAN REENEN, G. and HAKBIJL, T.
2003 "Environmental reconstruction of a Roman Period settlement site in Uitgeest (The Netherlands), with special reference to coprophilous fungi." *Journal of Archaeological Science* 30: 873–883.
VAN GEEL, B. and VAN DER HAMMEN, T.
1978 "Zygnemataceae in Quaternary Colombian sediments." *Review of Palaeobotany and Palynology* 25: 377–392.
VAN GROENEWOUD, H.
1992 "The robustness of Correspondence, Detrended Correspondence, and TWINSPAN analysis." *Journal of Vegetation Science* 3: 239–246.
VAN ZEIST, W.
1967 "Late Quaternary vegetation history of Western Iran." *Review of Palaeobotany and Palynology* 2: 301–311.
VAN ZEIST, W. and BOTTEMA, S.
1977 "Palynological investigations in Western Iran." *Palaeohistoria* 24: 19–85.
1991 *Late Quaternary Vegetation of the Near East.* Weisbaden: Verlag.
VAN ZEIST, W. and WOLDRING, H.
1978 "A postglacial pollen diagram from Lake Van in eastern Anatolia." *Review of Palaeobotany and Palynology* 26: 249–276.
VAN ZEIST, W., WOLDRING, H. and STAPERT, D.
1975 "Late Quaternary vegetation and climate of southwestern Turkey." *Palaeohistoria* 17: 53–143.
VAN ZEIST, W. and WRIGHT, H. E., JR.
1963 "Preliminary pollen studies at Lake Zeribar, Zagros Mountains, southwestern Iran." *Science* 140: 65–67.
VANCE, R. E. and MATHEWES, R. W.
1994 "Deposition of modern pollen and plant macroremains in a hypersaline prairie lake basin." *Canadian Journal of Botany* 72: 539–548.
VAVILOV, N. I.
1992 *Origin and Geography of Cultivated Plants.* Cambridge: Cambridge University Press.
VEKUA, A. and LORDKIPANIDZE, D.
1998 "The Pleistocene paleoenvironment of the Transcaucasus." *Quaternaire* 9: 261–266.
VEKUA, A. K.
1984 "Zhivotnyi mir Vostochnoi Gruzii v epokhu eneolita po materialam poseleniia Arukhlo," in *Chelovek i Okruzhaiushchaia Ego Sreda,* edited by T. N. Chubinishvili, pp. 91–119. Tbilisi: Metsniereba.

VERMEERSCH, S., DE GENST, W., VERMOESEN, F. and TRIEST, L.
2003 "The influence of transformations of an ordinal scale of a floristic gradient, applied on a TWINSPAN classification." *Flora* 198: 389–403.

VERMOERE, M., BOTTEMA, S., VANHECKE, L., WAELKENS, M., PAULISSEN, E. and SMETS, E.
2002 "Palynological evidence for late-Holocene human occupation recorded in two wetlands in SW Turkey." *The Holocene* 12: 569–584.

WALKER, D.
1971 "Quantification in historical plant ecology." *Proceedings of the Ecological Society of Australia* 6: 91–104.

WALTER, H.
1974 *Die Vegetation Osteuropas, Nord- und Zentralasiens.* Stuttgart: Fischer.

WANG, X., VAN DER KAARS, S., KERSHAW, A. P., BIRD, M. and JANSEN, F.
1999 "A record of fire, vegetation and climate through the last three glacial cycles from Lombok Ridge core G6–4, eastern Indian Ocean, Indonesia." *Palaeogeography, Palaeoclimatology, Palaeoecology* 147: 241–256.

WARTENBURG, D., FERSON, S. and ROHLF, F. J.
1987 "Putting things in order: a critique of Detrended Correspondence Analysis." *American Naturalist* 129: 434–448.

WASYLIKOWA, K.
1967 "Late Quaternary plant macrofossils from Lake Zeribar, western Iran." *Review of Palaeobotany and Palynology* 2: 313–318.
1986 "Plant macrofossils preserved in prehistoric settlements compared with anthropogenic indicators in pollen diagrams," in *Anthropogenic Indicators in Pollen Diagrams,* edited by K.-E. Behre, pp. 173–185. Rotterdam: A.A. Balkema.
2005 "Palaeoecology of Lake Zeribar, Iran, in the Pleniglacial, Lateglacial and Holocene, reconstructed from plant macrofossils." *The Holocene* 15: 720–735.

WATSON, P. J.
1995 "Explaining the transition to agriculture," in *Last Hunters – First Farmers,* edited by T. D. Price and A. B. Gebauer, pp. 21–37. Santa Fe: School of American Research Press.

WEBB, T., III
1973 "A comparison of modern and pre-settlement pollen from southern Michigan, U.S.A.." *Review of Palaeobotany and Palynology* 16.
1980 "The reconstruction of climatic sequences from botanical data." *Journal of Interdisciplinary History* 10: 749–772.

WHITTAKER, R. H.
1967 "Gradient analysis of vegetation." *Biological Reviews* 42: 207–264.

WICK, L., LEMCKE, G. and STURM, M.
2003 "Evidence of Lateglacial and Holocene climatic change and human impact in eastern Anatolia: high-resolution pollen, charcoal, isotopic and geochemical records from the laminated sediments of Lake Van, Turkey." *The Holocene* 13: 665–675.

WILLIAMS, J. W., SUMMERS, R. L. and WEBB, T., III
1998 "Applying plant functional types to construct biome maps from eastern North American pollen data: comparison with model results." *Quaternary Science Reviews* 17: 607–627.

WILLIS, K. J., SÜMEGI, P., BRAUN, M., BENNETT, K. D. and TÓTH, A.
1998 "Prehistoric land degradation in Hungary: who, how and why?." *Antiquity* 72: 101–113.

WOLDRING, H. and CAPPERS, R.
2001 "The origin of the 'wild orchards' of Central Anatolia." *Turkish Journal of Botany* 25: 1–9.

WRIGHT, H. E., JR.

1967 "The use of surface samples in Quaternary pollen analysis." *Review of Palaeobotany and Palynology* 2: 321–330.

1970 "Environmental changes and the origin of agriculture in the Near East." *BioScience* 20: 210–212.

1980 "Cores of soft lake sediments." *Boreas* 9: 107–114.

1993 "Environmental determinism in Near Eastern prehistory." *Current Anthropology* 34: 458–469.

WRIGHT, H. E., JR., AMMANN, B., STEFANOVA, I., ATANASSOVA, J., MARGALITADZE, N. A., WICK, L. and BLYAKHARCHUK, T. A.

2003 "Late-glacial and early-Holocene dry climates from the Balkan peninsula to Southern Siberia," in *Aspects of Palynology and Palaeoecology,* edited by S. B. Tonkov, pp. 127–136. Sofia – Moscow: Pensoft Publishers.

WRIGHT, H. E., JR. and BARNOSKY, C. W.

1984 "Introduction to the English edition," in *Late Quaternary Environments of the Soviet Union,* edited by A. A. Velichko, pp. xiii–xxii. Minneapolis: University of Minnesota Press.

WRIGHT, H. E., JR., MCANDREWS, J. H. and VAN ZEIST, W.

1967 "Modern pollen rain in western Iran, and its relation to plant geography and Quaternary vegetational history." *Journal of Ecology* 55: 415–443.

WRIGHT, H. E., JR. and PATTEN, H. L.

1963 "The pollen sum." *Pollen et Spores* 5: 445–450.

YANUSHEVICH, Z. V. and RUSISHVILI, N. S.

1984 "Novye paleoetnobotanicheskie nakhodki na eneoliticheskom poselenii Arukhlo I," in *Chelovek i okruzhaiushchaia ego sreda: materialy po arkheologii Gruzii i Kavkaza,* edited by T. N. Chubinishvili, pp. 21–33. Tbilisi: Metsniereba.

YASUDA, Y., KITAGAWA, H. and NAKAGAWA, T.

2000 "The earliest record of major anthropogenic deforestation in the Ghab Valley, northwest Syria: a palynological study." *Quaternary International* 73/74: 127–136.

YATSENKO-KHMELEVSKII, A. A. and KANDELAKI, G. V.

1941 "Drevesnye ugli iz raskopok Beshtashenskoi tsiklopicheskoi kreposti (Tsalka)." *Soobshcheniia Akademii Nauk Gruzinskoi SSR* 2: 441–448.

YAZVENKO, S. B.

1991 "Modern pollen-vegetation relationships on the Southeast Caucasus." *Grana* 30: 350–356.

1994 "Holocene pollen record from a peatland in the West Caucasus, Abkhasia, Black Sea region." *Journal of Paleolimnology* 12: 65–74.

YU, G., PRENTICE, I. C., HARRISON, S. P. and SUN, X.

1998 "Pollen-based biome reconstructions for China at 0 ka and 6 ka." *Journal of Biogeography* 25: 1055–1069.

ZHORZHIKASHVILI, L. G. and GOGADZE, E. M.

1974 *Pamiatniki Trialeti Epokhi Rannei i Srednei Bronzy.* Tbilisi: Metsniereba.

ZOHARY, D.

1996 "The mode of domestication of the founder crops of Southwest Asian agriculture," in *The Origins and Spread of Agriculture and Pastoralism in Eurasia,* edited by D. R. Harris, pp. 142–158. Washington D.C.: Smithsonian Institution Press.

ZOHARY, D. and HOPF, M.

1993 *Domestication of Plants in the Old World.* Oxford: Clarendon Press.

ZOHARY, D. and SPIEGEL-ROY, P.

1975 "Beginnings of fruit growing in the Old World." *Science* 187: 319–327.

ZOHARY, M.

1973 *Geobotanical Foundations of the Middle East.* Stuttgart: Verlag.

APPENDIX A

PRESENT AND PAST POLLEN MAPS OF GEORGIA

On the following pages I present a gazetteer of changing distribution and abundance for Georgia's main pollen types. Maps of present and past pollen distribution and abundance for the last 14,000 years are given, based on the mapping method described in Chapter Seven. Site-distribution maps are followed by maps of modern abundances of individual pollen taxa, then by maps of changing pollen abundance for each 1000-year time-slice. The maps are arranged in alphabetical order, with arboreal pollen types first, followed by non-arboreal types. Chenopodiaceae, *Betula*, *Fagus*, *Quercus* and *Picea* are presented separately in Chapter Seven.

Directly following the maps is a section detailing the ecology, pollen representation and Late Quaternary history of the individual taxa. This section is included primarily as an aid and reference to future research work in Georgia, and presents much of the ecological information (as applicable to the Caucasus) in English for the first time. It is not intended to be the last word on the reasons for the observed changes in distribution and abundance.

Hopefully these data will, as in other parts of the world, provide the basis for robust debate on the reasons that some plants are more common now than in the past while others are less common today.

ABIES

Ecology

Abies is represented by only one species in the Caucasus region: *Abies nordmanniana*. It is a conifer that can grow to 50 m or more in height and up to 1½ m in trunk diameter. It grows in the middle and upper forest belts from about 800 to 2000 m above sea level. It has a preference for cool, moist, north-facing slopes and often grows admixed with *Picea orientalis* and *Fagus orientalis*. It is widely distributed in Western Caucasus, extending eastward along the Main Caucasus Range as far as the Little Liakhvi River and along the Trialeti Range as far as the town of Manglisi in Eastern Georgia.[1] The mean annual temperature of *Picea-Abies* forests in the Caucasus is between 4 and 6 degrees, annual precipitation in excess of 800 mm and mostly above 1300 mm, and relatively high humidity (**Table A.1**).[2] This precludes *Abies* from entering much of the Eastern Caucasus where the climate is too dry. The soils beneath *Abies* forests are usually brown forest soils or podzols, with a slightly acidic pH.[3]

[1] Ketskhoveli 1971.
[2] Gulisashvili 1964; Walter 1974.
[3] Gulisashvili 1964.

Table A.1. The climate of two contrasting *Abies nordmanniana* habitats: Monthly climatic data for Mestia, 43° 3' N, 42° 44' E, 1320 m a.s.l., and Manglisi, 41° 42' N, 44° 23' E, 1100 m a.s.l. T: average monthly temperature; P: average monthly precipitation. Source: Gulisashvili 1964.

	J	F	M	A	M	J	J	A	S	O	N	D	Avg.
Mestia, Svaneti													
T (°C)	–7.6	–4.7	–0.6	5.6	11.1	14.2	16.8	16.8	12.8	7.9	1.8	–4.5	5.8
P (mm)	56	49	77	76	86	77	79	74	88	118	141	129	1333
Manglisi, Trialeti													
T (°C)	–2.2	–1.9	2.2	6.7	11.7	15.2	18.6	18.4	13.9	9.6	3.7	–0.1	8.0
P (mm)	16	34	32	79	123	103	71	49	59	54	44	26	690

The main co-dominants in *Abies nordmanniana* forests in Georgia are *Picea* and *Fagus*. A.G. Dolukhanov differentiates these forests on the basis of their undergrowth:[4] those without a developed understorey and those with 'Colchic' understorey. In the first group are forests with *Festuca drymeja, Luzula sylvatica,* dwarf shrubs, ferns, and subalpine meadow herbs. These are more typical of the drier parts of fir's distribution. Trees such as *Betula litwinowii* and *Sorbus aucuparia* are found with *Abies nordmanniana* in some tree line situations. The Colchic understorey group includes fir forests with *Rhododendron ponticum, Laurocerasus officinalis, Ilex colchica* and *Vaccinium arctostaphylos* in the understorey.[5] Many of these Colchic understorey species are evergreens.

Abies nordmanniana, apart from being drought susceptible, is prone to attack by fir beetles and is particularly intolerant of fire. Its exposure to beetle infestation has increased in recent times through overexploitation.[6] Palynological studies of *Abies alba*, the European equivalent of *A. nordmanniana*, have shown that fire usually results in the death of the tree.[7] The Greek species, *A. cephalonica*, is likewise killed by fire.[8]

The economic value of *Abies nordmanniana* is considerable. It is regarded as the best material for paper-making. It is also used for construction, in tanning and to obtain rosin. Essential oils are distilled from the foliage and turpentine from the branches.[9] It was logged heavily in the early nineteenth century by Caucasian industry, particularly for paper-making.[10]

Abies nordmanniana is a Euxine species, with a distribution spreading into northeastern Anatolia. In northwestern Anatolia it is replaced by the subspecies *A. nordmanniana* subsp. *bornmuelleriana*.

Pollen Representation

Abies is wind pollinated and produces very large (to 190 μm) saccate pollen grains with thick exines, posing no problems for identification. Its pollen is relatively heavy and does not disperse well, so large quantities of *Abies* pollen imply large populations of *Abies nordmanniana*

[4] Dolukhanov 1989.
[5] Dolukhanov 1989.
[6] Dolukhanov 1966.
[7] Tinner *et al.* 2000.
[8] Grove and Rackham 2001.
[9] Ketskhoveli 1971.
[10] Dolukhanov 1966.

trees. To a degree its pollen is carried by strong updraughts into treeless mountain areas, where it is over-represented only because of the lack of other wind pollinators in the alpine flora. However, this occurs to a much lesser degree than for *Pinus*, making the interpretation of *Abies* more straightforward.

The map of modern *Abies* percentages shows a clear geographical division between northwestern Georgia and the remainder of the study area. The highest *Abies* percentages (23–49 percent) are recorded from sites where *Abies* is a forest dominant or co-dominant, or sites just above an *Abies*-dominated tree line. High *Abies* percentages on the Black Sea coast are probably influenced by fluvially-transported pollen. Sites with 2.5–6.5 percent *Abies* pollen are those where fir is a minor component of forests, and less than 2.5 percent pollen indicates long-distance pollen transport. These data accord with that of Huntley and Birks,[11] *i.e.* that proportions of 5 percent or more indicate the presence of *Abies* in the surrounding forest.

Late Quaternary History of Abies *from Pollen Records*

Abies pollen is prevalent in late Quaternary sediments from some sites on the Black Sea coast, indicating that in mixed deciduous-coniferous low-altitude refugia, *Abies* was an important, if not dominant, tree. This was a position it occupied through much of the Pleistocene.[12]

In the late glacial, the data available indicate that *Abies* was quite frequent in the Black Sea lowlands, contributing up to 25 percent of arboreal pollen at Sukhumi. Fir was obviously a major component of the forest refugia there. The one inland site on the Tsalka Plateau, Aligol, has trace proportions of *Abies*, which are likely to be pollen carried by the wind from refugia in the river valleys and gorges bordering the South Georgian Uplands.

The representation of *Abies* decreased during the Younger Dryas, and at the beginning of the Holocene it began to expand in the highlands. By 9000 cal BP, *Abies* had moved outward from its glacial refugia into most of the areas for which we have pollen records. Its importance remained strong in the lowlands near Sukhumi, but it also became important on the Tsalka Plateau. Nowadays, *Abies* grows only on the fringes of the Tsalka Plateau, which is almost entirely deforested. Over the next few thousand years, *Abies* representation declined at Black Sea lowland sites. It remained an important component of the mountain flora, and expanded into new areas in the high mountains of the Caucasus and the South Georgian Uplands.

Then, during the mid Holocene, fir pollen experiences a slight resurgence. It increases along the Black Sea lowlands and becomes dominant at some northwestern mountain sites. Fir remained important in the South Georgian Uplands, as a constituent of mixed forests (its presence there independently confirmed by archaeo-botanical analysis.[13]) However, by 3000 cal BP, it had begun to decline in this area, and by 2000 cal BP was all but eliminated. Its decline appears to correlate very strongly with an increasingly intensive fire regime on the plateaux of Southern Georgia. The same may explain the decline of *Abies* pollen values generally throughout the study area during the late Holocene. *Abies* remained a significant member of the mountain forests of Abkhazia (NW Caucasus).

Fir's history in the study area is complex. It almost certainly responded to increasing precipitation at the beginning of the Holocene by expanding outwards and upwards from its maritime and riverine refugia. Its spread gradually extended as it migrated upslope, but at the same time it was outcompeted by deciduous species at lower altitudes. It remained a component of mixed forests, a modern analogue of which we are at a loss to find in the present day. It is difficult to discern the importance of the records from the Black Sea lowlands, because of

[11] Huntley and Birks 1983.

[12] Shatilova 1984; Shatilova and Mchedlishvili 1980.

[13] Yatsenko-Khmelevskii and Kandelaki 1941.

the possibility that the pollen deposited there was fluvially transported. It is possible that fir trees lingered in Colchic vegetation long after their main cohort had migrated to higher altitudes, especially considering that the Black Sea coastline was not always at its present high-stand.

The most obvious temporal pattern in its development is its decline from most sites in southern Georgia. As mentioned, it seems highly likely that this was a change brought on by human impact. The present-day distribution of fir in Caucasian forests therefore does not reflect merely climatic or edaphic preferences; it is determined to a significant extent by past human activity.

ACER

There are ten *Acer* species found in the study area. These are *Acer laetum, A. platanoides, A. campestre, A. velutinum, A. pseudoplatanus, A. trautvetteri, A. tataricum, A. hyrcanum, A. ibericum* and the Georgian endemic *A. sosnowskyi*. All are deciduous trees that tend to occur solitarily or in small groups within forest.[14]

Acer laetum grows 15–25 m in height amongst forests dominated by oak, oak-hornbeam, chestnut, hornbeam-beech and spruce-fir. It occurs from the lowlands up to about 1700 m above sea level in all forested parts of the study area. *Acer platanoides* is a taller tree to 30 m, occurring frequently in mixed broadleaf forests, as well as forests dominated by beech, hornbeam, chestnut and oak. Like *A. laetum*, it is widespread in the study area.

Acer campestre grows 15–20 m in montane broadleaf forests almost as far as the upper tree line. It is also found in gallery forest (oak-elm) in the lowlands; rarely in coniferous or beech forests. *Acer velutinum* is a large tree to 40 m high, growing in the eastern Georgian region of Kakheti as well as near the Caspian Sea. It occurs along mountain gorges, ravines and riversides amongst mixed broadleaf forests.[15]

Acer pseudoplatanus is widely distributed in Europe, the Mediterranean and Western Asia, and in the study area grows in various forest types up to 1600 m elevation, having a preference for river gorges and streamside terraces. *Acer trautvetteri* is a subalpine tree, mostly occurring alongside *Betula* woods and tall subalpine herbfields at 1700–2300 m elevation. *Acer tataricum* is a small tree or shrub growing in broadleaf forest understorey, in canopy gaps and at forest margins. *Acer hyrcanum* is a tree of dry forest slopes, confined to the eastern parts of the study area and eastwards. *Acer ibericum* is a small tree of arid forest, piedmont steppes and rocky slopes, and is distributed through Eastern Caucasus, Iran and Turkey. Finally, *Acer sosnowskyi* is found growing only in the limestone areas of Western Georgia.

Most *Acer* species are forest trees, but a few, like *A. ibericum* and *A. trautvetteri*, inhabit more open landscapes at the forest margins. All maples are valued in the Caucasus as honey trees and for their strong timber.[16]

Acer pollen is tricolpate with a characteristic, striated exine. Maples are pollinated by insects, so their pollen never occurs in great numbers. The presence of *Acer* pollen indicates the presence of maple trees within the surrounding vegetation. The map of modern pollen percentages demonstrates the scarcity of *Acer* pollen in the current environment. Values up to 0.9 percent indicate local presence of a few *Acer* trees, and values between 0.9 and 4 percent indicate a significant number of maples in the vegetation. Values above 4 percent have no analogue in the present-day, but it is assumed that they represent co-dominance in the vegetation. Sites in the modern pollen map with significant proportions of maple pollen are found

[14] Ketskhoveli 1983.
[15] Ketskhoveli 1983.
[16] Ketskhoveli 1983.

at the steppe-forest border near Tbilisi, at the upper tree line in the South Georgian Uplands and at forest and tree line situations on the Main Caucasus.

Acer pollen is well represented through the late glacial at Sagarejo in Eastern Georgia. This site appears to have been a refugium for thermophile broadleaf trees, and *Acer* was an important component. At the beginning of the Holocene its role at this site increases, and *Acer* pollen also appears near Bakuriani and Kazbegi, places where *Acer trautvetteri* probably grew. Maples became more prevalent in southern Georgia during the mid Holocene, especially around 5000 cal BP. By 4000 cal BP, their role appears to have declined. Late Holocene occurrences seem to be concentrated in Southern and Eastern Georgia, with one record from Svaneti on the Caucasus Range.

The maple record is problematic because it shows changes in abundance that are perhaps only a result of poor pollen dispersion and the choice of 1000-year intervals for data collection. The apparent decline of *Acer* in Southern Georgia around 4000 cal BP is only an artefact of sample selection. However, it can be said that *Acer* had a glacial refugium in Eastern Georgia during the late Pleistocene and does not appear to have been a major participant in refugia in the Black Sea lowlands. There it probably found refuge in mountain gorges. The Southern Georgian expansion of *Acer* during the mid Holocene follows the development of mixed forests, but it is interesting to note that *Acer* pollen is perhaps best represented during phases when these sites were above the tree line. Similarly, *Acer* pollen is best represented near Tbilisi in *Juniperus*-dominated pollen spectra, indicative of scrub vegetation. *Acer* pollen is simply more likely to be recorded in non-forest situations than in forests because of its poor dispersion and representation. In a forest its pollen signal is diluted by wind-pollinating trees and trapped by the canopy.[17] Hence *Acer* records from the Caucasus may indicate forest borders better than they do the forests themselves.

ALNUS

Ecology

There are two *Alnus* species in the study area: *Alnus barbata* and *Alnus incana*. The former is a large deciduous tree growing 30–40 m in height. It occurs from sea level up to 1500 m elevation, with a strong preference for moisture.[18] In the Colchis lowlands *Alnus barbata* once formed very extensive swamp forests, in which it was the dominant tree, but many of these habitats have been drained for agriculture.[19] Elsewhere it clings to riversides, waterways and other damp places. Because of its hygrophilous nature, it is far more widely distributed in Western Caucasus than in the East, though it is also frequent along the Alazani Valley in Eastern Georgia and in Lenkoran near the Caspian Sea. *Alnus barbata* is often found admixed with other deciduous trees in moist forests. Common associates include *Ulmus glabra, Tilia begoniifolia, Acer pseudoplatanus, A. platanoides, Fagus orientalis* and *Carpinus betulus*. It is also something of a pioneer of deforested and open ground. The geographer L. S. Berg reports that abandoned cornfields in Western Georgia have been known to vanish beneath six-metre high thickets of *Alnus barbata* and *Pterocarya fraxinifolia* in just two years.[20]

Alnus incana has a similar ecology but different distribution. It occurs in mountain ravines from about 1200–1950 m elevation, but mostly from 1500 to 1800 m. It too is hygrophilous, occurring usually on mountain floodplains and streamsides. *Alnus incana* is found in forests in

[17] Tauber 1967.
[18] Ketskhoveli 1975.
[19] Dolukhanov 1966.
[20] Berg 1950.

which beech prevails, in association with elm, hornbeam, lime and poplar. The distribution of *Alnus incana* is considerably smaller than that of *A. barbata*, though at the upper tree line *A. incana* is more likely to dominate.[21]

Both species produce timber that is light and weathers quickly. However, if kept waterlogged it is relatively rot-resistant, being used for mine shafts, water tanks, *etc.* Charcoal from its wood is used in gunpowder manufacture and the bark is used in tanning.[22]

Pollen Representation

Alnus species are wind pollinated, spreading their pollen over large distances. Alder pollen hence tends to be somewhat over-represented. Although alders are arboreal plants, their tendency to grow on peatbogs and floodplains means that they are sometimes part of the sediment-forming system, and as such are more akin to wetland plants. *Alnus* is sometimes excluded from the pollen sum where this is thought to have been the case.

Alder's affinity for damp places is shown clearly on the modern pollen map. High values are found along the central Colchis coastline where *Alnus*-dominated swamps occur. At these sites it is the dominant pollen type represented. The highest percentages occur in the Urukh basin in the northwest Caucasus, where *Alnus* comprises about 90 percent of the arboreal pollen sum, being a significant component of floodplain vegetation in that area. High values are characteristic of the entire northwestern Caucasus, due to high precipitation and mountainous topography. Several sites that are situated in treeless subalpine and alpine have *Alnus* values around 30 percent or more, derived from long-distance pollen transport from lower altitudes. The same applies in a vast number of sites in Southern Georgia, though here its proportions are significantly lower and generally do not exceed 10 percent. Sites with moderate values are those in the lowland steppe region, or those in close proximity to floodplains. Because of its wide dispersion and over-representation, the *Alnus* pollen record needs to be interpreted carefully. The broad correlation between precipitation and *Alnus* pollen in the study area suggests that it may be a palaeoecological proxy for rainfall.

Late Quaternary History of Alnus *from Pollen Records*

Alnus pollen is present in all the late glacial records examined. It appears that alder trees grew in refugia in the Black Sea lowlands and Eastern Georgia. *Alnus* declined through the Younger Dryas in Western Georgia. At the beginning of the Holocene it reacted quickly to rising moisture levels, becoming the dominant pollen type in the Sukhumi record. The opposite trend is observed in Eastern Georgia, where *Alnus* appears to have migrated away from the riverine lowlands in the early Holocene, perhaps as glaciers retreated, rivers stabilised and floodplain habitats contracted. The species concerned here was probably *Alnus incana*, while in Western Georgia it was almost certainly *A. barbata*. By 9000 cal BP, the importance of *Alnus* in the vegetation of the Colchis lowland can already be observed. By 6000 cal BP, *Alnus* was well established at most elevations in Western Georgia, while having largely retreated from the eastern parts of the study area. The late Holocene sees the extension of alder swamps in the Colchis coastal zone, the continued presence of *Alnus* in the mountain flora of the northwest Caucasus, and the sustained lack of *Alnus* pollen in the study area's drier, more continental eastern and southern parts. Throughout its late Quaternary history, *Alnus* has been strongly associated with the Black Sea lowlands and the Main Caucasus Range, places with a far greater concentration of hydrography than the east.

[21] Ketskhoveli 1975.
[22] Ketskhoveli 1975.

While alder's migration into the Caucasus Mountains during the early-mid Holocene was almost certainly tied to the network of glacier-fed rivers in that part of the study area, its expansion along the Black Sea lowlands may be linked to the paludification of the coastal plain by rising sea levels. Tvalchrelidze and colleagues have conducted a detailed sedimentological investigation of changes in the Black Sea level during the past 20,000 years.[23] Their data indicate that during the late Pleistocene and early Holocene the Black Sea level stood 40 or more metres below its present high-stand. An abrupt rise in sea level is recorded around 7000 cal BP, bringing the sea level to within ten metres of its present level. This was followed by further rises around 5500, 4000 and 2000 cal BP. A minor regression occurred between 5000 and 4000 cal BP, and a major drop 3000 to 2000 cal BP, when the sea level fell by as much as 10 metres.

All of these events are significant in the history of *Alnus* in the coastal lowlands. As sea levels rose, paludification of large tracts of formerly well-drained land led to the replacement of chestnut forests by alder swamps. The expansion of *Alnus* follows a north-south trend: sites in the northern coastal section (Sukhumi, Gagra) had alder swamps in the early Holocene; sites in the central section (Anaklia, Imnati) were paludified in the early-mid Holocene; while the southernmost sites (Ispani) became alder swamps only in the mid-late Holocene. This need not be a migrational lag, but merely the effect of differential rates of coastal subsidence.[24]

The history of *Alnus* in the study area is actually far more complex than the above synopsis suggests, with rapid expansions and contractions, especially in the coastal zone, caused by local hydrological changes. Its apparent preference for the floodplains of glacier-fed rivers may have caused its decline in Eastern Georgia at the beginning of the Holocene, but until this pattern is replicated at another site, conjecture will remain. And while history records that *Alnus* was adversely affected by the drainage of swamps in Colchis in the early twentieth century,[25] little can be said about its response to prehistoric human impact because of the almost complete lack of charcoal records and the neglect of anthropogenic indicators in pollen diagrams from Western Georgia.

Betula

Ecology

Five species of the genus *Betula* occur in the study area: *B. medwediewii, B. megrelica, B. raddeana, B. litwinowii* and *B. pendula*. The two last named species are the most widely distributed in the Caucasus region. All birches are characteristic of the upper tree line, and all are shade intolerant deciduous trees and rarely grow amongst other deciduous species. Birches are more common to the North Caucasus than the study area.[26]

Betula medwediewii is a creeping, suckering shrub or small tree that grows on moist subalpine soils (1900–2250 m above sea level) in Ajara in southwestern Georgia, sometimes with *Quercus pontica*, evergreen Colchic understorey plants and among subalpine tall herbfields. *Betula megrelica* is dwarf tree of the subalpine belt (1700–2000 m) endemic to the limestone mountains of the Samegrelo region of Western Georgia. *Betula raddeana* occurs in *Betula litwinowii* and *Sorbus aucuparia* subalpine woods at the tree line (1700 to 2100 m) in the central part of the Main Caucasus Range where the climate is more continental.[27]

[23] Tvalchrelidze *et al.* 2004.
[24] Kaplin and Selivanov 2004.
[25] Ketskhoveli 1959.
[26] Ketskhoveli 1975.
[27] Ketskhoveli 1975.

Betula litwinowii is common at the upper tree line in most parts of the study area, from 1900–2500 m elevation and higher where protected by late snow cover.[28] It is often associated with *Acer trautvetteri, Sorbus aucuparia, S. boissieri, Salix caprea, Pinus kochiana, Ribes biebersteinii, Vaccinium arctostaphylos, Sambucus nigra* and other species in subalpine krummholz forests, dwarf woodlands, *Rhododendron caucasicum* thickets and subalpine meadow vegetation, having a preference for north-facing slopes. It is distributed throughout Asia Minor and the Caucasus. *Betula pendula*, the common Silver Birch, has a similar distribution and ecology to *B. litwinowii* – the two species often grow together in the subalpine belt. Silver Birch, however, occurs at generally lower altitudes (to 2100 m) and is often seen in the middle mountain zone growing in forest gaps or colonising disturbed soils.

The growing season in the subalpine belt is 3–4 months,[29] with a mean annual temperature of 3–4 °C. Average July temperatures are around 13 °C, and below freezing from December til March. Rainfall ranges from 600 mm in the Anticaucasus Mountains to as much as 1700 mm on the Main Caucasus Range,[30] with a rainfall maximum around May. Most *Betula* habitats have relatively moist soils. In the continental parts of the study area, birches have an association with north-facing slopes, where snow cover lingers somewhat longer and soil moisture is consequently higher than on the drier, southern slopes. Deep snow in winter also protects these trees from low temperatures and desiccation.[31]

Birches are favoured by disturbance and are able to cope with heavy snowfall by assuming krummholz form.[32] In forest situations they rarely persist long once other, more light-competitive deciduous trees are established. As such they are something of a pioneer tree. Human impact has probably favoured birches in the forest belt, but extensive wood-cutting and clearing of subalpine birch thickets has reduced the extent of these considerably and they have been replaced by treeless subalpine meadows and *Rhododendron* thickets.[33] Birches are particularly valued in the Caucasus for firewood and charcoal.

Pollen Representation

Betula pollen is triporate with large, vestibulate pores. Birches are wind-pollinating and produce large amounts of pollen that is well dispersed over long distances. In the present-day, *Betula* pollen is only really well-represented in tree line situations along the Main Caucasus Range, with values above 4 percent. Moderate values of 1.5–4 percent are found in above-treeline sites in Southern Georgia and mountainous Abkhazia, but these are mostly the result of long-distance pollen transport. The same applies to occurrences in the semidesert-steppe area of Eastern Georgia. Only *Betula* values greater than 7 percent probably indicate birch-dominated woodlands; values of 4–7 percent indicate birch trees in the surrounding area; while values below 4 percent are probably indicative of long-distance pollen transport from distant sources. A surprising number of sites have no *Betula* pollen at all, especially in the Black Sea area, perhaps because of the restricted distribution of *Betula* compared to other deciduous, wind-pollinating trees.

[28] Ketskhoveli 1975; Nakhutsrishvili 1999.
[29] Walter 1974.
[30] Gulisashvili 1964.
[31] Nakhutsrishvili 1999.
[32] Dolukhanov 1978.
[33] Ketskhoveli 1975.

Late Quaternary History of Betula *from Pollen Records*

Betula pollen occurs through the late glacial in low, but significant, proportions. Values of around 5 percent from a site in the South Georgian Uplands demonstrate that birch was at least present on the surrounding mountainsides and had begun to expand during the lead up to the Holocene. *Betula* peaks at the beginning of the Holocene, but was soon replaced by other deciduous taxa. It was probably during the period 11,000 to 9000 cal BP that *Betula* achieved its maximum spread as it rapidly migrated up the mountains and retreating snow lines left new ground to be colonised. It remained during the mid Holocene at mountain sites along the Main Caucasus Range and in Javakheti, but elsewhere its role was insignificant. Around 4000 cal BP *Betula* values decline in the southern parts of the study area, while increasing in the north. This trend was then reversed in the late Holocene, as birch became again important in Colchis and Southern Georgia, a trend probably related to lower and middle altitude forest disturbance.

This *Betula* record is heavily influenced by the distribution of sites. The undated Nariani record from 2050 m elevation on the Tsalka Plateau shows that *Betula* pollen was dominant in early-mid Holocene spectra until *Abies* and *Pinus* later supplanted it.[34] The birch woods that migrated past the 1500-m elevation Aligol site at 11,000 cal BP established themselves in the subalpine zone from that time onward. Their mid-late Holocene decline at Nariani is almost certainly tied to deforestation, grazing and other human impacts in the subalpine zone beginning in prehistoric times.

The behaviour of *Betula* during the early Holocene is worthy of note. It does not appear to have resided in the refugia of either the Black Sea lowlands or Eastern Georgia. One would have to assume that dwarf birch woods were a feature of mountain landscapes during the glacial period. Birch's apparently rapid migration was no doubt assisted by its wind-dispersed seeds, but also by its ability to survive under extreme conditions. Pollen production of *Betula* is considerably lower at the upper limit of its range than at lower altitudes.[35] Large birch thickets may produce little or no pollen under these conditions, being therefore palynologically 'silent'. The question must be asked whether the early Holocene birch increase was the result of rapid migration and colonisation, or merely the mass flowering of stunted birch thickets already existing in the mountains. Obviously pollen data cannot adequately address this problem, which looks to macrofossil analysis for its resolution.

CARPINUS BETULUS

Ecology

Debate rages in the Caucasus region over the taxonomic status of *Carpinus caucasica*. Most botanists outside the region regard the Caucasian hornbeam as a minor variation of *Carpinus betulus,* which is widespread in Central and Eastern European forests. The *caucasica* epithet obviously has some resonance for Caucasian botanists and is retained there, as many such appellations are. The plant in question is a large tree to about 25 m that occurs frequently in forests of lower and middle altitudes. It is a thermophile usually associated with *Fagus orientalis, Castanea sativa* and *Quercus* spp., with which it is often co-dominant. Pure *Carpinus betulus* forests are somewhat less common. The Oriental Hornbeam, *Carpinus orientalis*, is discussed with *Ostrya* as it produces that pollen type.

[34] Margalitadze 1977.
[35] Kvavadze 1993.

Carpinus betulus prefers fertile soils with good drainage, but is adaptable to a wide range of conditions. Its Caucasian variant is distributed throughout Georgia and extends into Daghestan, Azerbaijan, Armenia, Anatolia and northern Iran. Within the study area it is concentrated in Western Georgia and, in the East, on the slopes of the Trialeti and Gombori Ranges and in the Alazani Valley.

The ecology of *Carpinus betulus* in Caucasian forests is not well studied, but of considerable interest. Beech is able to out-compete hornbeam for space and light, but on rough ground hornbeam is able to gain the upper hand.[36] The proportion of hornbeam in oak forests is related to the age of the forest:[37] young forests tend to have a much higher proportion of *Carpinus betulus* than do older, more established forests. It capitalises on forest gaps and other disturbances, keeping a ready store of seedlings in the understorey. Hornbeam forests are far more common at the lower forest border and in areas close to human habitation than elsewhere, and I suspect that most pure hornbeam forests are of secondary origin.

Hornbeam timber is valued for its hardness. Its wood makes an excellent fuel and is highly favoured for charcoal production.[38] As such it was one of the most commonly coppiced trees in the Caucasian vegetation.

Pollen Representation

Carpinus betulus pollen grains are large with usually three, four or more pores. Its pollen representation is relatively poor, as shown on the modern pollen map. Values of 4–8 percent represent sites where *Carpinus betulus* trees are present in small numbers. Values between eight and 13 percent indicate a considerable number of hornbeam trees. Values above 13 percent indicate co-dominance in the surrounding forests. However, long distance transport is also significant with this species, and some sites in the treeless highlands and lowland steppes of Eastern Georgia have *Carpinus* proportions up to 30 percent where no trees were in close proximity. Interestingly this form of over-representation in the mountains appears not to occur in Western Georgia.

Sites with high *Carpinus betulus* proportions are concentrated in the Tbilisi environs, Colchis, lowland Abkhazia and the Central Caucasus. This is in good agreement with the distribution of *Carpinus betulus* in the forests of the area. The high values from Lagodekhi in far eastern Georgia are the result of updraughts carrying pollen from extensive stands of *Carpinus betulus* on the slopes below.

Late Quaternary History of Carpinus betulus *from Pollen Records*

Carpinus betulus occupied a place in forest refugia in the Black Sea lowlands during the late Pleistocene, judging from *Carpinus* pollen values of up to 25 percent. It is not clear whether *Carpinus betulus* existed in Eastern Georgia at the same time – only single grains are recorded in the Sagarejo record prior to the Holocene. A single occurrence is found at 12,000 cal BP from the Tsalka Plateau. By 10,000 cal BP, however, *Carpinus betulus* was expanding throughout the study area, being recorded in Western, Southern and Eastern Georgia in small quantities. Over the next two thousand years it spread substantially. In Southern Georgia, maximum *Carpinus betulus* values are recorded at an elevation of 1100 m by 9000 cal BP and at 1600 m by 8000 cal BP. This suggests that *Carpinus betulus*-dominated forests ascended the mountain slopes from the east and west during this period. Values in Abkhazia decline slightly by

[36] Walter 1974.
[37] Dolukhanov 1966.
[38] Ketskhoveli 1975.

8000 cal BP, followed by Colchis in about 7000 cal BP, and finally Southern Georgia by 6000 cal BP.

The mid Holocene appears to have been an unfavourable time for *Carpinus betulus* across the study area, perhaps as *Fagus orientalis* and *Quercus* spp. cemented their positions as forest dominants. The late Holocene sees the return of *Carpinus betulus* to parts of the Black Sea littoral zone and Eastern Georgia. At 3000 cal BP, high values are recorded mostly in Eastern and Southern Georgia, with only one site in Colchis yielding substantial quantities of *Carpinus betulus*. During the remaining years *Carpinus betulus* continued to be important in Southern and Eastern Georgia, and increased in importance in Colchis. Its role in Abkhazia during the late Holocene appears to have been a minor one.

Hornbeam's post-glacial migration was a relatively slow process. It does not reach the mountains *en masse* until about 9000 cal BP. Huntley and Birks observed this same migrational lag in *Carpinus betulus* in Europe, hypothesising that it was delayed by edaphic factors.[39] I can offer no explanation here, except that the early Holocene in the Caucasus may have been relatively dry, confining hornbeam to moister lowland locations. Its mid Holocene decrease is probably the result of climatic changes, soil maturation and the replacement of mixed forests by monodominant *Fagus orientalis* and *Quercus iberica* forests. Even at this early date, however, the charcoal records from the Tsalka Plateau suggest that humans may have been active in burning vegetation. The adoption of metallurgy in the Bronze Age may have encouraged the coppicing of hornbeam trees for charcoal production, reducing their palynological signal. While this process could have been important locally, presumably it could not have occurred on a regional scale.

As for hornbeam's late Holocene expansion in Eastern Georgia, this was probably the result of human activity. Several pollen diagrams from Europe have shown a rapid increase in *Carpinus betulus* pollen coinciding with the Roman Period. Such an increase is not recorded universally in Georgia, but at Tsavkisi in the hills near Tbilisi, hornbeam pollen values increase dramatically from less than 10 percent to over 40 percent in the space of a few years. There seems no reasonable ecological explanation for this other than the abandonment of coppice woods and the subsequent mass flowering of regrowth during the Roman Period. While a climatic cause has been invoked to explain this pattern in Europe,[40] it cannot account for the direction and rapidity of vegetation change at Tsavkisi – its cause was almost certainly a change in the economic rather than atmospheric climate.

Castanea

Ecology

Castanea sativa is a thermophilous, deciduous tree, growing to a height of 30 metres. It is widely distributed in Colchis, with smaller populations in the east along the Alazani and Liakhvi valleys and near Borjomi. It occurs in the northwestern part of Azerbaijan as well. Chestnut is generally found at elevations from 500–1400 m above sea level, though in certain river canyons in Ajara and Abkhazia it can descend to 200 m and in Eastern Georgia it is known to climb to 1380 m. It is generally found in the lower to middle forest zones where it may occur in pure stands, or, more usually, admixed with *Fagus orientalis, Carpinus betulus, Quercus hartwissiana, Acer laetum, Pyrus caucasica* and others.[41]

[39] Huntley and Birks 1983.
[40] Huntley 1988.
[41] Gulisashvili 1964; Nakhutsrishvili 1999.

The understorey of chestnut forests is quite variable. Dolukhanov distinguishes seven chestnut understorey associations:[42] those with *Trachystemon orientalis* (Boraginaceae), those with ferns (*Dryopteris filix-mas*), those with forbs and *Hedera helix*, and those with various deciduous and evergreen shrubs (*Rhododendron luteum, R. ponticum, Vaccinium arctostaphylos* and *Laurocerasus officinalis*).

Chestnut forests occur on well-drained, slightly acidic, yellow-brown forest soil and have a preference for non-calcareous substrate.[43] The climate throughout chestnut's distribution is warm-temperate and relatively wet, conditions also favourable to species such as *Quercus iberica, Q. hartwissiana, Fraxinus excelsior* and *Zelkova carpinifolia*. The mean annual temperature ranges from about nine to 12 °C, with warm summers (around 20 °C) and less than one month where the temperature average is below freezing. The earliest frosts are experienced in September-October and the latest in April-May. Precipitation is moderately high, averaging between 900 and 2100 mm per annum. And although the seasonal distribution of this rainfall is uneven within *Castanea*'s area, rainfall during the growing period exceeds that of the colder months.[44]

Castanea timber has considerable economic value and because of this has been logged extensively in the Caucasus for a long time, exposing the remaining stands to the depredations of parasitic fungi.[45] Chestnuts are also prized for their nutritious fruits, which are eaten raw, boiled or baked, and prepared as a surrogate for coffee,[46] though their place at the Georgian table is minor compared to that of walnuts and hazelnuts. Zohary and Hopf note that the chestnut was probably domesticated in the Caucasus and northern Anatolia.[47]

Pollen Representation

Castanea is insect-pollinated and produces small amounts of pollen, which is poorly dispersed. Chestnut pollen is quite small (14–20 μm), tricolporate and prolate. It is difficult to identify with certainty, having many similarities to *Androsace alpina*-type and *Hypericum perforatum*-type.[48] Whilst the difference between these can be discerned when the pollen is well-preserved, poorly preserved *Castanea* pollen is almost impossible to distinguish from poorly preserved *Hypericum perforatum*-type. As such the records from the semi-arid parts of the study area may be treated with some degree of caution, as both taxa occur there frequently.

The modern distribution of *Castanea* pollen in lake and wetland sediments in the study area shows that *Castanea* values are higher in Colchis, consistent with the tree's distribution. Several of these sites are at lower altitudes than where *Castanea sativa* grows and the pollen may have arrived there by aeolian or fluvial processes. Long-distance transport of chestnut pollen is demonstrated by its presence at the Sögütlü and Pasinler sites in Eastern Anatolia where no chestnut trees occur. *Castanea* pollen values in Eastern Georgia do not exceed 2–3 percent and there is a possibility that some of this derives from cultivated trees. Values less than 8 percent represent sites where chestnut pollen arrived via long distance transport, while values between eight and 15 percent represent sites where chestnut forms a component of the vegetation. Values in excess of 15 percent, and especially those above 27 percent, represent the significant participation of chestnut in forest vegetation.

[42] Dolukhanov 1992.
[43] Gulisashvili 1964; Nakhutsrishvili 1999.
[44] Gulisashvili 1964.
[45] Dolukhanov 1966.
[46] Grossgeim 1945.
[47] Zohary and Hopf 1993.
[48] Beug 2004.

Late Quaternary History of Castanea *from Pollen Records*

In the late glacial, *Castanea* appears in low proportions in the Sukhumi record from Abkhazia, indicating that it was present amongst the mixed deciduous-coniferous suite of forest plants that found refuge there. It is not recorded at any other site during the late glacial. By 10,000 cal BP, it had apparently spread along the Black Sea lowlands of Abkhazia and became a more significant forest constituent. By 9000 cal BP, *Castanea* pollen is recorded along the length of the Black Sea lowlands and traces are found on the South Georgian Upland. It enters Eastern Georgia in a major way at 8000 cal BP, when moderately high values near Tbilisi suggest chestnut was present there in small numbers. At 7000 cal BP, chestnut's distribution continued to expand, especially in Abkhazia, where, it would seem, chestnut-dominated forests had begun to form. *Castanea* appears to have migrated northward from the mountains of Ajara onto the Colchis lowland during the early-mid Holocene, but was curtailed by rising sea levels around 4500 cal BP. Its role in Eastern Georgia also decreased after this time perhaps due to unfavourable climatic changes. By 3000 cal BP, chestnut had again established amongst forests in the Colchis lowlands. In Abkhazia, *Castanea* forests remained widespread throughout the mid Holocene, but declined quite dramatically in the late Holocene, especially during the last two millennia.

The present distribution of chestnut is much smaller than its mid Holocene extent (and its probable potential spread under the current climate.) Sea level changes and the resulting paludification of the Black Sea littoral almost certainly had a negative impact on lowland *Castanea*-dominated forests, as chestnut is very sensitive to changes in soil moisture.[49] Selective logging in the last few millennia has also taken its toll on *Castanea* populations in Colchis and elsewhere. This perhaps explains why the current distribution of chestnut forest is concentrated around steep ravines distant from human habitation. But climatic change cannot be ignored: a cooling climate would be unfavourable to *Castanea* populations, perhaps limiting their ability to regenerate in the aftermath of intensive human exploitation.

The history of *Castanea* in Southern Georgia is problematic because only late Holocene sediments are blessed with excellent pollen preservation in this area. Early-mid Holocene *Castanea* pollen is hence difficult to differentiate from *Hypericum perforatum*-type, which is produced by a plant widespread in Southern Georgian woodlands and grasslands. At Aligol, for instance, *Hypericum*-type pollen declines to absence at the same time as *Castanea*-type pollen does, about 1500 years ago. At Imera, *Hypericum*-type pollen declines at the beginning of peat formation, when *Castanea*-type pollen becomes suddenly more frequent.

If the two pollen types are taken to represent *Castanea sativa* alone, then it appears that chestnut migrated onto the Tsalka Plateau from lower altitudes amongst mid Holocene mixed woods with a high proportion of *Carpinus, Abies, Fagus, Ulmus, Tilia* and *Quercus*. It remained important until about 1500 years ago. *Castanea* could have been present there in the mid Holocene, as archaeological excavations have unearthed a considerable quantity of *Castanea* fruit dated to the Middle Bronze Age (early second millennium BC) at Tetri Tsqaro, 20 km south-east of Lake Imera.[50]

This history implies that *Castanea* was present in the forests around Tbilisi until around 7000 cal BP, when it moved upslope. A climatic trigger probably initiated this migration, but then early arboricultural activities may have also selectively conserved chestnuts long after the original mixed woods had been replaced by oaks or conifers. The most important late Holocene climatic changes seem not to have directly led to the chestnut's demise in Southern Georgia, so what then was the cause? If humans were to blame, why did chestnut timber

[49] Grove and Rackham 2001.
[50] Gobejishvili 1978.

become suddenly more valuable than its fruit? Since *Castanea* is intolerant of browsing,[51] did livestock grazing in chestnut coppices result in their slow death? Perhaps fungal diseases took their toll? Clearly much more palynological work is required to shed light on prehistoric climatic and human interactions with this interesting and difficult species.

CORYLUS

Ecology

Six *Corylus* species occur within the study area: *Corylus colurna, C. avellana, C. pontica, C. kachethica, C. imeretica* and *C. colchica*.[52] These last three are endemic to the region and all species are deciduous.

Corylus colurna is distinguished from the others by its arboreal habit, growing to a height of 30–50 m in broadleaf forests of the middle and upper mountain zones, especially in Eastern Georgia. It is a canopy tree distributed from northern Iran through the Caucasus, NW Anatolia and Romania to the Balkans. *Corylus avellana*, a species widespread in European forests, is a shrub or small tree common in the understorey of broadleaf forests and at forest edges throughout the study area. *Corylus pontica* grows in similar habitats, but is only found in the Black Sea basin and is somewhat less common. *Corylus kachethica* occurs in Eastern Georgian broadleaf forests and has a very restricted distribution. *Corylus imeretica* and *C. colchica* are associated with areas of rocky limestone terrain in Western Georgia, from the middle forest belt to the tree line and in the lower forest belt, respectively.[53] Of all these species, only *Corylus avellana* and *C. colurna* are widely distributed and significant in forest vegetation.[54]

The ecology of hazel has received little attention in Caucasian botanical literature because of its ubiquity in mixed forests. Most often it is associated with *Fagus orientalis, Quercus iberica, Carpinus orientalis, Acer laetum, A. pseudoplatanus, Tilia begoniifolia, Fraxinus excelsior* and other deciduous species. In the understorey it finds company with shrubs such as *Cornus mas, Swida australis, Euonymus europaeus, Lonicera caucasica, Viburnum lantana, Malus orientalis, Pyrus caucasica, Prunus* spp. and *Crataegus* spp. It is usual for *Corylus* to be accompanied by a diverse understorey and canopy, and it has a preference for calcareous soils.[55]

To a certain degree, *Corylus* species are favoured by disturbance. They flourish at forest edges, but are soon overtopped by other trees. They are especially common along tracksides, where animals readily spread their seeds. *Corylus* is also a favoured shrub for coppicing – many examples of this can be seen in forests near Tbilisi, although the practice has declined recently allowing these coppice stools to regrow. *Corylus* coppice wood was traditionally used for fencing, stakes, wattles, hoops, walking sticks and pipe-stems.[56] Hazelnuts are an important part of the Georgian diet, replacing walnuts as a source of fat in some regions.

Pollen Representation

Corylus pollen is triporate, around 30 μm in size, with a blunted triangular shape in polar view. Although it is a good pollen producer, its pollen production as an understorey shrub tends to be smothered by the forest canopy. As such, its pollen signal is far greater when it forms part of the canopy. *Corylus colurna* is a canopy tree and hence is unaffected by this

[51] Grove and Rackham 2001.
[52] Gagnidze 2005.
[53] Ketskhoveli 1959.
[54] Ketskhoveli 1975.
[55] Ketskhoveli 1975.
[56] Ketskhoveli 1975.

problem. Much of the *Corylus* pollen in the study area is derived from *Corylus avellana*, and to a lesser extent *C. colurna*; other species are so rare as to have only local significance, assuming that all *Corylus* species have a similar pollen productivity.

Corylus pollen makes its way to most places in the study area. It is more common along the southern slope of the Caucasus Range than it is in the south. The highest values are recorded in sites from the limestone areas of Racha and Imereti, where the concentration of *Corylus avellana* is highest.[57] Values greater than 3.5 percent almost certainly indicate the presence of *Corylus* within forests, while values greater than 7 percent and especially 14 percent represent a large proportion of hazel in the landscape. Significant long-distance transport occurs in areas with steep topography.

Late Quaternary History of Corylus *from Pollen Records*

Corylus pollen was present in small proportions through the late glacial in Abkhazia and Southern Georgia and increased through the Younger Dryas stadial. At the beginning of the Holocene its role in Abkhazia declined rapidly, whilst in Southern Georgia *Corylus*-dominated woods expanded with equal rapidity. By 10,000 cal BP, *Corylus* had spread throughout the study area and had been incorporated into mixed forest vegetation. Near Bakuriani the arrival of *Corylus* was delayed until about 9500 cal BP, which could be the result of dating irregularities or a migrational lag. By 9000 cal BP, *Corylus* was well established in the forests of Southern Georgia, declining somewhat in the Black Sea lowlands.

By the mid Holocene this decline had reversed – the role of *Corylus* in Southern Georgia decreases, while increasing in Western Georgia. Particularly high proportions are recorded at Cheliagele at 6000 cal BP. Mid-late Holocene fluctuations of *Corylus* pollen are difficult to interpret on a regional scale and were probably the result of local forest perturbations and succession. Hazel's part in the vegetation of the mid-late Holocene was obviously much less than during the early Holocene. In the last 2000 years there has been a slight increase in *Corylus* values across the study area in keeping with the present-day pattern. Most of the sites that saw this increase are today situated at the forest border.

The data show that *Corylus* had glacial refugia in a number of localities in the Caucasus, probably in mountain gorges as well as the lowlands. From these locations it expanded out rapidly at the beginning of the Holocene. As in Europe, however, its dominance in the woodland vegetation was short-lived. It seems probable that the species represented in this migration was *Corylus avellana*, if only because its subsequent decline in the early Holocene can be explained by the migration of other, taller, deciduous tree species that shaded out the early Holocene hazel woods. There is no evidence to show that the other shrubby species were not involved in the same way, although their preference for calcareous soils may have confined them to limestone areas. *Corylus colurna*, the arboreal species, probably moved in concert with *Quercus, Fagus* and *Carpinus*, but this is only a supposition based on its habit and present-day ecology, a supposition that does not hold for *C. avellana*.

The mid-late Holocene decline of *Corylus* in Southern Georgia raises questions about the possible role of humans in the landscape at that time. Hazel can be both positively and negatively impacted by human activity – it is encouraged by forest clearance, but reduced through grazing.[58] The pollen decline in Southern Georgia may merely be a function of increasing forest cover in that area and the consolidation of the forest canopy above hazel plants. Equally, the collection of hazelnuts, coppicing of *Corylus* stools for livestock fencing and housing, and browsing by domestic herds may have impacted upon *Corylus* populations.

[57] Grossgeim 1945.
[58] Rackham 2001.

Hazel's late Holocene increase across the study area relates very strongly to episodes of forest disturbance and fire. Logging and clearing augmented the extent of forest gaps and edges, the preferred habitat of *Corylus*. It is not clear what impact coppicing may have had on the pollen productivity of *Corylus* plants, as this depends greatly on the frequency of cutting, which in Georgia was usually between 2–5 years.[59] At least near Tbilisi, a reduction in grazing and an increase in forestry activity led to the increase of *Corylus*, which was coppiced widely there until recently.

Ephedra

Ecology

Ephedra is represented by two species in the study area: *Ephedra distachya* and *E. procera*. *Ephedra* species are gymnosperms that produce fleshy red fruits. They are scrambling, dioecious shrubs characteristic of dry environments, and have spreading root systems.[60] *Ephedra distachya* grows only about 20–50 cm high, occurring on dry, rocky slopes in semidesert-steppe vegetation in the lowlands and piedmonts. It is confined to the arid parts of Eastern Georgia within the study area, but extends through much of S and E Europe, W Siberia and the Mediterranean Basin. *Ephedra procera* (syn. *E. major* subsp. *procera*) grows larger, sometimes up to 2 metres, but more usually 0.5–1.5 m. It also grows on dry, rocky ground, eroded slopes and occasionally on saline soils. It occurs at higher altitudes than *E. distachya*, from the foothills to 1600 m or more. It is distributed from the Eastern Mediterranean to the Western Himalayas.[61]

Ephedra species are associated with other xerophytes. In shrubby, forest edge situations, *Ephedra procera* can coexist with *Quercus iberica, Carpinus orientalis, Paliurus spina-christi, Spiraea hypericifolia, Rosa canina* and xerophile *Pyrus* species. In low scrub, both species can be found amongst *Pistacia mutica, Juniperus polycarpos, J. foetidissima, J. oxycarpa, J. oblonga, Paliurus spina-christi, Rhamnus pallasii, Rhus coriaria, Lonicera iberica, Celtis caucasica, Atraphaxis spinosa, Cerasus incana, Jasminum fruticans* and many more.[62] They are commonly associated with steppe grasses such as *Bothriochloa ischaemum* and *Stipa capillata*, and herbs like *Teucrium polium, Helianthemum salicifolium, Artemisia lerchiana* and *Scleranthus annuus*. *Ephedra* does not presently dominate the vegetation of which it is a part, usually occurring solitarily or in small groups.

Ephedra produces the alkaloid ephedrine, which is used for medicinal purposes. The plant is also used for erosion mitigation in arid areas.

Pollen Representation

Ephedra pollen is elongated with numerous 'pseudocolpi'[63] running meridionally around the grain. Two morphological types are distinguished: *Ephedra distachya*-type and *E. fragilis*-type. The former has irregular fissures radiating out from each colpus, while *E. fragilis*-type is unornamented and has more colpi. It seems probable that both pollen types are produced by Caucasian *Ephedra* species,[64] but there is also a possibility that *E. fragilis*-type pollen comes

[59] Ketskhoveli 1957; Ketskhoveli 1959.
[60] Chikov 1983.
[61] Ketskhoveli 1971.
[62] Arabuli 2003; Ketskhoveli 1959.
[63] Beug 2004.
[64] Dr E.V. Kvavadze, Institute of Paleobiology, Tbilisi, pers. comm.

from the Eastern Mediterranean by aeolian transport. The plant has poor pollen representation – soil samples taken from beneath *Ephedra* bushes yielded spectra dominated by *Alnus* and *Pinus* pollen. *Ephedra* pollen values up to 17 percent were recorded.[65] In lake and wetland sediments, however, values greater than one or two percent are infrequent.

The modern pollen map shows a cluster of sites in the steppe region of Eastern Georgia where *Ephedra* pollen is recorded in low percentages, reflecting the prevalence of the plant there relative to other parts of the study area. Similar values occur in sites from Eastern Turkey (Lake Van and the Pasinler Valley) and in the mountains of the NW Caucasus. So, whilst occasional *Ephedra* pollen grains are probably the result of long-distance transport, even small percentages (>2 percent) represent the local presence of *Ephedra* plants in the surrounding vegetation.

Late Quaternary History of Ephedra *from Pollen Records*

Ephedra seems to have been one of the dominant arboreal plants in the late glacial steppes of the southern Caucasus, growing amongst vegetation dominated by *Artemisia* and Chenopodiaceae. Palaeolithic humans consumed *Ephedra* fruits along with other plants like *Celtis*, according to macrofossil finds.[66] Its maximum pollen values are recorded during the Younger Dryas, suggesting a cold, arid climate. These conditions were most extreme in Southern Georgia, though the effects were felt as far west as the Black Sea lowlands. *Ephedra* remained important during the first millennia of the Holocene, declining slowly in the South Georgian Uplands as the climate became gradually wetter and more mesic arboreal species migrated upslope. Its importance in Abkhazia seems to have increased at the same time, perhaps as the climate there took on a more Mediterranean character.

The mid Holocene was unkind to *Ephedra* species. Pollen values decline almost universally. Moderate *Ephedra* values are only recorded at one site in Abkhazia and another in southern Javakheti (South Georgian Uplands) by 6000 cal BP. Elsewhere, *Ephedra* became very uncommon and remained so throughout the late Holocene. It occurs at 5000 cal BP on the Tsalka Plateau, associated with an increase in *Quercus* and *Juniperus* pollen. High values at 4000 cal BP are recorded in Abkhazia, and at 2000 cal BP near Tbilisi and in the mountains of NW Caucasus. Very little indication of *Ephedra*'s current range is given by its complex history.

Ephedra was certainly a major contributor to the glacial flora of Southern Caucasus and Eastern Anatolia. The agreement between the Lake Van and Aligol records is very good, showing a definite *Ephedra* peak during the Younger Dryas, followed by the decline of *Ephedra* around 8000 cal BP (Chapter Six). But why did *Ephedra* linger well into the Holocene in many parts and become so scarce only in the mid-late Holocene? Wright *et al.*[67] have suggested that the Caucasus and the whole region from the Balkans to Southern Siberia experienced a much drier early Holocene climate compared to the mid and late Holocene. Such a climate would certainly have favoured *Ephedra*, and moreover have been unfavourable to forest trees. A mid Holocene shift to greater precipitation in the arid interiors of Western Asia allowed the expansion of mesic forest species and the displacement of *Ephedra*. Palaeoclimatic reconstructions in this study suggest that while the early Holocene climate was somewhat dry, there appears to have been a shift in the seasonal distribution of rainfall perhaps favourable to *Ephedra* (see Chapter Six).

In the mid-late Holocene, *Ephedra*'s distribution was almost certainly affected by human activities. In particular, grazing appears to be highly destructive to *Ephedra* plants. This would

[65] Kvavadze and Stuchlik 1993.
[66] Vekua and Lordkipanidze 1998.
[67] Wright *et al.* 2003.

explain why, in the long record of steppe, scrub and woodland vegetation from Tsavkisi wetland near Tbilisi, *Ephedra* pollen is very infrequent until the apparent cessation of grazing in the Classical Era. *Ephedra* becomes important, paradoxically, at the time of greatest forest regeneration. For a shrub intolerant of shade, this only makes sense if grazing is taken into account. The relationship between fire and *Ephedra* is less clear, masked by its early Holocene climatic signature.

The *Ephedra* record from the Caucasus has some similarities with that of the European and Mediterranean — its prevalence during the late glacial and an early Holocene decline — except that in the Caucasus *Ephedra* lingered several millennia later into the Holocene, perhaps as a result of slow afforestation and the continentality of the early Holocene climate.

FAGUS

Ecology

Fagus orientalis is the most prevalent forest tree in the Caucasus region. Exactly half of the study area's forests are classified as beech-dominated.[68] *Fagus orientalis* is closely related to its European counterpart, *F. sylvatica*, and fills essentially the same ecological niche. It is a tall, long-lived deciduous tree growing 20–40 m in height and is found in forests from the Eastern Mediterranean to the Caspian region, grading into *F. sylvatica* forests in the Balkans, Romania, Crimea and western Anatolia.[69] It usually dominates the forests in which it is found, especially wet forests of the middle to upper forest belts (1000–1600 m). However, it is found in almost all woodland vegetation from sea level up to 2200 m (2380 m in places), growing as a dwarf shrub in the subalpine belt. Beech forest is strongly influenced by precipitation and is lacking in the driest and most continental parts of the Caucasus.

The climate of Oriental Beech habitats is characterised by cold winters, warm summers and high atmospheric moisture (**Table A.2**). The average annual temperature is between 5.6 and 8 °C, with coldest month temperatures between –2.2 and –5.6 °C (absolute minimum –28.8 °C) and the warmest month between 16.5 and 18.6 °C (abs. max. 35 °C).[70] This makes *Fagus orientalis* somewhat more thermophile than the European species, *Fagus sylvatica*. Precipitation varies from >1700 mm per annum in the wet forests of Colchis to less than 700 mm in the drier forests of Eastern Georgia (**Table A.2**). At all locations, however, atmospheric moisture exceeds 70 percent.

Table A.2. Two extremes of beech forest climates: Monthly climatic data from *Fagus orientalis* habitats at Kharistvali, 42° 25' N, 43° 3' E, 1320 m a.s.l., and Manglisi, 41° 42' N, 44° 23' E, 1100 m a.s.l. T: average monthly temperature; P: average monthly precipitation. Source: Gulisashvili 1964.

	J	F	M	A	M	J	J	A	S	O	N	D	Avg.
Kharistvali, Racha													
T (°C)	–5.2	–3.2	0.6	5.2	11.3	14.0	17.0	17.1	13.0	8.0	2.7	–2.2	6.6
P (mm)	123	98	108	155	131	212	162	145	170	152	138	142	1736
Manglisi, Trialeti													
T (°C)	–2.2	–1.9	2.2	6.7	11.7	15.2	18.6	18.4	13.9	9.6	3.7	–0.1	8.0
P (mm)	16	34	32	79	123	103	71	49	59	54	44	26	690

[68] Kruchinin 1973.
[69] Dolukhanov 1989; Ketskhoveli 1975.
[70] Gulisashvili 1964.

Beech forest is almost always associated with acidic brown forest soils. It also occurs on brown forest podzols, humified karst soils and krasnozems.[71]

Although beech forests are often monodominant, the most common associate of *Fagus orientalis* is hornbeam, *Carpinus betulus* (*C. caucasica*). Hornbeam is more prevalent in beech forests that have been subjected to woodcutting and grazing.[72] Other associates include *Picea orientalis, Abies nordmanniana* and *Castanea sativa*. Pure beech forests are classified according to their understorey dominants: *Galium odoratum, Rubus hirtus, Festuca drymeja, Rhododendron* spp., *Laurocerasus officinalis, Vaccinium arctostaphylos, Viburnum orientale, Dryopteris filix-mas, Athyrium filix-femina, Ilex colchica, Ruscus colchicus, etc*. Some oriental beech forests, particularly on thin soils on the Main Caucasus Range, have no understorey at all.

Oriental Beech seedlings are very shade-tolerant. They can survive in the understorey of other forest types for decades, eventually overtopping and shading out other trees. In the absence of disturbance, individual *Fagus orientalis* trees can typically live for 160–200 years, and in extreme cases, 500 years or more.[73] Their longevity allows them to totally dominate forest biomes where climatic conditions are optimal. But *Fagus orientalis* have shallow root systems and are susceptible to wind-throw, allowing for gap-phase regeneration. Their fine-grained timber is also prized (for carpentry, furniture making, ship-building and musical instrument manufacture) and for this reason they have been felled extensively. Beech wood burns long and hot on the fire and the oil extracted from beech mast was used in the past for cooking and lighting.[74]

Pollen Representation

Fagus pollen is dispersed by the wind and has characteristically short colpi almost encircled by the large, diffuse pores. Being the most widespread tree in the study area, its pollen is present in almost all lake and wetland deposits. However, its pollen representation is lower than many other wind-pollinated trees.[75] Values less than 6 percent represent situations where *Fagus* pollen is derived from long-distance transport or where beech trees occur sporadically in the extra-local vegetation. Values between six and 12 percent represent long distance transport from beech-dominated vegetation at lower altitudes (in sites above the tree line) or the presence of beech in the extra-local vegetation. Values greater than 12 percent, and especially those >22 percent reflect the co-dominance to dominance of *Fagus orientalis* in the vegetation.

Sites with high values of *Fagus* pollen are located in Abkhazia, along the southern slope of the Main Caucasus Range, and in the mountains of Southern Georgia. Its pollen distribution more or less accurately portrays the prevalence of *Fagus* trees in the sampled areas of the study area.

Late Quaternary History of Fagus *from Pollen Records*

Fagus orientalis is supposed to have been the species responsible for *Fagus* pollen records during earlier interglacials in Europe, prior to the development of the more cold-adapted *Fagus sylvatica*.[76] Today its range centres on the Eastern Mediterranean, Black and Caspian Sea Basins.

[71] Dolukhanov 1989; Gulisashvili 1964.
[72] Dolukhanov 1989; Gulisashvili 1964.
[73] Dolukhanov 1989.
[74] Grossgeim 1945.
[75] Andersen 1970.
[76] Huntley and Birks 1983.

Fagus was present in refugia in the Black Sea lowlands during the late Pleistocene. Its role was initially minor, but increased significantly during the late glacial, if the pollen records we have are any indication. By 12,000 cal BP, its role in Abkhazia was beginning to decline and it makes a first appearance in Southern Georgia as long-distance transported pollen. Upon entering the Holocene, *Fagus* began to migrate upward from its Black Sea lowland refugia. Still, its presence as a forest dominant remains very doubtful until 9000 cal BP, when it its pollen occurs in proportions greater than 20 percent in lowland Abkhazia in the west, and in the east near Bakuriani. This suggests that *Fagus* had climbed to about 1000–1200 m elevation in the Trialeti Ranges.

Eight-thousand years ago, *Fagus* had migrated upward yet further and was probably co-dominant in woods surrounding sites at 1100–1900 m elevation. Interestingly, it appears that beech continued to play a subordinate role in Colchic vegetation long after its main cohort had migrated upslope. By 7000 cal BP, beech had consolidated its distribution in the middle and upper forest belt of Western and Southern Georgia, present as a component of mixed forests. Beginning around this time, however, was the slow decline of *Fagus* in Colchis and Southern Georgia, brought on by the increased role of oak, chestnut and other thermophilous trees in the middle forest belt. Beech remained prevalent through the mid Holocene in the upper (montane) forest belt, especially in high rainfall areas.

The distribution of beech throughout the late Holocene changed very little – it continued to dominate in parts of the middle and upper forest belt and remained relatively scarce elsewhere. In the last two millennia, however, beech distribution has fragmented and spread, possibly as forest disturbance opened up new areas for colonisation.

Oriental beech populations seem to have been largely controlled by climate changes and migration rates. The Younger Dryas stadial forced beech populations down to lower altitudes in the Black Sea lowlands. Probably *Fagus orientalis* refugia were also located in river valleys in the eastern Caucasus, but direct evidence for this is lacking. *Fagus* trees were slow to migrate out from these refugia during the early Holocene. It took some 3000–4000 years for beech to move upslope and establish at the elevations it currently favours. This lag allowed other deciduous trees such as *Corylus, Betula* and *Carpinus* to dominate montane forests until the arrival of the more competitive *Fagus orientalis*. This explains why early Holocene forests were so different to the ones we see today.

Beech's mid Holocene decline on the Tsalka Plateau occurred at the same time as fire became an important element there. Whether these fires were caused by an abrupt climatic change or by humans wishing to stave off rapid afforestation of pastures, their result was the same – a decline in mixed forests (including *Fagus*) and the spread of oak-dominated woodland. Forest fires in the mid-late Holocene everywhere in the Caucasus would have reduced the forest area of *Fagus*, which is fire-sensitive.[77] Widespread timber felling and deforestation in the late Holocene, however, may have increased beech's spread into other vegetation communities since it has difficulty invading closed forests.[78] Beech forest is known to have invaded areas of former agricultural land near Tbilisi.[79] Increasing precipitation in the last few centuries could have been to the benefit of *Fagus orientalis* populations across Georgia. So, although beech forest is presently the most widespread forest type in the study area, even 1000 years ago its distribution was probably much more focused and restricted.

[77] Tinner *et al.* 2000.
[78] Huntley and Birks 1983.
[79] Avakov 1982.

FRANGULA

Frangula alnus is a deciduous shrub or small tree growing about 6–7 metres high. It is distributed throughout Europe and stretches as far as the Caspian Sea. In the Caucasus its distribution is concentrated around the Black Sea coast.[80] It grows in the understorey of moist broadleaf and gallery forests from sea level up to about 2000 m. It is a common shrub, around the margins of *Sphagnum* bogs on the Black Sea coast, with *Rhododendron ponticum, R. luteum, Vaccinium arctostaphylos* and *Alnus barbata*. But it also grows in quite arid parts of the Caucasus, on dry rocky slopes.[81] The bark and roots of the plant are used in traditional medicine as a purgative. Dyes are extracted from its roots and ripened fruit.

Frangula pollen is similar to *Rhamnus*-type pollen and some grains of *Frangula* fall under the category of *Rhamnus*-type.[82] *Frangula* has rarely been recorded in pollen diagrams from the study area, perhaps due to its scarcity. The pollen is poorly represented; occurring where plants are present *in situ*, although long-distance pollen transport is possible. The only occurrence of *Frangula* pollen in recent sediments comes from Lake Kumisi near Tbilisi. These grains may have been carried in from the forest belt by inflowing streams.

The Holocene history of *Frangula* in the study area is difficult to deduce. *Frangula* pollen is present in the Ispani-II sediments from 5600 cal BP, becoming more frequent from 4500 cal BP, and peaking between 3200–2200 cal BP. Probably *Frangula alnus* has always been common in the swamps of the Colchis lowlands. Its dramatic increase in the late Holocene at Ispani-II correlates with the charcoal record from that site, and suggests that fire assisted its spread. Archaeological evidence from the Ispani area indicates a strong human presence in the area during these fires. A similar association – of high *Frangula* proportions and human impact – is evident in another record from Racha at 2000 cal BP. Hence the proliferation of *Frangula* during the first millennium BC might have been a consequence, witting or unwitting, of economic activities in Western Georgia.

FRAXINUS

Ecology

Three species of ash occur in the study area: *Fraxinus excelsior, F. coriariifolia* and *F. oxycarpa*.[83] All of them are small- to medium-sized deciduous trees with compound leaves.

Fraxinus excelsior is the most common and widespread species and has a distribution that encompasses Central Europe, the Mediterranean, Balkans and Western Asia. It grows in most mixed and broadleaf forests of the lower and middle altitudes. It is far more widespread in the more arid forests of Eastern Georgia than the moist forests of Western Georgia, where it is most often found in ravines, gullies and canyons. *Fraxinus excelsior* is rarely dominant but is very cosmopolitan: it occurs in *Quercus iberica, Q. iberica-Carpinus betulus* and *Q. iberica-C. orientalis* forests, semi-arid *Juniperus* scrub and *Pistacia mutica* savanna, *Paliurus* shibliak, gallery forests, mixed forests, *Alnus barbata* swamps and subalpine *Quercus macranthera* woods.[84] *Fraxinus excelsior* is very commonly associated with shrubs from the Rosaceae and Rhamnaceae families. It favours relatively moist, slightly alkaline soils and does not thrive where the forest canopy casts deep shade. For these reasons it is generally absent from coniferous and *Fagus orientalis*-dominated forests.

[80] Browicz 1982 ff.
[81] Ketskhoveli 1983.
[82] Moore *et al.* 1991.
[83] Gagnidze 2005.
[84] Ketskhoveli 1959.

The remaining two species, *Fraxinus coriariifolia* and *F. oxycarpa*, are much less common than *F. excelsior*. The former is especially rare, occurring only in a handful of locations in the easternmost sector of Eastern Georgia and extending into Azerbaijan. *F. oxycarpa* is more widespread, but still confined in the study area to Eastern Georgia. It grows in open woods and scrub vegetation on dry piedmont slopes.[85]

Fraxinus species have light, flexible timber suitable for building and furniture-making. The bark is used in tanning and yields black, brown and blue dyes.[86] Ash trees respond well to coppicing, but since their foliage is poisonous to livestock they are not used as fodder plants.[87]

Pollen Representation

Fraxinus pollen is usually 3–4 colpate with a reticulate exine. *Fraxinus* species are poor pollen producers. Their pollen grains are poorly dispersed in large quantities, although the light grains can be transported considerable distances by the wind and may be over-represented in treeless environments. Because of its poor representation, *Fraxinus* is not always recorded in pollen diagrams. The highest values currently occur near Tbilisi and in Southern Georgia. Values greater than 5 percent indicate a considerable cover of *Fraxinus* trees in the surrounding forest canopy, as is the case with *Fraxinus excelsior* on many foothills around Tbilisi. Values of 1–5 percent reflect the occurrence of *Fraxinus* in the vegetation to varying degrees. Values below 1 percent represent only regional presence. However, due to wind transport, high values of *Fraxinus* pollen are recorded at several treeless sites both in highland grasslands and lowland steppes. At other sites where *Fraxinus* trees are known to occur, the pollen signal has been drowned out by more prolific pollen producing trees and herbs. The *Fraxinus* pollen record must therefore be interpreted carefully in relation to vegetation and pollen source area.

Erika Gobet (unpublished data) and I have identified *Fraxinus ornus*-type pollen at several sites in the study area. It only ever occurs in trace proportions, and is probably the result of long-distance pollen transport from the Mediterranean region.

Holocene History of Fraxinus *from Pollen Records*

The history of *Fraxinus* represents something of a palynological blind spot. It was almost undoubtedly present in the study area through the late Pleistocene but is not recorded until 9000 cal BP on the pollen maps. By 8000 cal BP, it was present and abundant throughout Southern Georgia admixed with *Corylus*, *Carpinus* and other deciduous trees. But by the mid Holocene it had been enveloped by other deciduous taxa and was unimportant except at the lower forest limit near Tbilisi. There it was co-dominant with oak, but by 4000 cal BP its importance had begun to decline. Then comes a period in which *Fraxinus* does not seem to have been an important part of the vegetation at any of the study sites. Surprisingly, a thousand years ago it expanded considerably in southern Colchis and Southern Georgia. Nowadays *Fraxinus* seems to be more important in Southern Georgia than at any time in the last 4000 years.

The Holocene history of *Fraxinus* is complex. It shows a triple expansion – first between 8000 and 7000 in the South Georgian Uplands, then between 6000–5000 cal BP on the foothills near Tbilisi, and finally in both these areas during the last millennium. The late Holocene expansion of *Fraxinus* on the Tsalka Plateau can probably be attributed to the deforestation of the plateau in the last two thousand years. This allowed *Fraxinus* pollen to be

[85] Gagnidze 1985.
[86] Gagnidze 1985.
[87] Grove and Rackham 2001; Polunin 1969.

blown in by updraughts from lower altitudes. Some of this contribution was also more local, as *Fraxinus excelsior* does grow around the disturbed forest margins of the Plateau even today.

Ash's expansions and contractions on the foothills are a bit more difficult to explain. The mid Holocene was a very favourable time for *Fraxinus excelsior* near Tbilisi, where pollen percentages indicate that it was co-dominant in forest vegetation with oak, hornbeam and hazel (and initially elm and lime as well). Such vegetation can be found today between Tbilisi and Mtskheta: in the environs of Chili Lake, above Mukhatgverdi Cemetery and in Karsnis-Khevi (see Chapter Four). These oak-ash forests are quite open, now heavily grazed but coppiced in the recent past. They differ from secondary woodlands in similar positions by their distinct lack of oriental hornbeam (*Carpinus orientalis*).

It is difficult to explain why oak-ash forest occurs where it does. Perhaps these forests were managed coppices, from which livestock were excluded. Then again, the climate near Tbilisi is in many places too dry for *Quercus iberica* to grow into a canopy tree, allowing canopy gaps to be colonised by *Fraxinus*, *Juniperus*, *Celtis* and other light-demanding but less competitively-dominant trees. The Eocene sedimentary rocks that underlay the hills between Tbilisi and Mtskheta are alkaline and may favour *Fraxinus* over *Quercus*. *Fraxinus* are often found in gullies and along small streams where soil moisture is somewhat higher, so increased precipitation might favour *Fraxinus* expansion.

Clearly the mid Holocene ash expansion at Tsavkisi was climatically driven, but it is unclear whether ash remained dominant in the vegetation because of climate, disturbance or some other factor. Its subsequent decline with oak was rapid, brought on by a simultaneous drop in temperature and precipitation and an increase in grazing pressure about 3500 years ago. Its late Holocene increase might well relate to anthropogenic forest disturbance (gap creation), but an increase in precipitation is also a plausible explanation. I cannot offer a more conclusive exploration of this topic because the precipitation records in this study are derived from pollen data.

HIPPOPHAË

The widespread Western Eurasian species, *Hippophaë rhamnoides*, is the only representative of the *Hippophaë* genus in the Caucasus region.[88] It is a many-branched spiny shrub growing 3–5 m tall, but sometimes occurs as a small tree. The plants are either male or female (dioecious) and have distinctive greyish deciduous foliage.

Hippophaë rhamnoides is occurs widely throughout the Caucasus, although its distribution is quite fragmented. Its fragmented distribution is a consequence of its habitat preferences, which include seaside sand dunes, river terraces, rocky slopes, mountain gorges, cliffs and screes. The soils in these places are usually poor, thin or unstable. *Hippophaë rhamnoides* has nitrogen-fixing root nodules, which allow it to grow where other woody plants struggle.[89] In the gallery forests of Eastern Georgia ('chala' in Georgian), *Hippophaë rhamnoides* is associated with *Populus × canescens, P. nigra, Alnus barbata, Salix* spp., *Quercus pedunculiflora* and *Ulmus minor*, shrubs such as *Tamarix ramossissima, Daphne caucasica, Lonicera caprifolium* and lianas.[90] It also occurs as one of the only woody plants along ephemeral waterways in the semi-desert-steppe zone. By the seashore, it forms scrub on sand dunes, and, to a lesser extent, grows in *Alnus*-dominated swamp forests. *Hippophaë rhamnoides* is also seen colonising disturbed places and forest openings from sea level up to the subalpine belt. It appears to be quite shade-intolerant.

[88] Gagnidze 2005.
[89] Huntley and Birks 1983.
[90] Ketskhoveli 1959.

The plant's leaves are used for pigmenting paint, its fruit as flavouring for jam, cordial and ice-cream, and as a traditional medicine for pain relief. Its spiny, twiggy branches are sometimes used to make brush fencing.[91]

Hippophaë pollen is tricolporate with distinctive, beaked apertures, which make identification relatively straightforward. *Hippophaë* pollen always occurs in low percentages because *Hippophaë rhamnoides* rarely covers large areas.[92] Abundances greater than 1 percent probably indicate that *Hippophaë rhamnoides* plants are present in the surrounding area. Modern *Hippophaë* pollen has been recorded in several places, with a clear geographic relationship with the Black Sea coastline and the treeless steppic areas of Eastern Georgia.

Not much can be deduced about the history of *Hippophaë* from pollen records because of its scattered distribution and poorly-represented pollen. Nonetheless, there are several occurrences of *Hippophaë* worth noting. Unlike the Lake Van record from Eastern Anatolia, where *Hippophaë* appears to have been important in the late-glacial, the available data from the study area have no late Pleistocene occurrences of *Hippophaë*. One of the earliest Holocene occurrences is at Gomnis Lake, coincident with the early Holocene *Corylus* peak at that site. *Hippophaë* was clearly prevalent in the subalpine belt near Kazbegi during the mid Holocene. Pollen values up to 22 percent are recorded in a mid Holocene buried peat at 1980 m elevation. *Hippophaë rhamnoides* plants were obviously growing very close to this site, probably as streamside vegetation, so the record is representative of only a tiny area.

Hippophaë occurs throughout the Tsavkisi record, if only in low percentages. There is however, one considerable peak (4 percent of dryland pollen) at 210 cm depth, corresponding to the disappearance of *Secale*-type pollen and an increase in carbonised particles. In this instance, it appears that *Hippophaë* acted as a pioneer species, occupying abandoned fields that were later overtaken by secondary forest.

There has been an apparent *Hippophaë* increase during the past two millennia. It occurs in sediments from Sukhumi on the Black Sea coast, on the Tsalka Plateau, and near Tbilisi. Around 1000 cal BP it is recorded again in these places, as well as at Lake Bazaleti. It is difficult to say whether this increase is an artefact of the greater number of sites for the late Holocene, or a reflection of real increases in *Hippophaë* populations due to land clearing, erosion and fire.

HEDERA

Three ivy species grow in the study area: *Hedera helix, H. pastuchowii* and *H. colchica*. All are evergreen lianas which scramble along the ground or climb up trees. *Hedera helix* is the most widespread, growing in low to middle altitude forests throughout the study area. *H. pastuchowii* is only found in Eastern Georgia in broadleaf forests, at forest edges and in fields. *H. colchica* has a Colchic distribution and is predominantly found in broadleaf and coniferous forests in Western Georgia from sea level up to an altitude of 1500 m.[93]

All three species are commonly found in 'relict' Colchic forests in Eastern and Western Georgia. These include lowland mixed broadleaf forests and gallery forests dominated by *Alnus barbata, Pterocarya fraxinifolia, Populus* × *canescens, Quercus pedunculiflora, Q. iberica, Q. hartwissiana, Ulmus minor, U. glabra, Diospyros lotus, Zelkova carpinifolia, Carpinus betulus, Castanea sativa, Fagus orientalis, Pinus kochiana, etc.*[94] The understorey of *Hedera* habitats is always species-rich and usually contains other lianas (*Clematis vitalba, C. orientalis, Smilax*

[91] Ketskhoveli 1959; Ketskhoveli 1984.
[92] Huntley and Birks 1983.
[93] Ketskhoveli 1984.
[94] Ketskhoveli 1959.

excelsa, Vitis sylvestris, Periploca graeca and *Humulus lupulus*). In Western Georgia, *Hedera* often grows alongside other evergreen shrubs (*Rhododendron ponticum, Laurus nobilis, Laurocerasus officinalis, Ilex colchica, Buxus colchica* and *Ruscus colchica*).[95]

Hedera species produce tricolporate pollen with heavy reticulation. The pollen is dispersed by insects and is therefore under-represented in pollen spectra. This is compounded by the fact that most ivy plants do not flower until they reach the crown of their host tree. *Hedera* pollen usually indicates the presence of ivy plants nearby, but long distance transport is also possible in treeless environments. Currently, *Hedera* pollen is found at three sites: on the Black Sea coast, along the central Caucasus Range, and in Southern Georgia.

Hedera pollen occurs very occasionally during the late Pleistocene at Sukhumi, indicating the presence of ivy in low-altitude forest refugia. Not until 8000 years ago does *Hedera* pollen again become important in Western Georgia. Five thousand years ago it only occurs along the Ajara coast (southwestern Georgia), and, apart from occasional occurrences in Southern and Eastern Georgia, it remains associated with Ajara for the remainder of the Holocene. Whilst the ivy maps only give a patchy indication of the species' history, it seems that ivy had a wider early-mid Holocene distribution, and then retreated southward during the mid-late Holocene.

Ivy is strongly influenced by climate and demands warm-to-hot, drought-free summers.[96] If ivy's southward retreat and eastward spread can be interpreted in climatic terms, then it indicates that the early-mid Holocene in Western Georgia had warm, moist summers, but Eastern Georgia was probably too dry for *Hedera* (except in restricted riparian habitats). In the mid-late Holocene, cooler and perhaps drier summers in Abkhazia pushed *Hedera* to the south. This explanation seems to fit with the increasingly continental rainfall regime hypothesised for the mid Holocene. A mid-late Holocene shift towards more Mediterranean (*i.e.* summer-dry) conditions may have been to the detriment of *Hedera* populations.

Ilex

Ilex colchica is the only holly in the study area. It is an evergreen shrub of 2–6 m stature, distributed throughout the mixed broadleaf forests of Colchis and northern Anatolia, with isolated populations in Eastern Georgia and the Iskenderun area near the Turkish-Syrian border.[97] *Ilex colchica* is shade-tolerant and grows slowly in the understorey of mixed forests dominated by *Fagus orientalis, Carpinus betulus, Castanea sativa, Abies nordmanniana* and *Picea orientalis*. At the upper limits of its range, which is from about 200 to 2300 m, it occurs in subalpine *Betula-Acer* woods and amongst evergreen Colchic understorey vegetation.[98] It spreads well vegetatively, but is intolerant of drought. It is a Euxine species closely related to the more widespread and cold-tolerant European species, *Ilex aquifolium*.

Ilex pollen is tricolpate and distinctly clavate-baculate.[99] It is insect-dispersed and very poorly represented in pollen spectra. The presence of *Ilex* pollen indicates the presence of *Ilex colchica* plants in the surrounding vegetation; its absence does not however indicate the plants' absence. The map of its present and past distributions demonstrates the rarity of *Ilex* pollen now and in the past. *Ilex* pollen is recorded at 10,000 cal BP in Abkhazia, but its greatest period of expansion was during the mid Holocene. At 7000 cal BP, *Ilex* was present at several locations in Colchis. By 5000 cal BP, its spread had grown to include most of the coastal lowlands. Later occurrences (4000, 3000 and 1000 cal BP) come from Devis Tba on the

[95] Ketskhoveli 1959.
[96] Huntley and Birks 1983.
[97] Browicz 1982 ff.; Ketskhoveli 1983.
[98] Ketskhoveli 1983.
[99] Moore *et al.* 1991.

Racha Range and the Anaklia site on the Black Sea coast. *Ilex* pollen occurs sporadically at several other sites through the late Holocene, but was missed in the 1000-year intervals chosen for mapping.

The European holly, *Ilex aquifolium*, is strongly influenced by climatic factors, particularly temperature.[100] It seems likely that *Ilex colchica* has similar requirements for relatively mild winters and cool summers. Kvavadze and Rukhadze have argued that tree lines in Abkhazia were pushed 300 m upslope during the mid Holocene by a warming climate.[101] While warmer winters may have increased the spread of *Ilex colchica*, warmer summers might be expected to increase its drought susceptibility. Like *Hedera* (above), *Ilex* may have have found mid Holocene conditions of summer-dominated rainfall to its benefit.

JUGLANS

Juglans regia is a large deciduous tree, 20–35 m in height, distributed throughout Western Asia. In the study area it grows from sea level to about 1400-m elevation. Walnut is common in gallery forests and piedmont forests, but is far more common in cultivation. Because of its long association with humankind and its tendency to run wild, it is difficult to determine where (if at all) it occurs naturally. Ketskhoveli suggests that it may be indigenous to the gallery forests of Kakheti (Eastern Georgia),[102] where it grows in association with *Populus* × *canescens, P. nigra, Alnus barbata, Salix excelsa, S. alba, S. caprea, Quercus pedunculflora, Ulmus glabra* and *U. minor*. These forests are species-rich and unusual for Eastern Georgia in their representation of lianas (*Clematis vitalba, C. orientalis, Smilax excelsa, Hedera helix, H. pastuchowii, Vitis sylvestris, Periploca graeca* and *Humulus lupulus*). Little is known of walnut's ecology because botanists tend to regard it as an adventive species.

It is highly valued for its beautifully decorative timber, which is used for cabinet making and furniture. In parts of Western Georgia, in particular, walnuts are the predominant source of fat in people's diet.[103] No Georgian garden or orchard can claim that title without at least one walnut tree. The nuts are ground with spices as a marinade and the green fruits are made into an unusually delicious, sweet preserve.

Walnut pollen is large (~50 μm) with many pores. Although similar to *Pterocarya* pollen in polar view, *Juglans* pollen has pores all over the grain surface, not just around the equator. The pollen is wind-dispersed, but is heavy and seems not to travel far from its source. *Juglans* pollen values less than 1.7 percent occur in treeless environments where the trees are absent locally but abundant regionally. Values above 1.7 percent indicate local presence and values of 4 percent or more occur where walnuts are planted extensively. *Juglans* percentages are high in the Tbilisi area, and also in Racha in central Western Georgia. Both these areas are well-populated and at suitable altitudes for walnut cultivation.

Juglans pollen was almost ubiquitous at the late Pliocene and early Pleistocene boundary in Western Georgia, in pollen spectra dominated by *Abies, Picea, Tsuga, Pterocarya, Carya* and Taxodiaceae.[104] During the late glacial, *Juglans* pollen was quite prevalent in the Sagarejo section on the Iori River floodplain, Eastern Georgia, suggesting that walnut grew there then in gallery forests.[105] *Juglans* is found more sporadically in the Sukhumi record from Abkhazia at the same time. By 9000 cal BP, however, *Juglans* had dwindled in both of these presumed

[100] Roberts 1998, p. 35.
[101] Kvavadze and Rukhadze 1989.
[102] Ketskhoveli 1957.
[103] Ketskhoveli 1957.
[104] Shatilova 1984; Shatilova and Mchedlishvili 1980.
[105] Gogichaishvili 1984b.

glacial refugia. Throughout the early Holocene, walnut pollen is found in trace proportions near Bakuriani, where it may have been wind-transported from populations at lower altitudes in the Borjomi Gorge. During the mid Holocene *Juglans* pollen continued to appear at this site and several locations in Abkhazia, but not in any quantity that would suggest local presence. High proportions at 2150 m elevation around 4000 cal BP on the northern slope of the Main Caucasus Range are almost certainly the product of wind transport. Not until 2000 cal BP do we see the widespread occurrence of *Juglans* pollen in the study area. By 1000 cal BP it was well established in many regions, with local plantations in Racha and Kartli. The situation is much the same today.

Was *Juglans regia* indigenous to the study area? Ketskhoveli draws attention to the geobotanical similarities between the Talysh region in Azerbaijan, where the walnut is assumed to be indigenous, and the gallery forests of Eastern Georgia. 'Hence it is possible that walnut is native to the forests of Kakheti. But, despite this, in most cases their existence in forests is secondary'.[106] Archaeological finds are few. Walnut shells dating to the fifth millennium BC have been identified in Georgia and from the second millennium BC in Azerbaijan, but subsequent archaeological finds post-date Greek colonisation.[107]

Palynological evidence from Sagarejo and Sukhumi certainly suggests that *Juglans* trees were present in lowland gallery forests in the late glacial, but where did they go during the early and mid Holocene? It seems paradoxical that a warmth-loving tree would be less common during the Holocene than the glacial, but perhaps the late glacial climate of Western Asia was, as suggested by some,[108] relatively warm but extremely arid. At Sagarejo, *Juglans* and other thermophile gallery forest elements may have survived this aridity by hugging close to rivers fed by glacial meltwater. Once river channels began to stabilise in the Early Holocene, *Juglans* populations along rivers draining the Main Caucasus Range were forced out by other, more light-competitive trees. Perhaps *Juglans* continued to grow elsewhere during the early Holocene, in the Borjomi Gorge, the Alazani Valley or in places in Western Georgia, but the data do not furnish solid evidence of this.

What is certain is that *Juglans* populations increased markedly during the last 2000 years, since Greek and Roman colonisation. A similar expansion is observed through much of Europe at the same time, possibly triggered by favourable climatic conditions.[109] The same climatic conditions may have also allowed *Pterocarya* to expand out of its Holocene refuges (see *Pterocarya*, this appendix). The appearance of *Olea europaea* pollen at the same time suggests a strong human hand in the appearance of walnut during Classical times (*Olea*, this appendix). The Tsavkisi record from near Tbilisi indicates a change in land-use from pastures to orchards as economic or cultural influences arrived from the West. Could a cultural change manifest in so many places at once? But equally, what kind of climatic change could allow *Juglans* to expand so rapidly and widely in the Caucasus region with such mountainous relief and such a variety of microclimates? And, from a cultural perspective, did *Juglans* have the economic and culinary importance it currently enjoys in Georgia prior to the Greeks?

The history of *Juglans regia* is perplexing but fascinating. Its decline in the early Holocene seems to contradict the idea that the glacial period was cold. The cause of its late Holocene expansion will continue to be debated. Hopefully future research will shed much needed light on the reasons for the past behaviour of this most valuable tree.

[106] Ketskhoveli 1957, p. 325.
[107] Lisitsyna and Prishchepenko 1977.
[108] Davis *et al.* 2003.
[109] Huntley and Birks 1983.

JUNIPERUS

Ecology

Seven juniper species inhabit the study area: *Juniperus oblonga, J. oxycedrus, J. foetidissima, J. polycarpos, J. depressa, J. pygmaea* and *J. sabina*.[110] The first four species can grow into trees about 10 m high and of great age under ideal conditions, while the last three are all small shrubs.

Juniperus oblonga, J. oxycedrus, J. foetidissima and *J. polycarpos* are often found together in xerophyte scrub and savanna vegetation and on dry, rocky, southern slopes in Eastern Georgia. *Juniperus oblonga* and *J. oxycedrus* grow both in Eastern and Western Georgia, in forest understorey, forest edges and open woodlands. *J. foetidissima* and *J. polycarpos* only occur in open vegetation between Mtskheta and the Azerbaijan border.[111] Although juniper woodlands are rare in Georgia, their area in Armenia and Azerbaijan is considerable – 1.7 and 6.2 percent of the total forest cover, respectively.[112]

The climate of juniper woodlands is variable. All are continental, having cold winters below freezing and summers that vary from mild (**Table A.3**) to hot (*e.g.* Azerbaijan, Eastern Georgia). Juniper woodlands occur over an extraordinary altitudinal range – from less than 300 m (Azerbaijan) to over 2200 m (Armenia). The constant characteristic of juniper woodlands is low rainfall – as little as 400 mm per annum. In areas with higher rainfall, junipers are confined to steep slopes and other places where soil moisture is insufficient to sustain other tree species.

Table A.3. Monthly climatic data from *Juniperus*-dominated woodland north of Lake Sevan, Armenia. T: average monthly temperature; P: average monthly precipitation. Source: Gulisashvili 1964.

	J	F	M	A	M	J	J	A	S	O	N	D	Avg.
T (°C)	–5	–4.6	–1.4	2.6	7.5	12.3	15.3	16.0	13.0	8.8	2.9	–2.6	6.4
P (mm)	4	13	10	45	77	100	37	39	46	42	21	8	442

The remaining species (*J. depressa, J. pygmaea* and *J. sabina*) are mountain species growing amongst rocks and screes in the subalpine and alpine belts. They occur in tragacanth vegetation on the drier parts of the Main Caucasus and Anticaucasus ranges in both Western and Eastern Georgia.

The arboreal species are commonly associated with *Quercus iberica, Carpinus orientalis, Paliurus spina-christi, Rhamnus pallasii, Pistacia mutica, Ephedra procera, Spiraea hypericifolia, Lonicera iberica, Acer ibericum, Celtis caucasica, Pyrus salicifolia,* xerophilous herbs and steppe grasses.[113] At higher altitudes they occur with the highland oak, *Quercus macranthera*. The alpine shrubby species are associated with tragacanthic *Astragalus* spp., *Artemisia* spp., *Berberis vulgaris, Ephedra procera, Allium* spp. and a variety of grasses.[114] In subalpine forests, they can be associated with *Pinus kochiana*.[115]

[110] Ketskhoveli 1971.
[111] Arabuli 2003; Ketskhoveli 1959.
[112] Kruchinin 1973.
[113] Gulisashvili 1964; Ketskhoveli 1959.
[114] Nakhutsrishvili 1999.
[115] Ketskhoveli 1959.

Juniperus species are drought tolerant, but intolerant of waterlogging and fire.[116] Fire kills junipers to the ground. Junipers tend to proliferate under grazing because the livestock avoid eating them and open pastures are good habitats for light-demanding shrubs. Graziers often set fire to individual shrubs where they threaten to overtake pastures.

Juniper timber is strong and durable, used for carving, furniture, cabinet making, *etc*. Several species yield essential oils for medicinal purposes and pigments for dyeing.[117]

Pollen Representation

Juniperus pollen is produced in abundance, but poorly dispersed by the wind.[118] The modern distribution of *Juniperus* pollen fairly accurately reflects the abundance of *Juniperus* plants in the vegetation. The highest values come from the Tbilisi area, where junipers are common in scrub and shibliak vegetation at the border between forest and steppe. Values of 10 percent or more indicate that junipers form an important part of the vegetation, while proportions between five and 10 percent indicate the presence of a few junipers. Values below these may represent long-distance pollen transport.

Late Quaternary History of Juniperus *from Pollen Records*

Juniper occupied an important place in the arid desert-steppe vegetation of the South Georgian Uplands during the late Pleistocene. Its role declined in the late glacial, as *Ephedra* became more abundant. During the early Holocene it was still present in Southern Georgia and relatively common near Tbilisi in scrub vegetation. Its role changed little until 5000 cal BP, when it became more abundant near Tbilisi and also on the Tsalka Plateau as a component of oak woodland.

Juniper had declined by 3000 cal BP, except near Tbilisi where very high proportions suggest a juniper-dominated landscape at that time. Juniper woodland was gradually replaced by steppe-phrygana vegetation, reflected in the lesser importance of *Juniperus* pollen at this site two thousand years ago. At 1000 cal BP, juniper had once again increased in steppic vegetation on the Tsalka Plateau and around Tbilisi, coincident with a dry lake phase at Lake Imera (Chapter Six).

The history of *Juniperus* is very closely linked to climate. It expands especially during drier phases and retracts in wetter phases. It is less clear whether it has any relationship to temperature. Junipers were widespread in the South Georgian Uplands in the late Pleistocene, just as they were in Western Europe at the same time,[119] and as they are today in parts of Eastern Georgia, southern Armenia and western Azerbaijan. On geobotanical grounds it seems likely that it was the arboreal rather than alpine species that prevailed in Southern Georgia during the late Pleistocene. This is in keeping with the hypothesis that the late glacial vegetation of Western Asia was controlled more by extreme aridity than temperature.[120]

The *Juniperus*–Chenopodiaceae association that appears in the Aligol pollen record between 14,000 and 12,000 cal BP is probably analogous to the juniper-saxaul communities found presently in parts of western Azerbaijan, where rainfall averages around 400 mm per annum. Lemke and Sturm have estimated a 30 percent drop in regional precipitation during the late glacial.[121] Aligol presently receives approximately 600 mm precipitation in an average year – a

[116] Grove and Rackham 2001; Huntley and Birks 1983.
[117] Ketskhoveli 1971; Polunin 1969.
[118] Huntley and Birks 1983.
[119] Huntley and Birks 1983.
[120] Roberts and Wright 1993; Wick *et al.* 2003.
[121] Lemke and Sturm 1997.

30 percent decrease brings that figure very close to 400 mm. As for the *Ephedra-Artemisia* association that replaced *Juniperus* and Chenopodiaceae during the Younger Dryas, I have argued that this came about through a reduction in winter temperatures. One might have to look to the *Ephedra* thickets of Tien Shan (Kirgizia/Kirgistan) for an equivalent of Southern Georgia's vegetation during the Younger Dryas.

Juniper's place in the Holocene vegetation of the study area was small and confined to Southern Georgia. A reduction in fires and a warming climate with perhaps more frequent drought helped *Juniperus* re-establish on the Tsalka Plateau during the mid Holocene. The same factors may explain its expansion there 1000 years ago. Near Tbilisi, its expansion between 4000 and 3000 cal BP occurred as the climate became colder and drier. After 3000 cal BP, the climate near Tbilisi became wetter, and it seems that grassland expanded at juniper's expense. In successional terms, it is unusual that trees could be displaced by grasses, but perhaps graziers had a hand in this process through selective burning. Today extensive juniper woodlands are found only in the most remote and waterless parts of the Caucasus where pyromaniac herders seldom set foot.

OLEA EUROPAEA[122]

Olea europaea is an evergreen small tree or large shrub native to the circum-Mediterranean area. In the study area it grows on dry, limestone slopes around the Gagra and Bichvinta areas of Abkhazia, where the climate is more Mediterranean than elsewhere.[123] Olives are not grown in Eastern Georgia today, but historical sources indicate that they once were. The early eighteenth century geographer, Vakhushti Bagrationi, wrote of the lower Khrami valley: "This place, compared to elsewhere, produces more pomegranates, figs, olives and other fruits".[124]

Olive pollen has been identified in the study area only recently. The Kumisi, Jvari, Tsavkisi, Imera, Bareti and Ispani records all contain *Olea europaea* pollen. Occasional pollen grains are recorded through much of the early and mid Holocene. After 2000 cal BP, finds become much more frequent and at many sites the *Olea* pollen curve becomes a continuous one. The sites with the greatest representation of olive pollen are Kumisi, Tsavkisi and Imera, the three sites closest to the location Vakhushti described. It seems certain olive cultivation began in the warmer lowlands of Eastern Georgia around the first century BC, as a result of contacts with Greece and Rome.

Ketskhoveli hypothesises that historical events, such as catastrophic frosts in the late 18th century and the effects of war, saw that olives were no longer planted in Eastern Georgia.[125] According to the pollen records, *Olea europaea* began to decline around 1000 years ago. Olive pollen is very scarce during the last 500 years. A similar history can be seen in the Lake Van pollen diagram.[126]

As well as the Mediterranean region, olive populations are reported from the Chorokh Valley, NE Anatolia, and from the northern shores of Lake Van.[127] All of these places have a Mediterranean climate, *i.e.* cool, wet winters followed by hot, dry summers. It is frost that determines where olive trees can be grown: a frost of –13 °C will kill olive trees outright.[128] In Lower Kartli, where Vakhushti described olives growing, the annual average absolute minimum temperature is –12 °C.[129] Even during a relatively mild winter, this is at the very limit of olive

122 *Olea* pollen has not been mapped due to a scarcity of finds.
123 Gagnidze 1985; Ketskhoveli 1957.
124 Vakhushti Bagrationi 1983.
125 Ketskhoveli 1957.
126 Wick *et al.* 2003.
127 Zohary 1973.
128 Grove and Rackham 2001.
129 AN GSSR 1964.

cultivation. If winters during the Classical Era less severe, however, olives could have been viable. The Tsavkisi record (Chapter Five) seems to indicate a warmer, wetter climate during the Classical Era.

The appearance of *Olea europaea* in the vegetation history of the study area is fleeting, but gives us a very detailed insight into the climate and economy of the last two millennia. A more Mediterranean climate during the Classical Era not only allowed olives and walnuts to be cultivated in great numbers, but may have also initiated the late expansion of thermophilous *Pterocarya*. Among the many possible reasons for olive's decline in recent centuries, the evidence supports Ketskhoveli's hypothesis,[130] that olives were repeatedly killed by heavy frost (during the 'little ice age', ca 1590–1850 AD) until their cultivation became uneconomic.

OSTRYA-TYPE

Ostrya-type represents two deciduous tree species native to the study area: *Ostrya carpinifolia* and *Carpinus orientalis*. The former is a small to medium-sized tree found in low and middle altitude forests, mostly in Western Georgia. The latter is far more widespread, occurring as a small tree or bush throughout the study area from the lower to the upper tree line. Both species grow in the understorey of mixed deciduous forests, but *Carpinus orientalis* is also a major component of shibliak and scrub vegetation. Here it is associated with species like *Paliurus spina-christi, Rhamnus pallasii, Juniperus* spp., *Quercus iberica* and other xerophytes.[131] This vegetation type is very similar floristically to sub-Mediterranean scrub.[132]

Carpinus orientalis is very common where forest (particularly oak) has been cut down.[133] It seems to tolerate grazing much better than other deciduous trees, quickly converting the landscape from pasture to woodland if livestock are removed. It is ubiquitous around Tbilisi, where it forms one of the penultimate successional stages in Avakov's model of reafforestation.[134] *Carpinus orientalis* and *C. betulus* were formerly coppiced extensively around Tbilisi for charcoal making. Most of the hornbeam forests around the city today are overgrown coppices.

Ostrya-type pollen is poorly represented in sediments. Even where it forms a substantial cover, values higher than 10 percent are rare. The modern distribution of *Ostrya*-type pollen shows a clear affinity for the Tbilisi area and Southern Georgia, where values between 11 and 18 percent represent its significant participation in the vegetation. Near Tbilisi values of 25–30 percent are also recorded where *Carpinus orientalis* are often the only trees or shrubs in deforested landscapes. Values less than 11 percent occur where *Carpinus orientalis* plants are present regionally. In Western Georgia, *Ostrya*-type pollen is relatively uncommon and confined to the forest belt.

Ostrya-type pollen occurs throughout the late Pleistocene levels of Sagarejo in Eastern Georgia, indicating its presence in the 'Colchic' forest refugium there. It increases during the early Holocene and is then recorded sparsely in Southern Georgia and Abkhazia by 10,000 cal BP. By 7000 cal BP, it had spread throughout the study area, and was well represented in the lowlands of Western Georgia. A thousand years later, however, *Ostrya*-type values had declined across the region and it appears to have disappeared from large areas. *Ostrya*-type began a slow increase 4000 years ago, when it regained some of its former distribution and remained a minor element in the vegetation. Its abundance magnified considerably between 2000–1000 years ago in Eastern Georgia, and *Ostrya*-type pollen there is nowadays more prevalent than at any other time.

[130] Ketskhoveli 1957.
[131] Ketskhoveli 1959.
[132] Walter 1974.
[133] Grossgeim 1945.
[134] Avakov 1982.

The changing abundances of *Ostrya*-type pollen can best be interpreted in terms of landscape openness. It is interesting to note that *Ostrya*-type shares a similar history to *Rumex,* a weedy herb found predominantly in open vegetation (see *Rumex,* below). *Ostrya carpinifolia* and/or *Carpinus orientalis* participated in the early Holocene expansion of forests, but after 7000 cal BP were subordinated by canopy trees such as *Fagus, Quercus* and *Carpinus betulus*. Forest disturbances after 5000 cal BP allowed *Ostrya* and *C. orientalis* to colonise woodland gaps, clearings and edges.

Their late Holocene history is the result of a complex interaction between deforestation, burning, grazing, coppicing and climate and warrants more detailed palaeoecological investigation. Because *Carpinus orientalis* has an extraordinary ability to survive intense browsing without flowering, it is difficult to tell whether its late Holocene increase is real or just the result of profuse flowering as heavily grazed pastures gave way to woodland coppices.

PICEA

Ecology

Picea orientalis is the only *Picea* species native to the study area. It is a coniferous tree that can grow to 50 metres high with a trunk diameter of 1½ metres. It is distributed in the western Caucasus and northeast Anatolia and is closely related to *Picea omorika* of the Drina Basin in the Balkans.[135] Spruce is much less common in the Eastern Caucasus where the climate is drier, but spreads along the Trialeti Range as far east as Tbilisi.[136] *Picea orientalis* occurs on shaded slopes from about 300 m elevation up to 2200 m, usually admixed with *Abies nordmanniana, Fagus orientalis* or *Pinus kochiana*. It forms pure montane forests between 1000 and 1800 m on moist, slightly-acidic soils.[137] *Picea orientalis* habitats have relatively cool temperatures, winter snow cover and high rainfall, similar to that of *Abies nordmanniana,* its associate species (see *Abies*, this appendix).

The understorey of *Picea* and *Picea-Abies* forests depends greatly on moisture availability. In the driest *Picea* forests, where *Quercus iberica* or *Pinus kochiana* may also occur in the canopy, grasses and herbs (*Poa nemoralis, Brachypodium sylvaticum* and *Oxalis acetosella*) are characteristic. In wetter situations in Colchis, the understorey may include shrubs such as *Vaccinium arctostaphylos, Daphne pontica, Viburnum orientale, Ilex colchica, Rhododendron ponticum* and herbs such as *Trachystemon orientalis, Oxalis acetosella, Sanicula europaea, Galium rotundifolium* and *Festuca drymeja*.[138]

Picea orientalis is tolerant of freezing temperatures, but not of drought. It appears to be less fire sensitive than *Abies*. It has been extensively exploited in the Caucasus for centuries, finding use in construction, paper-making, dyeing and the manufacture of turpentine and pitch.[139]

Pollen Representation

Picea pollen is large and saccate, characterised by much finer patterning on the air bladders than *Pinus* or *Abies*. It makes its way into the pollen spectra of most places in the study area, but is best represented where *Picea* trees dominate the vegetation. Values above 40 percent in the Bakuriani and Akhaltsikhe areas indicate the dominance of *Picea orientalis* in forests there. *Picea* pollen is also frequent in the subalpine zone along the Main Caucasus Range, but here the importance of *Abies* is greater on account of higher precipitation. Values less than 12 percent,

[135] Browicz 1982 ff.
[136] Nakhutsrishvili 1999.
[137] Gulisashvili 1964; Ketskhoveli 1971.
[138] Nakhutsrishvili 1999.
[139] Ketskhoveli 1971.

and especially below 4 percent, are often the result of long distance pollen transport from lower altitudes.

Picea was an important component of early Pleistocene forests, based on its prevalence in Chaudian deposits.[140] During the late Pleistocene, *Picea* pollen from the Black Sea lowlands indicates that spruce was present in the Colchic refugia. High early Holocene values in the Bakuriani area also suggest that *Picea* survived the glaciation in river gorges, such as the middle and upper reaches of the Kura River. Mostly though, *Picea* was not a major component of the vegetation near any of the study sites during the early and middle Holocene. It increases modestly in Abkhazia and Southern Georgia around 7000 cal BP and long-distance transported pollen also becomes more widely distributed. However, it is not until 3000 cal BP that *Picea* begins to gain ground in Western and Southern Georgia. *Picea* populations have since exploded – especially during the last millennium – in the Bakuriani area and at the upper treeline along the Main Caucasus Range. Spruce is more common today than at any other time during the Holocene.

Several hypotheses may help explain spruce's population explosion during the late Holocene. The first is that the climate became more suitable for *Picea* in the last 2–3 millennia. Declining temperatures and increasing rainfall could favour spruce expansion. There is some evidence to suggest that the late Holocene was somewhat wetter than the early-mid Holocene: after a dry spell 3000 years ago, lake levels began to increase in Eastern Anatolia and the Tsalka Plateau; and pollen based moisture curves (Chapter Seven) indicate higher moisture levels in the last 3000 years.

It could also be hypothesised that edaphic changes, perhaps caused by higher rainfall and soil maturation, increased the extent of podzols in the study area. This hypothesis, however, fails to explain why spruce's expansion was so rapid in many areas.

A third hypothesis is that the late Holocene *Picea* expansion was aided by human activities. Clearing of high altitude dwarf deciduous woods, burning and grazing of vegetation and cultivation of formerly forested areas could have, directly or indirectly, favoured *Picea* over competing deciduous species such as *Fagus orientalis*. This hypothesis is well supported by palynological evidence. At Gomnis Lake, *Picea* expanded around 1000 years ago onto open pastures and fields that were cleared of beech forest centuries earlier.[141] In Western Georgia spruce's expansion occurred after forest disturbance.[142] Disturbance may have been enough to halt the 'inertia'[143] of dwarf beech woods that were at their upper altitudinal limits and make way for spruce trees. While humans may have provided clearings and gaps for these conifers to colonise, perhaps *Picea* was only able to expand so rapidly and dominate so comprehensively because of a wetter, cooler late Holocene climate.

Pinus

Ecology

This pollen type is representative of three *Pinus* species distributed in the Caucasus: *Pinus kochiana, P. eldarica* and *P. pityusa*. Pine forests are very characteristic of the North Caucasus, especially in river gorges with nutrient-poor, rocky soils and a rather continental climate. Pines have a much lesser and very fragmented distribution in the South Caucasus, possibly occupying the same locations where they survived the last glacial period.[144]

140 Shatilova 1984; Shatilova and Mchedlishvili 1980; Shatilova *et al.* 2003.

141 Margalitadze 1971.

142 Kvavadze and Efremov 1996; Kvavadze and Rukhadze 1989.

143 *Sensu* Smith 1965.

144 Dolukhanov 1966; Gulisashvili 1949.

Pinus kochiana (syn. *P. hamata, P. sosnowskyi*; from the *P. sylvestris* group) is the commonest pine in the South Caucasus, found in all but the most arid areas. It also occurs in Crimea and Northern Anatolia.[145] Although it has a fragmentary distribution in the South Caucasus, there are some significant natural stands near the headwaters of the Kura River and in the Tana River basin. Here, the climate is dry and continental, with less than 600 mm of annual precipitation, mild summers and winters below freezing (**Table A.4**). *Pinus kochiana* is an evergreen tree growing to 20–35 m, but under extreme conditions is dwarfed to 3–5 m. It is usually associated with dry, rocky slopes where other, more competitive tree species are unable to establish. Its altitudinal distribution goes from about 700 to 2400 m, forming either pure stands, forest associations with *Fagus, Picea* and *Abies*, or high altitude open woodlands with *Betula*. The understorey of *Pinus kochiana* forests is extremely variable. In Southern Georgia (*e.g.* the Trialeti Range), they have sparse undergrowth with grasses, tragacanths and *Juniperus* shrubs. In the largely-treeless Javakheti region, scattered pines occur amongst diverse subalpine meadow vegetation, creating a park-like landscape.[146] In moister areas, undergrowth is suppressed by low light and the toxic leaf litter pines produce, although deciduous shrubs are common. A detailed description of Caucasian pine forest undergrowth is given by Walter.[147]

Table A.4. Monthly climatic data from *Pinus kochiana* habitat at Ormotsi, 41° 51'N, 43° 57'E, 1200 m a.s.l., Tana Basin, Trialeti Range, Georgia. T: average monthly temperature; P: average monthly precipitation. Source: Gulisashvili 1964.

	J	F	M	A	M	J	J	A	S	O	N	D	Avg.
T (°C)	–3	–2.4	2.1	7.1	12.3	15.3	18.3	18.6	13.9	9.0	3.0	–1.0	7.8
P (mm)	24	26	25	50	73	67	58	50	46	38	38	27	524

The two other pine species found in the Caucasus are the Pitsunda Pine (*Pinus pityusa*) and the Eldar Pine (*P. eldarica*). The former is a relict, seaside species confined to a few locations in Abkhazia, where it grows 20–35 m tall on rocky limestone soils fronting the Black Sea. Its habitat is from sea level to about 300 m elevation and characterised by mild temperatures and an abundance of rainfall. It is related to *Pinus eldarica* and pines from the *P. brutia* group.[148]

Pinus eldarica is a xerophile pine occurring in arid steppes near the border between Azerbaijan and Georgia. It grows 12–15 m tall in savanna with *Juniperus* species and steppe grasses. The elevation of these stands is 450–700 m above sea level and the climate is dry, with less than 400 mm annual rainfall and hot summers. Both *Pinus eldarica* and *P. pityusa* are currently so restricted in distribution that their contribution to pollen records in the Caucasus might be only of local significance.

Pines have considerable economic value in the Caucasus, reflected in the abundance of pine plantations (mostly of *Pinus kochiana*) established during the Soviet period. Its timber is easily worked and used extensively in construction. Turpentine, pitch and tar are obtained from the sap. Pines are also valued in the mountainous Caucasus for their ability to stabilise denuded slopes.[149]

One significant characteristic of pines from a palaeoecological standpoint is their ability to halt vegetation succession through the litter they produce. Pine needle litter is toxic to most

[145] Nakhutsrishvili 1999.
[146] Nakhutsrishvili 1999.
[147] Walter 1974.
[148] Ketskhoveli 1959.
[149] Dolukhanov 1966; Ketskhoveli 1971.

other plants and quite acidic, leading to the formation of podzol soils beneath pine forests. Pines are also adapted to fire.[150] Their resinous bark and flammable leaf litter encourage fire to periodically remove fire-sensitive competitors such as *Abies* and *Fagus*, which are otherwise able to shade out light-demanding pine seedlings.

Pollen Representation

Pine pollen is saccate and produced in enormous quantities. Individual pollen grains are wind-transported over long distances, allowing pines to cross-pollinate despite their disjunct distribution. Pine pollen percentages are difficult to interpret because of their over-representation, especially in treeless mountain areas. Pollen spectra from treeless sites often contain up to 50 percent *Pinus* pollen, derived from long-distance transport. Huntley and Birks estimate that when pollen values exceed 50 percent, pines are present locally and may be dominant.[151] Values of 25–50 percent in forested landscapes indicate that pines are present locally. These results are in general accord with data from Southern Georgia: *Pinus* pollen comprises up to 36 percent of arboreal pollen even when pine trees are absent from the surrounding vegetation and exceeds 55 percent when locally dominant or regionally significant (see Chapter Four). The interpretation of *Pinus* pollen requires consideration of the entire pollen spectrum to determine the character of the landscape at the time of pollen deposition.

In this study, I have not separated *Pinus* pollen into Diploxylon and Haploxylon types, as the latter were so rare as to be insignificant. *Pinus* appears sporadically as redeposited pollen in lake sediments near Tbilisi, where it is presumably eroded from Eocene sedimentary rocks. These redeposited grains are clearly identifiable by their smaller size, crumpled appearance and dark colour when stained, and are not included in the *Pinus* percentage. In some sediments, the prevalence of *Pinus* pollen may be the result of differential pollen preservation – pine pollen is easy to identify even when broken and corroded.

Pinus is perhaps the most ubiquitous pollen type in the area, occurring in most deposits regardless of their genesis or proximity to pine trees. The highest percentages occur in the highlands of Southern Georgia and in the treeless alpine belt of the Main Caucasus Range. Intermediate values occur in the mountains bordering these areas, while the lowest percentages appear in the Colchis lowlands where dense broadleaf and swamp forests predominate. Its pollen representation, therefore, does not rely on the presence of pine trees, but perhaps more on the absence of other trees. This presents a significant obstacle to reconstructing past vegetation, especially distinguishing pine forest from pine savanna and treeless grassland.

Late Quaternary History of Pinus *from Pollen Records*

Pinus pollen is an important component of late Pleistocene deposits in the study area. It is dominant in the Sagarejo palaeosols analysed by Gogichaisvili.[152] This suggests that pine forests descended to quite low altitudes in parts of Eastern Georgia in the late Pleistocene. Perhaps these were extensive stands of *Pinus eldarica*. At other sites, however, pine pollen occurs in proportions around 10–20 percent of arboreal pollen, increasing somewhat at all sites during the Younger Dryas.

In the early Holocene, pine began to emerge from its glacial refugia. This is shown in increasing pine percentages at sites such as Gomnis Lake (above the Kura gorge near Borjomi) and at Sukhumi (on the Black Sea coast). Most other sites exhibit a slight decline in pine

[150] Grove and Rackham 2001.
[151] Huntley and Birks 1983.
[152] Gogichaishvili 1984b.

percentages, but it would be difficult to attribute this to a decline in the areal extent of pine trees; rather, it could merely be a percentage dilution caused by the expansion of other species. By 9000 cal BP, most sites have what could be called 'background' levels of pine pollen. Pine had already migrated upslope to inhabit the marginal areas avoided by most broadleaf trees. By 7000 cal BP, pine was important at several sites along the Black Sea coast and at high altitudes in the South Georgian Uplands and along the Main Caucasus Range.

Pinus pollen is slightly less prevalent in mid Holocene sediments in comparison to the early Holocene. This was the time of maximum afforestation in many parts of the Caucasus. Five thousand years ago pine again began to increase on the Tsalka Plateau, in the Main Caucasus and along the Black Sea coast. This process continued throughout the late Holocene, with most sites in Eastern Georgia registering an increase. Although climatic factors clearly played a role in the expansion of pines in the mid to late Holocene, the most dramatic and sudden increases in pine pollen were almost certainly the product of late Holocene deforestation. These events were almost always associated with fire. It is difficult to say whether the late Holocene *Pinus* increase is a real reflection of expanding pine populations, or merely reflects a reduction in forest cover. At several tree line sites (*e.g.* Luganskoe, Gomnis Lake, Tseros Tbis), *Pinus* pollen declines in the late Holocene due to the rapid expansion of other coniferous species (*Picea* and *Abies*). Judging from existing pollen records, however, pine pollen is generally more prevalent in the present day than it has been at any other time in the Holocene.

Black Sea coast sites have a *Pinus* pollen record that stands apart from sites inland, due in part to the exceptional climate in Colchis. *Pinus* pollen in these humid lowlands is more likely to indicate local occurrences of pine trees than in other parts of the Caucasus because of the generally forested landscape. The Colchis records show considerable variation over space and time. Pine pollen was clearly less prevalent, however, at 6000 and 3000 cal BP, times characterised by the expansion of *Alnus* swamps. Nowadays, Pitsunda Pine (*P. pityusa*) has been pushed into a 200–hectare area on the Pitsunda Peninsula, but in the past it may have had a considerably wider distribution along the eastern Black Sea littoral. Sea level changes and coastal subsidence, processes that led to the paludification of the Black Sea coastline and the expansion of *Alnus* swamps, likely had a deleterious impact on seaside pine populations. Humans may have also exerted pressure on coastal pines; in antiquity Colchis was prized for "everything that pertains to ship-building; for it not only produces quantities of timber but also brings it down on rivers. And the people make linen in quantities, and hemp, wax, and pitch".[153] It is perhaps no coincidence that the Georgian word for the pine tree, *pich'vi*, echoes the Latin *pix* and its English equivalent, *pitch*.

In conclusion, pine's role in the Holocene vegetation of Georgia has always been a marginal one, occupying marginal habitats along the seashore, at the upper tree line or on soils too poor to support other types of vegetation. Its pollen record is difficult to interpret because of the wide dispersion of pine pollen and its over-representation in treeless environments. Nonetheless, it does appear that pine has been favoured by late Holocene human impacts and has probably benefited from the increasing role of fire in the landscapes of the Caucasus.

PISTACIA

Pistacia mutica is a small, spreading deciduous tree that grows only in the lowlands and foothills of Eastern Georgia. It usually occurs on dry, south-facing slopes and neutral, slightly saline soils. *Pistacia* is the principal tree of a kind of arid savanna vegetation[154] found most

[153] Strabo 1928, 11: 2: 17.

[154] Variously described as 'light forest', 'savannoid arid open woodland', *etc.* in Nakhutsrishvili 1999.

extensively in the Vashlovani Reservation near the Georgia-Azerbaijan border. Here it grows intermixed with *Bothriochloa* steppes and *Artemisia* semideserts, as well as various species of *Juniperus, Paliurus spina-christi, Rhamnus pallasii, Celtis caucasica, Pyrus salicifolia, Ephedra procera, Berberis iberica* and other xerophytes.[155] It is also a minor component of shibliak and scrub vegetation in the foothills of Eastern Georgia.

The climate of *Pistacia* habitats is dry. Rainfall averages 400–500 mm per annum, peaking in early summer. The average temperature is around 10–12 °C, with hot summers and cool winters.[156] In this regard it differs from highland *Juniperus*-dominated woodland. *Pistacia* is most prevalent today in the remotest parts of the arid lowlands. In part, this is due to its extreme sensitivity to grazing.

Pistacia pollen is very delicate, with small ruptures scattered over its thin exine. It is greatly underrepresented and its presence is usually a good indication of the occurrence of *Pistacia* trees in the landscape. The map of modern pollen frequencies shows that this pollen type is only recorded in the eastern part of Southern Georgia. Thence it is blown by the wind from *Pistacia mutica* stands in the lowlands of Eastern Georgia, Azerbaijan and Iran.

Pistacia pollen was present in Southern Georgia 8000 and 1000 years ago. Its appearance 8000 years ago corresponds to the maximum extent of *Pistacia* woodlands in the Lake Van area.[157] It also coincides with the '8.2 kyr' climatic event identified in the Greenland ice cores and several European pollen records. Wick *et al.* have suggested that this climatic event manifested in the Van area as increased summer drought, favouring *Pistacia* over *Quercus*. Only after the 8.2 event did oak woodland begin to expand in the Lake Van area.[158] The Georgian pollen records suggest that mesic woodland had begun to expand beforehand, but there is evidence to suggest droughtier conditions around 8,200 years ago: a transient increase in *Ephedra, Acer* and arboreal Rosaceae in the Aligol record. The *Pistacia* increase 1000 years ago likewise relates to a drying phase in the Tsalka Plateau lakes. In the highlands, the Mediaeval seems to have been slightly warmer, but with more pronounced late summer drought. The history of *Pistacia* in the study area is very much incomplete and awaits further investigation in the arid lowlands where its populations are concentrated today.

Pterocarya

Pterocarya fraxinifolia is a thermophilous, hygrophilous, Tertiary-relict tree native to the Caucasus, the Caspian coast of Iran, northeastern Anatolia, northwestern Anatolia and the Iskenderun area.[159] It is a deciduous tree, growing to a height of 10 metres or more, with long, compound leaves and winged seeds. It occurs in gallery, riparian and swamp forests in the lowlands and piedmonts of Colchis and the Kakheti region of Eastern Georgia.[160] In Turkey and Iran, *Pterocarya* is recorded at elevations up to 1160–1200 m.[161] The species is only found in places with moist soils and high watertables – swamps, floodplains, gullies and streamsides – but will not tolerate constant inundation.[162] The climate of its distribution is mild, with warm to hot summers and winter temperatures just above freezing.[163]

155 Ketskhoveli 1959.
156 Gulisashvili 1964.
157 Wick *et al.* 2003.
158 Wick *et al.* 2003.
159 Browicz 1982 ff.; Davis 1982.
160 Ketskhoveli 1975.
161 Davis 1982; Noirfalise and Djazirei 1965.
162 Denk *et al.* 2001.
163 Stuchlik and Kvavadze 1998.

Pterocarya is seldom dominant in the forest canopy. It usually co-dominates with *Alnus barbata, Populus × canescens, Salix australior, Carpinus betulus, Fraxinus excelsior* or *Quercus hartwissiana* in Colchis, and *Populus × canescens, P. nigra, Alnus barbata, Salix* spp., *Quercus pedunculiflora* or *Ulmus minor* in Eastern Georgia. *Morus alba, Ficus carica, Smilax excelsa* and *Vitis sylvestris* are shrubs and lianas often associated with *Pterocarya* habitats.[164]

Pterocarya fraxinifolia has a very restricted distribution. Denk *et al.*[165] classify it as an 'opportunistic stenoecious' species. Its behaviour is certainly opportunistic, rapidly invading deforested land and quickly attaining great height.[166] But, being stenoecious, its foothold in the region's vegetation is tenuous – several other Tertiary-relict taxa with a similar niche strategies died out during the Pleistocene: *Carya, Nyssa* and *Taxodium*.[167] Through selective felling along lowland rivercourses, humans have perhaps helped preserve *Pterocarya*, whose wood is light and not considered valuable.

Pterocarya pollen has 5–6 pores arranged equatorially around the grain. In this regard it differs from its close relative, *Juglans*, whose pollen has pores all over. *Pterocarya* pollen is dispersed by the wind, but only appears in high proportions where plants are present locally.[168] No pollen sites in the present-day have *Pterocarya* pollen percentages high enough to suggest local presence. However, the distibution of sites (along the Black Sea coast and in Eastern Georgia) is more-or-less in accordance with the distribution of *Pterocarya* trees. Its pollen is absent from the mountain regions, except when it is carried there by wind. Values between four and 10 percent suggest local presence, while values greater than 10 percent suggest abundance.

Pterocarya has been present in the study area since the Miocene.[169] It was widespread during the early Pleistocene interglacials, just as it was in Europe.[170] During the late Pleistocene, *Pterocarya* was present on lowland floodplains in Eastern Georgia amongst gallery forests of *Salix, Alnus, Populus* and other hygrophiles. This plant community was displaced in the early Holocene for reasons as yet unknown, although it may be that river stabilisation after deglaciation caused *Pterocarya* and *Juglans* to migrate away from their glacial refugia (see *Juglans,* this appendix).

Early Holocene records of *Pterocarya* come from the Black Sea lowlands, both at Sukhumi in the north and the Supsa River in the south. *Pterocarya* persisted in these places throughout the early Holocene and appears not to have spread any distance. At 6000 cal BP, *Pterocarya* suddenly becomes very abundant along the Supsa River, where it may have been favoured by rising water tables linked to sea level rise and coastal subsidence. However, this increase is very localised and is not replicated anywhere else. In fact, at 4000 cal BP, *Pterocarya* pollen was all but absent elsewhere in the study area. By 3000 cal BP, it appears to have begun a slow increase in both east and west, appearing as long-distance transported pollen on the Tsalka Plateau, and probably as a local population of *Pterocarya* trees near Gagra in Abkhazia.

Two thousand years ago *Pterocarya* began to move away from its 'Holocene refugium' on the Supsa River, apparently migrating slowly downstream to colonise the Black Sea coast near Kobuleti, a process that was almost complete a thousand years ago. This may have been aided by milder winters during the Classical Period, producing a more Mediterranean climate, but widespread human impact could also be invoked. Since that time, it seems that *Pterocarya* has

[164] Denk *et al.* 2001; Ketskhoveli 1959.
[165] Denk *et al.* 2001.
[166] Berg 1950; Walter 1974.
[167] Denk *et al.* 2001.
[168] Stuchlik and Kvavadze 1998.
[169] Stuchlik and Kvavadze 1998.
[170] Godwin 1975; Shatilova and Ramishvili 1990.

expanded its range throughout the study area and is much more widespread now than it was throughout the Holocene. In light of its current behaviour, it would seem that humans have assisted the spread of *Pterocarya* by clearing other, more competitive vegetation, through draining swamps and altering groundwater hydrology, and also by planting this Tertiary-relict curiosity in parks and gardens.

QUERCUS

Ecology

The genus *Quercus* has seven representatives that number amongst the most important trees in the study area: *Quercus macranthera, Q. pontica, Q. hartwissiana, Q. pedunculiflora, Q. imeretina, Q. iberica* and *Q. dshorochensis.*[171]

Quercus macranthera, the highland oak, grows 10–20 m tall from the middle forest belt to the subalpine zone (~1450–2400 m elevation), mostly in the drier parts of the Eastern Caucasus, especially Southern Georgia and Armenia. It prefers dry, south-facing slopes, where it forms open, park-like woodlands of itself or with *Acer trautvetteri, Betula litwinowii, Carpinus betulus* and *Fraxinus excelsior*, and usually an understorey of grasses or xerophytic herbs.[172]

Pontic oak, *Q. pontica*, is a bizarre creeping shrub with unusual serrated leaves relict to limestone substrates in the moist subalpine zone of Western Georgia. *Quercus hartwissiana* is also confined to Western Georgia, where it is found in Colchic mixed deciduous forests from the Black Sea lowlands to the middle mountain zone (200–1200 m). It prefers rich, moist soils and is often accompanied by *Zelkova carpinifolia, Castanea sativa, Carpinus betulus* and other *Quercus* species, especially the Georgian endemic *Quercus imeretina.*

Quercus pedunculiflora is a tree common to the gallery forests of the Eastern Caucasus, especially along the lower Kura, Aragvi, Alazani and Iori rivers. It is closely related to *Quercus robur* and similarly requires moist soil, although *Q. pedunculiflora* is very tolerant of low atmospheric moisture. It is associated with *Populus* × *canescens, P. nigra, Alnus barbata, Salix* spp., *Ulmus minor, Acer* spp., *Tilia begoniifolia, Carpinus betulus, Pterocarya pterocarpa* and lianas such as *Smilax excelsa, Periploca graeca, Clematis vitalba, Vitis sylvestris* and *Hedera helix.*[173]

Quercus iberica, the Georgian oak, is the most common and widespread oak in Caucasian forests. Its distribution stretches from the Caspian Sea to the Balkans; in Europe it is substituted by its close relative, *Quercus petraea*. Although Georgian oak often appears in the mixed deciduous forests of Western Georgia, it is under the more continental climate of Eastern Georgia that its distribution is concentrated. Oak-dominated forests extend from the *Quercus iberica* lower tree line to the *Q. macranthera* upper tree line in parts of Southern Georgia. *Quercus macranthera* intergrades with and eventually replaces *Quercus iberica* at elevations above 1450 m. In this area beech trees are confined to shaded northern slopes.

The most common associates of *Quercus iberica* are hornbeams: *Carpinus betulus* in mountain forests and *C. orientalis* at the lower tree line in scrub and woodland vegetation. *Acer, Ulmus, Tilia* and *Fraxinus* species are also frequent. Oaks cast a lighter shade than many other deciduous trees and consequently the understorey of oak forests is quite diverse. Shrubs like *Cornus mas, Swida australis, Ostrya carpinifolia, Euonymus europaeus, Ligustrum vulgare, Corylus avellana, Rhus coriaria, Lonicera caucasica, Viburnum lantana* and *Juniperus oblonga* are regularly encountered beneath an oak canopy. Rosaceous shrubs also occupy an important place (see Rosaceae, this appendix) and the ground cover is often grassy or xerophytic.

171 Gagnidze 2005.
172 Grossgeim 1945; Nakhutsrishvili 1999.
173 Ketskhoveli 1959; Nakhutsrishvili 1999.

Chorokhi oak, *Quercus dshorochensis*, is a relict species very closely related to *Q. iberica* and found only in a few locations in Ajara.[174]

The climate of *Quercus iberica* forests is moderately warm and relatively dry and continental. The average temperature varies between 9 and 11 °C, with winters just a few degrees below freezing and warm to hot summers (**Table A.5**). Rainfall averages between 430 and 744 mm per annum and is concentrated in early summer. *Quercus macranthera* habitats have colder winters, but the summers are warm to hot. The precipitation regime is similar to that of *Q. iberica* forests (**Table A.5**).

Table A.5. The climates of *Quercus iberica* and *Q. macranthera* habitats: Tsqneti, 41° 41' N, 44° 42' E, 1000 m a.s.l., and Aparan (Armenia), 40° 35' N, 44° 22' E, 1891 m a.s.l. T: average monthly temperature; P: average monthly precipitation. Source: Gulisashvili 1964.

	J	F	M	A	M	J	J	A	S	O	N	D	Avg.
Quercus iberica													
T (°C)	–2.2	–0.5	3.0	8.1	13.6	17.6	20.6	20.4	15.7	11.1	4.7	–0.8	9.3
P (mm)	15	27	36	75	116	101	61	48	58	57	47	30	671
Quercus macranthera													
T (°C)	–9.8	–8.1	–3.4	4.0	10.0	14.0	17.4	17.6	13.4	7.4	1.4	–6.0	4.8
P (mm)	34	29	42	57	87	69	70	60	36	54	57	30	625

Oak forests are most commonly associated with nutrient-rich brown earths. These soils were also favoured in the past for agricultural development and for this reason oaks were cleared extensively in prehistory.[175] The timber of oak is also highly valued, for construction, carpentry, ship-building, furniture and flooring. The acorns are eaten greedily by pigs and can be used as a coffee substitute. Oak galls have been used for ink making since ancient times. Joseph de Tournefort, passing through Erzerum in 1702, noted the considerable effort expended obtaining oak galls and the legislative protection given the oak trees that produced them:

> They also bring to Erzeron abundance of Gall-Nuts, five or six days Journey from the Town, and they preserve the Oaks with great care, by the Bassa's order; the Wood besides would be too dear, if 'twere carry'd thither for burning.[176]

Pollen Representation

Oak pollen is tricolpate with very distinctive and complex patterning on the exine. Oaks are wind pollinated and produce prodigious amounts of pollen. Despite this, the distribution of *Quercus* pollen in the present day is very closely linked to the main stands of *Quercus iberica* and *Q. macranthera* west of Tbilisi. Pollen values below 5 percent reflect long distance transport (in mountain areas) or the presence of *Quercus* (in lowlands). Values of 5–23 percent occur where *Quercus* is subdominant in the vegetation. Values greater than 23 percent occur where oak trees are co-dominant (usually with *Carpinus* spp.) and proportions of 39 percent or more occur in pure oak forest or oak-dominated woodland.

[174] Ketskhoveli 1975.
[175] Dolukhanov 1966.
[176] Tournefort 1718, p. 196.

Late Quaternary History of Quercus *from Pollen Records*

Quercus pollen is recorded during the late Pleistocene mostly at sites in Eastern Georgia. There oak existed as one of the few tree species present in the late glacial steppe, but its distribution was probably disjunct: scattered trees occurring in the few places able to support arboreal vegetation. *Quercus* began to expand on the Tsalka Plateau in the first millennia of the Holocene. Between 9000 and 8000 cal BP, oaks became co-dominant trees in much of Southern Georgia's open woodlands. The role of *Quercus* in Western Georgia was minor, but nevertheless more important than today. Oaks retained their primacy in the vegetation of Southern Georgia from the early Holocene until 3500–3000 cal BP. At this time oak was displaced from the Tsalka Plateau by conifers and grasslands. It also became considerably less frequent in the vegetation of Western Georgia. In the east, oak woods remained at lower altitudes, especially around Tbilisi.

Between 2000 and 1000 years ago the situation again changed – this time the lower forest belt near Tbilisi underwent a dramatic change in composition, greatly to the detriment of *Quercus*. Simultaneously it seems that scattered oaks returned to the deforested parts of the South Georgian Uplands. These progressively vanished during the last millennium and today only exist around the Tsalka Plateau margins and on isolated rocky slopes.

Quercus pollen is difficult to interpret because of the number of species contributing to it, each with their own distinct ecological preferences. It seems parsimonious to assume, however, that the altitudinal range that each occupies today was similar in the past. Therefore the Tsalka Plateau records represent the expansion of *Quercus macranthera* woods in the early and mid Holocene, followed by their late Holocene decline. *Quercus iberica* may have always dominated oak forests near Tbilisi, and *Q. hartwissiana* was probably the prevailing oak in Colchic forests.

This assumption is supported by the apparent conservatism of oak's distribution, which in spatial terms has changed very little since the early Holocene. The most significant temporal event in the history of *Quercus* throughout the study area was its dramatic decline between 3500 and 3000 years ago. The Lake Van oxygen isotope record demonstrates that this was a time of climatic cooling and drying across the region. Interestingly, European pollen records show a similar late Holocene decline, explained by climatic 'deterioration' and human impact.[177] Caucasian oaks have long been adversely affected by agricultural activities.[178] Clearing and logging enabled other tree species, like beech and pine, to penetrate oak forests and, when the climate was favourable, to displace them completely. Oak woods on the Tsalka Plateau were replaced by *Pinus*, *Abies* and grassland; oaks near Tbilisi were supplanted by *Carpinus* species; and in Western Georgia oaks yielded to the growing contribution of *Fagus orientalis* and *Carpinus betulus*.

Oak woods in Southern Georgia achieved their maximum extent at all elevations between 5000 and 3500 years ago, during the Early to Middle Bronze Age. Obviously the climatic conditions of that time were favourable to oak, which is better adapted to continentality than other deciduous trees. We could also entertain the idea that Bronze Age people applied the axe selectively to mould forest composition, or that livestock were removed from forest grazing to conserve trees for metallurgical production.

Before people had the wherewithal to produce metal tools, fire was the main agent of deliberate vegetation change. Fires burned often during the Chalcolithic, as sedimentary charcoal records show. But fires ceased abruptly 5000 years ago. The Bronze Age ushered in a change, a change that consolidated oak's dominant position in the woodlands of Southern Georgia.

[177] Huntley and Birks 1983, p. 354.
[178] Dolukhanov 1966.

Trees may have held a symbolic significance to Bronze Age people in the Caucasus through their depictions of the 'tree of life' and forest hunting scenes.[179] Oak timber was preferred in the construction of burial chambers ('kurgans') and the ceremonial chariots that were often placed within them.[180] Metallurgy perhaps allowed people to develop a new relationship with their environment. Pastoralists and farmers of the Neolithic and Chalcolithic had at best a limited ability to control the vegetation. If forests began to encroach on pastoral land, people might respond by either burning the vegetation or by moving their livestock elsewhere. It seems inevitable that both of these strategies would lead toward conflicts over land tenure.

If, however, agriculturalists were able to judiciously apply the axe to the forest, they could thin out unwanted trees and thus sustain their agricultural activities *in situ*. Oak trees, because of the light shade cast by their crowns, are much more amenable to pastoral agriculture than many other trees. The Bronze Age oak woodlands of Southern Georgia may have, in a functional sense, resembled Mediterranean savannas, which "are highly valued by their owners for fattening pigs, keeping stock over the difficult summer months, providing timber and firewood, and requiring a minimum of attention".[181] The Spanish savannas in question were developed most fully around 4000 years ago. It could be argued that the impressive Bronze Age wealth of Southern Georgia was aided by an increase in agricultural productivity as agriculture adapted to afforestation and its unexpected benefits.

Oak, more than most other wild trees, has a history that is intertwined with humankind. Oaks were standing by as Neolithic agriculturalists sowed their first crops and Chalcolithic herders drove their stock to highland pastures. Bronze Age agriculture flourished amongst the oak trees, and when the Anatolian Dark Age brought political instability to Southern Georgia, oaks and people together abandoned the mountain plateaux. Not until the Mediaeval did they both return in numbers, bringing political and economic stability to the once ravaged highlands.

RHAMNUS-TYPE

The *Rhamnus* pollen type represents two genera: *Rhamnus* and *Paliurus*. *Rhamnus* species include *R. imeretina* (occurring in the forest belt up to 2000 m elevation, mostly in Western Caucasus), *R. cordata* (on limestone cliffs in Western Georgia), *R. microcarpa* (on dry limestone slopes throughout the Caucasus, 1000–2300 m elevation), *R. depressa* (on dry limestone slopes from 2000–3000 m), *R. cathartica* (in dry places amongst scrub and broadleaf forest, widespread), *R. pallasii* (on dry piedmont slopes amongst xerophyte vegetation) and *R. spathulifolia* (in arid scrub vegetation).[182]

Paliurus is represented by only one species: *P. spina-christi*, a prickly shrub that grows on dry, deluvial clayey soils on piedmont slopes from sea level to 1500 m elevation. It is often associated with *Rhamnus pallasii, Spiraea hypericifolia* and xerophyte *Pyrus* species. *Paliurus* is the principal species of shibliak vegetation and forms extensive thickets at the lower forest border in Eastern Georgia. It is very much favoured by intensive livestock grazing:[183] cattle seem particularly averse to its well-armoured branches. It is difficult to remove from pastures once established.[184]

[179] Lang 1966.
[180] Gobejishvili 1980; Kuftin 1941.
[181] Stevenson and Harrison 1992, p. 227.
[182] Ketskhoveli 1983.
[183] Avakov 1982.
[184] Walter 1974.

It does, however, provide excellent fencing material for livestock, and the wood is used for burning and carpentry. *Rhamnus* species are used similarly, and also to produce dyes and pigments.[185] Both *Rhamnus* and *Paliurus* have a role in traditional medicine. Avakov describes the beneficial ecological role of *Paliurus* in initiating natural secondary succession in deforested areas.[186]

Rhamnus-type pollen is tricolporate with a distinct pore structure. According to Beug,[187] it represents both *Rhamnus* and *Paliurus* species. However, *Rhamnus* is more obviously reticulate, while the reticulum on *Paliurus* pollen is faint to absent.[188] Most of the *Rhamnus*-type pollen that I have seen in Holocene material from Georgia lacks reticulation and probably represents *Paliurus*. This accords with the wider distribution of *Paliurus spina-christi* in the landscape. Both *Paliurus* and *Rhamnus* are poor pollen producers, being largely insect-pollinated. For example, *Rhamnus*-type pollen comprised only 1 percent of pollen in surface sediments at a site (Lake Jvari) surrounded by oak-hornbeam scrub, *Spiraea hypericifolia* and extensive *Paliurus* thickets. When *Rhamnus*-type pollen occurs in any quantity above single pollen grains, it signifies the considerable contribution of *Rhamnus* or *Paliurus* shrubs to the vegetation.

Rhamnus-type pollen is recorded in less than a handful of sites. Its history suffers from poor pollen representation just as it does for other Rhamnaceous shrubs: *Hippophaë* and *Frangula*. It is worth noting here that *Rhamnus*-type pollen is found in most levels of the Tsavkisi sediments near Tbilisi. It also occurs in the Ispani-II sediments from the Black Sea coast. Both sites contain *Rhamnus*-type pollen in significant proportions through the mid Holocene, with an apparent decline in the last 2000 years. In the case of Ispani-II, this relates to paludification and afforestation; the latter cause also accounts for declining *Rhamnus*-type values near Tbilisi. Other records are too patchy to provide a clear view of this taxon's history, but it is certain that *Rhamnus* spp. and *Paliurus spina-christi* have long been a part of scrub and shibliak vegetation in the study area. The available palynological evidence does not support the contention of several botanists[189] that the area of *Paliurus spina-christi* has increased dramatically in recent centuries. Very likely its area was even greater in past millennia when livestock grazing occupied a more significant place in the region's economy than it does today.

Rhododendron

There are six *Rhododendron* species native to the study area, growing in a variety of habitats. *Rhododendron luteum* is the only deciduous species. It occurs in open landscapes from sea level to the subalpine belt throughout Georgia, but mostly in Colchis and along Turkey's NE coast. *R. smirnovii* and *R. ungernii* are tall-growing species of very restricted distribution, growing in the understorey of the middle and upper forest zones in Ajara and the Artvin district of Turkey. *R. ponticum* grows in the Colchis lowlands and has a disjunct distribution along the southern and eastern shores of the Black Sea, the mountains of Lebanon and in Portugal. *R. caucasicum* grows in large swathes on north-facing slopes at high altitudes (1600–3000 m) in the subalpine and alpine belts. *R.* × *sochadzeae* is a hybrid between *R. ponticum* and *R. caucasicum*, growing in the mountainous karst districts of northeastern Georgia.

All *Rhododendron* species prefer relatively moist environments (*R. ponticum* and *R. luteum* are common on peatbogs along the Black Sea coast) and are therefore more common in Western

185 Ketskhoveli 1983.
186 Avakov 1982.
187 Beug 2004.
188 J. van Leeuwen, Institute of Plant Sciences, Bern, pers. comm.
189 Ketskhoveli 1959; Sosnovskii 1915.

Georgia than the east. With the exception of the alpine *R. caucasicum*, the evergreen rhododendrons are the quintessential understorey plants of Colchic forest. Deciduous *R. luteum* is less shade-adapted and has a preference for open vegetation.[190]

Rhododendron pollen is produced in tetrads and dispersed by insects. Its occurrence in sediments is a reliable indication that *Rhododendron* plants occurred at or very near the site of sediment deposition. *Rhododendron* pollen is most abundant today in subalpine and alpine sites along the Main Caucasus Range, where *R. caucasicum* is most prevalent. *Rhododendron* pollen is probably therefore a good indicator of treeless environments.

This history of *Rhododendron* is brief, beginning seven thousand years ago. *Rhododendron* pollen appeared in sites in the Colchis lowlands and continued to occur there sporadically until a thousand years ago. These occurrences probably represent *Rhododendron ponticum*. In the highlands, *Rhododendron* did not become important until 1000 years ago, when it suddenly increased at several sites.

Because of *Rhododendron*'s poor pollen dispersion it is difficult to interpret these changes on a regional scale, but the apparent decline of *Rhododendron ponticum/luteum* along the Black Sea coastline over the past 2000 years could be explained by the expansion of *Alnus* swamps and dense *Fagus-Carpinus* forests in the coastal zone.

The explosion of *Rhododendron caucasicum* in the subalpine zone in the last millennium is almost certainly the product of human impacts at the upper tree line. Subalpine dwarf woods almost everywhere along the Main Caucasus Range have been affected by recent wood-cutting and intensive grazing.[191] *Rhododendron caucasicum* thickets have proliferated in the last thousand years because they are poisonous to livestock.

Rosaceae (Arboreal)

The arboreal Rosaceae are well represented in the study area, comprising the genera *Amelanchier, Amygdalus, Armeniaca, Cerasus, Cotoneaster, Crataegus, Cydonia, Laurocerasus, Malus, Mespilus, Padus, Pentaphylloides, Persica, Prunus, Pyracantha, Pyrus, Rosa, Rubus, Sorbus* and *Spiraea*.[192] While many of these genera are represented by one or two species, *Rubus, Rosa, Pyrus, Sorbus, Crataegus* and *Cotoneaster* each have a number of species, including several endemics.[193] It is pointless to describe the ecology of each of the species in the arboreal Rosaceae category, since this group is essentially a palynological one and may not represent all of the genera listed above.

I have included this group because rosaceous shrubs are important constituents of the oak scrub vegetation of Eastern Georgia. Were they always as common as they are now? Pollen analysis can only provide this question an answer in broad terms because Rosaceae pollen is poorly dispersed. Nevertheless, the distribution of Rosaceae (arboreal) pollen in the present day corresponds well to places where scrub vegetation is widespread or where deforestation of mixed forests has occurred. The highest proportions (>1 percent) occur around Tbilisi, in Southern Georgia and in Racha. Values less than 1 percent appear be mostly the result of wind-blown pollen deposition.

Between 8000 and 6000 cal BP, arboreal Rosaceae pollen occurs around Tbilisi, on the Tsalka Plateau and also in Racha. Then, 5000–3500 years ago, Rosaceae declined in the east and is only recorded along the Black Sea coast near Kobuleti. At 3000 cal BP, arboreal Rosaceae returned to Eastern Georgia and Racha. They became once again uncommon 2000 years ago, and returned 1000 years ago in numbers similar to today's.

190 Gagnidze 1985.
191 Dolukhanov 1978.
192 Ketskhoveli 1980.
193 Gagnidze 2005.

One striking feature of the arboreal Rosaceae record, apart from its rapid fluctuations, is that it occurs only at a handful of sites, mostly at middle and lower altitudes. Rosaceous shrubs appear to be no more common now than they were 8000, 6000 or 3000 years ago. In between these times arboreal Rosaceae declined considerably. The periods from 5000 to 3500 and 2000 to 1000 years ago were the times of maximum afforestation in Eastern Georgia (and perhaps on the foothills in the west also). So, while the arboreal Rosaceae have been established in the region since at least 8000 cal BP, they have only been abundant at times of partial forest cover. Heavy grazing appears to maintain scrub vegetation in Eastern Georgia today, and the Tsavkisi record (Chapter Five) indicates that this was true in the past also. When grazing is removed, the vegetation rapidly reverts to woodland.

The Caucasus is considered a possible 'hearth' of fruit tree domestication in Western Asia.[194] One of the first botanists to visit the region, Joseph de Tournefort, was clearly impressed by the diversity of 'wild' fruits he saw:

> All the rest [of the land] is taken up with natural Vineyards and Orchards, in which the Nut, Apricot, Peach, Plumb, Pear, and Apple-trees grow of themselves... we cannot doubt it being one of those parts of Georgia, wherein, according to Strabo, abound in all manner of Fruits, which the soil produces without Culture.[195]

The people of the Caucasus have, since ancient times, grafted scions of more-or-less domestic fruit varieties onto individual wild trees.[196] It is more than likely this apparently wild landscape greeted Tournefort as he journeyed along the Kura valley in 1702. Perhaps grafting was practised even in Strabo's time. Archaeo-botanical remains suggest that rosaceous fruits (*i.e. Persica, Amygdalus, Cerasus* and *Prunus*) were collected in the Caucasus at least as early as the third and second millennia BC.[197]

The pollen data presented here support the conclusion of Woldring and Cappers that the wild orchards of Western Asia developed alongside oaks but not within oak forest.[198] Rosaceous shrubs were established well before Bronze Age oak forests blanketed the landscape. They are quite prevalent now thanks to cattle grazing, deforestation and the careful protection of grafted wild fruit trees.

SALIX

In all, sixteen *Salix* species are found in the study area (see **Table A.6**). All *Salix* species are deciduous, moisture-loving plants, growing near watercourses, peatbogs and lakes. Several species are found in lowland gallery forests (*e.g. S. excelsa, S. wilhelmsiana*); others are more widespread (*e.g. S. caprea, S. elburzensis*); whilst most of the shrubby, Caucasian endemic species are found only at high altitudes in the subalpine and alpine belts. All *Salix* species are light-demanding, rarely occurring within closed forest. As well as being hygrophiles, they are often the first trees to colonise disturbed or denuded ground.

194 Zohary and Hopf 1993.
195 Tournefort 1718, p. 227.
196 Sosnovskii 1915.
197 Lisitsyna and Prishchepenko 1977.
198 Woldring and Cappers 2001.

Table A.6. *Salix* species from the study area. Sources: Ketskhoveli 1973 and Grossheim 1945.

Species	Habit	Caucasus distribution (world)	Habitat and altitudinal range
S. pentandroides	Small tree, 8–10 m	Mostly along Main Caucasus Range (Anatolia and Caucasus)	Streamsides and damp forests. Upper forest belt and subalpine to 2300 m.
S. triandra	Small tree, 4–5 m	Eastern and Western Georgia (Europe to Central Asia)	Riparian habitats, lake shores. Lowlands to 1300–1800 m.
S. alba	Tall tree, 20–30 m	Eastern Georgia, Black Sea coast (across Eurasia)	Riversides. From lowlands to 1600 m.
S. micans	Tall tree, 20–25 m	Mostly Western Georgia (Black Sea coast)	Beside rivers, reservoirs and canals. Lower forest belt.
S. excelsa	Tall tree to 20 m	Eastern Georgia (Caucasus and Iran)	Characteristic species of lowland gallery (riparian) forests.
S. apoda	Shrub, 1–1.5 m	Along the length of the Main Caucasus Range (endemic)	Birch woods and peatbogs. Subalpine and alpine belts.
S. caucasica	Shrub, 0.5–1.5 m	Mostly Western Georgia, also Kartli and Kakheti (Caucasus and Anatolia)	Damp places. Upper forest belt to subalpine (1500–2400 m).
S. caprea	Shrubby tree, 6–10 m	Widespread (most of Eurasia)	Streamsides, forest edges, dwarf woods. Sea level to 2300 m.
S. pseudomedemii	Shrubby tree, 3–4 m	Widespread in forested areas (Caucasus and Anatolia)	Forest edges, gullied slopes, on streamlets. Lower to middle forest belt.
S. kuznetzowii	Low shrub	Eastern part of Main Caucasus (endemic)	Birch woods and forest edges. Subalpine belt.
S. kikodseae	Shrub, 1–3 m	Western Georgia only (Georgian endemic)	Forest edges, slopes, moraines, fens. Subalpine belt.
S. kazbekensis	Shrub to 1.5 m	Widespread on Main Caucasus Range (endemic)	Screes, moraines, fens, lake edges, *Rhododendron* thickets. Subalpine and alpine 1700–3300 m).

Species	Habit	Caucasus distribution (world)	Habitat and altitudinal range
S. armeno-rossica	Tall shrub	Kartli and Javakheti (S. Caucasus and Anatolia)	Streamsides and ravines. Upper forest belt (1200–2000 m).
S. pantosericea	Low shrub	Western part of Main Caucasus (endemic)	Edges of dwarf woods, moraines, screes, denuded slopes. Subalpine (2100–2500 m).
S. elburzensis	Shrubby tree, 8–10 m	Abkhazia, Imereti, Kartli, Gare-Kakheti, Kvemo Kartli (Caucasus and Anatolia)	Riparian habitats (rivers, streams, gullies). From sea level to 2000 m.
S. wilhelmsiana	Twiggy shrub	Kartli and Gare-Kakheti (S. Caucasus, Anatolia, Iran, Central Asia)	Swamps; dry, rocky slopes; gallery forests. Lowlands, rarely to 1500 m.

In the present-day, *Salix* pollen is best represented in the central part of Georgia. It is all but absent from Colchis and Abkhazia, as well as much of Eastern Georgia. This distribution corresponds most closely to the areas where *Salix caprea*, the most widespread species in the study area, is prevalent. But, to a certain extent, the palynological signal is mixed by long-distance transport, so it is impossible to assign *Salix* pollen to a particular species. Pollen values below 1 percent probably represent long-distance transport, while values above 2.5 percent indicate local presence of *Salix* plants. Values above 7 percent indicate that willows are common in the nearby vegetation. Values of 17 percent or more have no modern analogue, but occur at sites that were floodplains or fens in the past.

Salix pollen was present in the late Pleistocene at Sagarejo in Eastern Georgia. At this site, *Salix* had its greatest abundance in the late glacial and early Holocene. It began to decline at Sagarejo around 9000 years ago, when it also began to appear in pollen records from elsewhere in the study area. By 8000 cal BP, trace amounts of *Salix* pollen were present in Racha, Eastern and Southern Georgia. *Salix* was probably locally dominant in fens near Mt Kazbegi. At 5000 cal BP, *Salix* trees were present near Tbilisi and also in the mountains of Abkhazia, followed by occurrences in Racha and near Bakuriani a thousand years later. Three thousand years ago *Salix* seems to have declined in mountainous Abkhazia, and occurred locally near only a handful of sites – in Racha, Javakheti and at Sagarejo. During the last two millennia, willow seems to have become more widely distributed in the study area.

The *Salix* record is very patchy and is therefore difficult to interpret. *Salix* must have been important in gallery forests at Sagarejo through the Younger Dryas. Perhaps, like *Juglans* and *Pterocarya* (this appendix), its prevalence there was assisted by channel instability. As glaciers began to melt and discharge large volumes of water into the river systems of Eastern Georgia, the seasonality of this discharge may have made floodplains quite unstable. Large, unstable floodplains would benefit *Salix* and *Pterocarya* to the detriment of other tree species, which increased only once riverbanks stabilised. During the Holocene, *Salix* has capitalised on disturbances and climatic and edaphic changes, but only on a very local scale. It appears to have been more widely distributed during the early and late Holocene, compared to its more restricted distribution in the wooded landscapes of the mid Holocene.

TILIA

Two lime species grow in the study area: *Tilia cordata* and *Tilia begoniifolia*. The former is a common species in the forests of Europe, but is less common in the Caucasus. It is a deciduous tree that grows to about 25 m high in mixed forests of the middle and upper forest belts (usually 1200–2000 m). It is found in drier, somewhat sparsely forested areas in the eastern part of the Main Caucasus Range. *Tilia begoniifolia*, the Caucasian lime, is larger in all its parts. It grows to 40 m high in mixed forests, at forest edges and in ravines from the lowlands to the middle forest belt throughout the Caucasus and into Anatolia, Iran and the Crimea.[199]

Tilia begoniifolia prefers deep, rich soils and is often associated with *Carpinus caucasica, Quercus iberica, Acer platanoides, A. campestre, A. laetum, Ulmus scabra, U. glabra, Castanea sativa, Fagus orientalis* and *Taxus baccata* in mixed forests.[200] It is rarely dominant, and is frequently accompanied by a rich understorey of *Corylus avellana, Ostrya carpinifolia* and other deciduous shrubs (or evergreens in Western Georgia).

Tilia is insect pollinated and always underrepresented in pollen spectra from lakes and wetlands.[201] Its presence is usually an indication that lime trees occur locally. The current distribution of *Tilia* pollen is generally in line with *Tilia begoniifolia*'s contribution to plant communities. However, *Tilia* pollen is present in several treeless, mountain sites, suggesting that long distance transport is possible. Perhaps only proportions greater than 2 percent can be interpreted as indicating the presence of *Tilia* in the local vegetation. The highest values come from Abkhazia and at the lower forest line near Tbilisi.

Tilia was an important component of late Pleistocene refugial forests at low altitudes in both Eastern and Western Georgia. At the beginning of the Holocene, *Tilia* appears to have vanished from the Black Sea lowlands. Lime pollen is recorded at several sites 10,000 years ago, but not in especially high proportions. By 9000 cal BP, it seems to have advanced onto the Tsalka Plateau and was again prevalent in the Sagarejo record. *Tilia* expanded in Western Georgia between 8000 and 7000 cal BP, and in Southern Georgia 6000 years ago.

By 5000 cal BP, *Tilia* had become less common in Southern Georgia, but remained very important in Abkhazia. It was present in southern Colchis lowlands by 4000 cal BP, at which time it experienced a sharp decline in Abkhazia. By 2000 cal BP, *Tilia* had almost vanished from Eastern Georgia. It began returning to Abkhazia around the same time. Its distribution has changed little in the last millennium.

The pollen records demonstrate convincingly that *Tilia* survived the glacial period in low altitude forests. At the beginning of the Holocene it began to migrate outwards from these refugia, and by 9000 cal BP had colonised much of Southern Georgia. *Tilia* was amongst the first cohort of trees to invade the highlands after rainfall increased in the early Holocene. In Abkhazia it appears to have been set back by paludification at low altitudes in the early Holocene, but recovered by 8000 cal BP and spread widely. According to Kvavadze *et al.*,[202] this was a time of rapid upward migration of tree lines as the climate became warmer and wetter.

Tilia became less common in Southern Georgia as mixed woodlands yielded to oak woodland. Even as early as 5000 years ago, human activities may have affected *Tilia* populations on the highland plateaux, for limes occupied soils that were valuable for agriculture. *Tilia* may have also declined due to cooler summers or the decline of mixed forest and scrub habitats. In Abkhazia, *Tilia*'s history is closely linked to climatic fluctuations – *Tilia* expanded as rainfall increased in the early Holocene (*i.e.* 8000 cal BP), declined sharply when the climate became drier (4500–3500 cal BP), and recovered once the climate returned to its former state.

[199] Ketskhoveli 1983.
[200] Ketskhoveli 1959.
[201] Huntley and Birks 1983.
[202] Kvavadze *et al.* 1992.

Ulmus and Zelkova

The Ulmaceae botanical family has three genera represented in the study area: *Ulmus, Zelkova* and *Celtis*. *Ulmus* and *Zelkova* produce pollen with similar morphology and so are grouped together in this section.

Ulmus is represented by *U. glabra* and *U. minor*, both of which have local forms that some botanists regard as separate species.[203] *Zelkova* is only represented by *Z. carpinifolia*, a Tertiary relict species with close relatives in East Asia, Crete and Sicily.

Ulmus glabra (syn. *U. elliptica*) is a large deciduous tree to 45 m in height, growing in mixed forests of the middle forest belt up to about 1500–2000 m above sea level. It is often associated with somewhat moister situations and hence occurs along streamsides, in gallery forest and in gullies.[204] Its distribution extends throughout the Caucasus, the Mediterranean Basin, Europe and Anatolia. *Ulmus minor* (syn. *U. suberosa, U. georgica*) is smaller (6–25 m) and grows in more arid situations up to an elevation of about 1000 m. This species is found in xerophytic vegetation dominated by *Quercus iberica, Pyrus salicifolia, Pistacia mutica, etc.*[205] It is distributed mostly in Eastern Georgia, but occurs also in the west.

Zelkova carpinifolia is a long-lived, mesophilous tree to 25–30 m in height. It prefers relatively moist, humus-rich soils and warm, sunny slopes. It can grow at altitudes up to 1600 m, but only reproduces from self-sown seed between 100 and 400 m above sea level.[206] *Zelkova* has a very restricted distribution in the lowlands of Georgia, where it occurs admixed with *Quercus* and *Carpinus*. Its main populations are along the coastlines of the Caspian Sea, in northern Iran and southeastern Azerbaijan.

Ulmus and *Zelkova* species produce very highly valued, extremely durable timber, used for furniture, machinery, ship-building, carts and other uses.[207] The pollen of *Ulmus* and *Zelkova* is very similar, usually with four or five pores and a characteristic rugulate exine. *Zelkova* can often be identified by its larger, thicker, elliptical pores and the lack of patterning around these,[208] but not all analysts make this morphological distinction.

Ulmus-type pollen is presently best represented in sites from the lowlands of the Black Sea coast and around Tbilisi, areas where *Ulmus* species are commonplace in the vegetation. *Zelkova* pollen only makes up a tiny propotion of *Ulmus*-type pollen in the present-day: single *Zelkova* pollen grains are recorded from very few sites in Southern Georgia and Colchis, probably as a result of wind transport. *Ulmus*-type proportions greater than 2 percent indicate the local presence of *Ulmus* in the vegetation, while values of 4–8 percent occur where *Ulmus* trees are relatively frequent. Higher percentages have no modern analogue, but it can be safely assumed that these indicate sub-dominance and co-dominance of *Ulmus* or *Zelkova* in forests.

Because some analysts have differentiated *Zelkova* pollen, the history of *Ulmus*-type pollen is presented in two ways: one including both *Ulmus* and *Zelkova* and the other with *Zelkova* alone (**Table A.7**).

[203] Czerepanov 1995; Gagnidze 2005.
[204] Ketskhoveli 1975.
[205] Ketskhoveli 1975.
[206] Kvavadze and Connor 2005.
[207] Grossgeim 1945.
[208] Kvavadze and Connor 2005; Nakagawa *et al.* 1998.

Table A.7. Pollen-based history of *Ulmus* and *Zelkova* during the late Quaternary. Dates given are timeslices (cal BP) and do not indicate the exact age at which events occurred.

Ulmus and *Zelkova* combined	*Zelkova* alone
14000–12000 BP – quite abundant in refugia of Abkhazia and Eastern Georgia (Sagarejo)	14000–12000 BP – present in Abkhazian refugium
11000 BP – appears locally in Southern Georgia; slight decline in Abkhazia	11000–10000 BP – absent from studied sites
10000 BP – increase in Abkhazia and Southern Georgia	
9000 BP – major increase across all Georgia	9000 BP – appears locally (>0.5%) in much of Southern Georgia
8000 BP – maximum percentages in Southern Georgia and Colchis; slight decline in Abkhazia	8000 BP – present in southern Colchis, slight decline in Southern Georgia
7000 BP – slight decline on Tsalka Plateau; stable or increasing elsewhere	7000 BP – maximum percentages (1–5%) in Southern Georgia and Colchis
6000–5000 BP – relatively stable populations	6000–5000 BP – slight decline in Colchis and Southern Georgia; present in Abkhazia
4000 BP – major decline in Southern Georgia; stable or increasing in Western Georgia	4000 BP – major decline, especially in Southern Georgia
3000–2000 BP – relative stability; slight increase in Western Georgia	3000–2000 BP – further decline; present only in southern Colchis and near Tbilisi
1000 BP – slight decrease in Western Georgia; slight increase in Eastern Georgia	1000 BP – major regional increase in Abkhazia, Southern Georgia, Colchis and near Tbilisi.
Present – slight increase in Abkhazia and near Tbilisi; major decrease in Colchis	Present – major decline

The differences between the history of *Ulmus* and *Zelkova* are salient. *Zelkova* vanishes during the Younger Dryas, but *Ulmus* does not. *Ulmus* seems to have migrated earlier and was established as a forest co-dominant everywhere 8000 years ago. *Zelkova* declined slightly at this time and achieved its maximum spread a thousand years later. The two records agree on the major decline of *Ulmus* and *Zelkova* in Southern Georgia between 5000 and 4000 years ago. Subsequently, *Zelkova* all but disappeared (apart from a brief resurgence 1000 years ago), while *Ulmus* remained stable or increased in Western Georgia until the twentieth century.

The causes of elm's catastrophic decline in European forests beginning around 5800 years ago – climate change, increased competition, fungal disease, human impact[209] – could also be invoked for the Caucasus. But here it is *Zelkova*, not *Ulmus*, that declines so dramatically between 5000 and 4000 years ago. Elm itself declines much more incrementally, beginning 7000 years ago. It is true that in Southern Georgia the two types decrease at more-or-less the same time, and this perhaps provides a clue to their decline. Around

[209] Peglar and Birks 1993; Roberts 1998.

4000 years ago, the so-called Trialeti culture emerged on the Tsalka Plateau. More than its predecessors, this was characterised by increased social stratification.[210] Monumental burial mounds ('kurgans') were erected as testament to the wealth and status of the dead. Within each of these kurgans was a burial chamber lined with wood and roofed with a thatch of branches. Dr Eliso Kvavadze has examined the pollen content of several of these kurgans and found *Zelkova* pollen in considerable abundance.[211]

Not only were *Ulmus* and *Zelkova* branches used in these and other constructions, they were probably also used as cattle fodder. Elm leaves are the most nutritious kind of fodder in European forests and have hence been coppiced and pollarded since the European Neolithic.[212] Similarly, *Zelkova carpinifolia* trees in Eastern Georgia were pollarded in the past to provide fodder for cattle grazing in the forest (see Chapter Eight).

The vast wealth of the Trialeti culture is thought to have come from an economy centred on cattle breeding.[213] It is possible that the Middle Bronze Age people of the Tsalka Plateau heavily exploited *Ulmus* and *Zelkova* to maintain their herds, leading to the apparently rapid decline of these tree species. As Troels-Smith[214] proposed, fodder production may not have killed the elms, but reduced their pollen production by defoliation. This in turn could have exposed them to attack by fungal disease or made them more susceptible to adverse climatic changes. Other forms of agricultural activity could also lead to an elm-zelkova decline.

Climate change cannot be overlooked in *Zelkova*'s history. It is a thermophilous, Tertiary relict species like *Pterocarya* and *Juglans*. These taxa expand in the Classical Period as the climate became more Mediterranean (see *Olea*, this appendix). So it is possible that *Zelkova* expanded during the early-mid Holocene because winters were relatively warm. If this hypothesis were tenable, one would expect the other Tertiary-relict thermophiles to react similarly, which is not the case.

In conclusion, the histories of *Ulmus* and *Zelkova* are as yet poorly understood. Both taxa were present in the region during the late Pleistocene and became abundant in forests during the early to middle Holocene. They began to decline in the mid to late Holocene, perhaps due to human interference in the vegetation.

Vitis

There are three *Vitis* species in the study area: *Vitis labrusca, V. vinifera* and *V. sylvestris*. The first of these is the Isabella grape, a North American species, which has run feral in the piedmont forests of Colchis and Abkhazia. The second is also a cultivated species, the common vine, which is grown widely in all but the most mountainous regions of the study area, but is also found wild in forests. *Vitis sylvestris*, the wild vine, is considered by many botanists to be a subspecies of *Vitis vinifera*.[215] It is scrambling deciduous liana growing about 6 m high and found in gallery forests, ravines and lowland forests with other lianas from sea level to 1500 m elevation.[216] Its indigenous range extends through mesic forests surrounding the Caspian, Black and Mediterranean seas.[217]

[210] Kushnareva 1997; Puturidze 2003.
[211] Kvavadze and Connor 2005.
[212] Austad 1988; Gardner and Willis 1999; Pott 1992.
[213] Puturidze 2003.
[214] 1960, cited in Roberts 1998.
[215] *Vitis sylvestris* is adopted here to conform to the standard nomenclature of Czerepanov 1995.
[216] See *Hedera* in this appendix and Ketskhoveli 1983.
[217] Zohary and Spiegel-Roy 1975.

Vines are amongst the most important agricultural plants in the study area. Their fruit is used in wine making, eaten raw, boiled to a paste or dried. Wild grapes in the Caucasus are sometimes cross-pollinated with domestic varieties to produce more disease-resistant cultivars.[218]

Vines are insect-pollinated. Cultivated vines are self-fertile, whereas wild vines are dioecious.[219] All produce very distinctive, heavily reticulated pollen grains that are poorly represented palynologically. Any occurrence of *Vitis* pollen is probably an indication of the presence of vines in cultivation or in the vegetation.[220] Today, *Vitis* pollen is best represented near the lower tree line in Eastern Georgia, but also occurs in the treeless South Georgian Uplands, where it appears because of wind updraughts. It is poorly represented in Western Georgia only because there are few pollen diagrams from the bacchanalian Imereti region. Because of vine's poor pollen representation, the following history can only be indicative of abundances at a local scale.

Vitis pollen first appears in the Bakuriani area of Southern Georgia in the early Holocene, amongst advancing *Corylus* woods. It is consistently represented near Tbilisi from 8000 to 6000 years ago and appears briefly on the Tsalka Plateau at 7000 cal BP. Five thousand years ago it was less frequent everywhere in the study area and appears only in Abkhazia at 4000 cal BP. Subsequently vines returned to prominence in Southern Georgia and Colchis, and increased during the next two millennia. Human activities in Racha saw *Vitis* and *Juglans* peak there 1000 years ago.

More than any other fruit, grapes are central to Georgian culture and life. Wine plays an almost sacramental role at feasts and its consumption is surrounded by ancient and elaborate rituals that differ markedly from those of other cultures. Even the word 'wine' is thought to have its roots in the Georgian *ghvino*. Strabo remarked on the bountiful fruit of Caucasian vines in spite of the little care given them besides pruning once every five years.[221] Joseph de Tournefort enumerated vines amongst the 'wild' fruits he saw growing in Kartli in 1702 AD.[222] Undoubtedly Tournefort observed vines grown in the traditional Caucasian way – in trees – as they still are in Western Georgia. Over 500 grape varieties are grown in Georgia alone.[223] Grape pips are common in early agricultural (Neolithic) sites of the Shulaveri type and number amongst the earliest finds of domestic grapes in the world.[224] All of these disparate lines of evidence point to a very ancient culture of viticulture and winemaking in the Caucasus.

Unfortunately pollen records are not much use when trying to determine whether indeed Georgia is the birthplace of viticulture. Certainly grapes are endemic to the study area and *Vitis* pollen is found throughout the Neolithic period, but differentiating wild and domestic vines is practically impossible using palynological methods.

ARTEMISIA

In all, twelves species of *Artemisia* grow in the study area. Most of these are xerophytic annuals, perennials or subshrubs growing in dry steppes and semideserts in lowland Eastern Georgia (*e.g. A. lerchiana*); others are found at higher altitudes on dry, rocky ground (*e.g.*

[218] Ketskhoveli 1983.
[219] Zohary and Hopf 1993.
[220] Turner and Brown 2004.
[221] Strabo 1928, 11: 4: 3.
[222] Tournefort 1718.
[223] Ketskhoveli 1957.
[224] Lisitsyna 1984; Lisitsyna and Prishchepenko 1977; Miller 1991; Zohary and Hopf 1993.

A. splendens); and a few species are quite weedy (*e.g. A. absinthium*).[225] *Artemisia* species are almost universally distributed through the Caucasus region, but are only prevalent in treeless environments in the arid south and east, and locally on dry, rocky slopes elsewhere.

Artemisia lerchiana is the dominant species of wormwood semideserts and grows on dry, slightly saline clay soils throughout Eastern Georgia, Azerbaijan and Armenia. It is associated with *Salsola dendroides, S. ericoides, Caragana grandiflora,* a number of ephemeral plants and steppe grasses such as *Bothriochloa ischaemum*.[226] *Artemisia* semideserts occur where the climate is dry, with cool winters, hot summers and pronounced late-summer drought (**Table A.8**).

Table A.8. Climatic data from an *Artemisia lerchiana* habitat near Yerevan, Armenia. T: average monthly temperature; P: average monthly precipitation. Source: Gulisashvili 1964.

	J	F	M	A	M	J	J	A	S	O	N	D	Avg.
T (°C)	–4.6	–2.5	4.7	12.1	17.2	21.6	25.3	25.2	20.6	14.0	6.6	–0.2	11.2
P (mm)	21	22	28	48	56	22	13	7	12	22	27	23	301

Artemisia species are wind-pollinated and produce tricolpate or tricolporate pollen, distinguished by its thickened mesocolpia. Because the percentages used in mapping were calculated from a sum including arboreal pollen and *Artemisia*, the values themselves are only useful for comparison. *Artemisia* pollen is best represented in the semidesert-steppe area southeast of Tbilisi, with values greater than 24 percent. Proportions between 13 and 24 percent are found at the steppe-forest border, where *Artemisia* is a minor element in scrub vegetation, and also at treeless sites in the mountains, where long-distance pollen transport is important.[227] Values below 13 percent are so widely distributed that they have little palaeoecological significance.

The Aligol record from Southern Georgia (Chapter Six) suggests that *Artemisia*-Chenopodiaceae semideserts were widespread during the late Pleistocene. *Artemisia* did not participate in lowland refugial forests in Abkhazia and Sagarejo. During the early Holocene, *Artemisia* values began to decline in Southern Georgia. High values occurred in parts of lowland Abkhazia at this time. *Artemisia* values then remained more or less stable in all parts of the study area until 1000 years ago, when they increased in Southern Georgia. In the last millennium, *Artemisia* declined in Southern Georgia. It has recently increased in the mountains of Abkhazia, perhaps as a result of deforestation and grazing in the subalpine belt.

These results show that *Artemisia* has always been an important component of woodland, scrub, steppe and semidesert vegetation in Southern and Eastern Georgia. Its continuous presence through the Tsalka Plateau records is a good indication that the tree cover there was never dense enough to be considered forest. *Artemisia* was most widespread and abundant in the region during the Younger Dryas, increasing in tandem with *Ephedra* under conditions of intense aridity. Like *Ephedra*, its retreat during the early Holocene was relatively unhurried. The climate of the Southern Caucasus and Eastern Anatolia may have remained seasonally dry until the mid Holocene, or early human activities worked to stall forest expansion.[228]

[225] Ketskhoveli 1952.
[226] Nakhutsrishvili 1999.
[227] Kvavadze 1993.
[228] Roberts 2002.

Chenopodiaceae

The Chenopodiaceae family includes some 24 genera in the study area.[229] They are generally halophytic plants found in association with dry, saline soils or salt lakes. Only in the semidesert-steppe regions of Eastern Georgia do Chenopodiaceae contribute significantly to the vegetal cover, but they are widespread in Azerbaijan. Some of the most important species are *Anabasis aphylla, Salsola ericoides, S. dendroides, Halothamnus glaucus, Gamanthus pilosus, Suaeda microphylla, Petrosimonia brachiata* and *Kalidium caspicum*.[230] These form pure communities, or are associated with *Artemisia*, steppe grasses and ephemeral plants. Where conditions are unstable, annual Chenopodiaceae species prevail over perennials, and on saline flats and dry salt lake beds, *Salicornia europaea* is often dominant.[231] Outside the semidesert-steppe area, Chenopodiaceae are found on disturbed soils and in grasslands.

Presently, Chenopodiaceae pollen is best represented in southeastern Georgia, especially in the semidesert-steppe region. It also appears in the South Georgian Uplands and at several other upland sites where long-distance pollen transport is important. Values below 15 percent are widespread throughout the region and probably represent both long-distance transport and scarce occurrence of Chenopodiaceae in the local weed flora.

Chenopodiaceae were widespread in Southern Georgia and Eastern Anatolia 14,000 years ago along with *Artemisia, Juniperus* and *Ephedra*. In Southern Georgia, Chenopodiaceae declined during the cold, dry conditions of the Younger Dryas. They appear in Abkhazia at the same time, indicating that aridity at this time extended even to the refugia of the Black Sea lowlands. By 10,000 cal BP, Chenopodiaceae were not especially common anywhere in the study area (except, presumably, in the semidesert lowands, from which we have no records). They began to increase in Southern Georgia about 7000 years ago, and more markedly by 5000 cal BP. Elsewhere in the study area Chenopodiaceae were present but unimportant. Chenopodiaceae increased again about 1000 years ago in Southern Georgia and thereafter declined slightly.

Chenopodiaceae's history is almost identical to *Artemisia*'s (this appendix). It was a major component of late Pleistocene arid landscapes. Unlike *Artemisia* though, Chenopodiaceae was less common during the Younger Dryas and early Holocene in Southern Georgia. A Chenopodiaceae-*Juniperus* association was pushed out by an *Artemisia-Ephedra* association at the time of greatest winter cold. This suggests an ecological relationship between Chenopodiaceae and *Juniperus* that is not clearly evident in the present-day.

Increases in Chenopodiaceae at 7000, 5000 and 1000 cal BP are probably related to human impact. The earlier increases occur at a time when burning and cereal cultivation became more frequent in Southern Georgia; the most recent one is accompanied by the intensification of livestock grazing.

Poaceae and Cerealia-Type

The large Poaceae family has some 113 genera represented in the study area.[232] Grasses are ubiquitous, but best represented in treeless environments. This is clearly reflected in the modern distribution of Poaceae pollen. Poaceae pollen is most abundant in Southern and Eastern Georgia. These regions have little forest cover in comparison to Western Georgia. Montane grasslands are extensive in the South Georgian Uplands, and grade into the lowland steppes of Eastern Georgia. All of the sites in Western Georgia with a high proportion of grass pollen are situated in treeless or deforested landscapes.

[229] Gagnidze 2005.
[230] Nakhutsrishvili 1999.
[231] Dolukhanov 1966.
[232] Gagnidze 2005.

The montane grasslands of Southern Georgia are dominated by species such as *Agrostis planifolia*, *Phleum phleoides*, *P. pratense*, *Festuca valesiaca* and, in areas of intense pasturage, *Nardus stricta*.[233] Lowland steppe species include *Bothriochloa ischaemum*, *Stipa capillata*, *S. tirsa*, *S. lessingiana*, *Bromus japonicus*, *Dactylis glomerata*, *Festuca valesiaca*, *Hordeum leporinum* and many others. Grasses also include a number of wetland plants (*e.g. Phragmites australis, Calamagrostis neglecta*) that manifest in pollen records regardless of the extent of grassland around the pollen site.

In the late Pleistocene, grasses were present in all parts of the study area. Their representation in Southern Georgia was high, so it seems that they were an important component of glacial semidesert-steppe. During the early Holocene, grasses declined at Sagarejo, but continued to be abundant in the South Georgian Uplands. Grassy landscapes diminished near Bakuriani around 9000 cal BP, as forests began to advance. The mid Holocene brought relative stability in the extent of grassy landscapes – high values in Southern Georgia and low values elsewhere. Along the Black Sea coast, grasses expanded in some places during the late Holocene.

The Poaceae record is complicated by the presence of grasses in the wetland flora. This poses an especial problem in the shallow lakes of the Tsalka Plateau (Imera, Aligol, Bareti), because most of the wetland plants today are grasses. The situation was probably the same in the past. Therefore it is difficult to determine whether their Poaceae records reveal much about the past extent of grasslands. Palaeovegetation reconstruction relies therefore on the whole pollen spectrum, not simply grasses.

Cereals, both wild and cultivated, produce pollen that may be differentiated from other grasses.[234] Several workers have reported *Cerealia*-type pollen in the study area. A number of cereals are endemic to the Caucasus region.[235] In particular, *Triticum* (wheat) species are well represented, begging the question: did the presence of wild cereals lead to the adoption of agriculture in the Caucasus?

Cerealia-type pollen is nowadays most prevalent in lowlands and piedmonts of Eastern Georgia. The main areas of wheat cultivation in Georgia are in the areas between forest and steppe in Eastern Georgia: the Alazani Valley, Kvemo Kartli, Shida Kartli and, to a lesser extent, the Tsalka Plateau and Javakheti region.[236] The correspondence between cereal pollen and the areas of wheat cultivation is good, when considering the lack of sites in the Alazani Valley and Shida Kartli. On the maps that follow, values greater than 3 percent, and especially greater than 6 percent, indicate the occurrence of cereals nearby. *Zea mays* is the predominant cereal crop of Western Georgia, but produces heavy pollen grains that disperse poorly.

Cereal pollen is recorded in Southern Georgia through the late Pleistocene in small amounts. At the beginning of the Holocene, however, *Cerealia* increases quite conspicuously and remains important until about 6000 cal BP.[237] Near Tbilisi, cereals (especially *Triticum*-type) increase around 7000 years ago. At 4000 cal BP, cereal pollen increases in Southern Georgia. It also appears then in Western Georgia and expands over the coming millennia. Two thousand years ago, *Cerealia* seems to have declined in Southern Georgia and expanded in Abkhazia and Colchis. Its role in the west was the same at 1000 cal BP, at a time when cereals again expanded in Southern Georgia.

[233] Nakhutsrishvili 1999 and Chapter Four.

[234] Bottema 1992.

[235] Menadbe in Nakhutsrishvili 1999.

[236] AN GSSR 1964.

[237] The importance of late-glacial and early Holocene occurrences may be overstated due to the way percentages were calculated, *i.e.* from the arboreal sum plus *Cerealia*-type.

The pollen data therefore suggest that wild cereals were present in Southern Georgia prior to the adoption of agriculture. It may even be that these wild cereals lived in or around the shallow wetlands from which the pollen records come.[238]

The first evidence for cereal cultivation comes from the lowlands of Kvemo Kartli, where irrigation seems to have been used.[239] The Tsavkisi record shows that, although wild cereals were present prior to the adoption of agriculture, they proliferated during the late Neolithic. Among cereal macrofossils identified at early agricultural sites are endemic wheat (*Triticum carthlicum*) and a number of wild cereals (*Triticum boeoticum, Hordeum spontaneum, Aegilops*). The Neolithic site of Arukhlo yielded remains of no less than nine wheat species, five barley species, oats, rye, common millet and foxtail millet.[240]

This huge diversity of wild, endemic and domestic species suggests that cereal-based agriculture emerged in the Caucasus as a response to increased resource availability in the sixth millennium BC, rather than being imported wholesale from the Levant.

Fluctuations in *Cerealia* during the mid-late Holocene are closely tied to human occupation. Its increase 4000 years ago corresponds to the flowering of the Trialeti culture in Southern Georgia and the concurrent *Ulmus-Zelkova* decline (see this appendix). *Cerealia* declined during the Classical Period as the Tsalka Plateau was depopulated and returned during the more stable political climate of the Mediaeval. In times of political and economic turmoil, people simply abandoned their fields and took up a semi-nomadic existence herding livestock.[241]

PLANTAGO AND RUMEX

The final pollen taxa to be included in this analysis are *Plantago* and *Rumex.* These two genera are grouped together because in Europe they are regarded as good palynological indicators of pastoralism and disturbance.[242] In the study area, *Plantago* has eight species and *Rumex* twenty species.[243] All are low-growing annual or perennial herbs most often encountered in grasslands. Their pollen, although ubiquitous, is often poorly dispersed over large distances.[244]

Plantago pollen occurs in almost everywhere in Georgia. Its highest values are found in Eastern Georgia, where forests are least expansive and grasslands are widespread. To a certain extent, the percentages reflect the abundance of arboreal pollen used to calculate *Plantago* percentages. This means that sites with a high arboreal to non-arboreal pollen ratio have lower *Plantago* percentages as a result. In general, though, more than 5 percent *Plantago* on the maps indicates open vegetation, and more than 10 percent indicates localised pastoral activity or disturbance.

Plantago was present in the glacial steppe flora of Southern Georgia. It increased in that region with the replacement of *Artemisia*-Chenopodiaceae semideserts by grassy steppes during the early Holocene. It was important near Tbilisi 8000 cal BP in scrub vegetation, and again 6000 cal BP. *Plantago* entered Western Georgia in a minor way about 5000 years ago. In all regions, plantain increased quite conspicuously during the late Holocene. It appears to have become much more abundant during the last 3000 years than at any other time.

[238] *Cf.* Bottema 1995.
[239] Dzhanelidze 1984.
[240] Chataigner 1995; Kiguradze 1986.
[241] Thomas 1999.
[242] Behre 1986.
[243] Gagnidze 2005.
[244] Huntley and Birks 1983.

Rumex is typically less abundant in pollen spectra than *Plantago*. The highest *Rumex* values today are found across a range of vegetation types and altitudes. Unlike *Plantago*, the abundance of *Rumex* does not seem to depend primarily on the presence of grasslands around the pollen site. Many sites surrounded by grassland have low *Rumex* values. All of the sites with moderate *Rumex* percentages (>2 percent) are subject to grazing, and sites with higher values have intensive, localised grazing.

During the Younger Dryas, *Rumex* grew in the semidesert-steppes of the South Georgian Uplands. Eight thousand years ago it was frequent between the Tsalka Plateau and Tbilisi. It is recorded in Racha, Western Georgia, 6000 years ago and was still important near Tbilisi at that time. Four thousand years ago, *Rumex* was absent in pollen records from Eastern Georgia and present in small amounts in the West. Thenceforth it faded into the background, and not until 1000 years ago did *Rumex* return at levels that mimicked its early-mid Holocene abundance.

The differences between these histories of *Plantago* and *Rumex* can be explained ecologically. *Plantago* species, although present in most open habitats, are especially prevalent in secondary grasslands created through deforestation. Thus *Plantago* became more important in the last three millennia as forests, woodlands and scrub were cleared in many parts of the Caucasus. *Rumex*, on the other hand, is considered a more reliable indicator of treeless vegetation.[245] Its Holocene history indicates the persistence of relatively treeless landscapes in Southern and Eastern Georgia until the mid Holocene. The hypothesis that oak woodland reached its greatest cover on the Tsalka Plateau 4000 years ago is supported by the low *Rumex* values for that period. *Rumex* returned as forest cover subsequently diminished.

As an indicator of grazing, *Plantago* seems to be more reliable than *Rumex*. Dung fungi (*Sporormiella*) have been identified in several pollen records and correlate well with *Plantago*, especially *Plantago major*-type. *Rumex* correlates far better with other indicators of landscape openness. Based on this, it is difficult to explain why some grassland sites have low *Rumex* values today; but pasture 'improvement' via the application of fertilisers and manures cannot be ruled out.

[245] Huntley and Birks 1983.

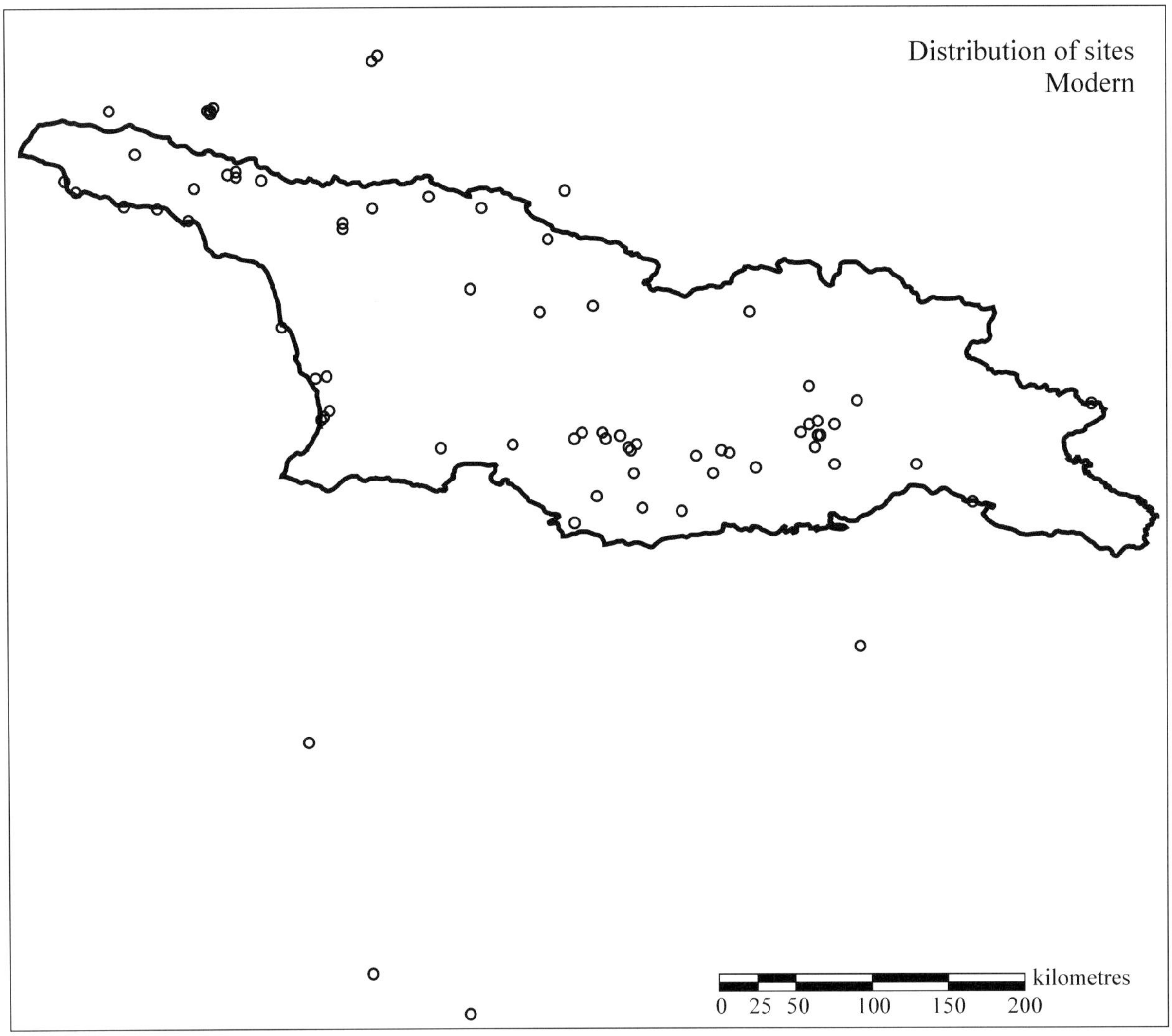
Distribution of sites
Modern
kilometres
0 25 50 100 150 200

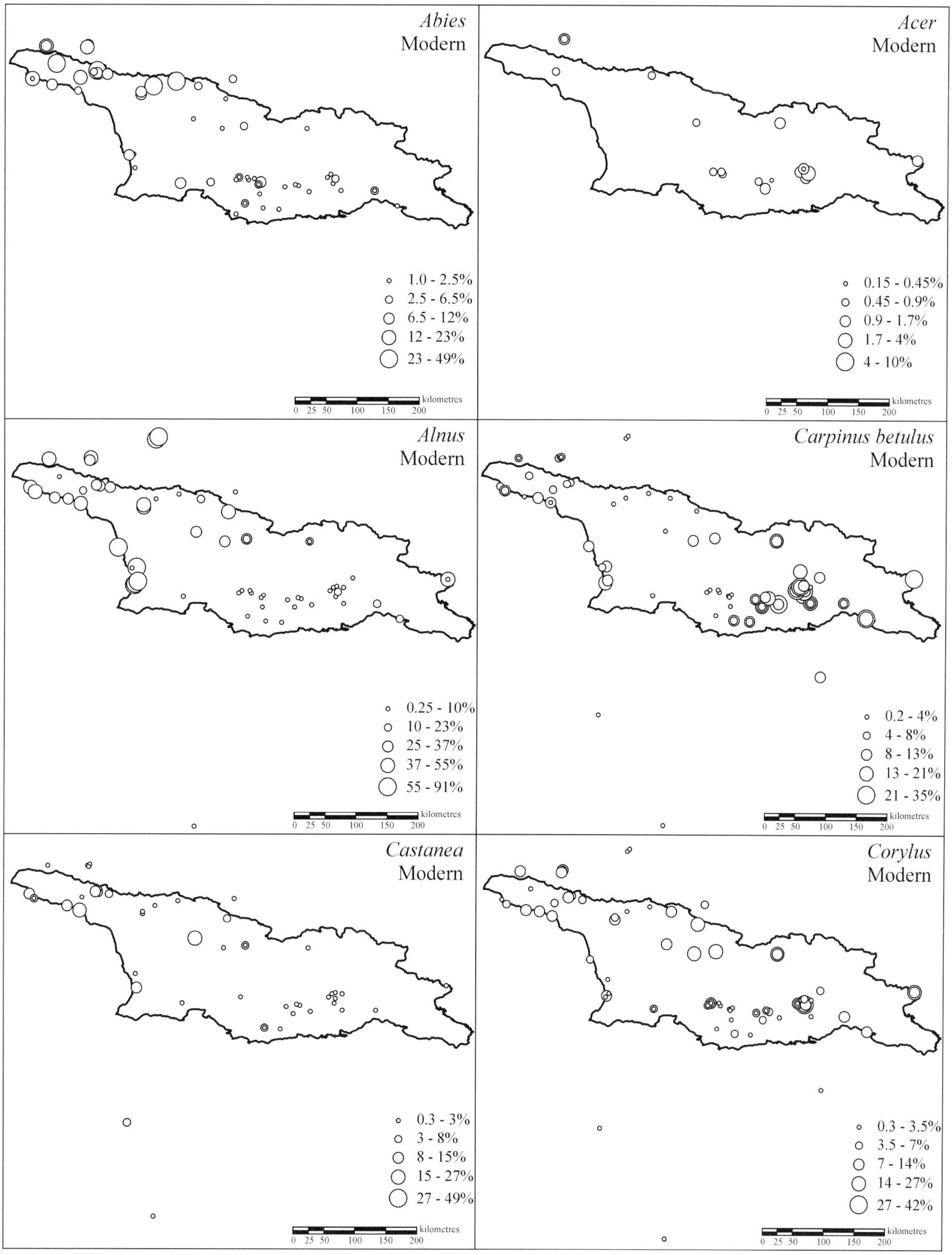
Abies
Modern
1.0 - 2.5%
2.5 - 6.5%
6.5 - 12%
12 - 23%
23 - 49%
kilometres
0 25 50 100 150 200
Acer
Modern
0.15 - 0.45%
0.45 - 0.9%
0.9 - 1.7%
1.7 - 4%
4 - 10%
kilometres
0 25 50 100 150 200
Alnus
Modern
0.25 - 10%
10 - 23%
25 - 37%
37 - 55%
55 - 91%
kilometres
0 25 50 100 150 200
Carpinus betulus
Modern
0.2 - 4%
4 - 8%
8 - 13%
13 - 21%
21 - 35%
kilometres
0 25 50 100 150 200
Castanea
Modern
0.3 - 3%
3 - 8%
8 - 15%
15 - 27%
27 - 49%
kilometres
0 25 50 100 150 200
Corylus
Modern
0.3 - 3.5%
3.5 - 7%
7 - 14%
14 - 27%
27 - 42%
kilometres
0 25 50 100 150 200

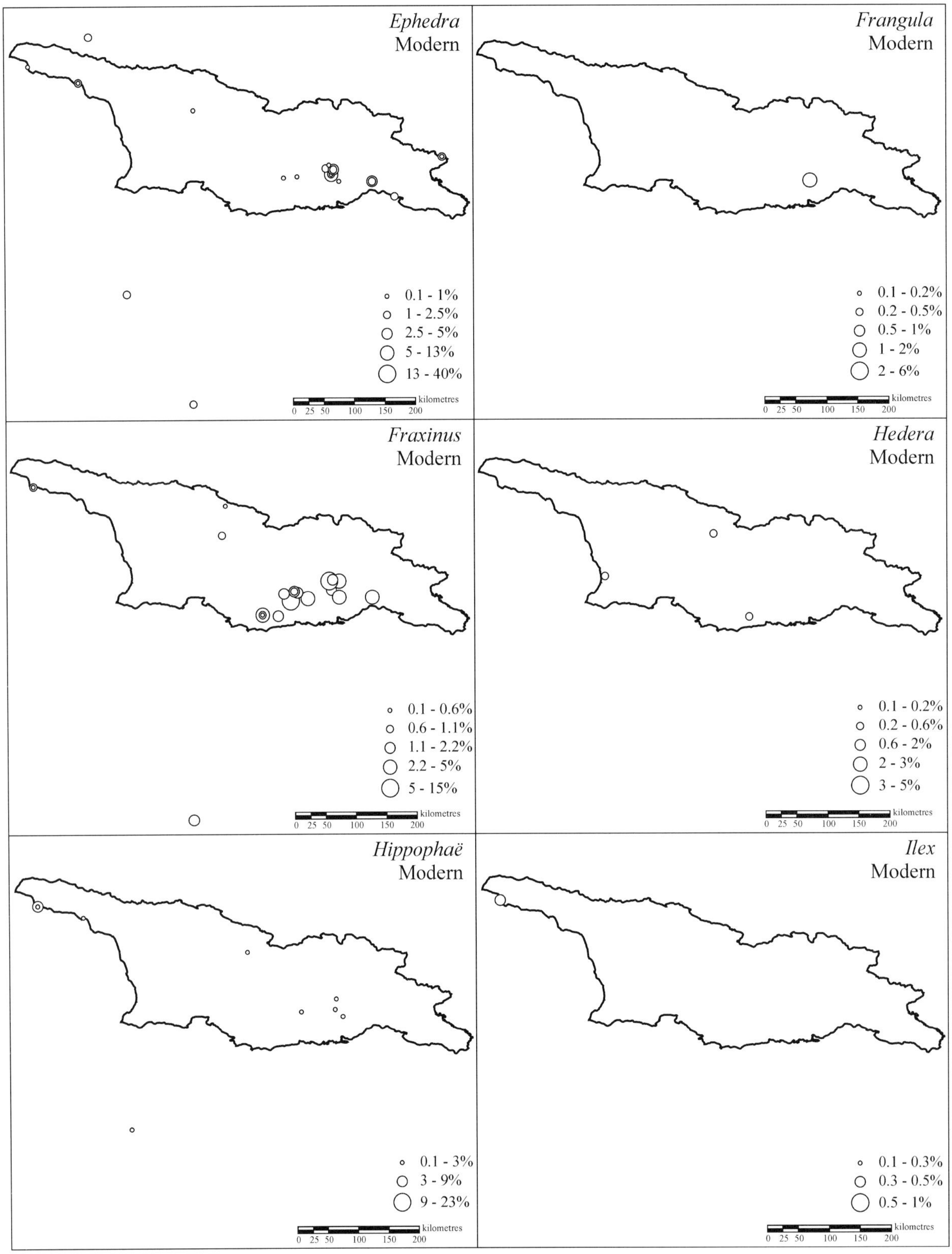
Ephedra
Modern
0.1 - 1%
1 - 2.5%
2.5 - 5%
5 - 13%
13 - 40%
kilometres
0 25 50 100 150 200
Frangula
Modern
0.1 - 0.2%
0.2 - 0.5%
0.5 - 1%
1 - 2%
2 - 6%
kilometres
0 25 50 100 150 200
Fraxinus
Modern
0.1 - 0.6%
0.6 - 1.1%
1.1 - 2.2%
2.2 - 5%
5 - 15%
kilometres
0 25 50 100 150 200
Hedera
Modern
0.1 - 0.2%
0.2 - 0.6%
0.6 - 2%
2 - 3%
3 - 5%
kilometres
0 25 50 100 150 200
Hippophaë
Modern
0.1 - 3%
3 - 9%
9 - 23%
kilometres
0 25 50 100 150 200
Ilex
Modern
0.1 - 0.3%
0.3 - 0.5%
0.5 - 1%
kilometres
0 25 50 100 150 200

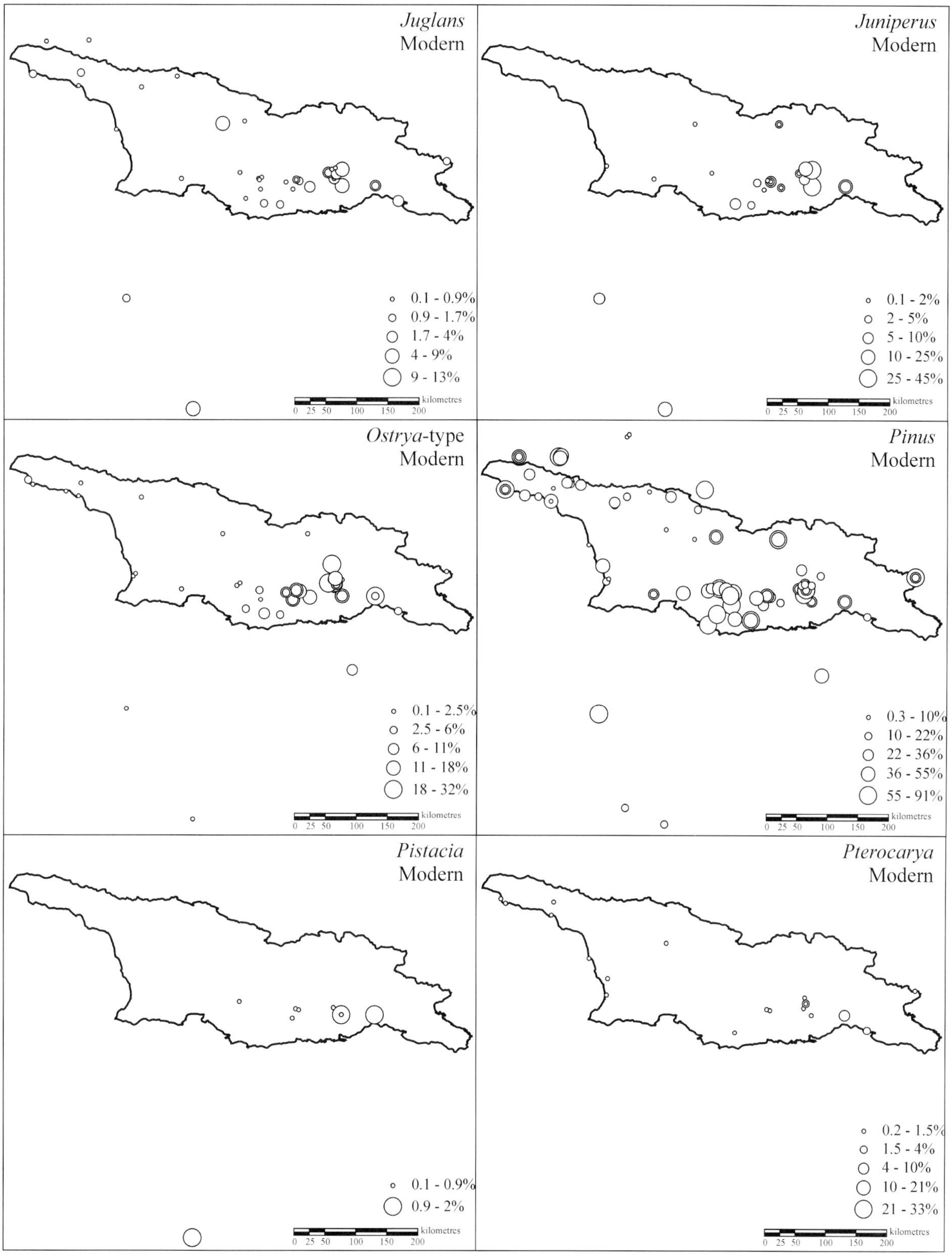
Juglans
Modern
0.1 - 0.9%
0.9 - 1.7%
1.7 - 4%
4 - 9%
9 - 13%
kilometres
0 25 50 100 150 200
Juniperus
Modern
0.1 - 2%
2 - 5%
5 - 10%
10 - 25%
25 - 45%
kilometres
0 25 50 100 150 200
Ostrya-type
Modern
0.1 - 2.5%
2.5 - 6%
6 - 11%
11 - 18%
18 - 32%
kilometres
0 25 50 100 150 200
Pinus
Modern
0.3 - 10%
10 - 22%
22 - 36%
36 - 55%
55 - 91%
kilometres
0 25 50 100 150 200
Pistacia
Modern
0.1 - 0.9%
0.9 - 2%
kilometres
0 25 50 100 150 200
Pterocarya
Modern
0.2 - 1.5%
1.5 - 4%
4 - 10%
10 - 21%
21 - 33%
kilometres
0 25 50 100 150 200

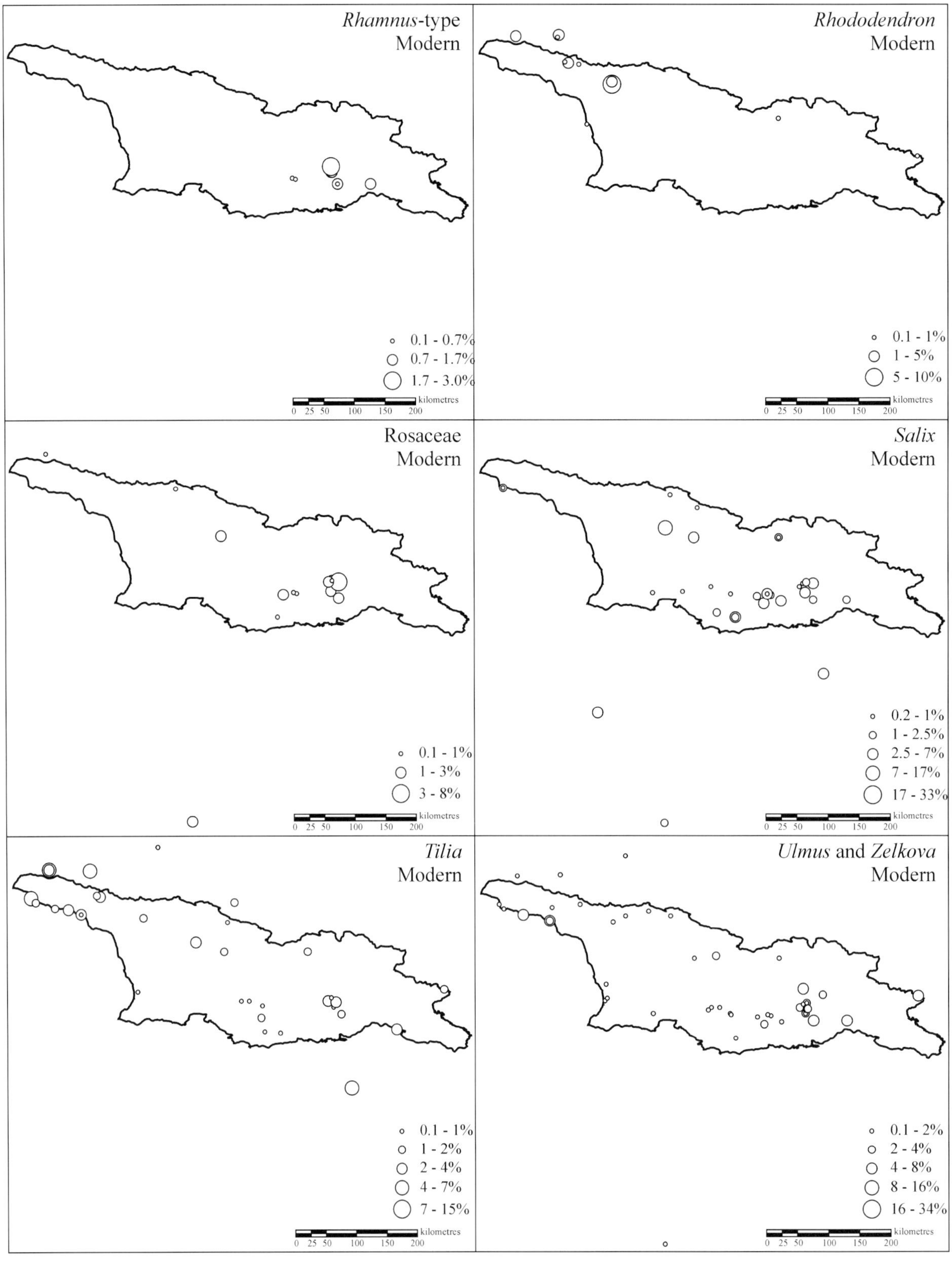
Rhamnus-type
Modern
0.1 - 0.7%
0.7 - 1.7%
1.7 - 3.0%
kilometres
0 25 50 100 150 200
Rhododendron
Modern
0.1 - 1%
1 - 5%
5 - 10%
kilometres
0 25 50 100 150 200
Rosaceae
Modern
0.1 - 1%
1 - 3%
3 - 8%
kilometres
0 25 50 100 150 200
Salix
Modern
0.2 - 1%
1 - 2.5%
2.5 - 7%
7 - 17%
17 - 33%
kilometres
0 25 50 100 150 200
Tilia
Modern
0.1 - 1%
1 - 2%
2 - 4%
4 - 7%
7 - 15%
kilometres
0 25 50 100 150 200
Ulmus and Zelkova
Modern
0.1 - 2%
2 - 4%
4 - 8%
8 - 16%
16 - 34%
kilometres
0 25 50 100 150 200

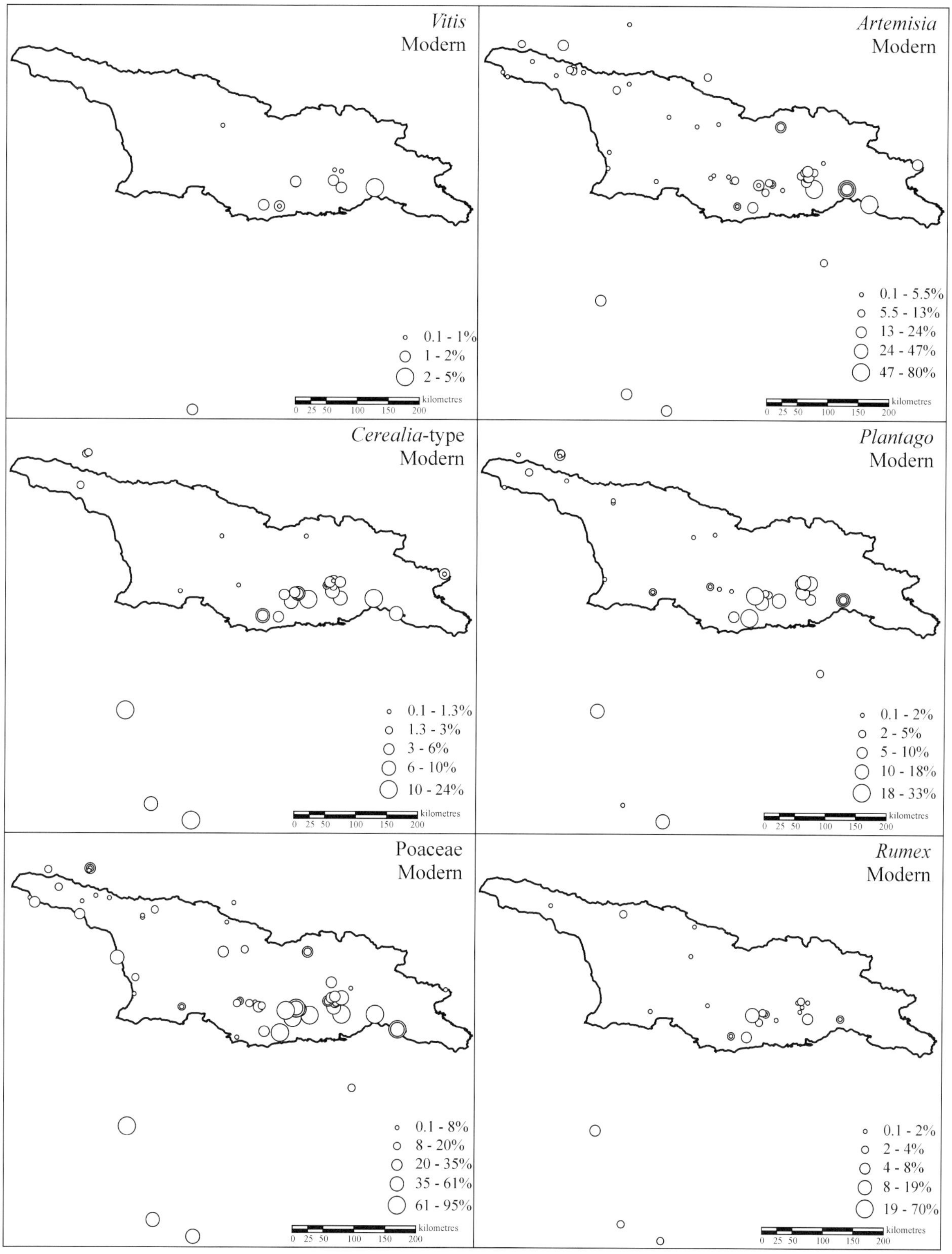
Vitis
Modern
0.1 - 1%
1 - 2%
2 - 5%
0 25 50 100 150 200 kilometres
Artemisia
Modern
0.1 - 5.5%
5.5 - 13%
13 - 24%
24 - 47%
47 - 80%
0 25 50 100 150 200 kilometres
Cerealia-type
Modern
0.1 - 1.3%
1.3 - 3%
3 - 6%
6 - 10%
10 - 24%
0 25 50 100 150 200 kilometres
Plantago
Modern
0.1 - 2%
2 - 5%
5 - 10%
10 - 18%
18 - 33%
0 25 50 100 150 200 kilometres
Poaceae
Modern
0.1 - 8%
8 - 20%
20 - 35%
35 - 61%
61 - 95%
0 25 50 100 150 200 kilometres
Rumex
Modern
0.1 - 2%
2 - 4%
4 - 8%
8 - 19%
19 - 70%
0 25 50 100 150 200 kilometres

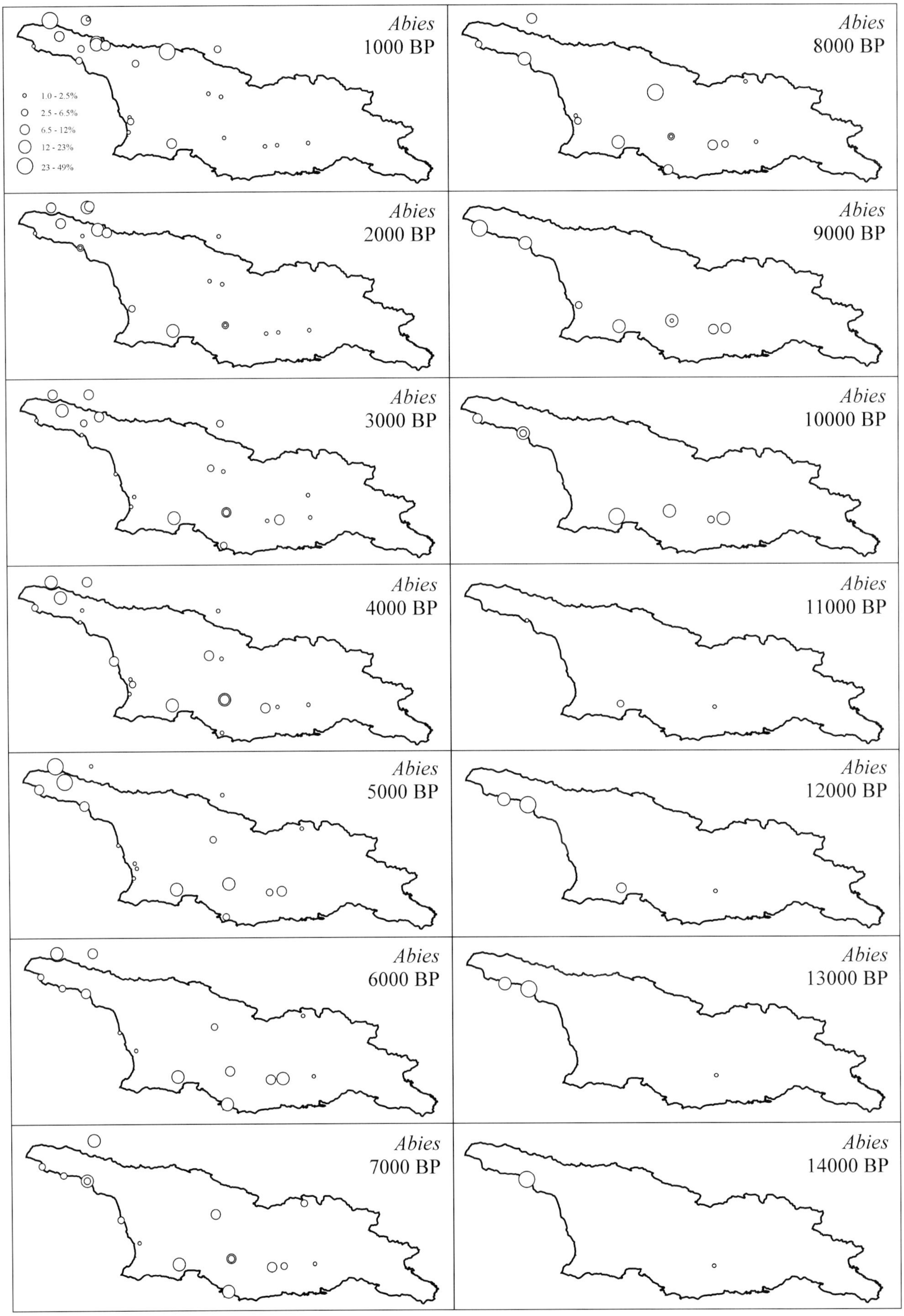

Abies
1000 BP
1.0 - 2.5%
2.5 - 6.5%
6.5 - 12%
12 - 23%
23 - 49%
Abies
8000 BP
Abies
2000 BP
Abies
9000 BP
Abies
3000 BP
Abies
10000 BP
Abies
4000 BP
Abies
11000 BP
Abies
5000 BP
Abies
12000 BP
Abies
6000 BP
Abies
13000 BP
Abies
7000 BP
Abies
14000 BP

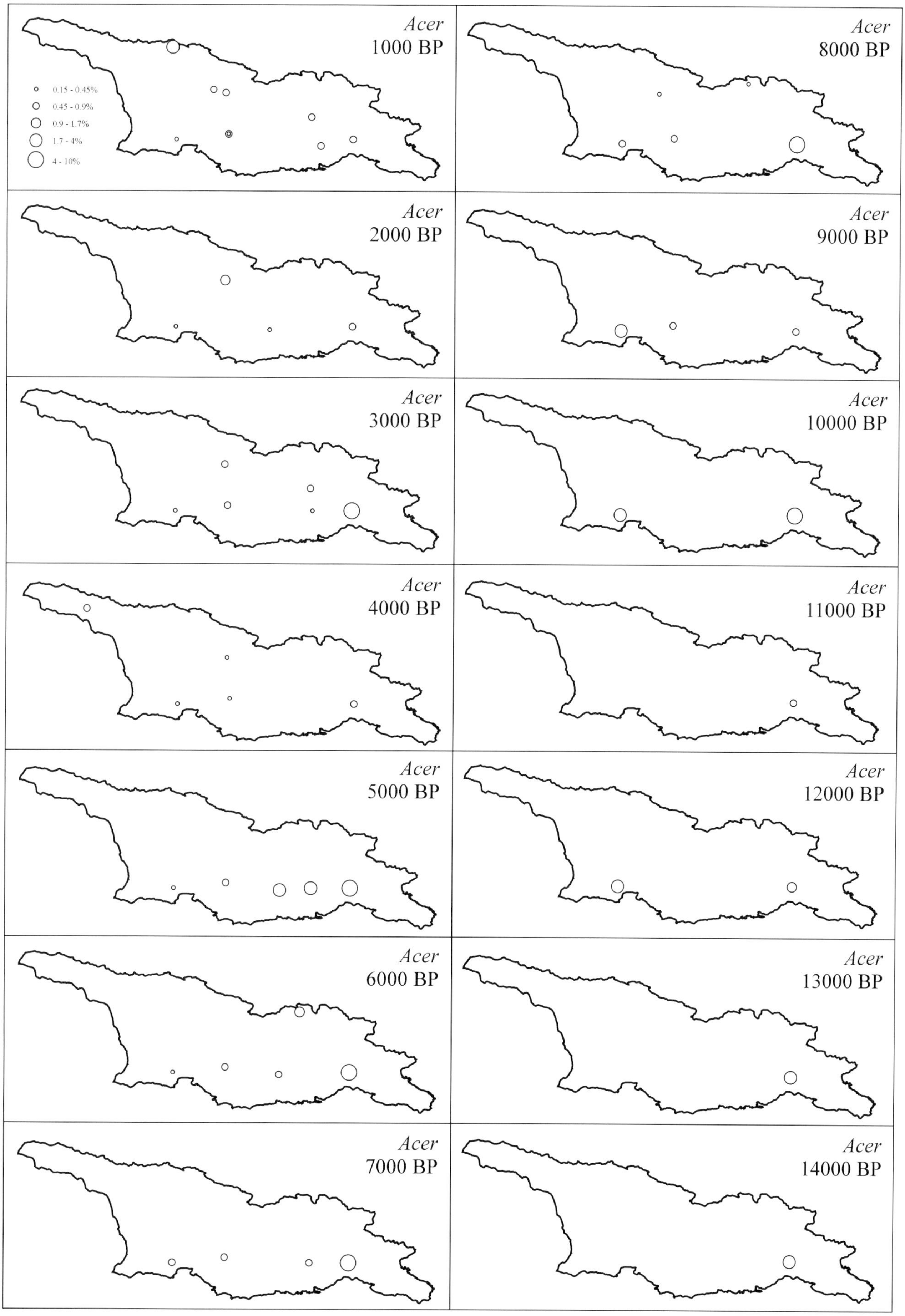
Acer
1000 BP
0.15 - 0.45%
0.45 - 0.9%
0.9 - 1.7%
1.7 - 4%
4 - 10%
Acer
2000 BP
Acer
3000 BP
Acer
4000 BP
Acer
5000 BP
Acer
6000 BP
Acer
7000 BP
Acer
8000 BP
Acer
9000 BP
Acer
10000 BP
Acer
11000 BP
Acer
12000 BP
Acer
13000 BP
Acer
14000 BP

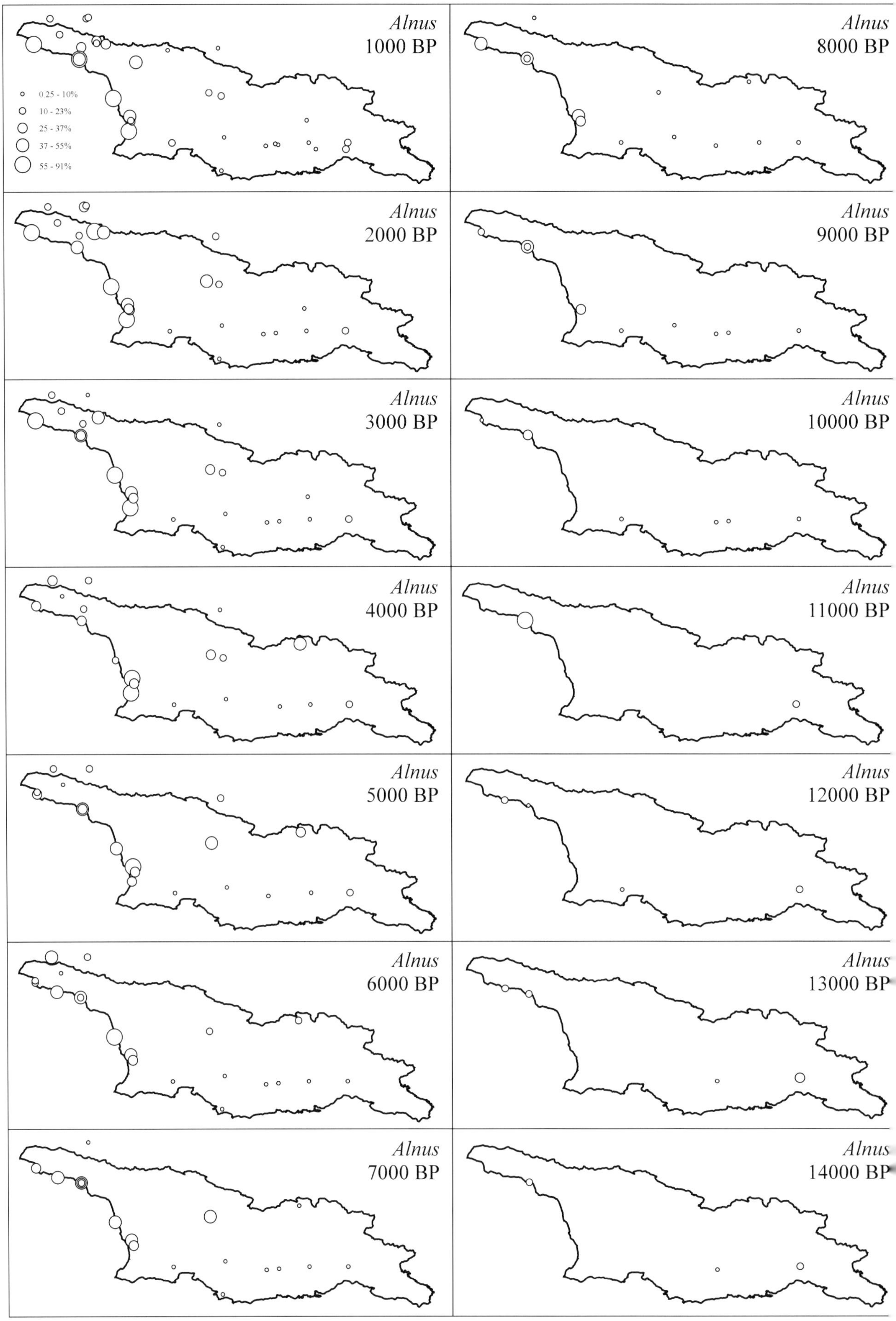

Alnus
1000 BP
0.25 - 10%
10 - 23%
25 - 37%
37 - 55%
55 - 91%
Alnus
8000 BP
Alnus
2000 BP
Alnus
9000 BP
Alnus
3000 BP
Alnus
10000 BP
Alnus
4000 BP
Alnus
11000 BP
Alnus
5000 BP
Alnus
12000 BP
Alnus
6000 BP
Alnus
13000 BP
Alnus
7000 BP
Alnus
14000 BP

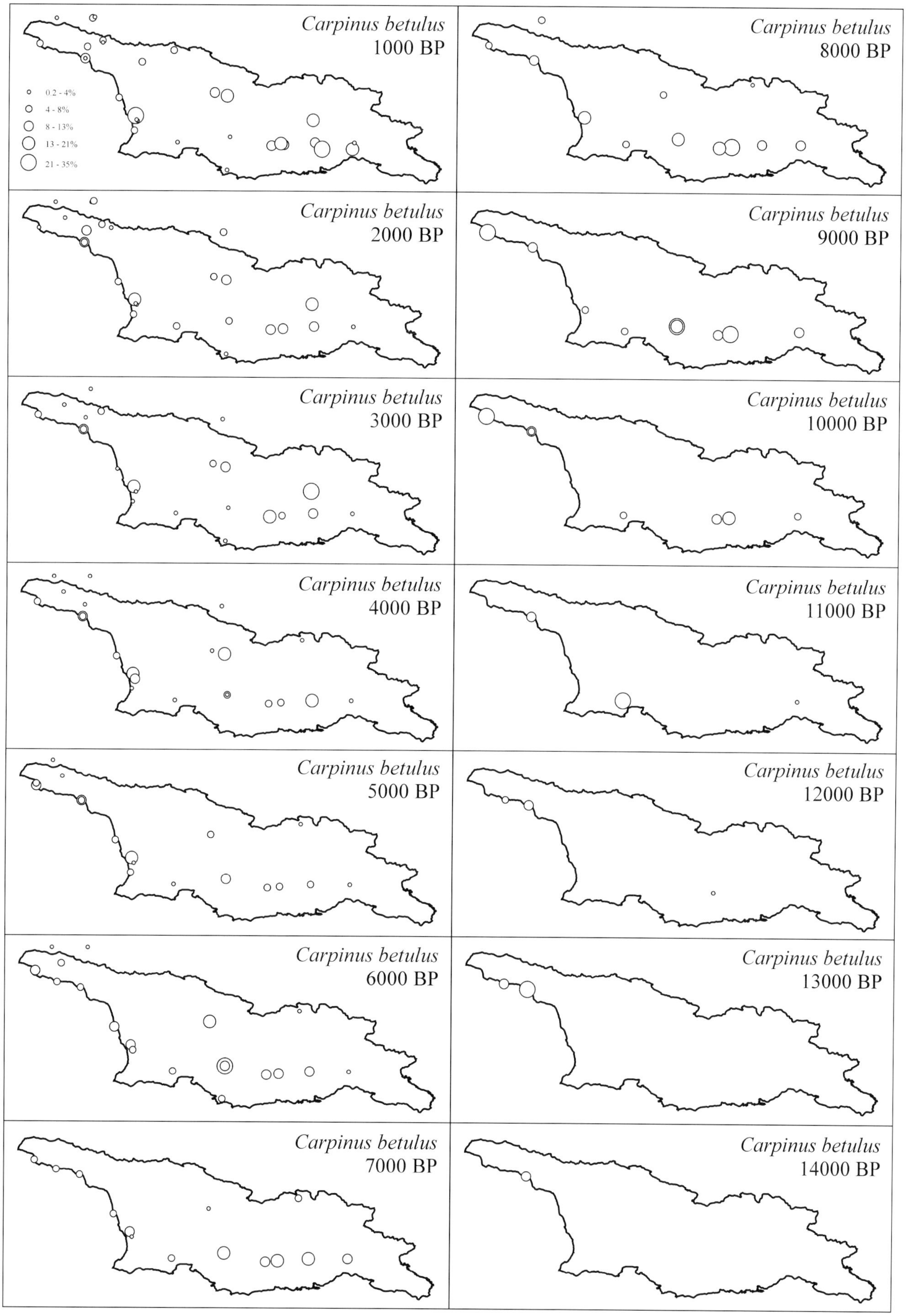
Carpinus betulus
1000 BP
0.2 - 4%
4 - 8%
8 - 13%
13 - 21%
21 - 35%
Carpinus betulus
8000 BP
Carpinus betulus
2000 BP
Carpinus betulus
9000 BP
Carpinus betulus
3000 BP
Carpinus betulus
10000 BP
Carpinus betulus
4000 BP
Carpinus betulus
11000 BP
Carpinus betulus
5000 BP
Carpinus betulus
12000 BP
Carpinus betulus
6000 BP
Carpinus betulus
13000 BP
Carpinus betulus
7000 BP
Carpinus betulus
14000 BP

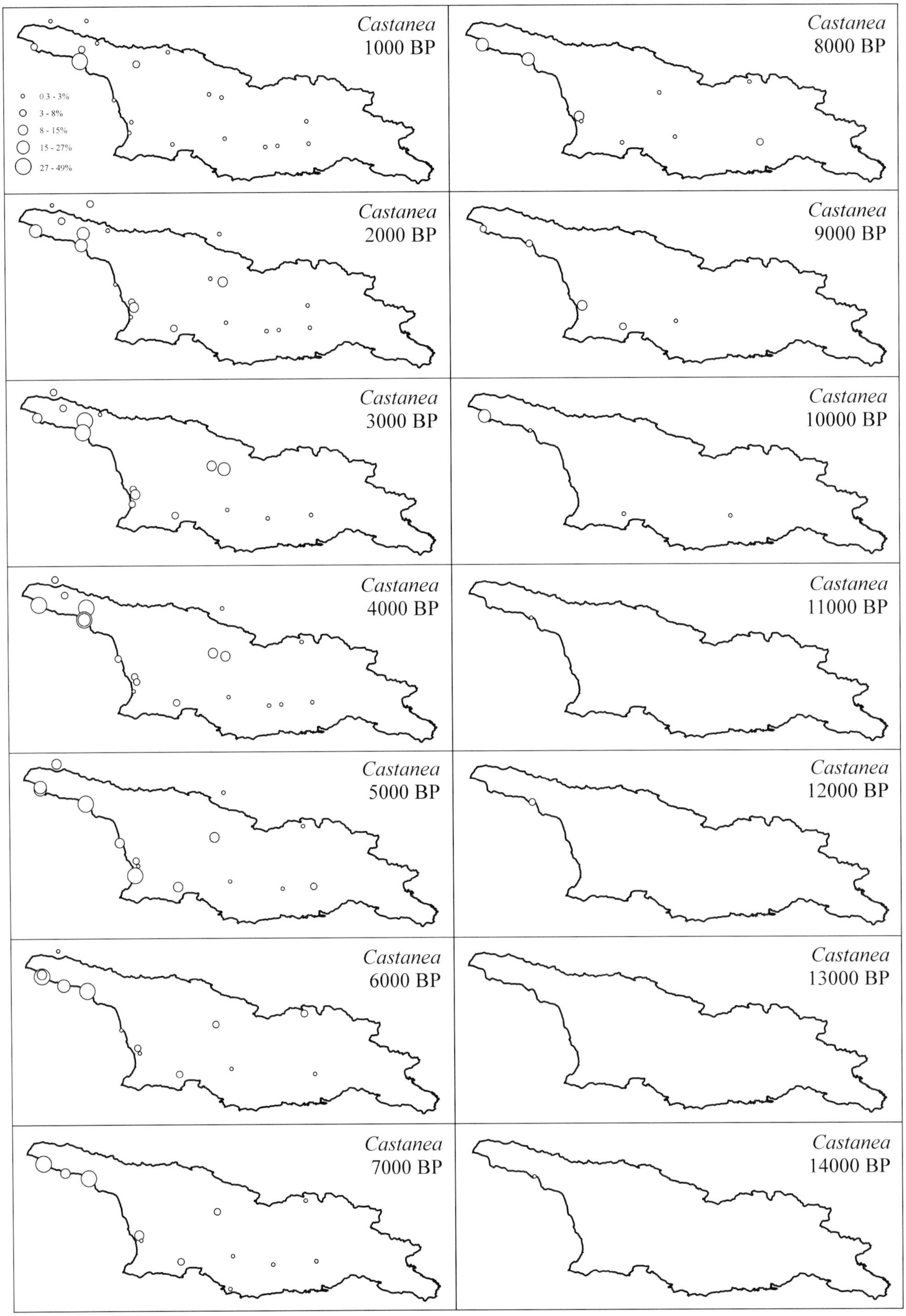

Castanea 1000 BP
0.3 - 3%
3 - 8%
8 - 15%
15 - 27%
27 - 49%
Castanea 8000 BP
Castanea 2000 BP
Castanea 9000 BP
Castanea 3000 BP
Castanea 10000 BP
Castanea 4000 BP
Castanea 11000 BP
Castanea 5000 BP
Castanea 12000 BP
Castanea 6000 BP
Castanea 13000 BP
Castanea 7000 BP
Castanea 14000 BP

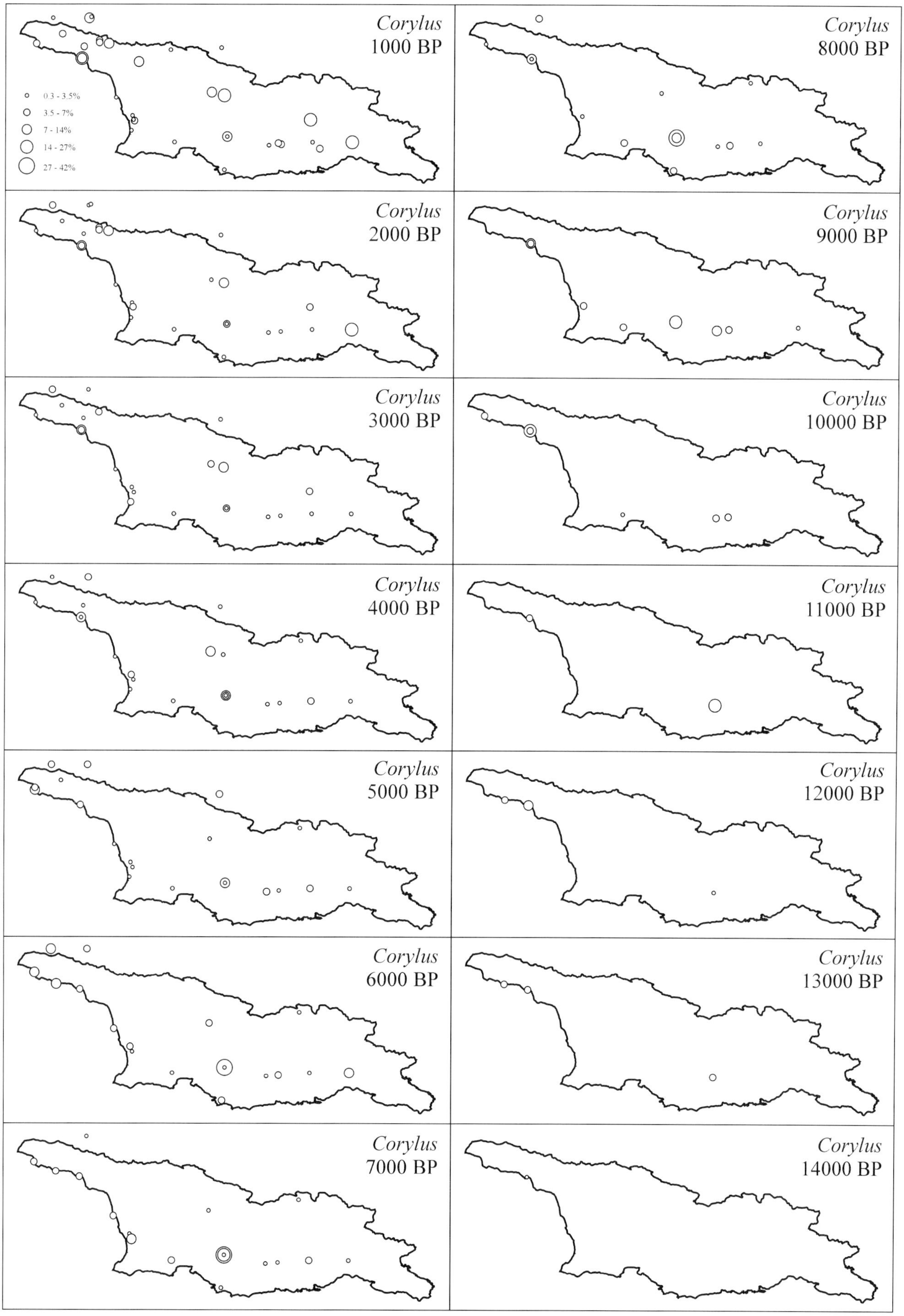
Corylus 1000 BP
0.3 - 3.5%
3.5 - 7%
7 - 14%
14 - 27%
27 - 42%
Corylus 8000 BP
Corylus 2000 BP
Corylus 9000 BP
Corylus 3000 BP
Corylus 10000 BP
Corylus 4000 BP
Corylus 11000 BP
Corylus 5000 BP
Corylus 12000 BP
Corylus 6000 BP
Corylus 13000 BP
Corylus 7000 BP
Corylus 14000 BP

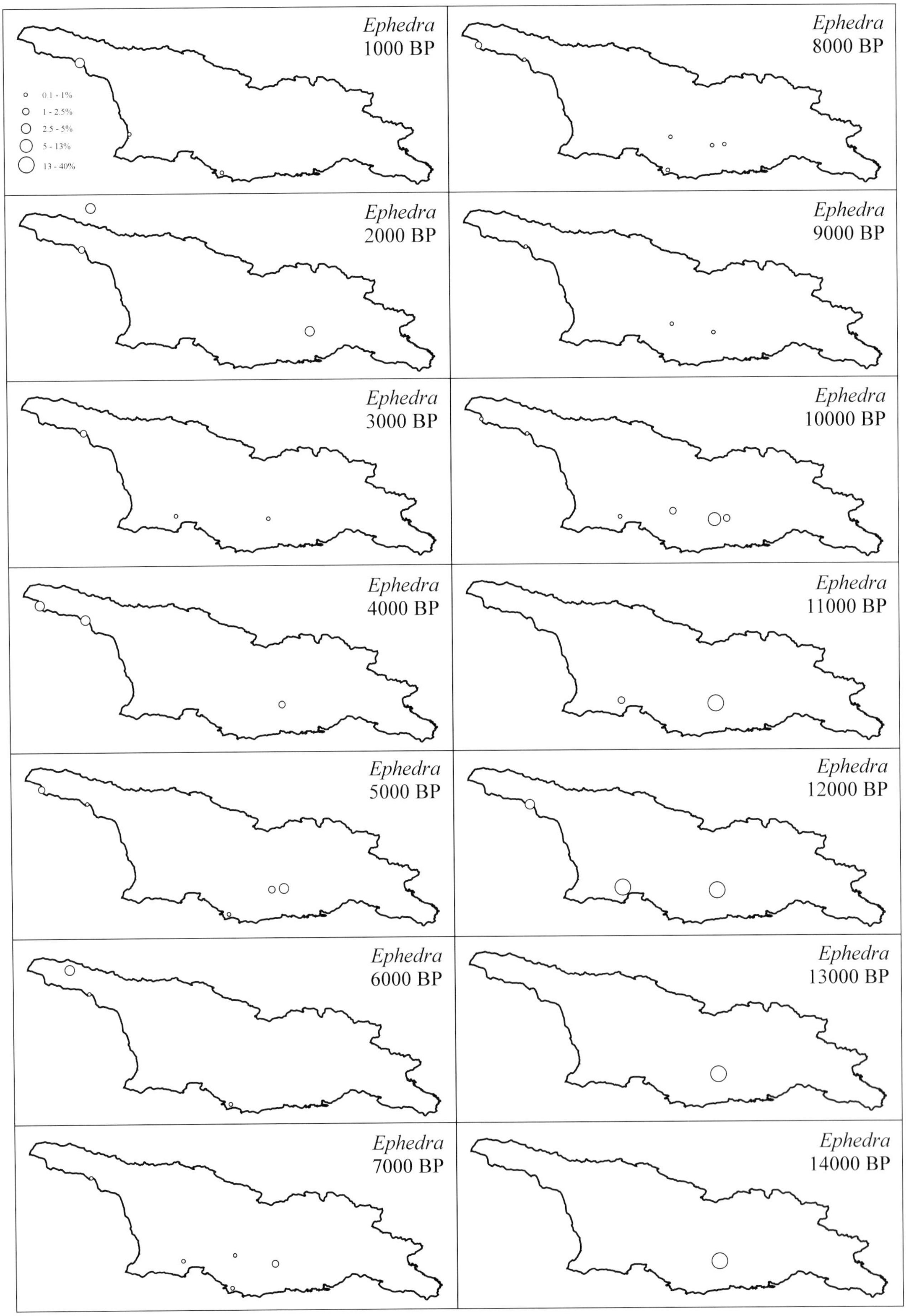

Ephedra
1000 BP
0.1 - 1%
1 - 2.5%
2.5 - 5%
5 - 13%
13 - 40%
Ephedra
8000 BP
Ephedra
2000 BP
Ephedra
9000 BP
Ephedra
3000 BP
Ephedra
10000 BP
Ephedra
4000 BP
Ephedra
11000 BP
Ephedra
5000 BP
Ephedra
12000 BP
Ephedra
6000 BP
Ephedra
13000 BP
Ephedra
7000 BP
Ephedra
14000 BP

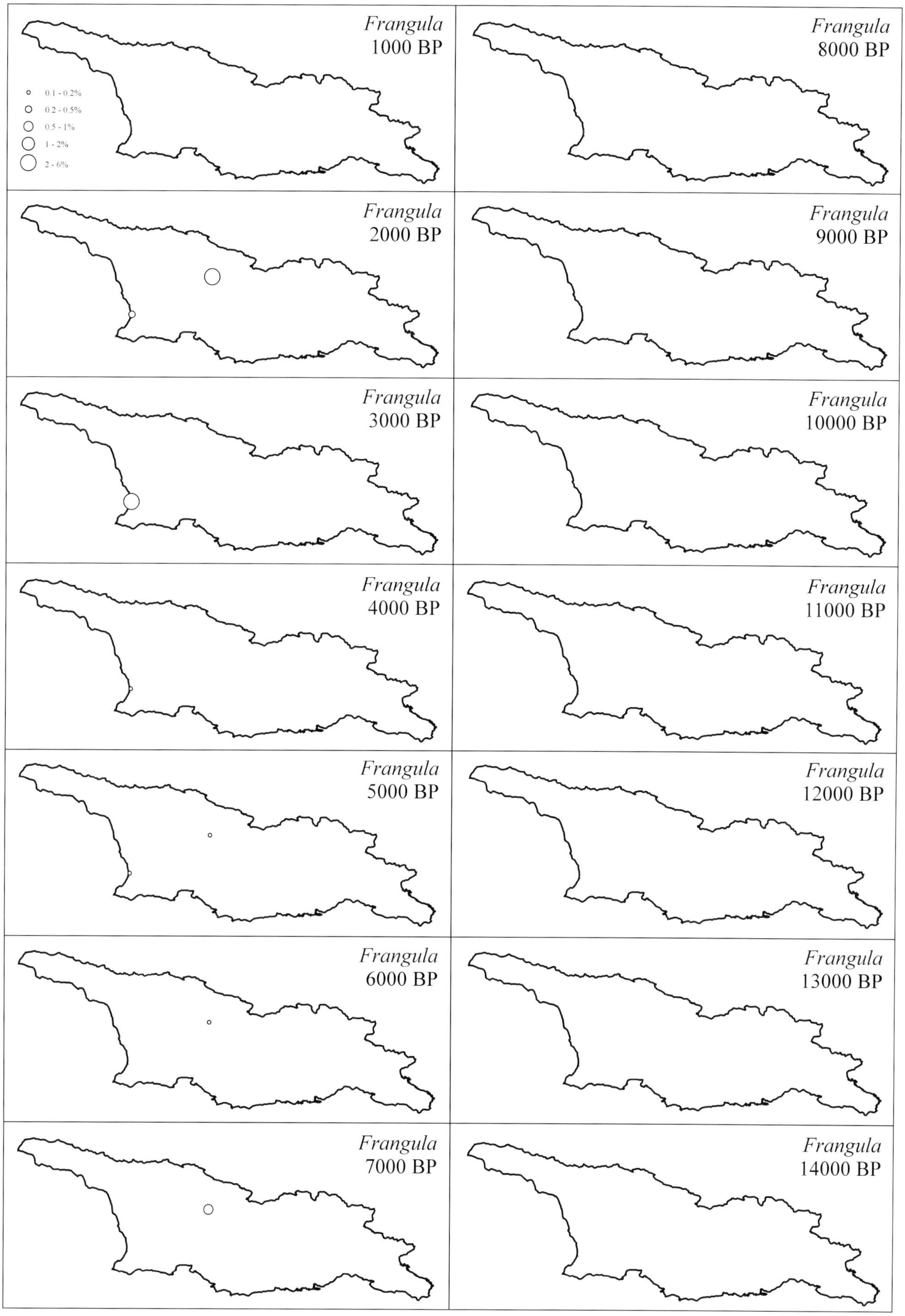

Frangula
1000 BP
0.1 - 0.2%
0.2 - 0.5%
0.5 - 1%
1 - 2%
2 - 6%
Frangula
8000 BP
Frangula
2000 BP
Frangula
9000 BP
Frangula
3000 BP
Frangula
10000 BP
Frangula
4000 BP
Frangula
11000 BP
Frangula
5000 BP
Frangula
12000 BP
Frangula
6000 BP
Frangula
13000 BP
Frangula
7000 BP
Frangula
14000 BP

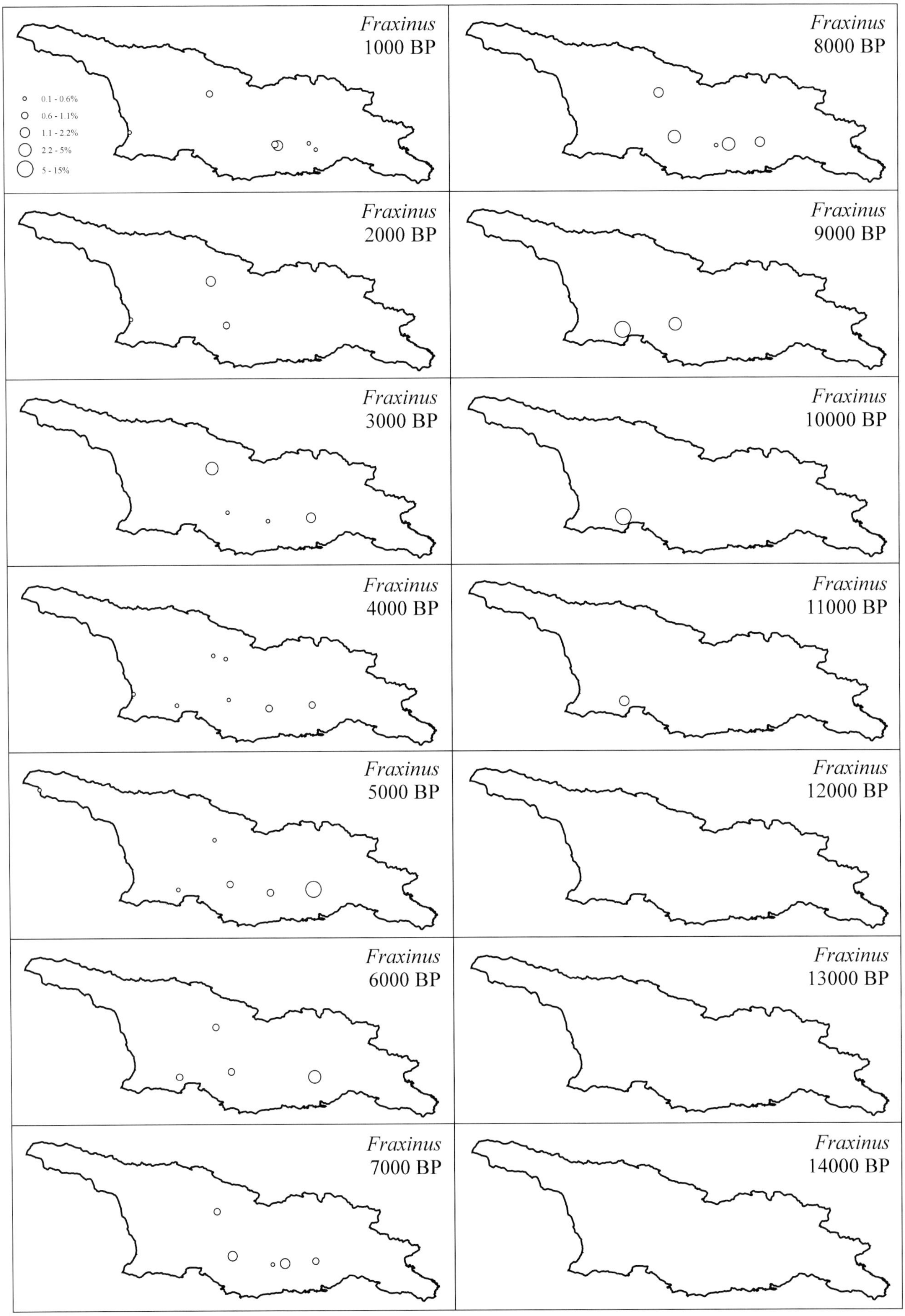
Fraxinus
1000 BP
0.1 - 0.6%
0.6 - 1.1%
1.1 - 2.2%
2.2 - 5%
5 - 15%
Fraxinus
8000 BP
Fraxinus
2000 BP
Fraxinus
9000 BP
Fraxinus
3000 BP
Fraxinus
10000 BP
Fraxinus
4000 BP
Fraxinus
11000 BP
Fraxinus
5000 BP
Fraxinus
12000 BP
Fraxinus
6000 BP
Fraxinus
13000 BP
Fraxinus
7000 BP
Fraxinus
14000 BP

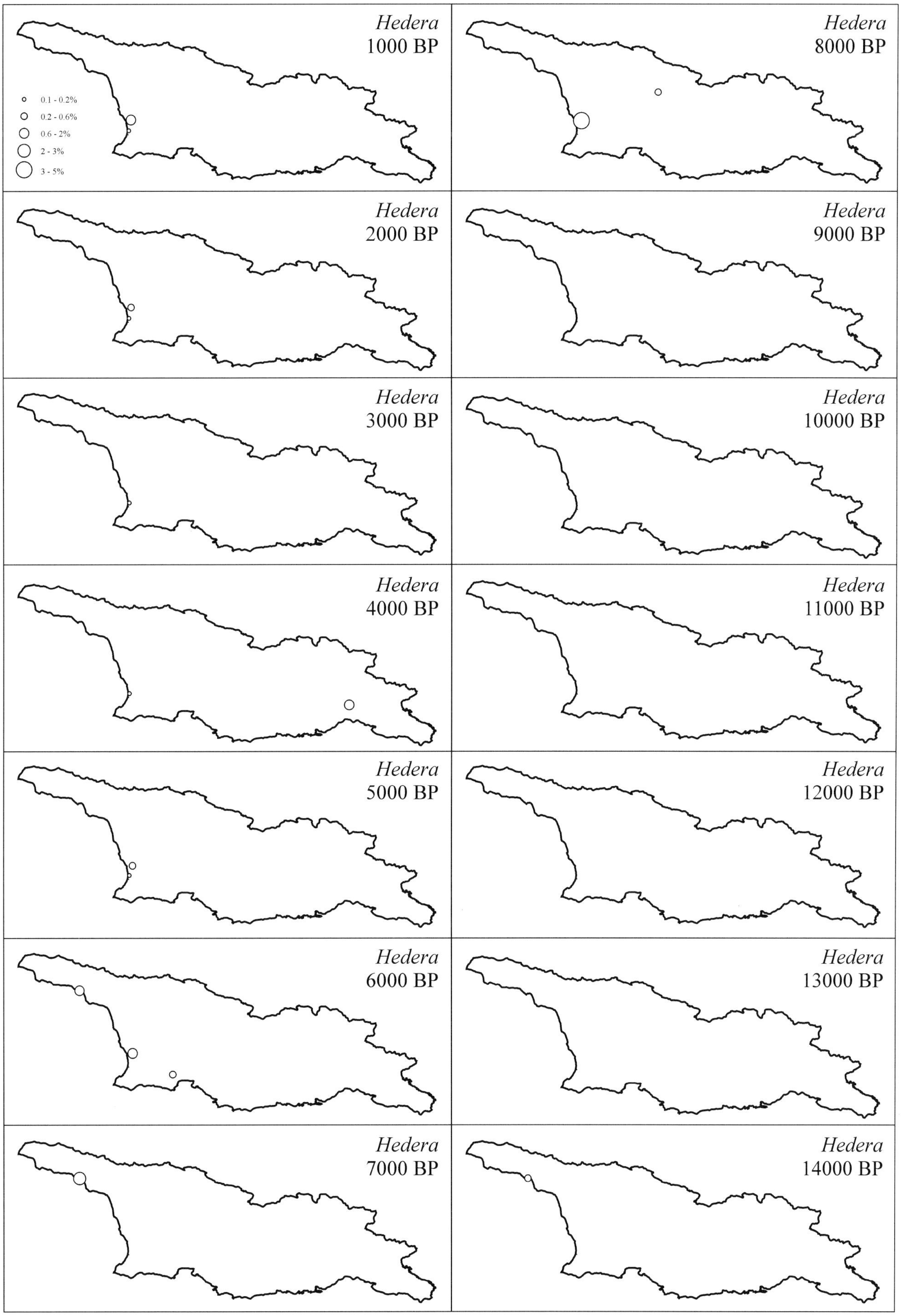

Hedera
1000 BP
0.1 - 0.2%
0.2 - 0.6%
0.6 - 2%
2 - 3%
3 - 5%
Hedera
8000 BP
Hedera
2000 BP
Hedera
9000 BP
Hedera
3000 BP
Hedera
10000 BP
Hedera
4000 BP
Hedera
11000 BP
Hedera
5000 BP
Hedera
12000 BP
Hedera
6000 BP
Hedera
13000 BP
Hedera
7000 BP
Hedera
14000 BP

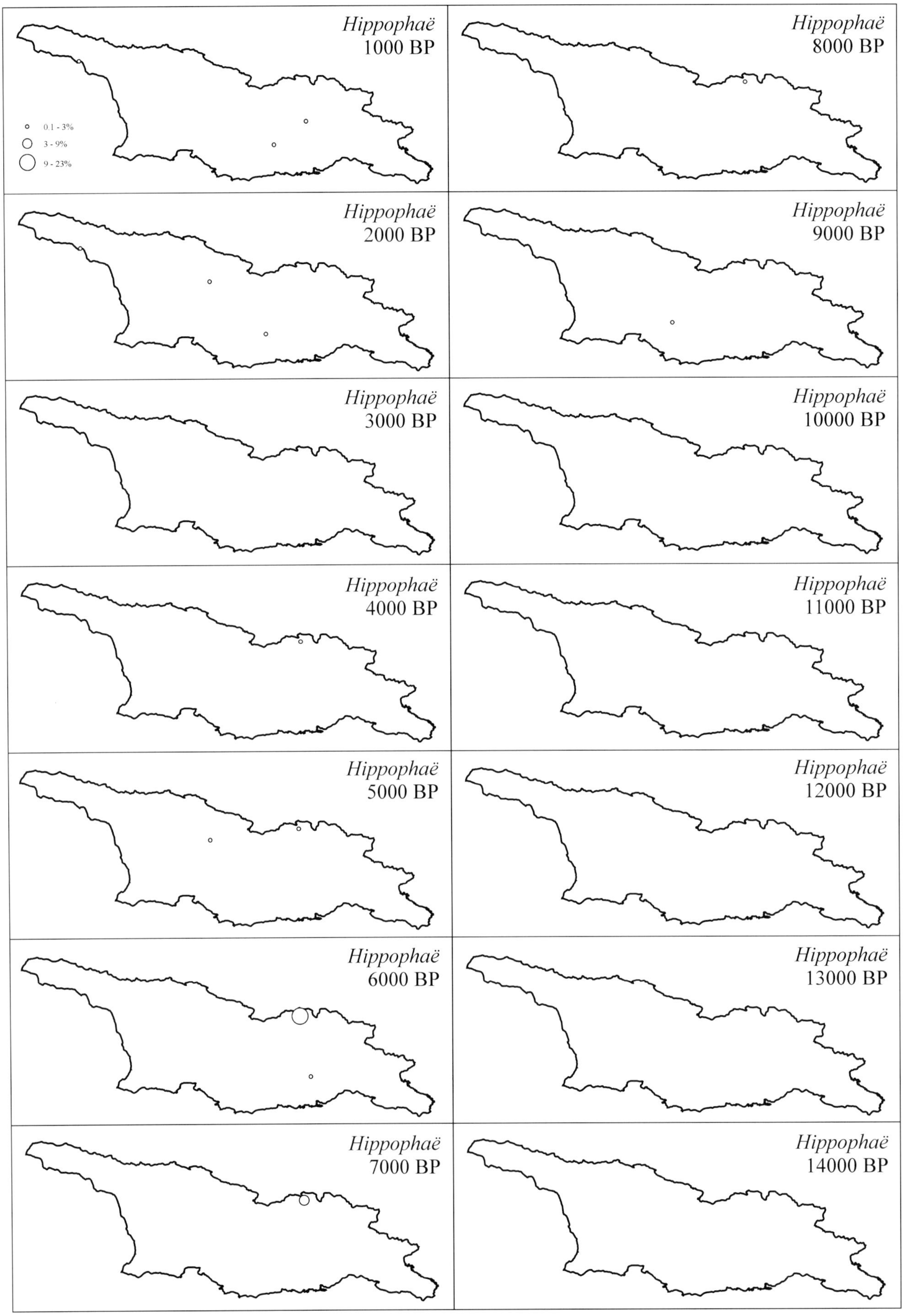

Hippophaë
1000 BP
0.1 - 3%
3 - 9%
9 - 23%
Hippophaë
2000 BP
Hippophaë
3000 BP
Hippophaë
4000 BP
Hippophaë
5000 BP
Hippophaë
6000 BP
Hippophaë
7000 BP
Hippophaë
8000 BP
Hippophaë
9000 BP
Hippophaë
10000 BP
Hippophaë
11000 BP
Hippophaë
12000 BP
Hippophaë
13000 BP
Hippophaë
14000 BP

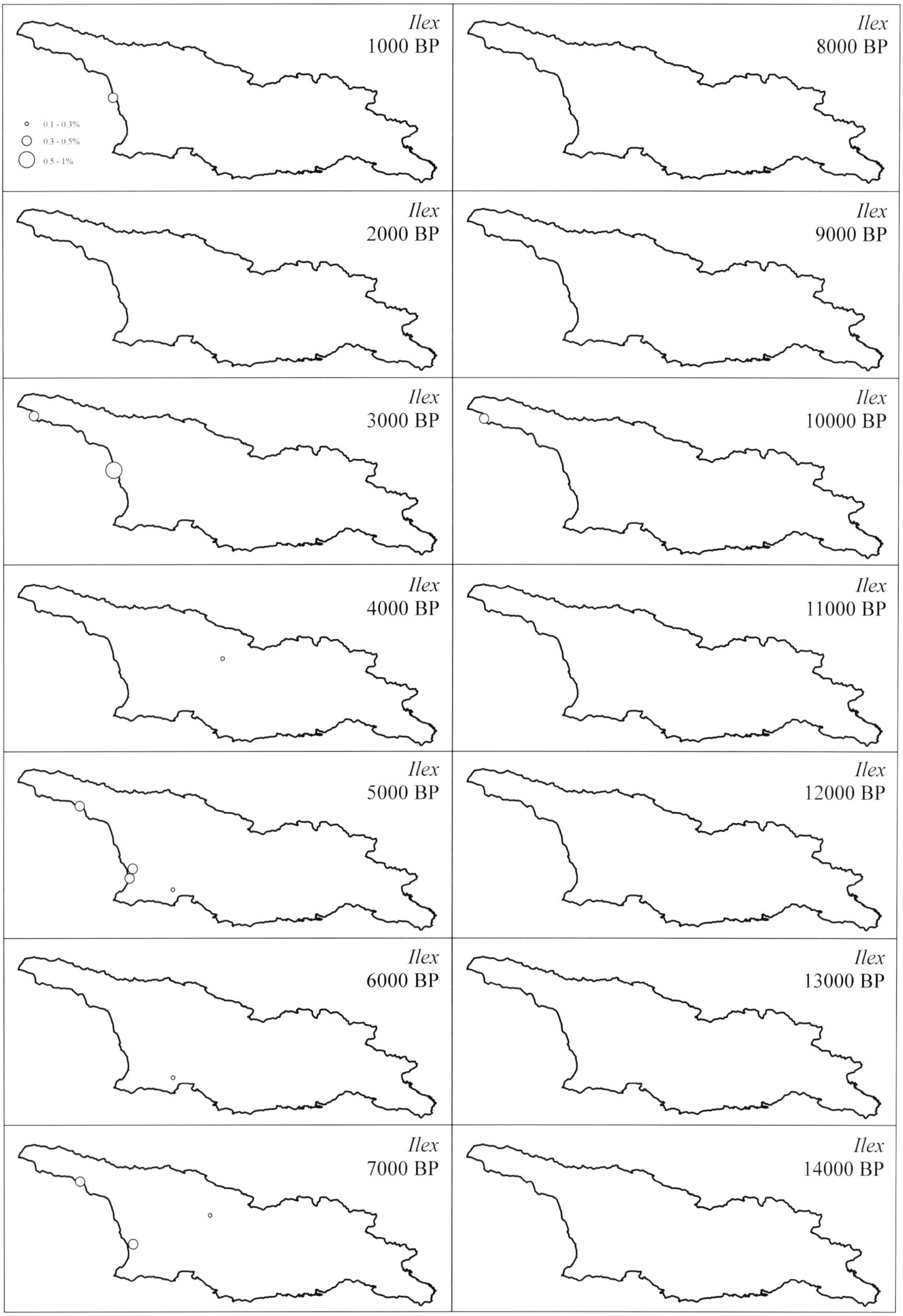
Ilex
1000 BP
0.1 - 0.3%
0.3 - 0.5%
0.5 - 1%
Ilex
8000 BP
Ilex
2000 BP
Ilex
9000 BP
Ilex
3000 BP
Ilex
10000 BP
Ilex
4000 BP
Ilex
11000 BP
Ilex
5000 BP
Ilex
12000 BP
Ilex
6000 BP
Ilex
13000 BP
Ilex
7000 BP
Ilex
14000 BP

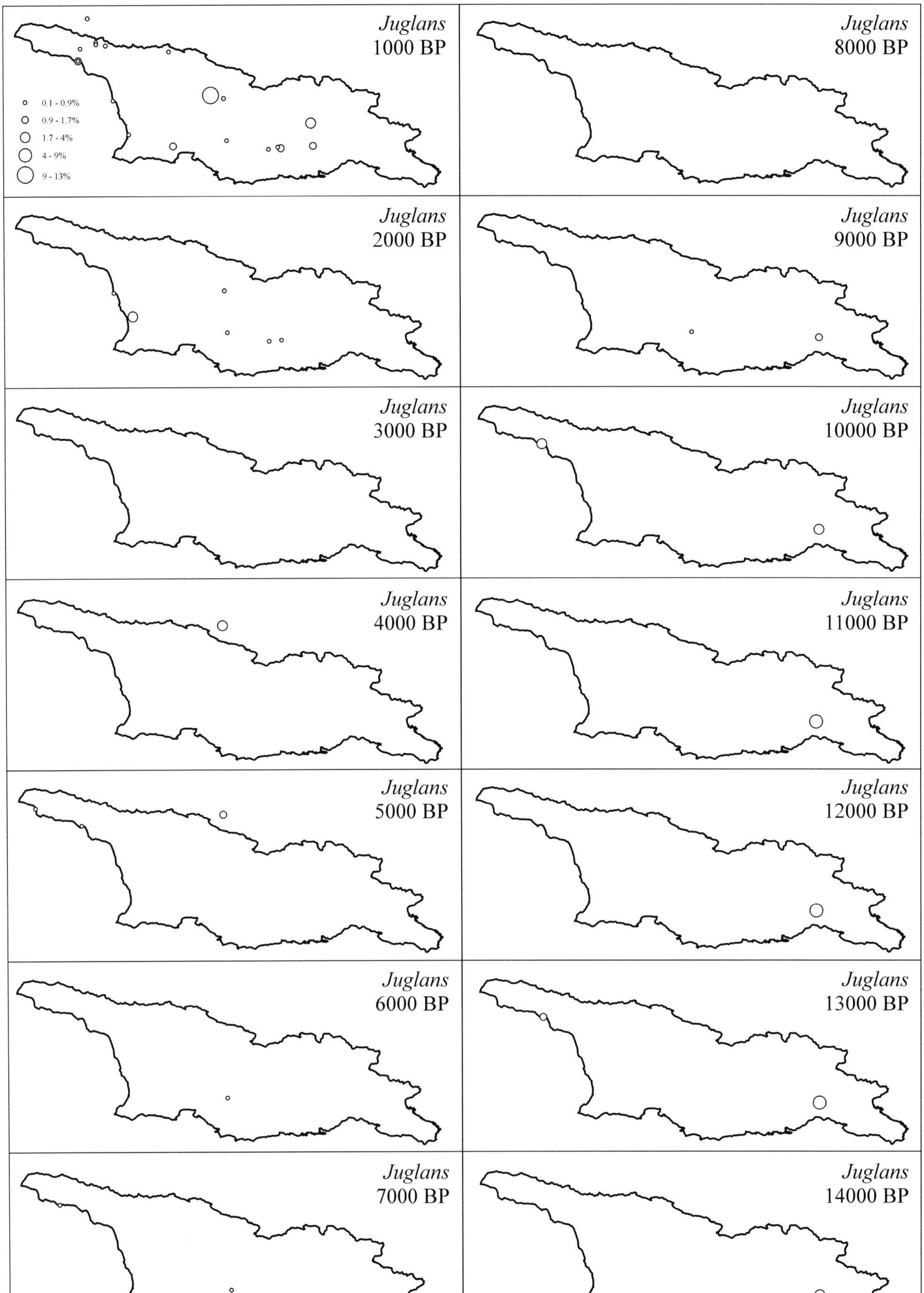

Juglans 1000 BP
0.1 - 0.9%
0.9 - 1.7%
1.7 - 4%
4 - 9%
9 - 13%
Juglans 8000 BP
Juglans 2000 BP
Juglans 9000 BP
Juglans 3000 BP
Juglans 10000 BP
Juglans 4000 BP
Juglans 11000 BP
Juglans 5000 BP
Juglans 12000 BP
Juglans 6000 BP
Juglans 13000 BP
Juglans 7000 BP
Juglans 14000 BP

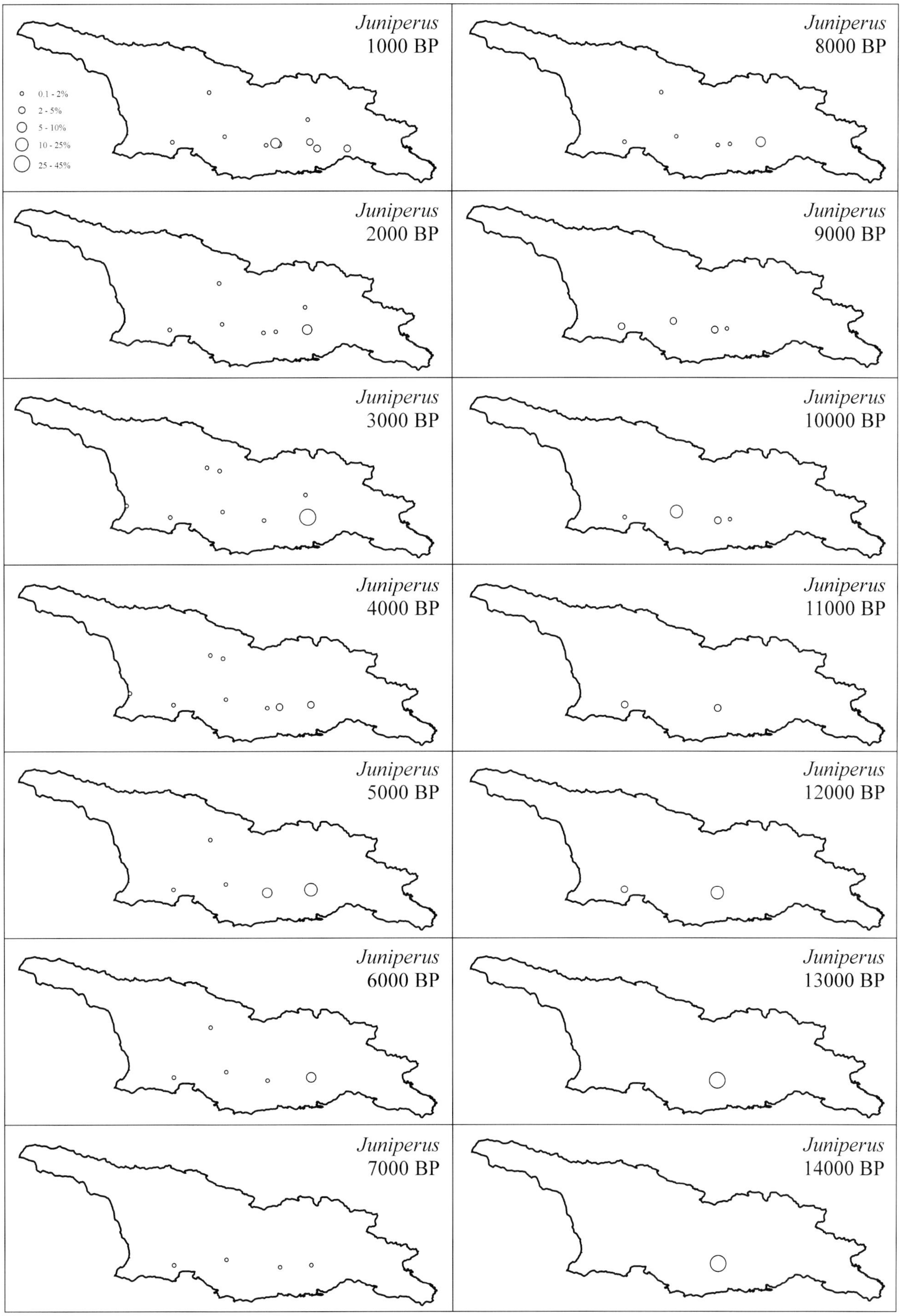

Juniperus 1000 BP
0.1 - 2%
2 - 5%
5 - 10%
10 - 25%
25 - 45%
Juniperus 8000 BP
Juniperus 2000 BP
Juniperus 9000 BP
Juniperus 3000 BP
Juniperus 10000 BP
Juniperus 4000 BP
Juniperus 11000 BP
Juniperus 5000 BP
Juniperus 12000 BP
Juniperus 6000 BP
Juniperus 13000 BP
Juniperus 7000 BP
Juniperus 14000 BP

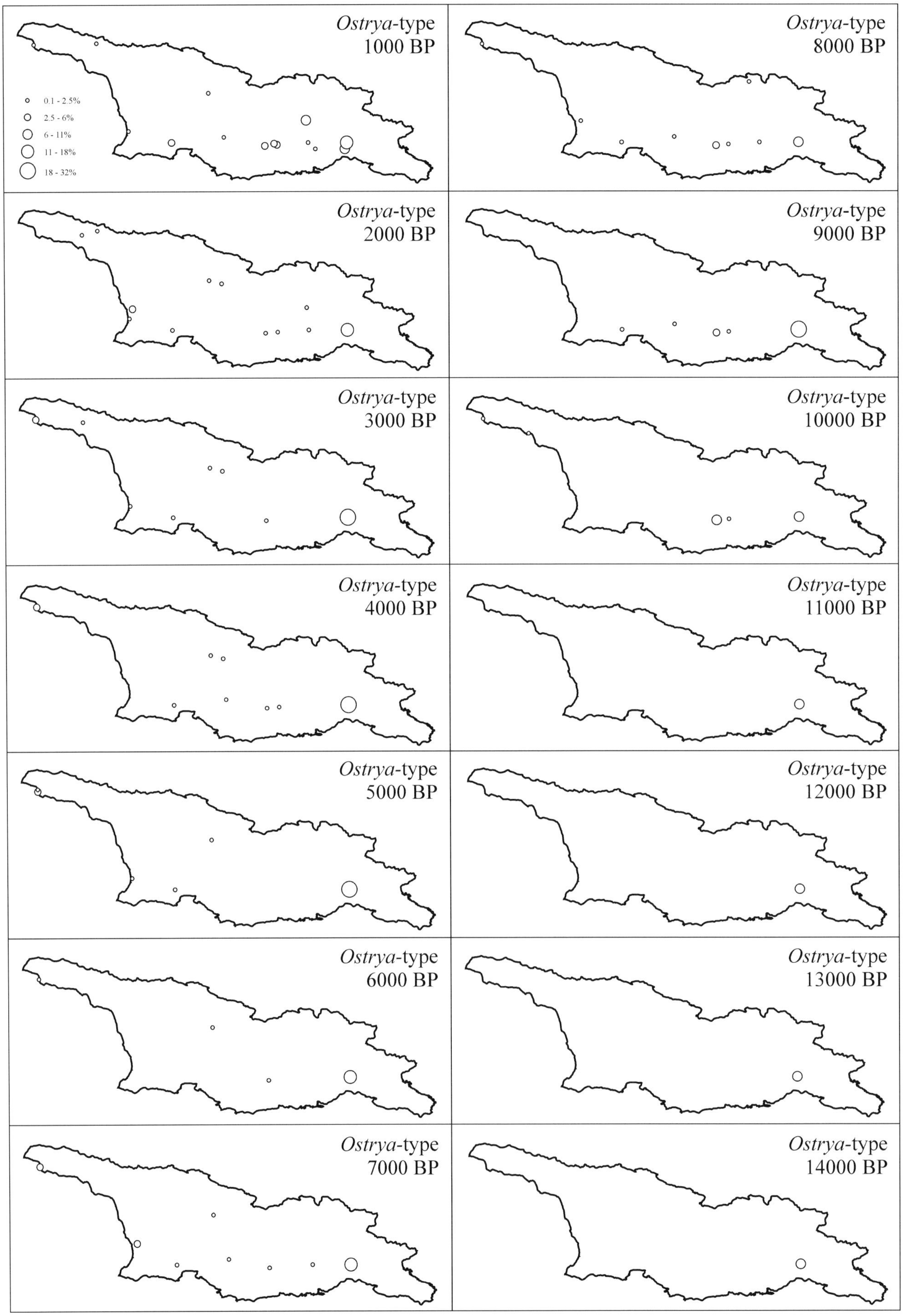
Ostrya-type
1000 BP
0.1 - 2.5%
2.5 - 6%
6 - 11%
11 - 18%
18 - 32%
Ostrya-type
2000 BP
Ostrya-type
3000 BP
Ostrya-type
4000 BP
Ostrya-type
5000 BP
Ostrya-type
6000 BP
Ostrya-type
7000 BP
Ostrya-type
8000 BP
Ostrya-type
9000 BP
Ostrya-type
10000 BP
Ostrya-type
11000 BP
Ostrya-type
12000 BP
Ostrya-type
13000 BP
Ostrya-type
14000 BP

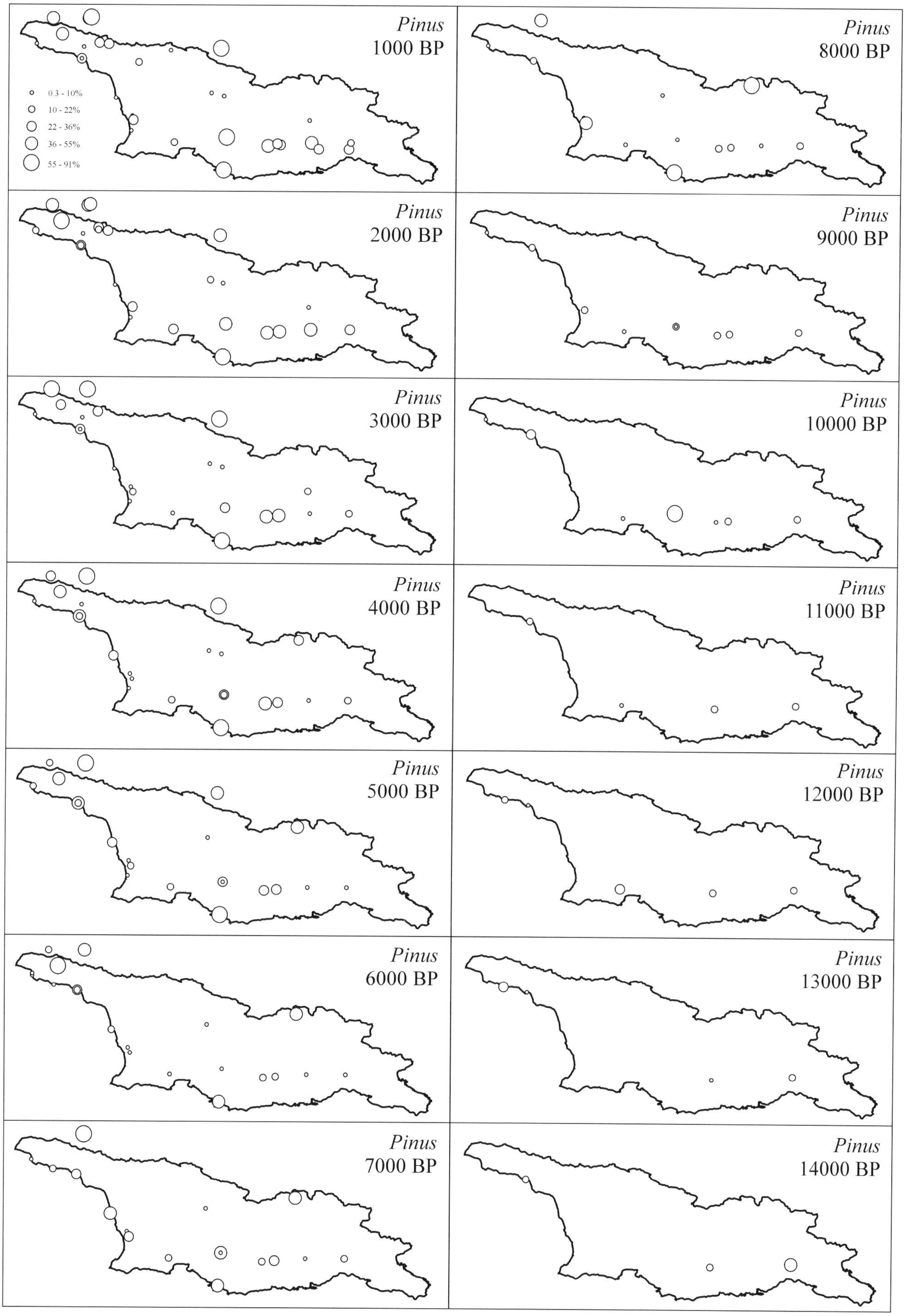
Pinus
1000 BP
0.3 - 10%
10 - 22%
22 - 36%
36 - 55%
55 - 91%
Pinus
8000 BP
Pinus
2000 BP
Pinus
9000 BP
Pinus
3000 BP
Pinus
10000 BP
Pinus
4000 BP
Pinus
11000 BP
Pinus
5000 BP
Pinus
12000 BP
Pinus
6000 BP
Pinus
13000 BP
Pinus
7000 BP
Pinus
14000 BP

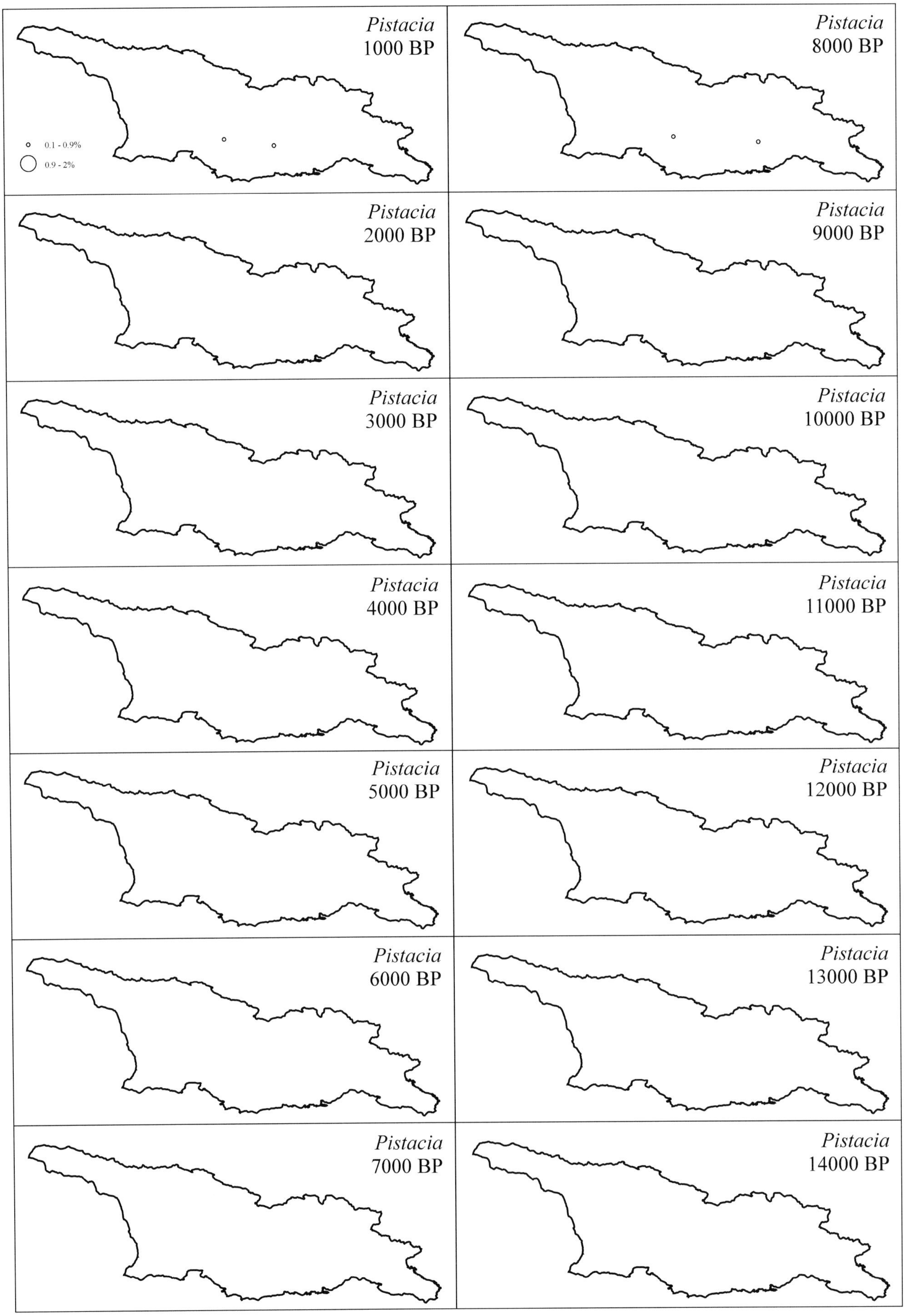
Pistacia
1000 BP
0.1 - 0.9%
0.9 - 2%
Pistacia
2000 BP
Pistacia
3000 BP
Pistacia
4000 BP
Pistacia
5000 BP
Pistacia
6000 BP
Pistacia
7000 BP
Pistacia
8000 BP
Pistacia
9000 BP
Pistacia
10000 BP
Pistacia
11000 BP
Pistacia
12000 BP
Pistacia
13000 BP
Pistacia
14000 BP

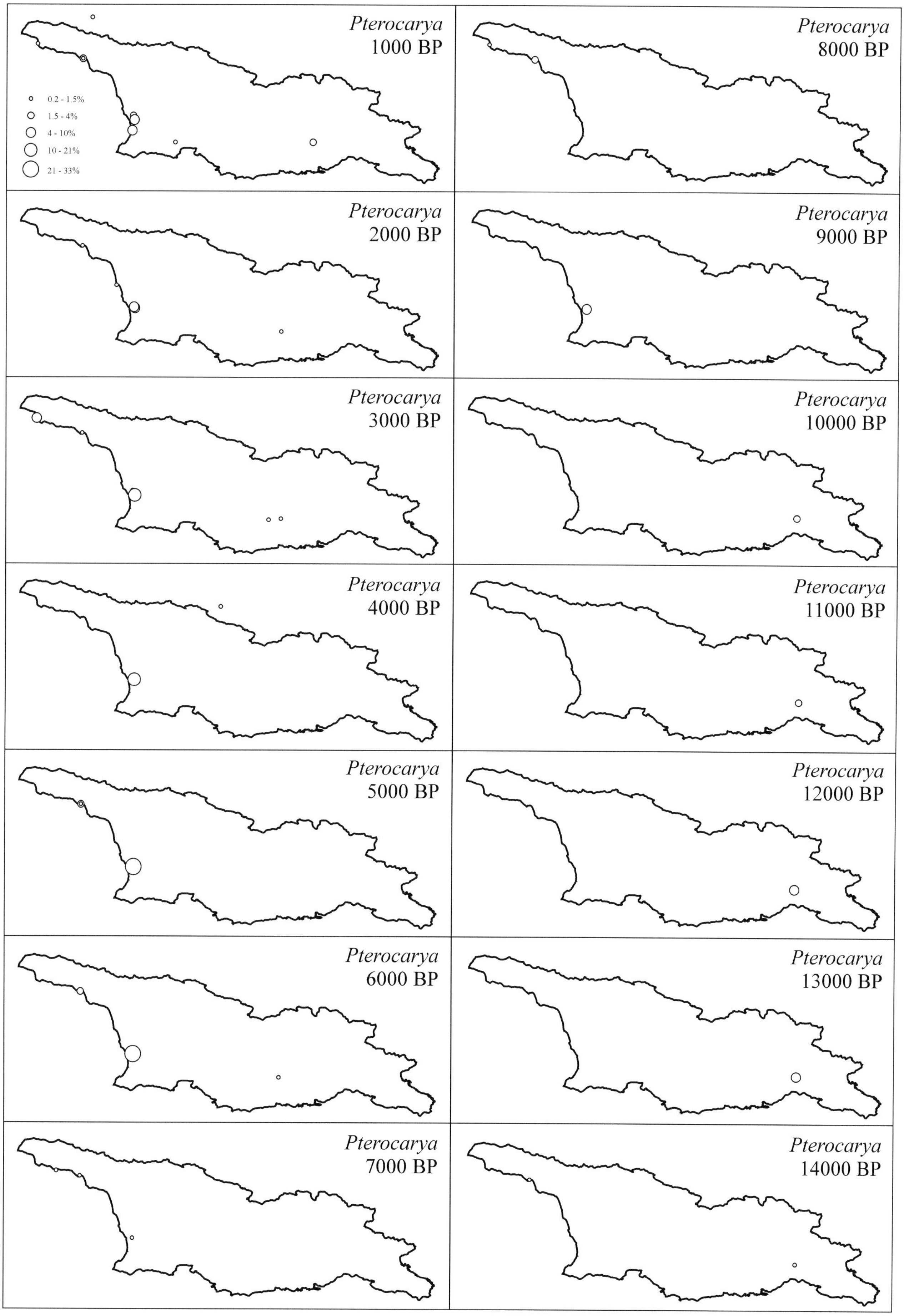
Pterocarya
1000 BP
0.2 - 1.5%
1.5 - 4%
4 - 10%
10 - 21%
21 - 33%
Pterocarya
8000 BP
Pterocarya
2000 BP
Pterocarya
9000 BP
Pterocarya
3000 BP
Pterocarya
10000 BP
Pterocarya
4000 BP
Pterocarya
11000 BP
Pterocarya
5000 BP
Pterocarya
12000 BP
Pterocarya
6000 BP
Pterocarya
13000 BP
Pterocarya
7000 BP
Pterocarya
14000 BP

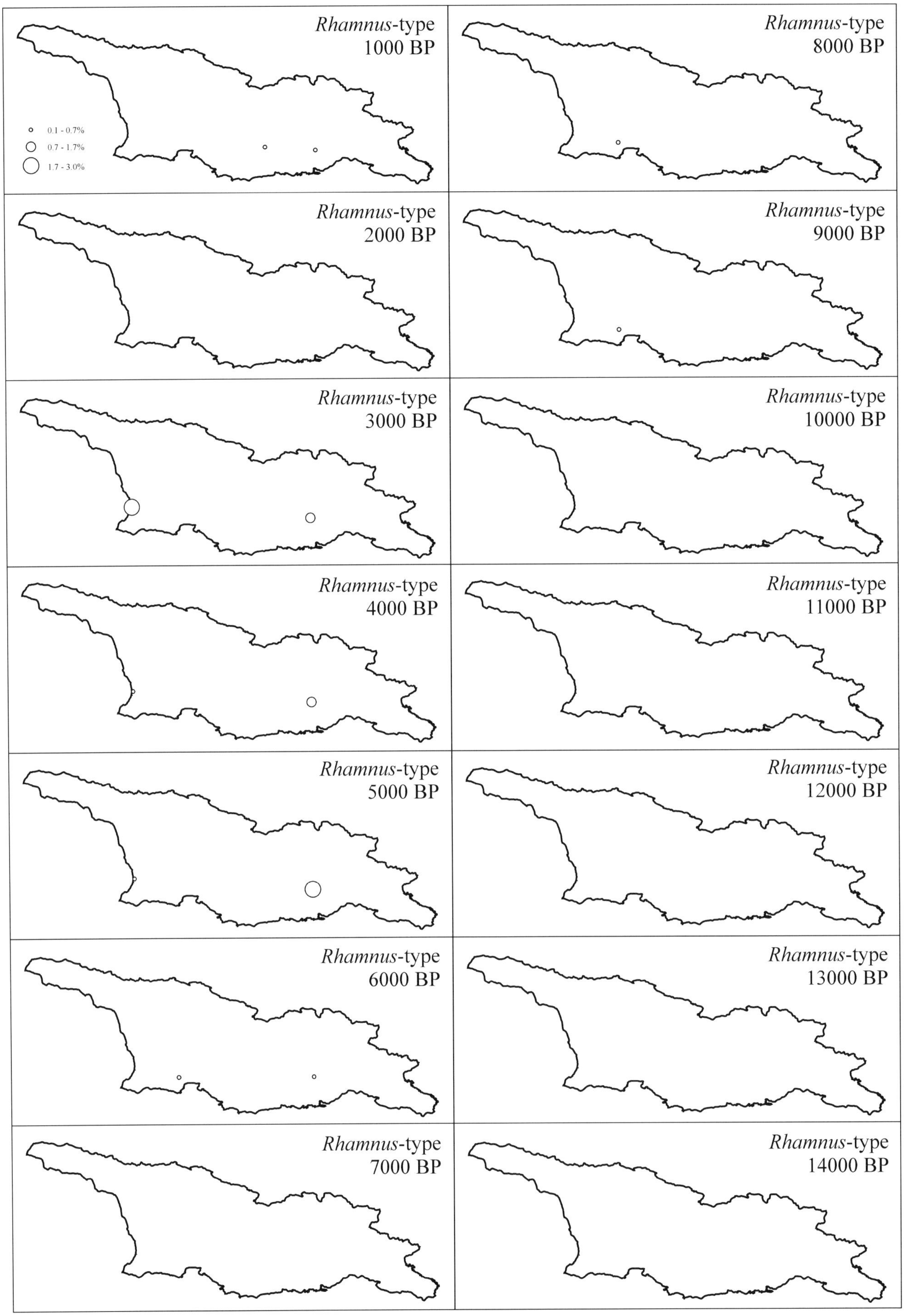
Rhamnus-type
1000 BP
0.1 - 0.7%
0.7 - 1.7%
1.7 - 3.0%
Rhamnus-type
2000 BP
Rhamnus-type
3000 BP
Rhamnus-type
4000 BP
Rhamnus-type
5000 BP
Rhamnus-type
6000 BP
Rhamnus-type
7000 BP
Rhamnus-type
8000 BP
Rhamnus-type
9000 BP
Rhamnus-type
10000 BP
Rhamnus-type
11000 BP
Rhamnus-type
12000 BP
Rhamnus-type
13000 BP
Rhamnus-type
14000 BP

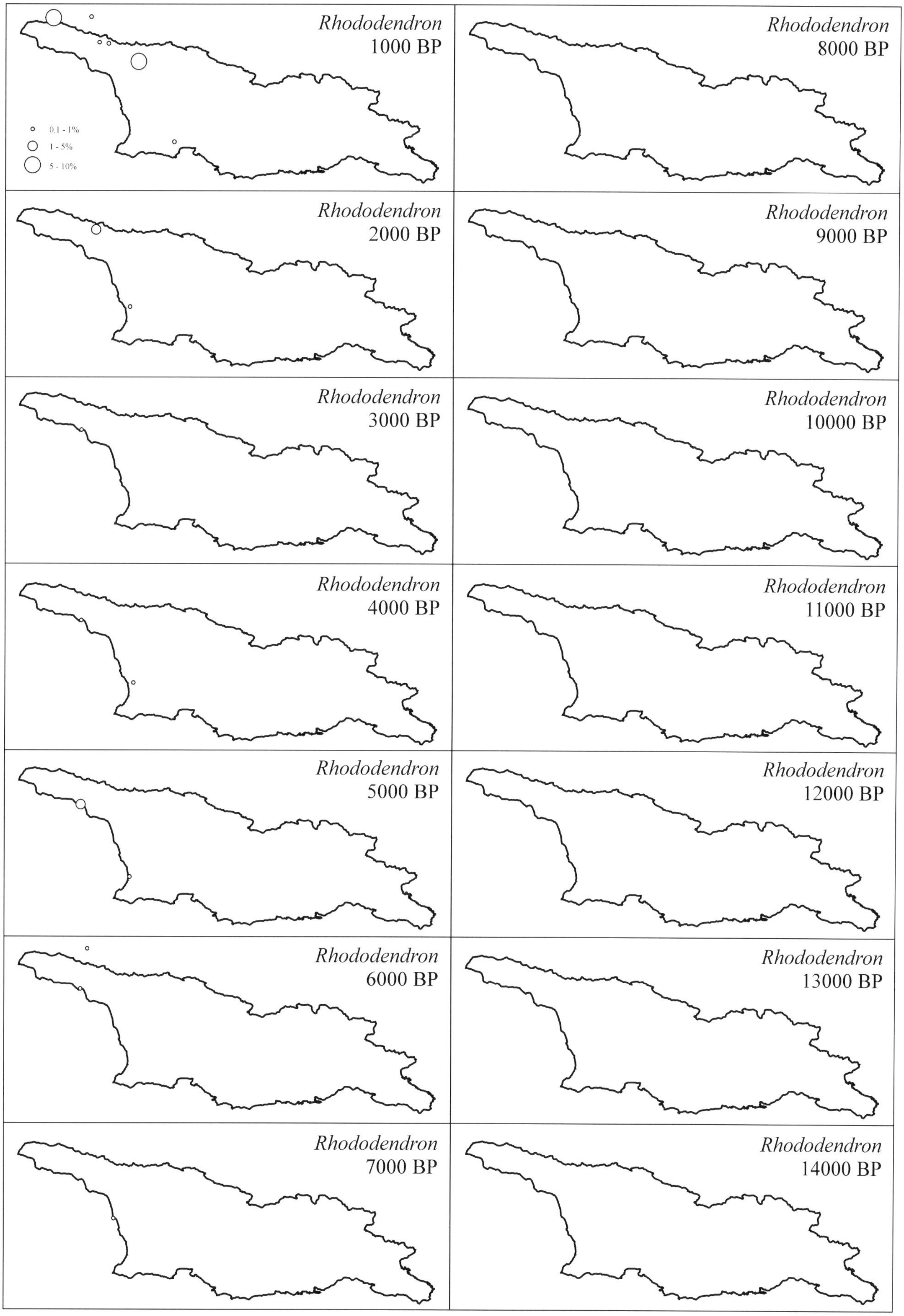
Rhododendron
1000 BP
0.1 - 1%
1 - 5%
5 - 10%
Rhododendron
8000 BP
Rhododendron
2000 BP
Rhododendron
9000 BP
Rhododendron
3000 BP
Rhododendron
10000 BP
Rhododendron
4000 BP
Rhododendron
11000 BP
Rhododendron
5000 BP
Rhododendron
12000 BP
Rhododendron
6000 BP
Rhododendron
13000 BP
Rhododendron
7000 BP
Rhododendron
14000 BP

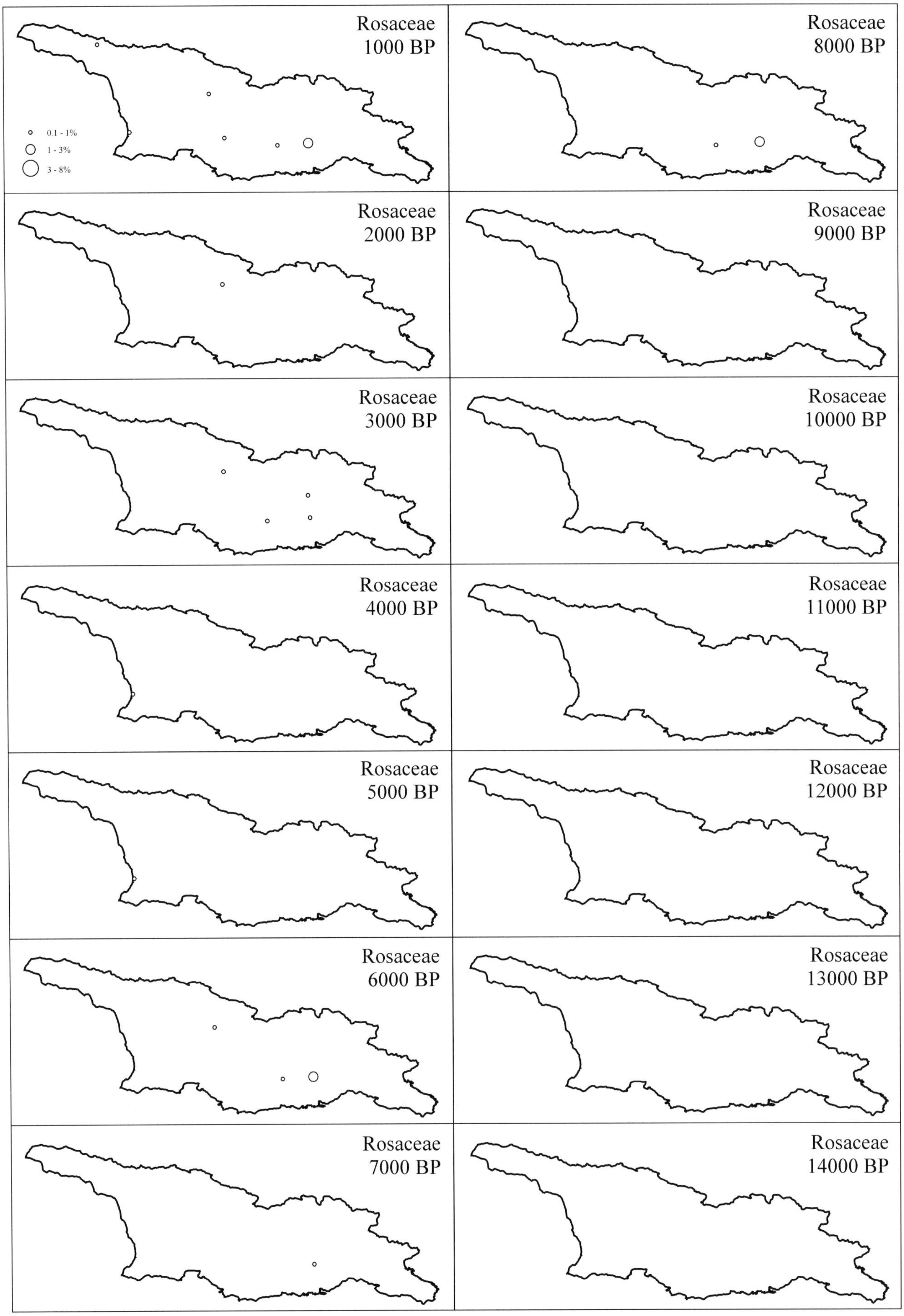
Rosaceae
1000 BP
0.1 - 1%
1 - 3%
3 - 8%
Rosaceae
2000 BP
Rosaceae
3000 BP
Rosaceae
4000 BP
Rosaceae
5000 BP
Rosaceae
6000 BP
Rosaceae
7000 BP
Rosaceae
8000 BP
Rosaceae
9000 BP
Rosaceae
10000 BP
Rosaceae
11000 BP
Rosaceae
12000 BP
Rosaceae
13000 BP
Rosaceae
14000 BP

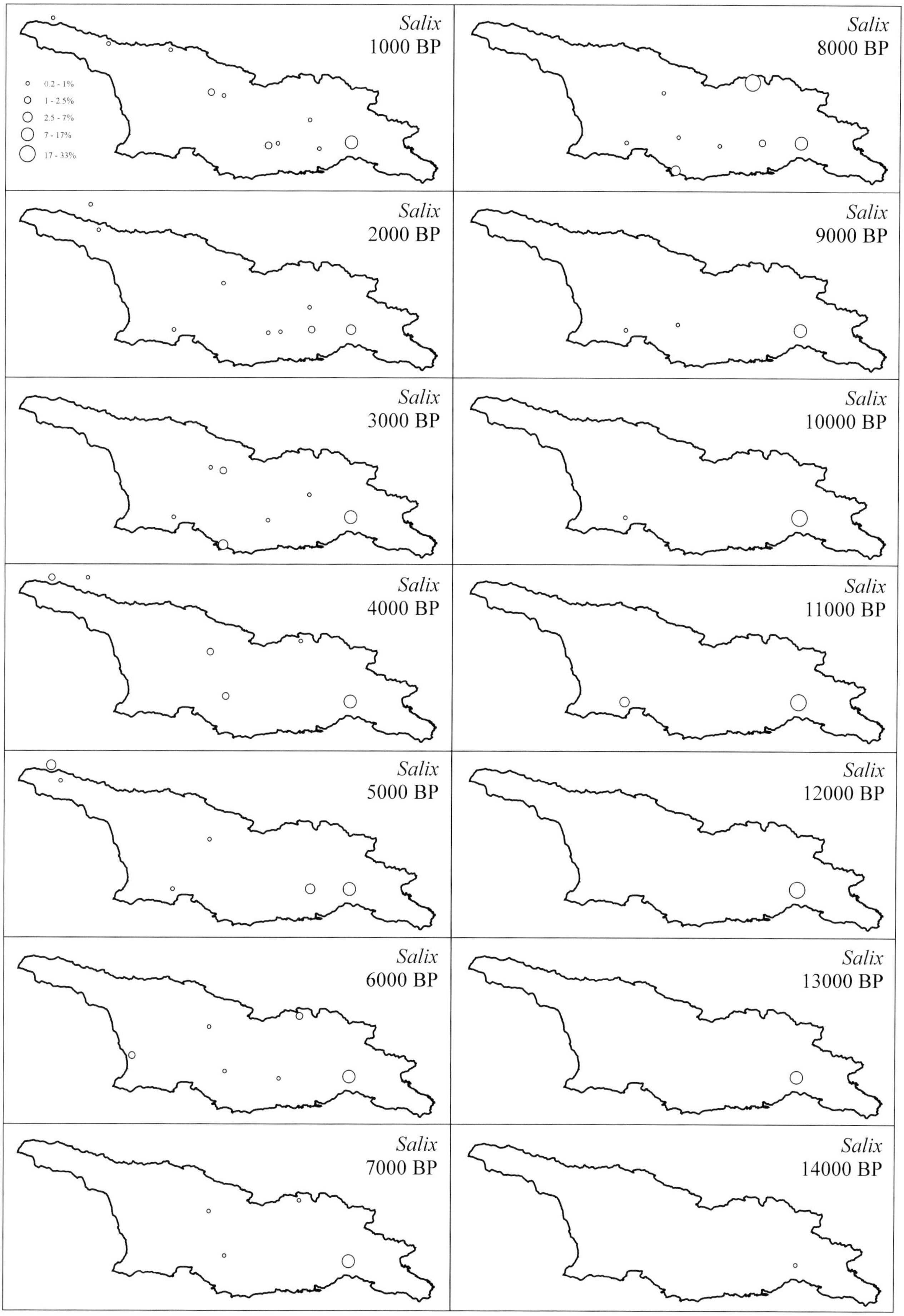
Salix
1000 BP
0.2 - 1%
1 - 2.5%
2.5 - 7%
7 - 17%
17 - 33%
Salix
8000 BP
Salix
2000 BP
Salix
9000 BP
Salix
3000 BP
Salix
10000 BP
Salix
4000 BP
Salix
11000 BP
Salix
5000 BP
Salix
12000 BP
Salix
6000 BP
Salix
13000 BP
Salix
7000 BP
Salix
14000 BP

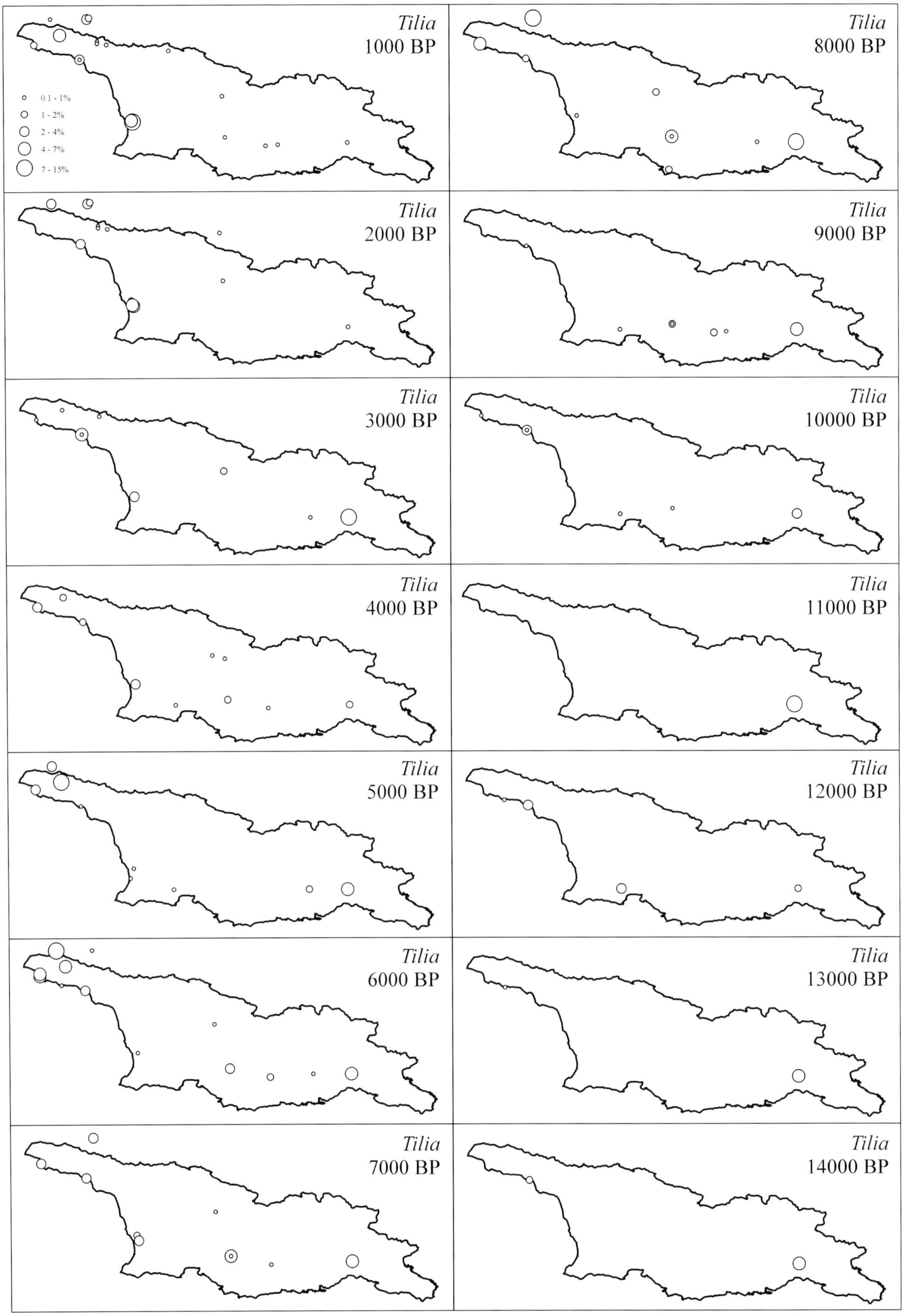
Tilia
1000 BP
0.1 - 1%
1 - 2%
2 - 4%
4 - 7%
7 - 15%
Tilia
8000 BP
Tilia
2000 BP
Tilia
9000 BP
Tilia
3000 BP
Tilia
10000 BP
Tilia
4000 BP
Tilia
11000 BP
Tilia
5000 BP
Tilia
12000 BP
Tilia
6000 BP
Tilia
13000 BP
Tilia
7000 BP
Tilia
14000 BP

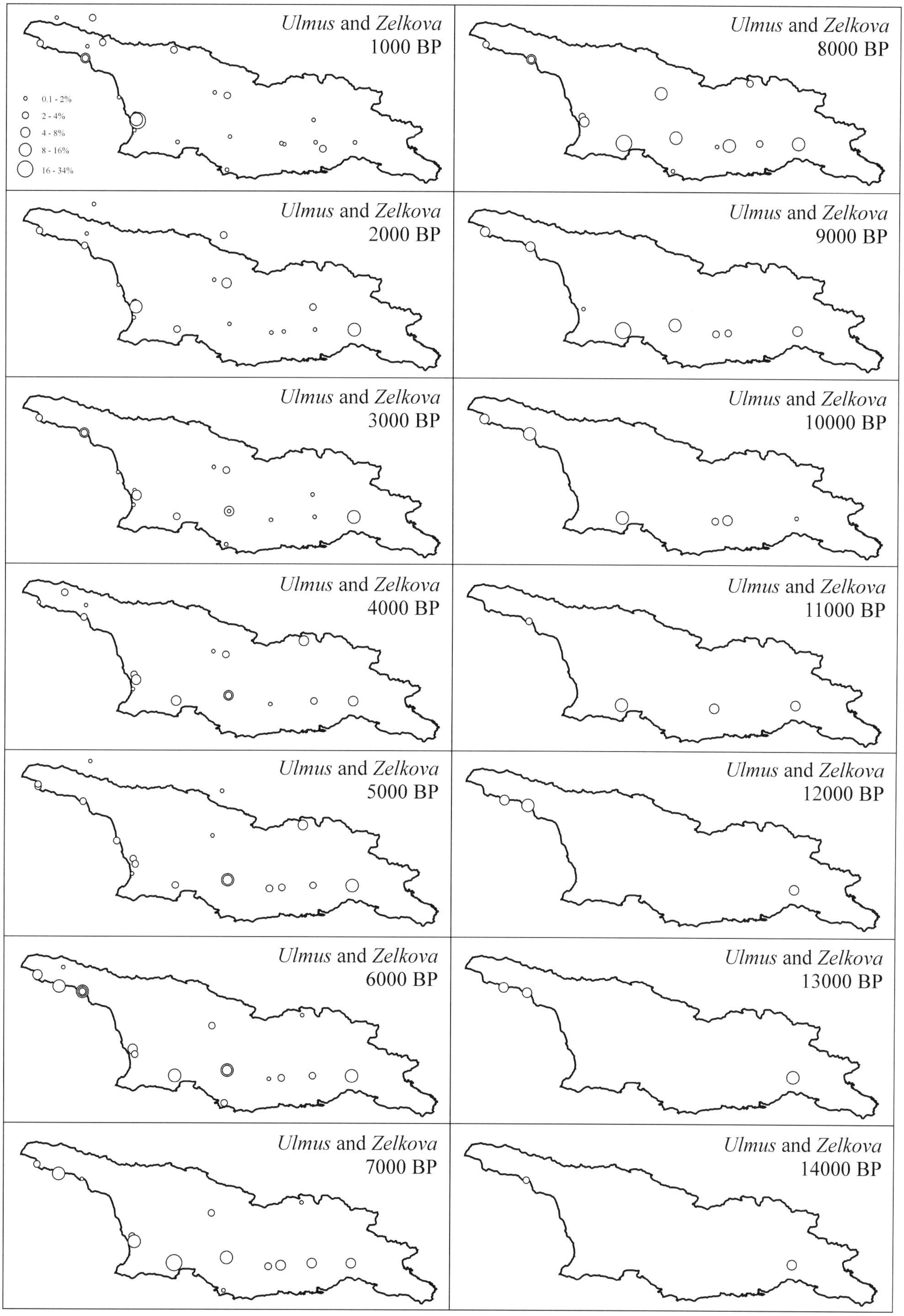
Ulmus and Zelkova
1000 BP
0.1 - 2%
2 - 4%
4 - 8%
8 - 16%
16 - 34%
Ulmus and Zelkova
2000 BP
Ulmus and Zelkova
3000 BP
Ulmus and Zelkova
4000 BP
Ulmus and Zelkova
5000 BP
Ulmus and Zelkova
6000 BP
Ulmus and Zelkova
7000 BP
Ulmus and Zelkova
8000 BP
Ulmus and Zelkova
9000 BP
Ulmus and Zelkova
10000 BP
Ulmus and Zelkova
11000 BP
Ulmus and Zelkova
12000 BP
Ulmus and Zelkova
13000 BP
Ulmus and Zelkova
14000 BP

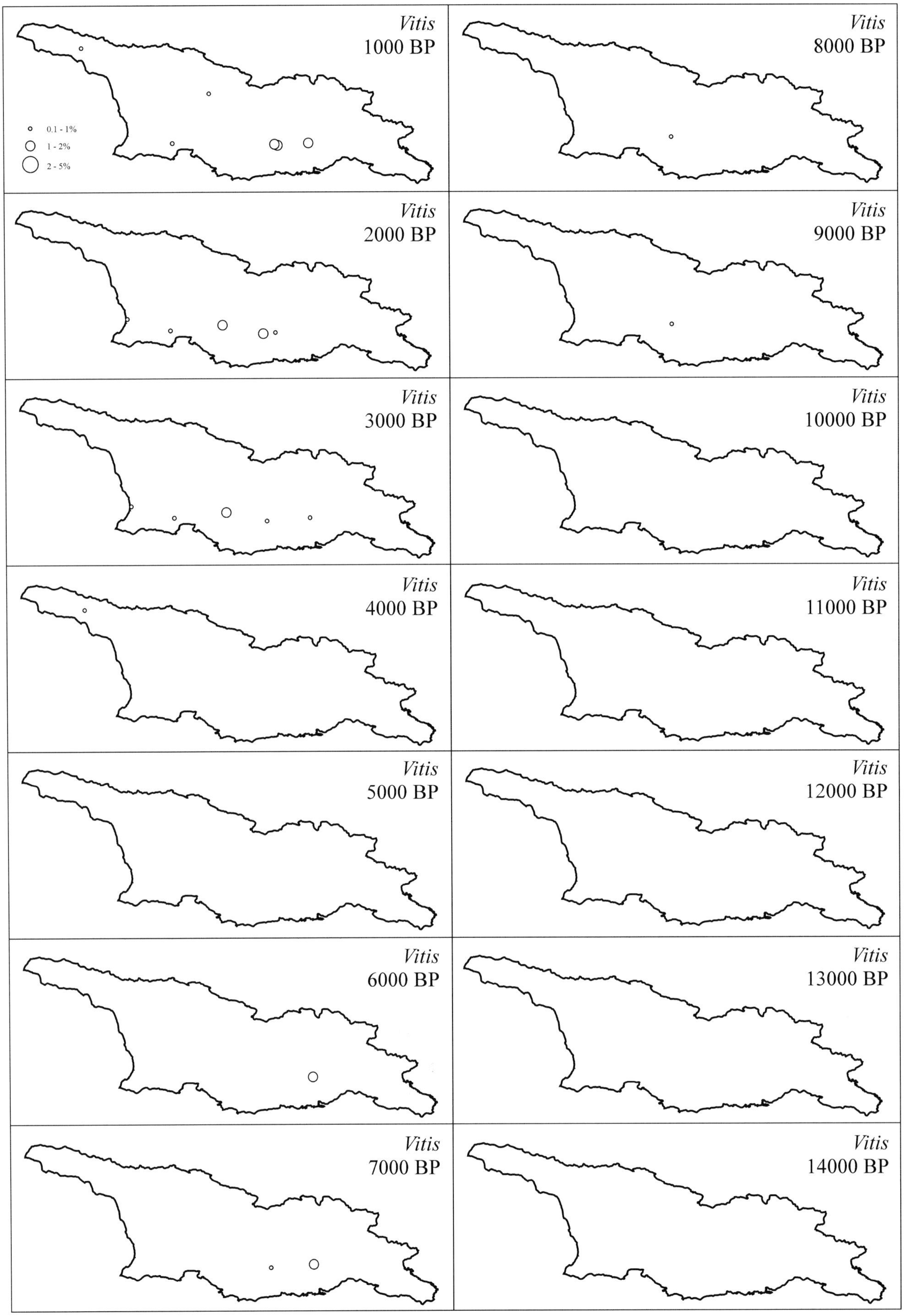

Vitis
1000 BP
0.1 - 1%
1 - 2%
2 - 5%
Vitis
2000 BP
Vitis
3000 BP
Vitis
4000 BP
Vitis
5000 BP
Vitis
6000 BP
Vitis
7000 BP
Vitis
8000 BP
Vitis
9000 BP
Vitis
10000 BP
Vitis
11000 BP
Vitis
12000 BP
Vitis
13000 BP
Vitis
14000 BP

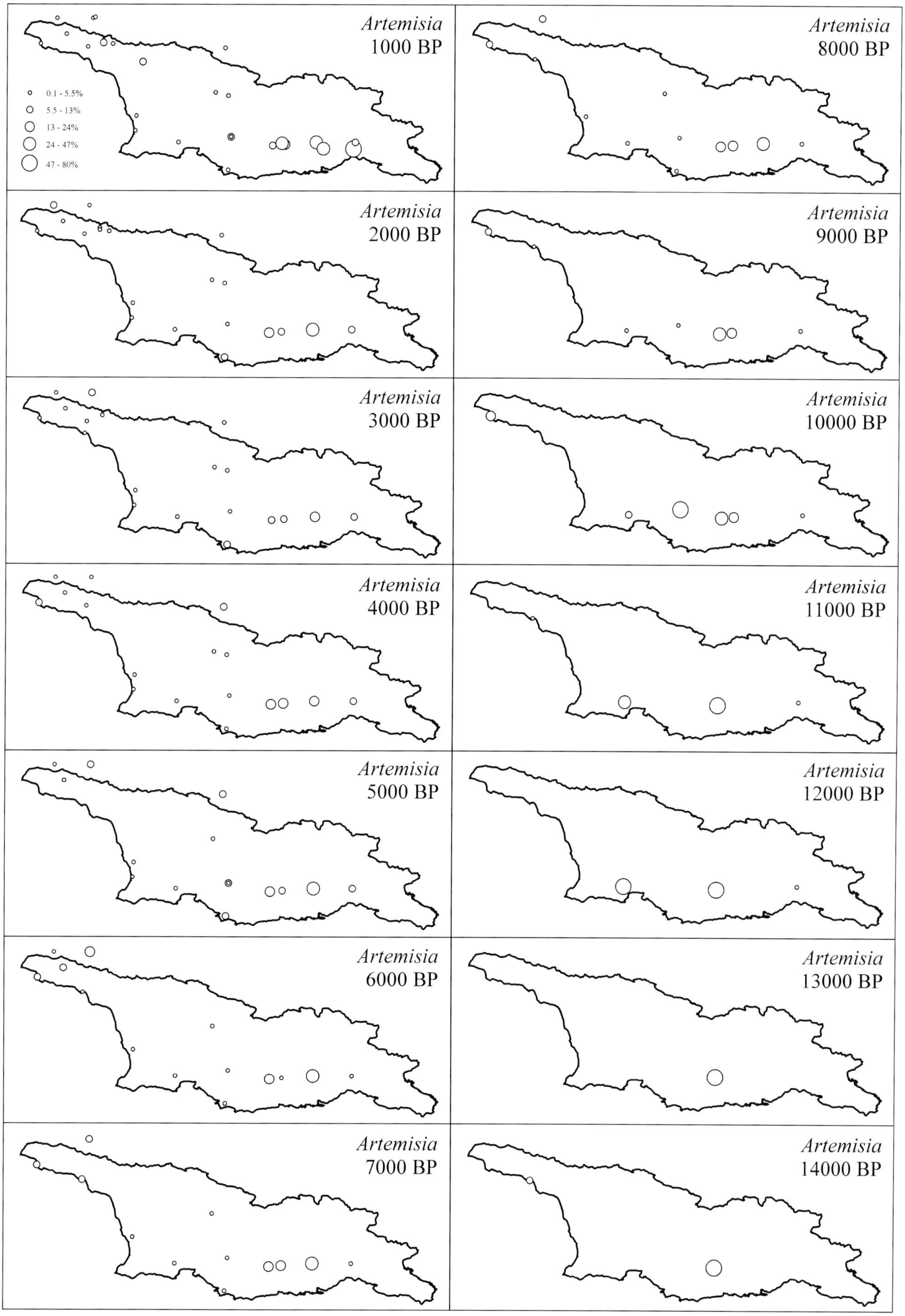
Artemisia 1000 BP
0.1 - 5.5%
5.5 - 13%
13 - 24%
24 - 47%
47 - 80%
Artemisia 8000 BP
Artemisia 2000 BP
Artemisia 9000 BP
Artemisia 3000 BP
Artemisia 10000 BP
Artemisia 4000 BP
Artemisia 11000 BP
Artemisia 5000 BP
Artemisia 12000 BP
Artemisia 6000 BP
Artemisia 13000 BP
Artemisia 7000 BP
Artemisia 14000 BP

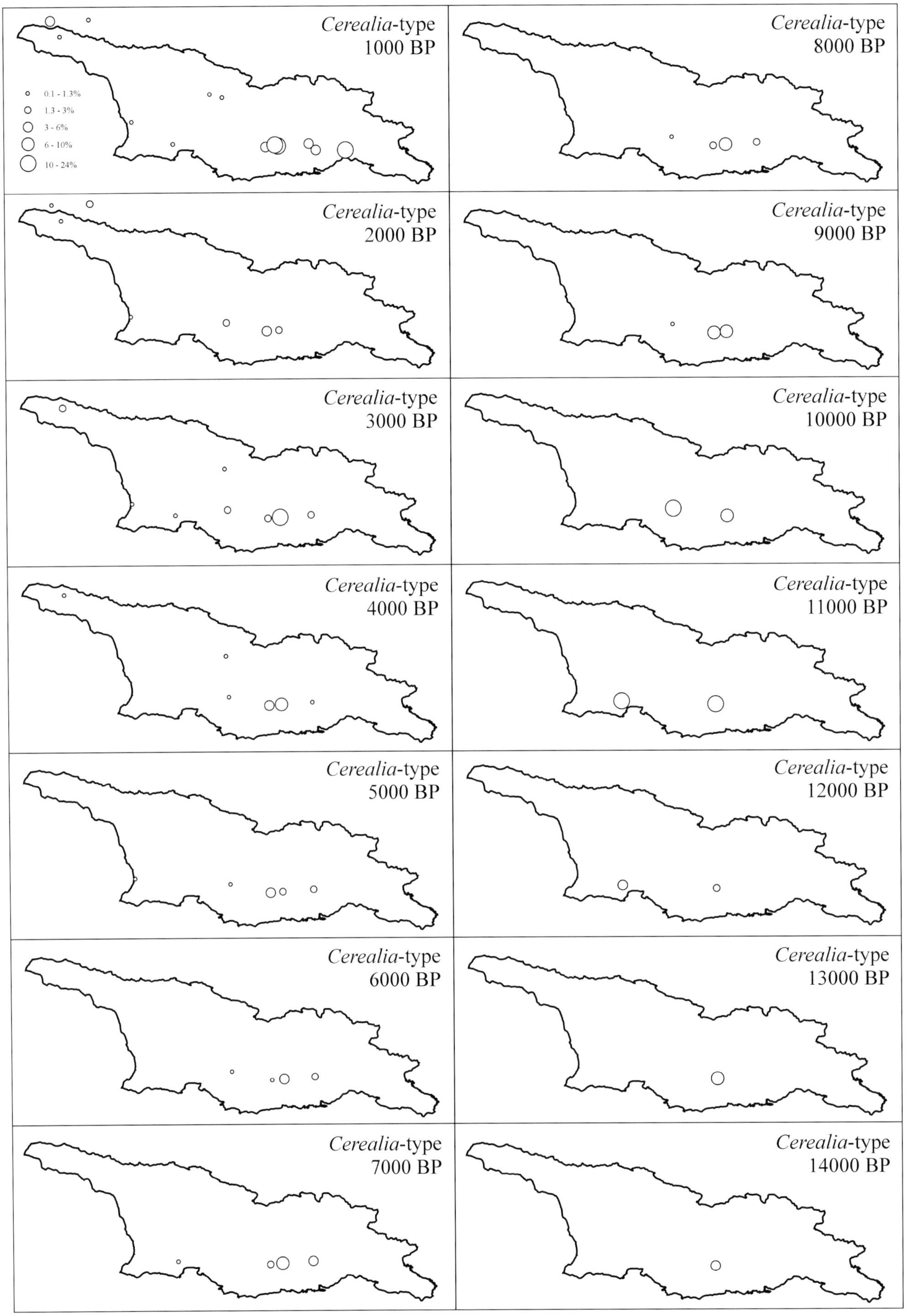
Cerealia-type 1000 BP
0.1 - 1.3%
1.3 - 3%
3 - 6%
6 - 10%
10 - 24%
Cerealia-type 8000 BP
Cerealia-type 2000 BP
Cerealia-type 9000 BP
Cerealia-type 3000 BP
Cerealia-type 10000 BP
Cerealia-type 4000 BP
Cerealia-type 11000 BP
Cerealia-type 5000 BP
Cerealia-type 12000 BP
Cerealia-type 6000 BP
Cerealia-type 13000 BP
Cerealia-type 7000 BP
Cerealia-type 14000 BP

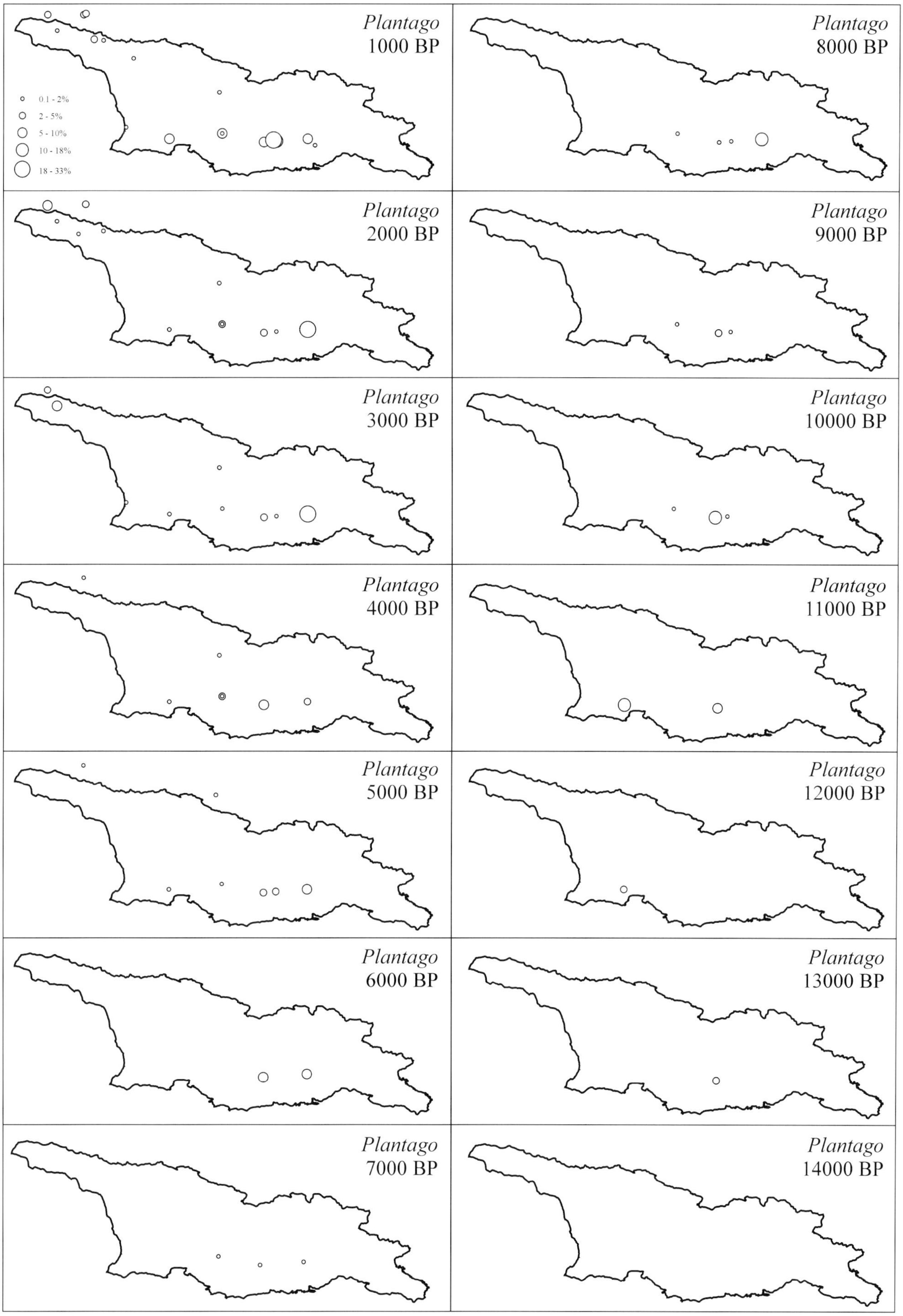

Plantago 1000 BP
0.1 - 2%
2 - 5%
5 - 10%
10 - 18%
18 - 33%
Plantago 8000 BP
Plantago 2000 BP
Plantago 9000 BP
Plantago 3000 BP
Plantago 10000 BP
Plantago 4000 BP
Plantago 11000 BP
Plantago 5000 BP
Plantago 12000 BP
Plantago 6000 BP
Plantago 13000 BP
Plantago 7000 BP
Plantago 14000 BP

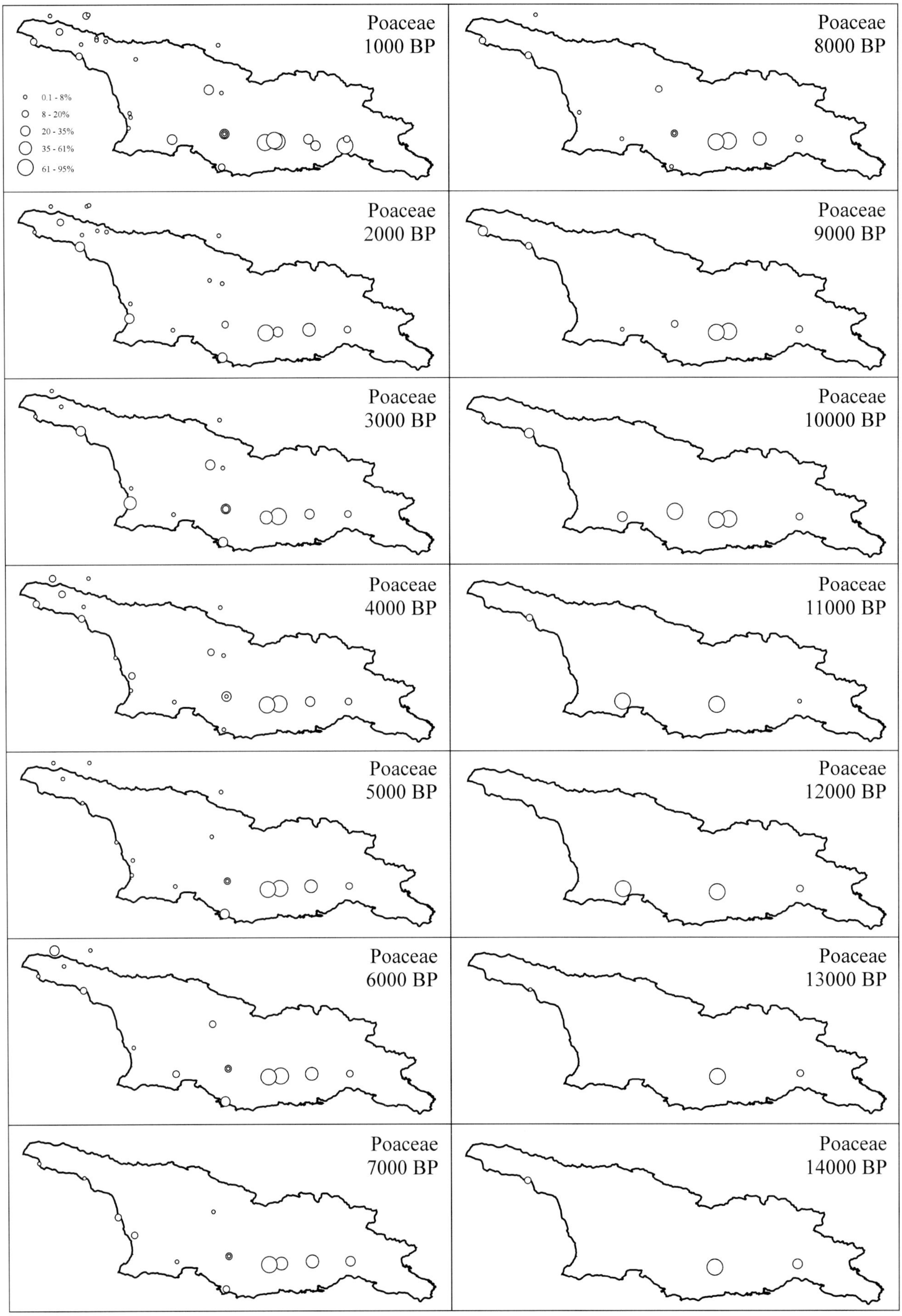
Poaceae 1000 BP
0.1 - 8%
8 - 20%
20 - 35%
35 - 61%
61 - 95%
Poaceae 8000 BP
Poaceae 2000 BP
Poaceae 9000 BP
Poaceae 3000 BP
Poaceae 10000 BP
Poaceae 4000 BP
Poaceae 11000 BP
Poaceae 5000 BP
Poaceae 12000 BP
Poaceae 6000 BP
Poaceae 13000 BP
Poaceae 7000 BP
Poaceae 14000 BP

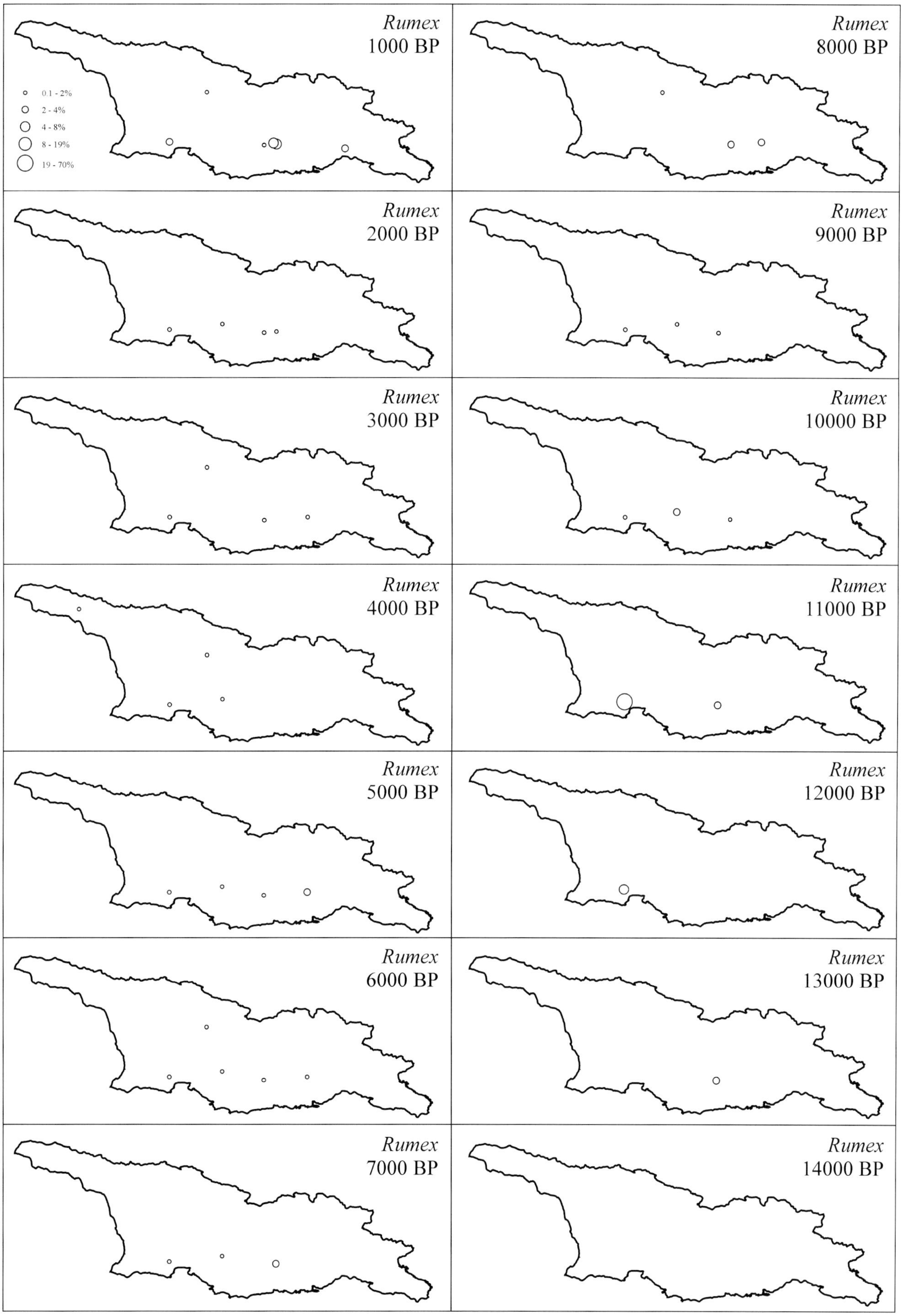
0.1 - 2%
2 - 4%
4 - 8%
8 - 19%
19 - 70%
Rumex 1000 BP
Rumex 2000 BP
Rumex 3000 BP
Rumex 4000 BP
Rumex 5000 BP
Rumex 6000 BP
Rumex 7000 BP
Rumex 8000 BP
Rumex 9000 BP
Rumex 10000 BP
Rumex 11000 BP
Rumex 12000 BP
Rumex 13000 BP
Rumex 14000 BP

APPENDIX B

CORRELATIONS WITH WEIGHTED AVERAGING TRENDS

This appendix presents the results of correlating palaeoenvironmental trends produced by weighted averaging (WA) with detrended correspondence analysis (DCA) ordination results from individual sites (using dryland pollen data from Appendix E). The aim of this exercise was to determine whether the dominant, site-specific trend (represented by DCA axis 1) was related to factors controlling pollen assemblages in the present landscape (here interpreted as 'precipitation' – WA axis 1; 'forest cover and winter temperature' – WA axis 2; and 'land-use' – WA axis 3).

By way of example, if the dominant trend in the Aligol data is related to precipitation, DCA axis 1 should have a close relationship with WA axis 1. Indeed this is the case, with a correlation of r^2=0.71. Where there is no relationship between the two sets of results, one may assume that a particular trend is related to other factors. Some assessment of these factors can be made by referring to the pollen types most correlated with the ordination axis.

The results are presented below, in alphabetical order by site. The figure following each DCA axis name is its eigenvalue, representing the amount of variance explained by that axis. The figure following each taxon or WA axis name is the correlation coefficient, with the direction of correlation indicated by a '+' or '–' sign.

ALIGOL

DCA axis 1 (0.25)	r^2 (±)	DCA axis 2 (0.07)	r^2 (±)	DCA axis 3 (0.03)	r^2 (±)
WA axis 1	0.71 +	WA axis 1	0.07 +	WA axis 1	0.06 –
WA axis 2	0.19 –	WA axis 2	0.07 –	WA axis 2	0.22 –
WA axis 3	0.37 –	WA axis 3	0.01 +	WA axis 3	0.36 –
Ephedra distachya	0.89 +	Poaceae (total)	0.73 –	*Pinus*	0.41 –
Artemisia	0.86 +	Chenopodiaceae	0.60 +	*Sinapis*-type	0.34 –
Anthemis-type	0.76 +	*Rumex*	0.57 +	*Campanula*-type	0.31 +
Indet. (cf. *Artemisia*)	0.75 +	*Sporormiella*	0.48 +	*Corylus*	0.29 +
Juniperus	0.65 +	*Urtica*	0.46 +	*Abies*	0.28 +
Degraded pollen	0.57 +	*Galium*-type	0.43 +	*Dipsacus*	0.25 +
Carpinus betulus	0.51 –	Cichoroiidae	0.39 +	*Fagus*	0.24 +

BARETI

DCA axis 1 (0.04)	r^2 (±)	DCA axis 2 (0.02)	r^2 (±)	DCA axis 3 (0.01)	r^2 (±)
WA axis 1	0.17 –	WA axis 1	0.05 –	WA axis 1	0.27 –
WA axis 2	0.63 +	WA axis 2	0.05 +	WA axis 2	0.09 –
WA axis 3	0.01 –	WA axis 3	0.02 –	WA axis 3	0.20 –
Alisma	0.76 +	*Galium*-type	0.59 +	*Secale*-type	0.61 –
Peucedanum-type	0.71 +	*Pediastrum*	0.47 –	*Trifolium pratense*	0.58 –
Cyperaceae	0.62 +	*Betula*	0.46 –	*Teucrium*-type	0.54 –
Pinus	0.60 –	Poaceae	0.44 +	*Onobrychis*-type	0.53 –
Typha-type	0.60 +	*Triticum*-type	0.43 –	*Sedum*-type	0.44 –
Artemisia	0.58 –	*Cannabis*-type	0.42 –	*Gentiana*	0.44 –
Oxyria-type	0.56 +	Chenopodiaceae	0.38 –	*Ammi majus*-type	0.44 –

IMERA

DCA axis 1 (0.16)	r^2 (±)	DCA axis 2 (0.05)	r^2 (±)	DCA axis 3 (0.02)	r^2 (±)
WA axis 1	0.32 –	WA axis 1	0.10 –	WA axis 1	0.13 +
WA axis 2	0.07 –	WA axis 2	0.06 +	WA axis 2	0.38 +
WA axis 3	0.04 +	WA axis 3	0.04 +	WA axis 3	0.03 +
Chenopodiaceae	0.79 +	*Campanula*-type	0.54 +	*Quercus*	0.46 +
Polygonum aviculare	0.73 –	*Silene*-type	0.50 –	*Pinus*	0.37 –
Cichoroiidae	0.68 –	*Zelkova*-type	0.48 +	*Alnus*	0.19 +
Plantago lanceolata	0.62 +	*Carpinus betulus*	0.47 +		
Rumex	0.62 +	*Arctium*-type	0.47 +		
Artemisia	0.62 +	*Thalictrum*	0.44 +		
Galium-type	0.57 +	*Pinus*	0.36 –		

JVARI

DCA axis 1 (0.09)	r^2 (±)	DCA axis 2 (0.03)	r^2 (±)	DCA axis 3 (0.02)	r^2 (±)
WA axis 1	0.71 –	WA axis 1	0.00 –	WA axis 1	0.12 +
WA axis 2	0.11 +	WA axis 2	0.38 –	WA axis 2	0.03 +
WA axis 3	0.01 –	WA axis 3	0.08 –	WA axis 3	0.07 –
AP	0.90 +	*Cirsium*-type	0.45 –	*Mercurialis*	0.51 +
NAP	0.85 –	*Artemisia*	0.38 +	*Scabiosa*	0.51 +
Pinus	0.64 +	*Centaurea nigra*	0.36 –	Chenopodiaceae	0.43 +
Cedrus	0.59 +	*Triticum*-type	0.35 –	Charcoal concentr.	0.40 +
Fraxinus	0.54 +	*Onobrychis*-type	0.32 –	*Carduus*-type	0.32 +
Gentianella	0.49 +	Cichoroiidae	0.31 –	*Fagus*	0.29 +
Quercus	0.45 +	*Hordeum*-type	0.30 –	Apiaceae	0.27 –

KUMISI

DCA axis 1 (0.07)	r^2 (±)	DCA axis 2 (0.02)	r^2 (±)	DCA axis 3 (0.01)	r^2 (±)
WA axis 1	0.93 –	WA axis 1	0.00 –	WA axis 1	0.01 +
WA axis 2	0.40 +	WA axis 2	0.03 +	WA axis 2	0.34 –
WA axis 3	0.88 –	WA axis 3	0.11 –	WA axis 3	0.01 –
AP	0.90 +	*Artemisia*	0.65 –	*Primula*	0.41 +
Chenopodiaceae	0.85 –	*Pinus*	0.62 +	*Ammi majus*-type	0.39 +
Humulus-type	0.67 +	*Carpinus betulus*	0.56 +	Cichoroiidae	0.38 –
Polygonum aviculare	0.55 –	*Quercus*	0.54 +	*Peucedanum*-type	0.36 –
Alnus	0.54 +	NAP	0.50 –	Fabaceae undiff.	0.32 +
Carpinus betulus	0.52 +	*Sporormiella*	0.47 –	*Falcaria*-type	0.30 –
Betula	0.51 +	*Trifolium repens*	0.47 –	*Convovulus*	0.29 +

SAKHARE

DCA axis 1 (0.04)	r^2 (±)	DCA axis 2 (0.01)	r^2 (±)	DCA axis 3 (0.01)	r^2 (±)
WA axis 1	0.10 +	WA axis 1	0.00 –	WA axis 1	0.02 +
WA axis 2	0.00 +	WA axis 2	0.09 +	WA axis 2	0.16 +
WA axis 3	0.47 –	WA axis 3	0.01 +	WA axis 3	0.02 +
Polygonum indet.	0.84 –	*Triticum*-type	0.54 +	Rosaceae undiff.	0.39 –
NAP	0.59 –	*Artemisia*	0.51 –	*Plumbago*	0.38 +
Chenopodiaceae	0.58 +	Poaceae	0.50 +	*Rumex*	0.37 –
Sinapis-type	0.48 +	*Anthemis*-type	0.37 –	*Hordeum*-type	0.28 –
Apiaceae undiff.	0.48 +	*Cirsium*-type	0.36 +	Poaceae	0.27 +
Fabaceae undiff.	0.37 +	*Fagopyrum*	0.35 +	Primulaceae	0.27 –
Astragalus-type	0.33 +	*Ostrya*-type	0.31 +	Charcoal concentr.	0.24 +

TSAVKIS

DCA axis 1 (0.14)	r^2 (±)	DCA axis 2 (0.07)	r^2 (±)	DCA axis 3 (0.04)	r^2 (±)
WA axis 1	0.15 –	WA axis 1	0.61 +	WA axis 1	0.06 –
WA axis 2	0.14 +	WA axis 2	0.72 –	WA axis 2	0.06 +
WA axis 3	0.81 –	WA axis 3	0.06 +	WA axis 3	0.01 +
AP	0.78 +	*Quercus*	0.59 –	Poaceae	0.67 –
Carpinus betulus	0.59 +	NAP	0.44 +	*Plantago lanceolata*	0.37 +
Ostrya-type	0.56 +	*Fraxinus*	0.42 –	*Juniperus*	0.33 +
Sporormiella	0.40 –	Degraded pollen	0.42 +	*Scleranthus*-type	0.23 +
Plantago coronopus	0.39 –	Chenopodiaceae	0.41 +		
Potentilla-type	0.38 –	*Pinus*	0.31 +		
Alnus	0.34 +	Polypodiaceae	0.30 +		

APPENDIX C

POLLEN PREPARATION METHOD

This section describes the method used to isolate pollen from sediment as adopted in this study. The aim of this method is to remove as much of the inorganic and organic contaminants from the sediment as possible, whilst retaining pollen, charcoal and other microfossils of interest. In most respects my method follows the standard method involving alkali treatment to remove humic material, strong acid digestion to remove carbonates and silicates, and acetolysis to remove dispersed organics (Moore et al., 1991). I have incorporated a number of small modifications to the standard method to account for the diversity of sedimentary environments in the study area and to overcome hazards encountered with the use of hydrofluoric acid in some samples.

As a first step, 1 cm^3 of sediment is added to a 50 ml centrifuge tube and suspended in 10% HCl solution. Two *Lycopodium* marker spore tablets (University of Lund) are also added at this stage and thoroughly combined with the suspended sediment. The tube is then centrifuged at 3000 rpm for 3 minutes and the solution is discarded. If abundant carbonates are present, the hydrochloric acid step may need to be repeated, otherwise the sample is resuspended in distilled water, centrifuged, decanted and repeated as necessary.

If the sample is peaty and organic, the tube is filled with hot 10% KOH solution and placed in a water bath to remove humic material. If the sample is predominantly silty, then the tube is filled with Calgon (sodium hexametaphosphate: $Na_2P_2O_7$), the sample resuspended and left overnight to deflocculate the compact sediments. A combination of these two treatments may be required for some samples. Samples are then sieved through a 150-micron mesh to remove large particles and transferred to 10 ml centrifuge tubes during the washing and centrifuging process. The empty 50 ml centrifuge tubes may be washed at this stage.

Next, the samples are gravity separated using sodium polytungstate ($3Na_3WO_4.9WO_3.H_2O$). The specific gravity of this 'heavy liquid' should be between 2.00 and 2.02, as measured with a hygrometer. If not, distilled water is added to decrease or the solution evaporated at 50 °C to increase specific gravity. The heavy liquid is added to the sample and mixed thoroughly. The tube is then centrifuged at low speed (2000–2500 rpm) for 20 minutes or more. The solution and any floating material is returned to 50 ml centrifuge tubes and diluted in distilled water. Material remaining at the bottom of the 10 ml tube can be checked for pollen content on a microscope slide and discarded if safe. The 10 ml tubes are then washed.

The 50 ml tubes containing the suspended pollen and diluted heavy liquid are centrifuged at high speed (3000 rpm) for 10 minutes or until a pellet forms at the bottom of the tube. The liquid can then be decanted for filtering and re-use, while the pellet should be transferred back into 10 ml tubes and washed repeatedly with distilled water in preparation for the acetolysis step. If substantial amounts of fine silica remain at this stage, 3 ml of 50% HF may be carefully added to the sample and heated, stirring with a rod of HF-resistant polymer or copper. Once silica has been digested, the tube is balanced with 10% HCl and centrifuged under the fume hood. The solution is decanted into an acid waste container for neutralisation.

Following washing in distilled water, the sample pellet is resuspended in 30% acetic acid, centrifuged and decanted, followed by the addition of glacial acetic acid (100%). Once this has been thoroughly mixed, centrifuged and decanted, 8 ml of acetic anhydride is added to the sample from a dry measuring cylinder. This is followed by the addition of 1 ml of concentrated sulphuric acid to produce the acetolysis reaction. The sample tube is placed in a heating block and stirred continuously with a glass rod for 1 minute (or longer if the reaction continues). The tube is then removed from heat; the reaction halted with the addition of a few drops of glacial acetic acid; centrifuged and decanted. Next, the sample is resuspended in decreasing concentrations of acetic acid: firstly 100%, then 30%, and finally distilled water. Following each step the tube is centrifuged and the acid solution decanted.

To neutralise any remaining acid and assist staining, a few drops of 10% KOH are added to the sample tube, then 2 ml of distilled water and 5 drops of safranin stain. The sample is thoroughly mixed to ensure complete staining. Distilled water is then added to balance the tube in the centrifuge. Once the stain solution is decanted, the sample is resuspended in a 30% solution of ethanol and distilled water, and transferred to 5 ml sample storage vials. The alcohol solution is mixed with the sample, centrifuged and decanted. Then 100% ethanol is added, centrifuged and decanted, ensuring that no liquid remains in the vial. The alcohol treatment removes water from the sample and greatly increases the pollen's shelf life. Next, 0.5 ml of glycerol is measured using a syringe and combined completely with the sample.

Finally, a labelled glass microscope slide is placed on a heating block at 50 °C, along with a small quantity of sealing wax in a separate Petri dish. A 15-μl drop of sample is placed on the slide and any residual alcohol allowed to evaporate. Each edge of a #0 cover-slip is dipped into the hot wax and placed over the sample drop on the slide. The wax should melt evenly to seal in the sample completely.

APPENDIX D

RADIOCARBON DATING

Radiocarbon dating is based on the radioactive isotope of carbon, namely ^{14}C. This isotope is formed by the interaction of cosmic rays and nitrogen in the earth's atmosphere and decays at a more-or-less uniform rate with a half-life of 5,730 years. Organisms absorb this radioactive ^{14}C during their lifetime; for plants this occurs as carbon dioxide is taken up through photosynthesis. When the organism dies, ^{14}C begins to decay, and so the abundance of ^{14}C compared to other, more stable carbon isotopes is a measure of the time since the death of the organism. However, the radiocarbon age of any given sample may be rendered erroneously old or young by contamination. The aim of pre-treatment methods is to minimise the impact of this contamination by older or younger material.

The most usual pre-treatment for organic sediments is the ABA (acid-base-acid) method.[1] Any visible contamination is removed (roots *etc.*) and the sample is washed in distilled, deionised water and dried at 60 °C. The dried sample is then ground to a powder using a mortar and pestle, then placed in a 50 ml centrifuge tube, resuspended in 2M HCl and placed in a water bath at 60 °C for 2 hours. This removes carbonate contamination and some infiltrated fulvic acids. The tube is then centrifuged and the liquid decanted; the sample is then resuspended in distilled water and centrifuged, and this process is repeated until the pH returns to 4. The alkali step removes contamination by humic and fulvic acids. The sample is suspended in 0.5–1.0% NaOH and placed in the water bath for up to 2 hours. The tube is again centrifuged and the alkali step repeated at the same or higher concentration until the solution is pale yellow in colour. After washing with distilled water, the sample is given a final acid treatment to remove atmospheric carbon absorbed during the alkali treatment. Once the sample has been washed in distilled water and dried overnight at 60 °C, it can proceed to the graphitisation step, which converts the carbon in the sample to carbon dioxide and thence to a 'target' of pure graphite, and it is this target that is placed in the accelerator mass spectrometer (AMS) for age determination.

Another pre-treatment technique employed in this study, albeit with limited success, is the isolation of pollen from sediment.[2] The standard technique for isolating pollen from sediments for pollen analysis is unsuitable for dating purposes because the use of acetic acid and acetic anhydride may introduce extraneous carbon to the sample. The method is, however, very similar. Sediments are deflocculated in Calgon, followed by sieving, treatment with 10% HCl, 10% KOH, heavy liquid separation and concentrated HF if required. The sample is then bleached using a 2% solution of NaOCl, allowed to stand for 5 minutes, centrifuged and decanted. The bleaching step is repeated as required. Next, the sample is thoroughly washed in distilled, deionised water, centrifuged, and the sample resuspended in concentrated sulphuric acid and left for several hours. After the acid step, the sample is again washed and the tube refilled with 10% NaOH to remove dissolved humic material. Finally, any remaining alkali is neutralised by the addition of 5 ml 2M HCl solution and the sample washed before being transferred to the drying oven for the preparation of the graphite target. This process is described in detail in Hua *et al.* (2001).

[1] The ABA method described here is that used at the Australian Nuclear Science and Technology Organisation (ANSTO): Environment Division Instruction Manual PMC-I-09-01-003.

[2] ANSTO Environment Division Instruction Manual PMC-I-09-01-004.

The results of these age determinations are presented in the chapters concerned, with particular emphasis on calibrating radiocarbon ages to calendar ages based on the tree-ring chronology. Table D.1 gives some additional details for the samples, including laboratory codes, sample types and $\delta^{13}C$ results.

Table D.1. Details of radiocarbon age determinations presented in this study.

N°	Site, core and sample	Lab. code	Sample type	^{14}C age (yr bp) and 1σ error	Calibrated age range (yr BP)	$\delta^{13}C$ ‰
1	Aligol 39 cm	OZH-397	G.S.O.	190 ± 40	228–135	–30.2
2	Aligol 71 cm	OZH-396	G.S.O.	930 ± 40	928–759	–32.4
3	Aligol 111 cm	OZH-395	G.S.O.	1510 ± 40	1423–1326	–29.8
4	Aligol 147 cm	OZH-394	G.S.O.	2700 ± 40	2866–2750	–29.5
5	Aligol 165 cm	OZH-393	G.S.O.	4030 ± 50	4647–4406	–27.8
6	Aligol 203 cm	OZH-392	G.S.O.	4660 ± 50	5485–5299	–26.8
7	Aligol 219 cm	OZH-391	G.S.O.	4450 ± 40	5089–4957	–27.4
8	Aligol 252 cm	OZH-390	G.S.O.	9540 ± 160	11224–10467	N/A
9	Aligol 285 cm	OZH-389	G.S.O.	10250 ± 90	12387–11552	–25.5
10	Aligol 310 cm	OZH-388	G.S.O.	12430 ± 90	15420–14134	–21.1
11	Bareti 123 cm	OZH-070	G.S.O.	900 ± 40	916–736	–28.0
12	Imera 54 cm	OZG-619	Bulk	990 ± 40	964–880	–27.4
13	Imera 115 cm	OZG-642	Bulk	1600 ± 40	1566–1389	–29.6
14	Imera 146 cm	OZH-067	G.S.O.	2360 ± 40	2488–2320	–28.9
15	Imera 198 cm	OZG-623	Bulk	4030 ± 110	4830–4234	N/A
16	Imera 230 cm	OZH-399	G.S.O.	4290 ± 50	4976–4812	–27.1
17	Imera 271 cm	OZH-398	G.S.O.	6590 ± 60	7582–7417	–25.8
18	Ispani 525 cm	Beta-209353	Macro	1940 ± 40	1991–1812	–28.7
19	Ispani 700 cm	Beta-209354	Macro	4060 ± 40	4641–4421	–28.7
20	Ispani 950 cm	OZH-068	G.S.O.	4900 ± 50	5736–5582	–23.3
21	Kumisi 246 cm	OZG-616	Bulk	1180 ± 40	1179–977	–20.4
22	Sakhare 87 cm	OZG-614	Bulk	1450 ± 80	1525–1237	N/A
23	Tsavkisi-2002 100 cm	OZG-617	Bulk	1880 ± 40	1892–1715	–26.4
24	Tsavkisi-2003 80 cm	OZH-867	Pollen	1370 ± 70	1412–1139	–31.3
25	Tsavkisi-2003 115 cm	OZH-866	Pollen	2140 ± 50	2209–1993	–25.7
26	Tsavkisi-2003 175 cm	OZH-865	Pollen	3225 ± 50	3564–3355	–25.9
27	Tsavkisi-2003 305 cm	OZH-864	Pollen	5510 ± 60	6412–6183	–25.3
28	Tsavkisi-2003 435 cm	OZH-863	Pollen	5040 ± 60	5908–5658	–26.2
29	Tsavkisi-2003 625 cm	OZH-862	Pollen	6890 ± 60	7836–7610	–27.9
30	Tsavkisi-2003 780 cm	OZH-069	G.S.O.	5340 ± 60	6209–5989	–25.3
31	Tsavkisi-2005 1055 cm	Wk-18593	Bulk	5003 ± 31	5757–5653	–32.0
32	Tsavkisi-2005 1065 cm	Beta-209355	Bulk*	4890 ± 40	5712–5584	–28.5

Notes: Calibrated age ranges are given following Method B of Stuiver and Reimer (1993), using the age range with the highest statistical probability at two standard deviations. Key to sample types: 'G.S.O.' = gravity-separated organics; 'Bulk' = bulk organics; 'Macro' = macroscopic plant remains; 'Pollen' = pollen fraction. Samples were pretreated according to the standard acid-base-acid technique except for pollen fraction samples and the acid-washed sample denoted by an asterisk (*). 'OZG' and 'OZH' samples were processed at ANSTO in Lucas Heights, New South Wales; 'Wk' sample at the University of Waikato in New Zealand; 'Beta' samples at Beta Analytic in Florida, U.S.A.

APPENDIX E

COMPLETE POLLEN DIAGRAMS

Complete pollen diagrams (*i.e.* comprising all identified pollen types, spores, fungal remains, *etc.*) are provided on the following pages. They are arranged in the same order as they appear in the text, and the diagrams from the relevant chapters should be consulted for details such as pollen sum calculations, radiocarbon dating locations, sediment logs and the like.

Pollen sums for each sample are given below in alphabetical order by site. For each entry, the site code (A- for Aligol, B- for Bareti, *etc.*) is followed by sample depth in centimetres, then the dryland pollen sum, and the total pollen sum (including aquatics). For example, 'A-4, 838, 1078' means Aligol 4-cm depth, 838 dryland pollen sum, 1078 total pollen sum. Original data may be obtained from the author on request.

Aligol: A-0, 1502, 1666; A-4, 838, 1078; A-8, 949, 1114; A-12, 1292, 1574; A-16, 852, 1365; A-20, 800, 1222; A-24, 1011, 1197; A-28, 438, 624; A-32, 830, 1035; A-36, 703, 1390; A-40, 1050, 1368; A-44, 851, 1032; A-48, 790, 1127; A-52, 1085, 1291; A-56, 788, 896; A-60, 708, 887; A-64, 903, 1139; A-68, 1016, 1342; A-72, 1091, 1391; A-76, 1295, 1615; A-80, 1011, 1287; A-84, 928, 1246; A-88, 979, 1302; A-92, 1355, 1729; A-96, 810, 1079; A-100, 1363, 1719; A-104, 1320, 1675; A-108, 1629, 2023; A-118, 891, 1091; A-125, 679, 855; A-136, 1380, 1688; A-147, 1142, 1320; A-156, 968, 1475; A-165, 1572, 2136; A-172, 1055, 1322; A-179, 1529, 1924; A-187, 1153, 1431; A-195, 1134, 1409; A-203, 976, 1285; A-209, 811, 1047; A-214, 820, 1038; A-219, 588, 854; A-224, 772, 1097; A-230, 1407, 1798; A-236, 1266, 1591; A-244, 1145, 1463; A-252, 236, 305; A-260, 469, 535; A-267, 514, 604; A-276, 624, 723; A-285, 377, 482; A-293, 508, 657; A-300, 553, 707; A-310, 781, 949.

Bareti: B-2, 585, 945; B-12, 458, 701; B-22, 918, 1299; B-32, 758, 821; B-42, 971, 1063; B-52, 931, 1021; B-62, 1015, 1072; B-72, 784, 814; B-82, 761, 797; B-92, 782, 823; B-102, 551, 584; B-112, 815, 918; B-122, 813, 1005.

Imera: I-0, 466, 644; I-5, 403, 517; I-10, 480, 622; I-15, 514, 646; I-20, 541, 667; I-24, 633, 841; I-28, 552, 673; I-31, 442, 636; I-38, 498, 1035; I-43, 472, 883; I-48, 985, 2145; I-54, 834, 1223; I-59, 545, 854; I-65, 671, 819; I-70, 559, 678; I-74, 555, 728; I-79, 666, 751; I-83, 395, 511; I-89, 747, 881; I-95, 849, 1012; I-98.5, 964, 1247; I-102, 572, 1014; I-106, 822, 1097; I-108, 588, 712; I-111, 682, 797; I-114, 727, 909; I-116, 657, 757; I-120, 660, 808; I-123, 435, 606; I-127, 588, 885; I-130, 657, 948; I-133, 923, 1230; I-136, 555, 763; I-140, 532, 779; I-144, 668, 1321; I-148, 778, 1510; I-150, 870, 1617; I-152.5, 929, 1427; I-155, 555, 699; I-160, 1090, 1710; I-163.5, 803, 988; I-165, 1469, 2241; I-170, 1064, 1669; I-175, 1024, 1622; I-180, 1104, 1592; I-185, 1094, 1511; I-190, 1523, 1959; I-198, 862, 1144; I-208, 901, 1242; I-217, 970, 1546; I-223, 396, 594; I-230, 449, 556; I-238, 792, 1085; I-246, 653, 1272; I-253, 549, 799; I-260, 535, 653; I-266, 507, 617; I-271, 612, 740; I-279, 748, 1049; I-286, 1344, 1636; I-294, 1452, 1736; I-300, 912, 1113; I-317, 1165, 1425.

Jvari: J-0, 494, 968; J-12.5, 657, 1026; J-25, 707, 2971; J-37.5, 588, 3209; J-50, 778, 2526; J-62.5, 549, 778; J-75, 614, 1433; J-87.5, 611, 858; J-100, 711, 861; J-112.5, 738, 1323; J-125, 582, 1838; J-137.5, 597, 699; J-150, 603, 680; J-162.5, 638, 870; J-175, 716, 1269; J-187.5, 855, 1073; J-200, 881, 1117; J-212.5, 872, 994; J-225, 886, 1035; J-237.5, 813, 879; J-250, 745, 1154.

Kumisi: K-0, 1376, 1554; K-10, 1132, 1305; K-20, 1032, 1232; K-30, 547, 744; K-40, 808, 1446; K-50, 980, 1112; K-60, 1220, 1358; K-70, 916, 1266; K-80, 1088, 1811; K-90, 759, 1089; K-100, 604, 1286; K-110, 880, 1001; K-120, 1097, 1248; K-130, 1323, 1434; K-140, 1106, 1795; K-150, 1443, 1926; K-155, 1265, 1980; K-160, 1104, 1195; K-165, 1545, 4504; K-170, 1566, 2504; K-175, 2026, 2165; K-180, 612, 706; K-185, 965, 1856; K-190, 1373, 1515; K-200, 1664, 2127; K-210, 1374, 1534; K-220, 451, 517; K-230, 1161, 1545; K-240, 1285, 1623; K-250, 1834, 1960.

Sakhare: S-0, 701, 822; S-4, 872, 1042; S-9, 817, 989; S-14, 943, 1025; S-19, 785, 868; S-25, 895, 965; S-29, 766, 876; S-34, 795, 875; S-39, 773, 864; S-45, 721, 815; S-51, 1066, 1125; S-57, 844, 918; S-62, 781, 838; S-66, 754, 841; S-70, 685, 755; S-75, 716, 784; S-79, 694, 746; S-86, 793, 853; S-90, 851, 925.

Tsavkisi: T-0, 352, 476; T-10, 513, 625; T-15, 592, 735; T-20, 498, 672; T-25, 625, 795; T-30, 752, 895; T-35, 829, 967; T-40, 1013, 1261; T-45, 383, 525; T-50, 680, 792; T-55, 450, 751; T-60, 360, 680; T-65, 399, 839; T-70, 332, 525; T-75, 499, 686; T-80, 736, 1122; T-85, 626, 811; T-90, 625, 819; T-95, 542, 711; T-100, 732, 949; T-105, 588, 810; T-110, 843, 1122; T-115, 600, 796; T-120, 470, 532; T-125, 520, 585; T-130, 591, 731; T-135, 741, 802; T-140, 631, 710; T-145, 740, 850; T-150, 709, 805; T-155, 884, 957; T-160, 514, 865; T-165, 479, 703; T-170, 369, 497; T-175, 423, 641; T-180, 352, 607; T-185, 303, 392; T-190, 570, 691; T-200, 446, 535; T-210, 752, 950; T-220, 453, 715; T-230, 816, 921; T-240, 1037, 1201; T-250, 775, 918; T-260, 735, 853; T-270, 741, 871; T-280, 567, 640; T-290, 526, 628; T-300, 666, 782; T-310, 579, 713; T-320, 467, 534; T-330, 426, 488; T-340, 643, 697; T-350, 620, 689; T-360, 721, 836; T-370, 662, 751; T-380, 676, 732; T-390, 724, 779; T-400, 809, 867; T-410, 866, 949; T-420, 757, 832; T-430, 641, 728; T-440, 348, 409; T-450, 595, 635; T-460, 594, 652; T-470, 637, 713; T-480, 623, 669; T-490, 555, 597; T-500, 666, 708; T-510, 617, 667; T-520, 685, 738; T-530, 653, 712; T-540, 750, 839; T-550, 594, 679; T-560, 493, 541; T-590, 642, 758; T-600, 774, 1605; T-610, 542, 1010; T-620, 485, 822; T-630, 431, 529; T-640, 490, 603; T-650, 488, 583; T-660, 644, 720; T-670, 675, 759; T-680, 609, 767; T-690, 583, 829; T-700, 676, 827; T-710, 581, 675; T-720, 675, 779; T-730, 711, 875; T-740, 436, 728; T-750, 697, 1100; T-760, 545, 899; T-770, 786, 1187; T-780, 689, 981; T-782, 584, 778; T-790, 666, 916; T-800, 716, 964; T-810, 690, 950; T-820, 851, 1178; T-830, 626, 822; T-840, 589, 785; T-850, 860, 983; T-860, 920, 1152; T-870, 460, 490; T-880, 743, 922; T-890, 511, 608; T-900, 921, 1061; T-910, 824, 1038; T-920, 765, 868; T-930, 1173, 1406; T-940, 712, 866; T-950, 818, 907; T-960, 944, 1099; T-970, 807, 969; T-980, 955, 1076; T-990, 711, 832; T-1000, 950, 1100; T-1010, 762, 933; T-1020, 871, 1022; T-1030, 731, 894; T-1040, 1035, 1146; T-1050, 845, 1011; T-1060, 849, 887; T-1070, 897, 1014; T-1080, 949, 1083; T-1090, 925, 1143; T-1100, 1118, 1308; T-1110, 1200, 1535.

Modern Pollen

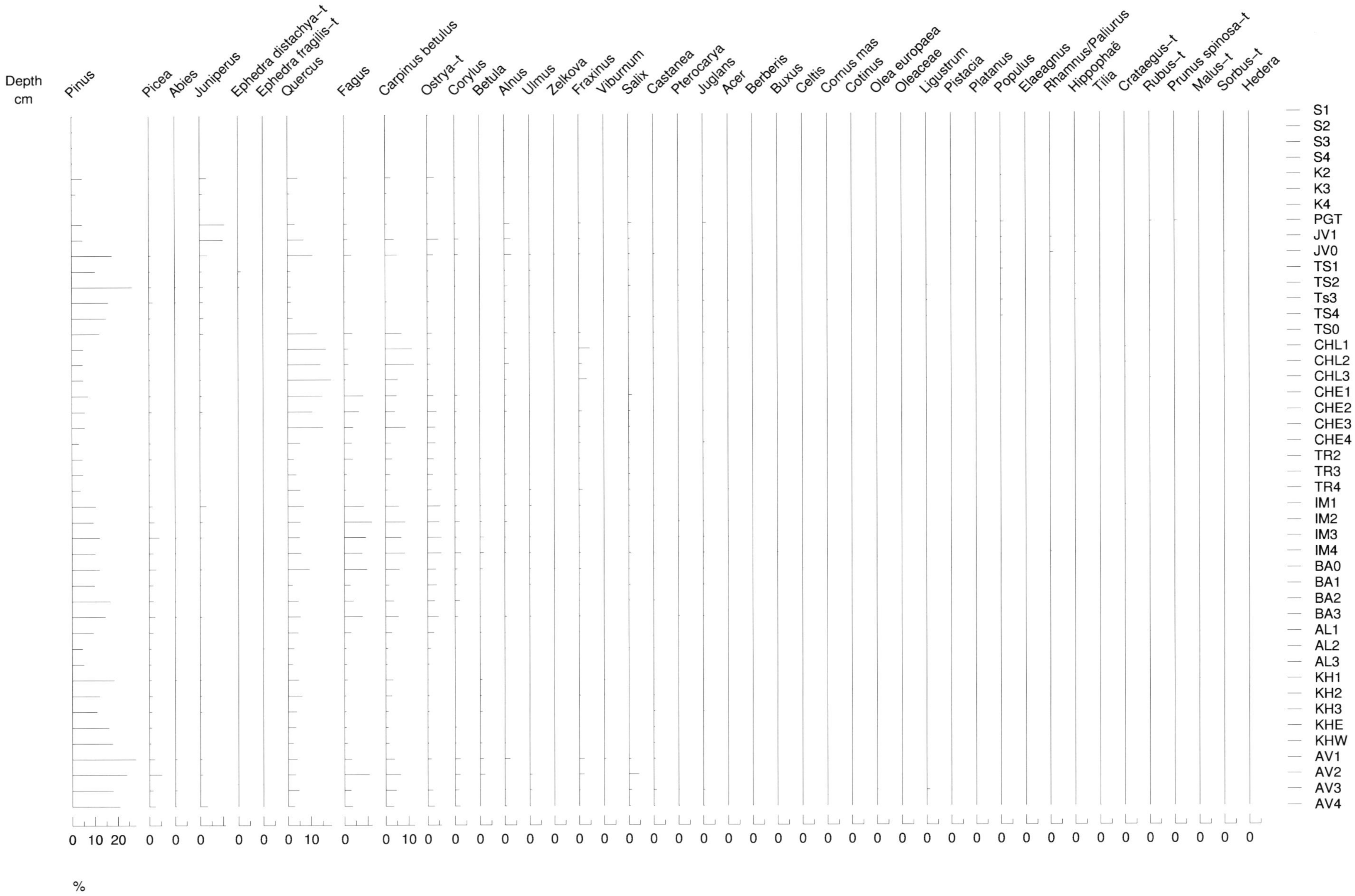

Modern Pollen

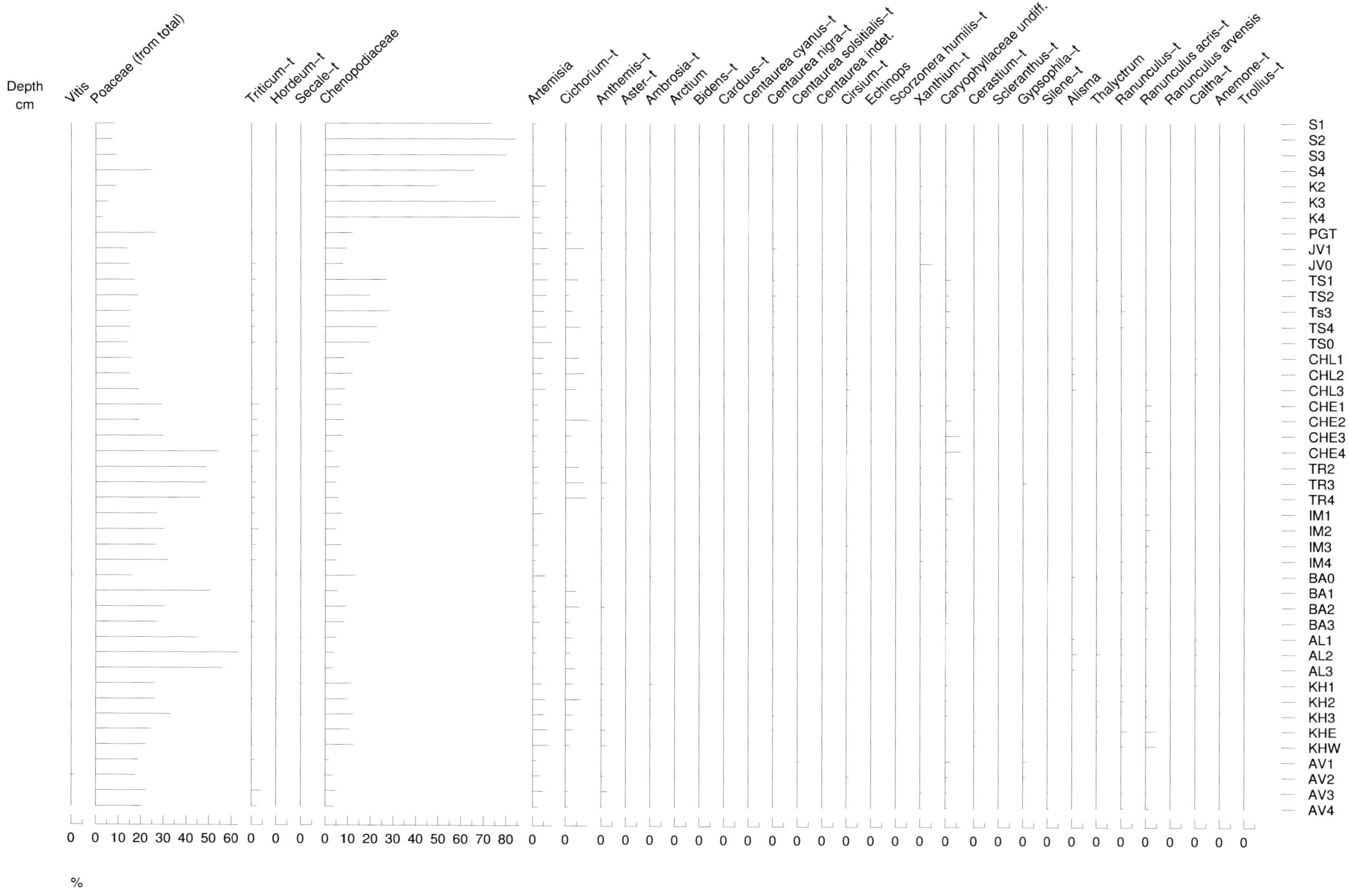

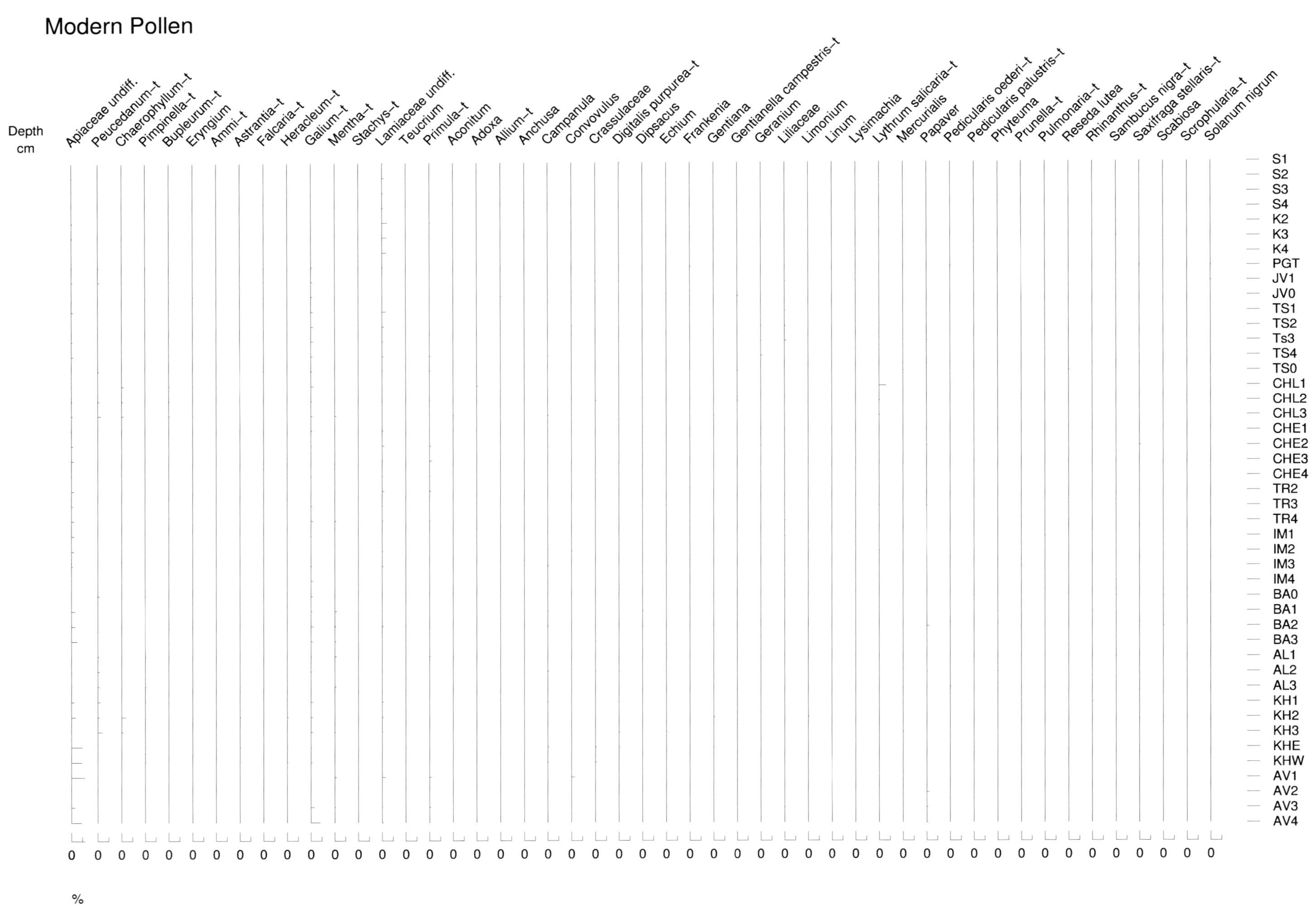
Modern Pollen
Depth
cm
Apiaceae undiff.
Peucedanum–t
Chaerophyllum–t
Pimpinella–t
Bupleurum–t
Eryngium
Ammi–t
Astrantia–t
Falcaria–t
Heracleum–t
Galium–t
Mentha–t
Stachys–t
Lamiaceae undiff.
Teucrium
Primula–t
Aconitum
Adoxa
Allium–t
Anchusa
Campanula
Convovulus
Crassulaceae
Digitalis purpurea–t
Dipsacus
Echium
Frankenia
Gentiana
Gentianella campestris–t
Geranium
Liliaceae
Limonium
Linum
Lysimachia
Lythrum salicaria–t
Mercurialis
Papaver
Pedicularis oederi–t
Pedicularis palustris–t
Phyteuma
Prunella–t
Pulmonaria–t
Reseda lutea
Rhinanthus–t
Sambucus nigra–t
Saxifraga stellaris–t
Scabiosa
Scrophularia–t
Solanum nigrum
S1
S2
S3
S4
K2
K3
K4
PGT
JV1
JV0
TS1
TS2
Ts3
TS4
TS0
CHL1
CHL2
CHL3
CHE1
CHE2
CHE3
CHE4
TR2
TR3
TR4
IM1
IM2
IM3
IM4
BA0
BA1
BA2
BA3
AL1
AL2
AL3
KH1
KH2
KH3
KHE
KHW
AV1
AV2
AV3
AV4
0 0 0 0 0 0 0 0 0 0 0 0 0 0 0 0 0 0 0 0 0 0 0 0 0 0 0 0 0 0 0 0 0 0 0 0 0 0 0 0 0 0 0 0 0 0 0 0 0
%

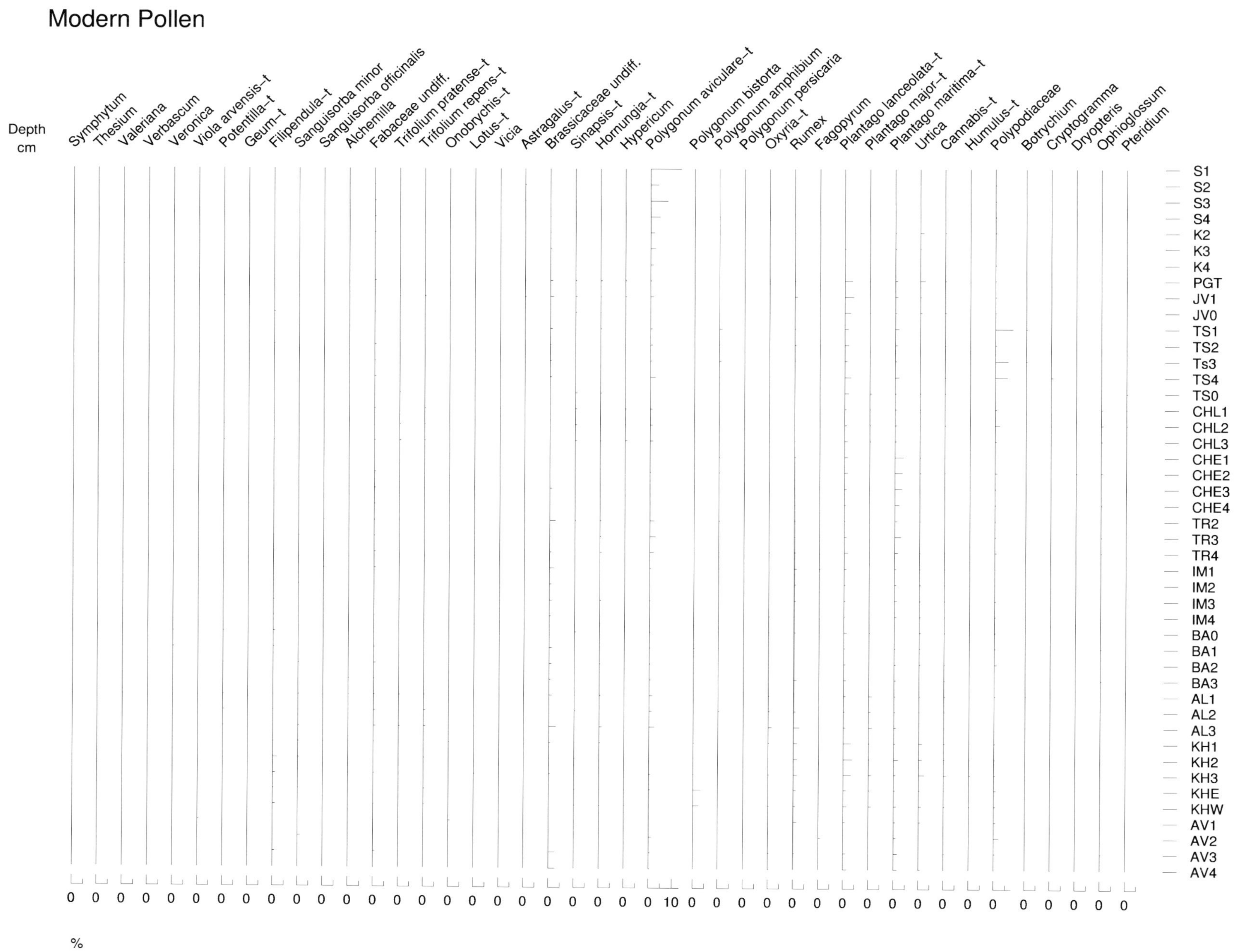
Modern Pollen
Depth cm
Symphytum
Thesium
Valeriana
Verbascum
Veronica
Viola arvensis–t
Potentilla–t
Geum–t
Filipendula–t
Sanguisorba minor
Sanguisorba officinalis
Alchemilla
Fabaceae undiff.
Trifolium pratense–t
Trifolium repens–t
Onobrychis–t
Lotus–t
Vicia
Astragalus–t
Brassicaceae undiff.
Sinapsis–t
Hornungia–t
Hypericum
Polygonum aviculare–t
Polygonum bistorta
Polygonum amphibium
Polygonum persicaria
Oxyria–t
Rumex
Fagopyrum
Plantago lanceolata–t
Plantago major–t
Plantago maritima–t
Urtica
Cannabis–t
Humulus–t
Polypodiaceae
Botrychium
Cryptogramma
Dryopteris
Ophioglossum
Pteridium
S1
S2
S3
S4
K2
K3
K4
PGT
JV1
JV0
TS1
TS2
Ts3
TS4
TS0
CHL1
CHL2
CHL3
CHE1
CHE2
CHE3
CHE4
TR2
TR3
TR4
IM1
IM2
IM3
IM4
BA0
BA1
BA2
BA3
AL1
AL2
AL3
KH1
KH2
KH3
KHE
KHW
AV1
AV2
AV3
AV4
0 0 0 0 0 0 0 0 0 0 0 0 0 0 0 0 0 0 0 0 0 0 0 0 10 0 0 0 0 0 0 0 0 0 0 0 0 0 0 0 0 0 0
%

Lake Kumisi, Tbilisi area

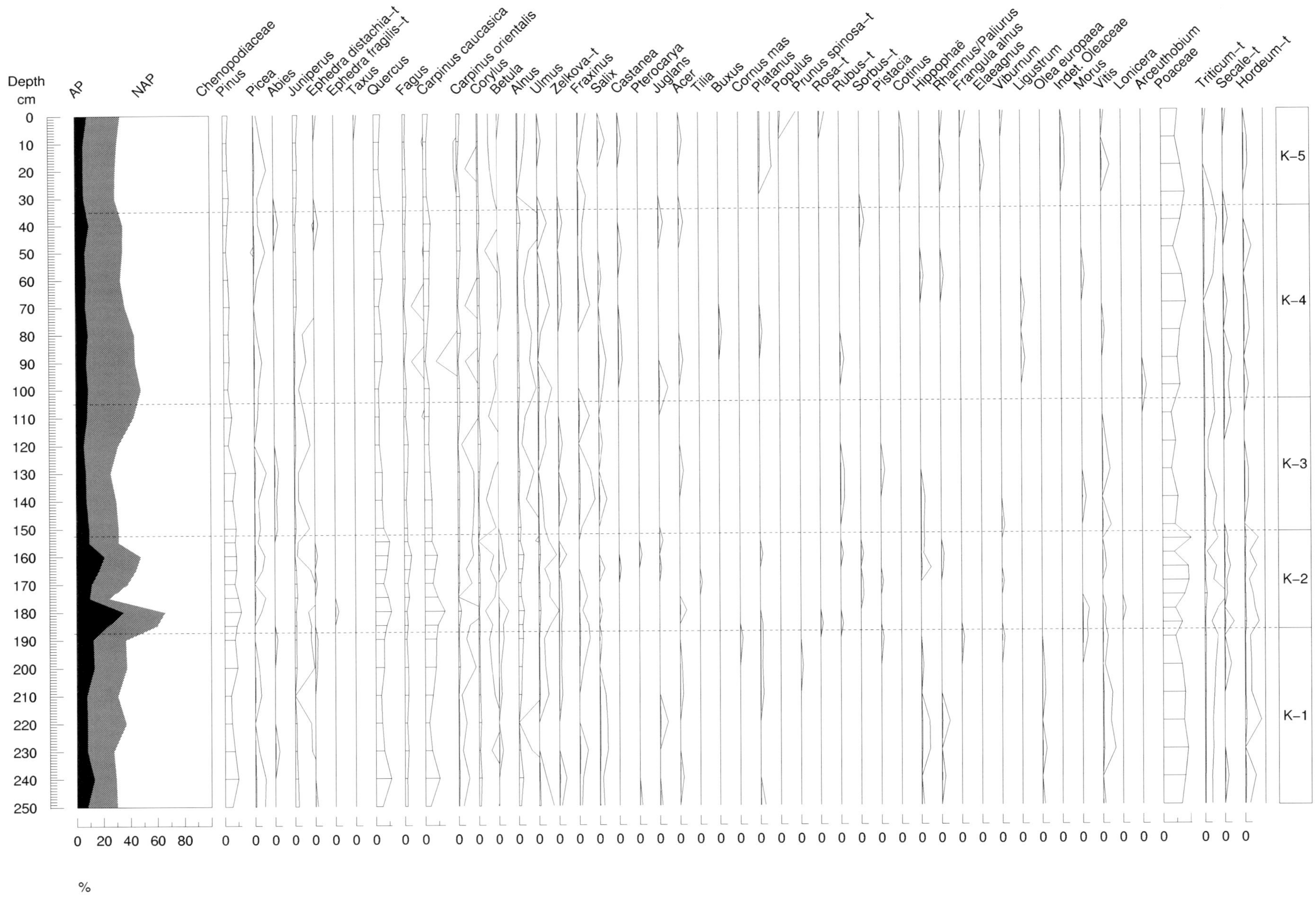

Lake Kumisi, Tbilisi area

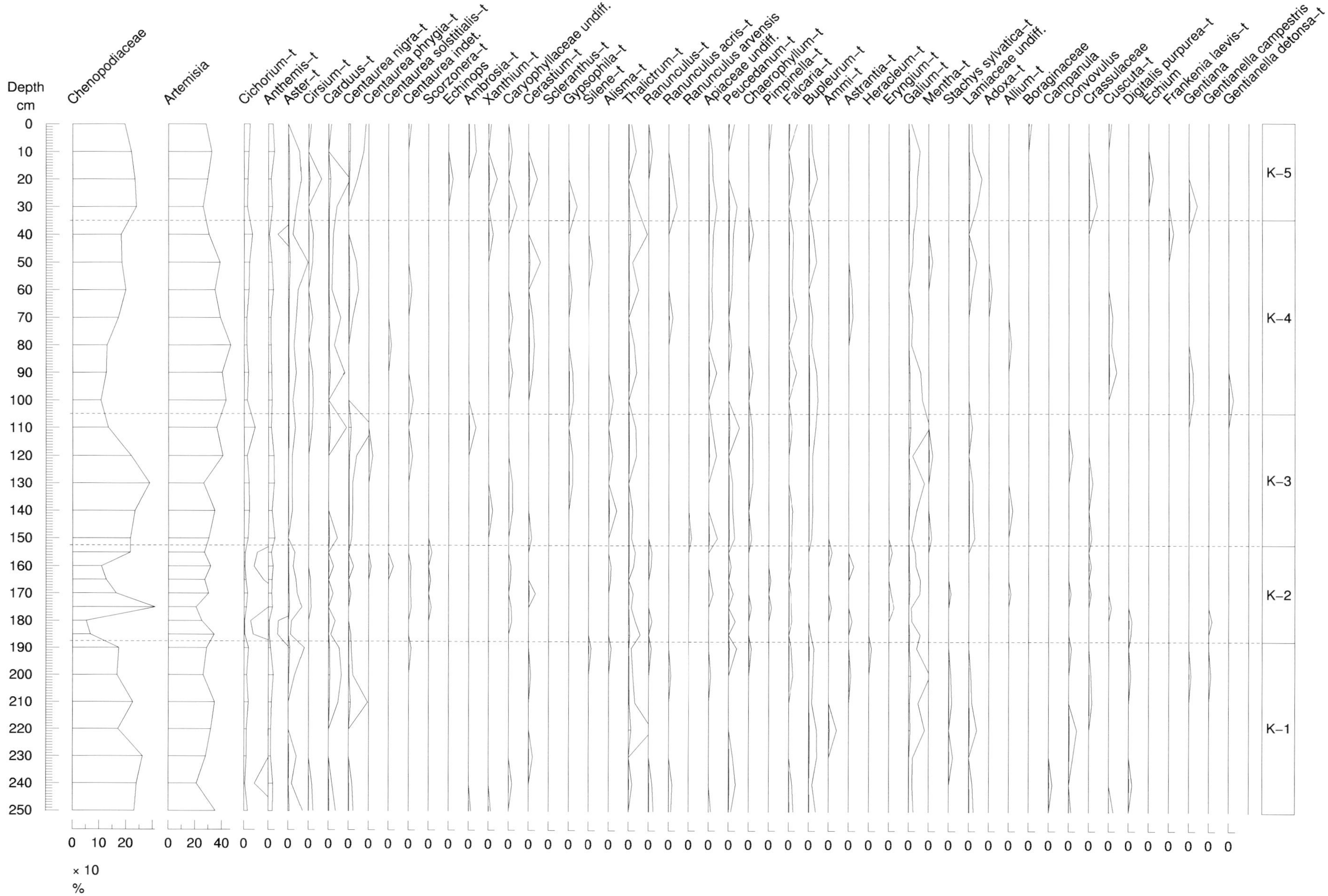

Lake Kumisi, Tbilisi area

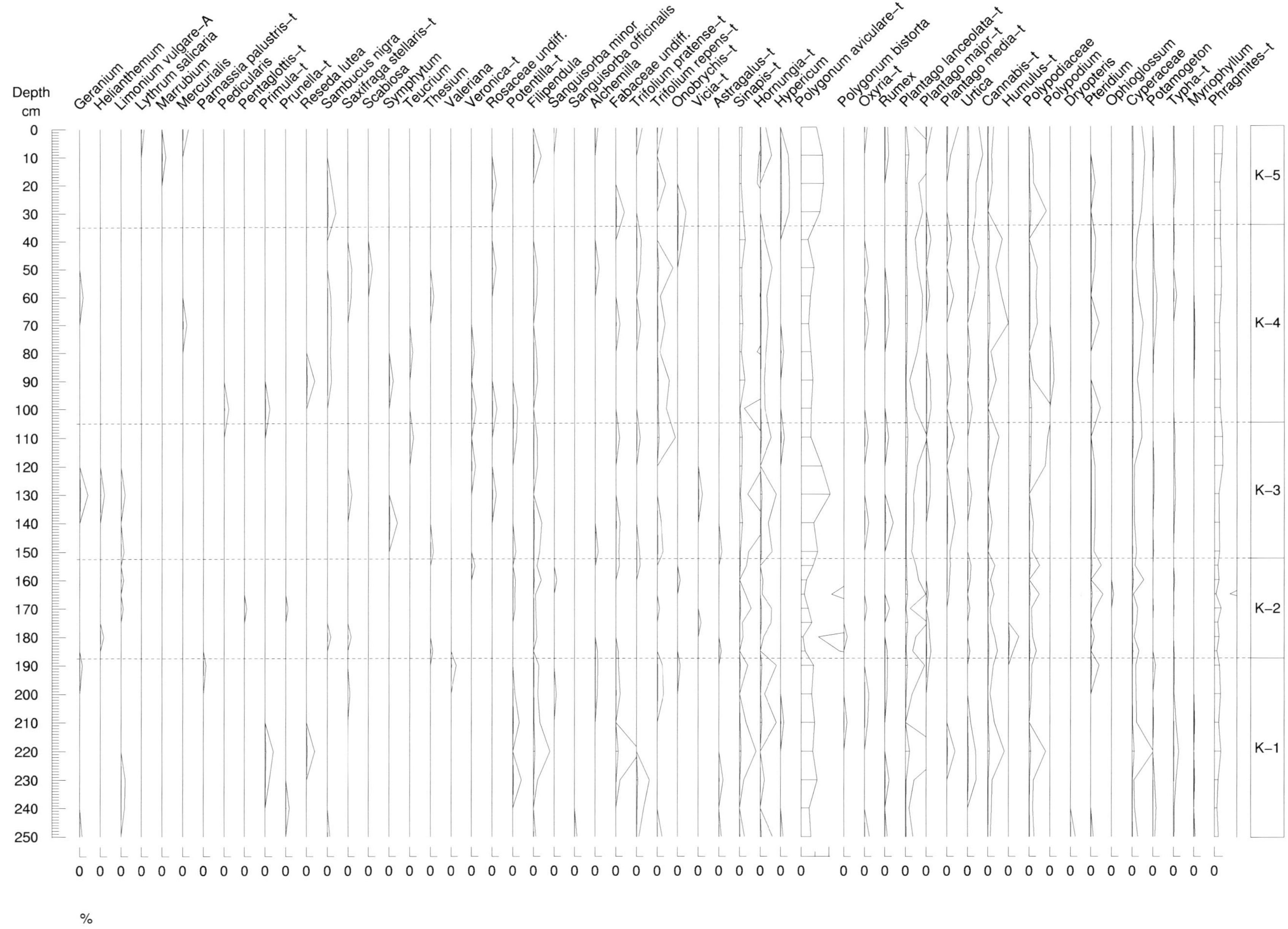

Lake Kumisi, Tbilisi area

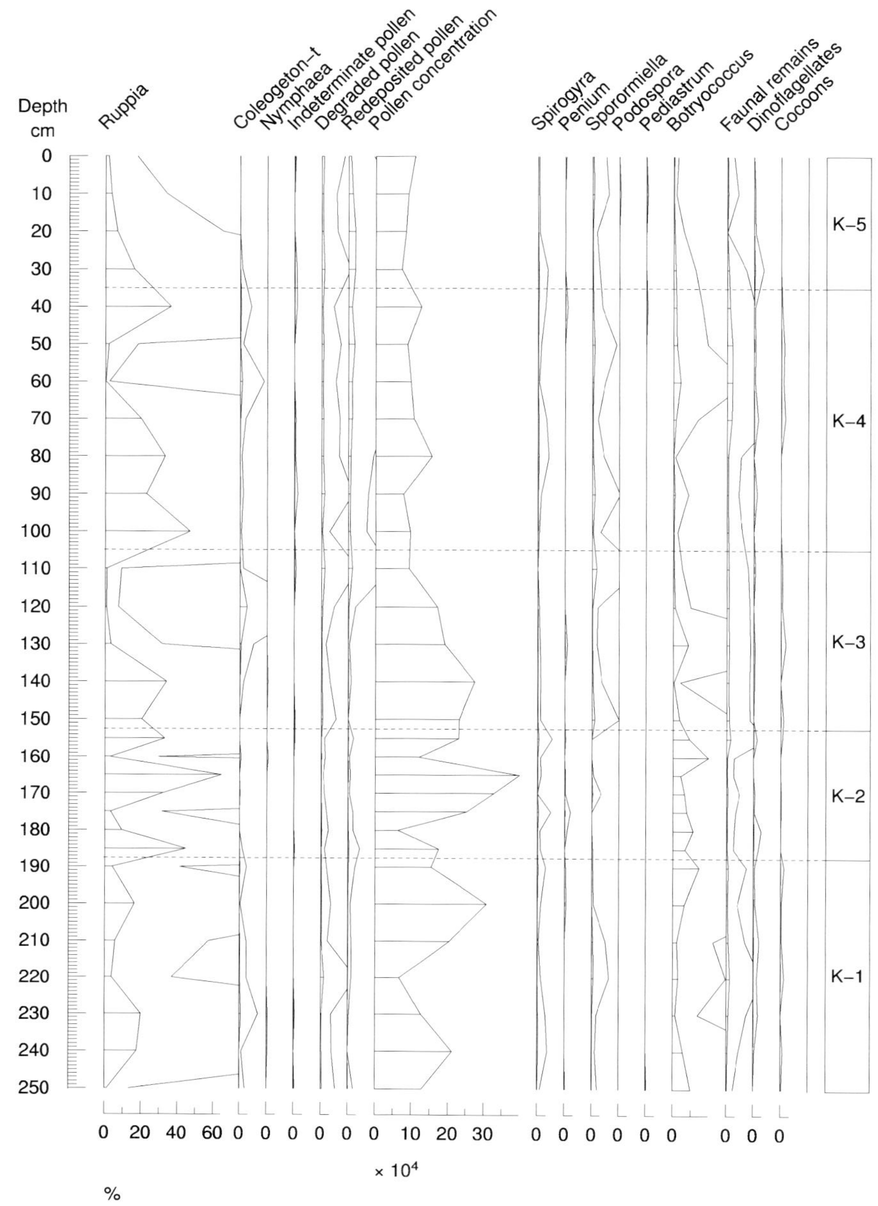

Sakhare Lake, Tbilisi area

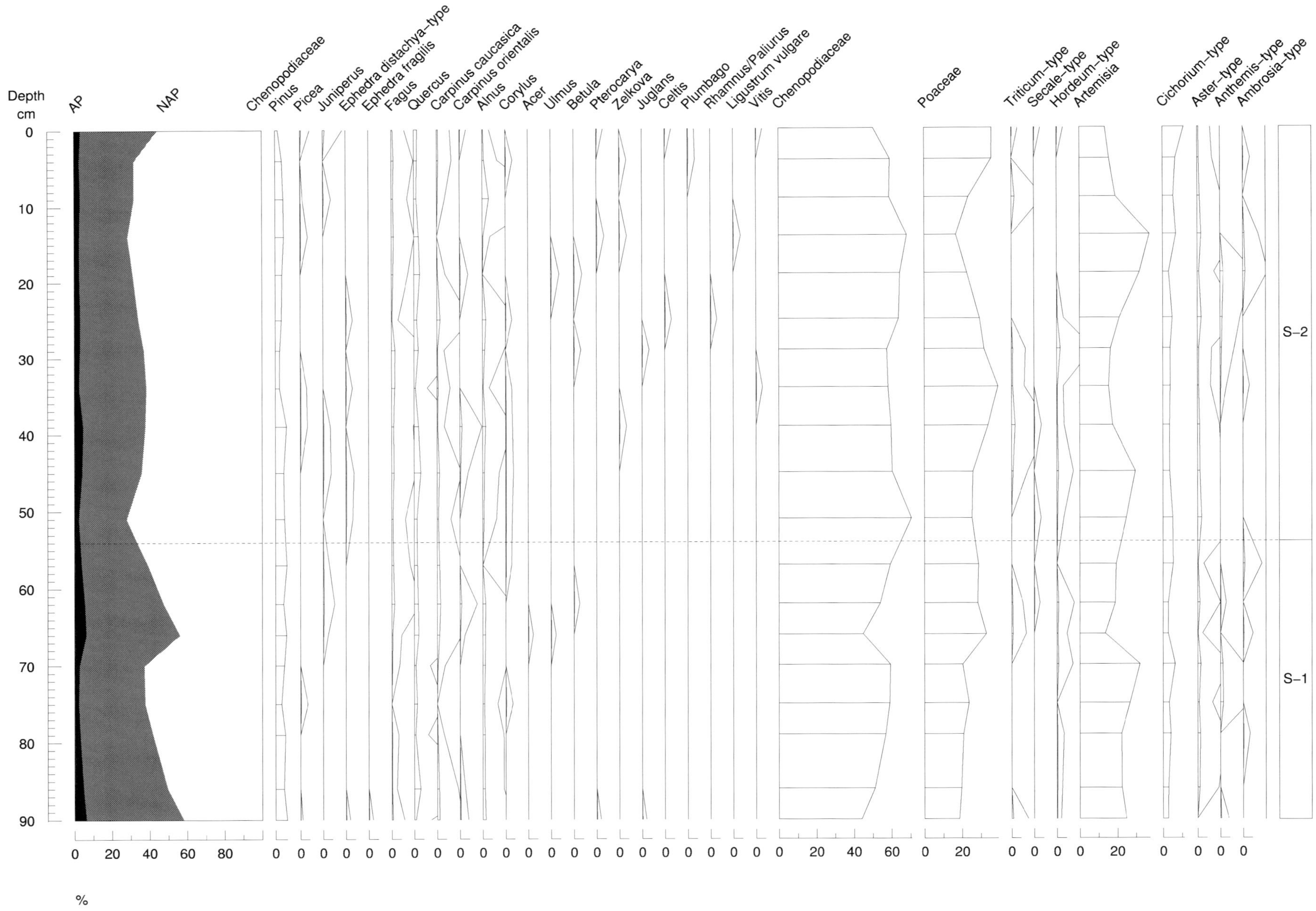

Sakhare Lake, Tbilisi area

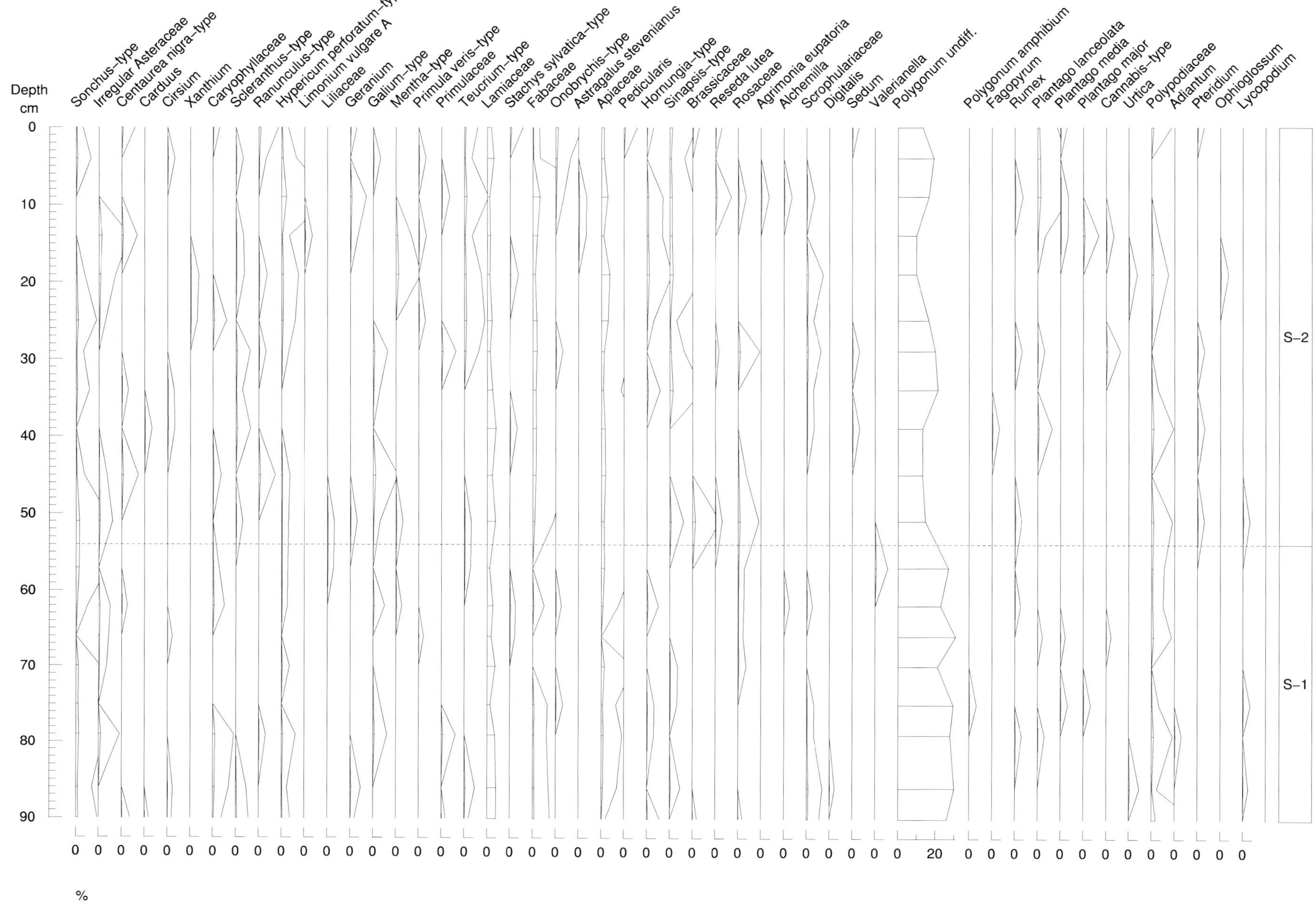

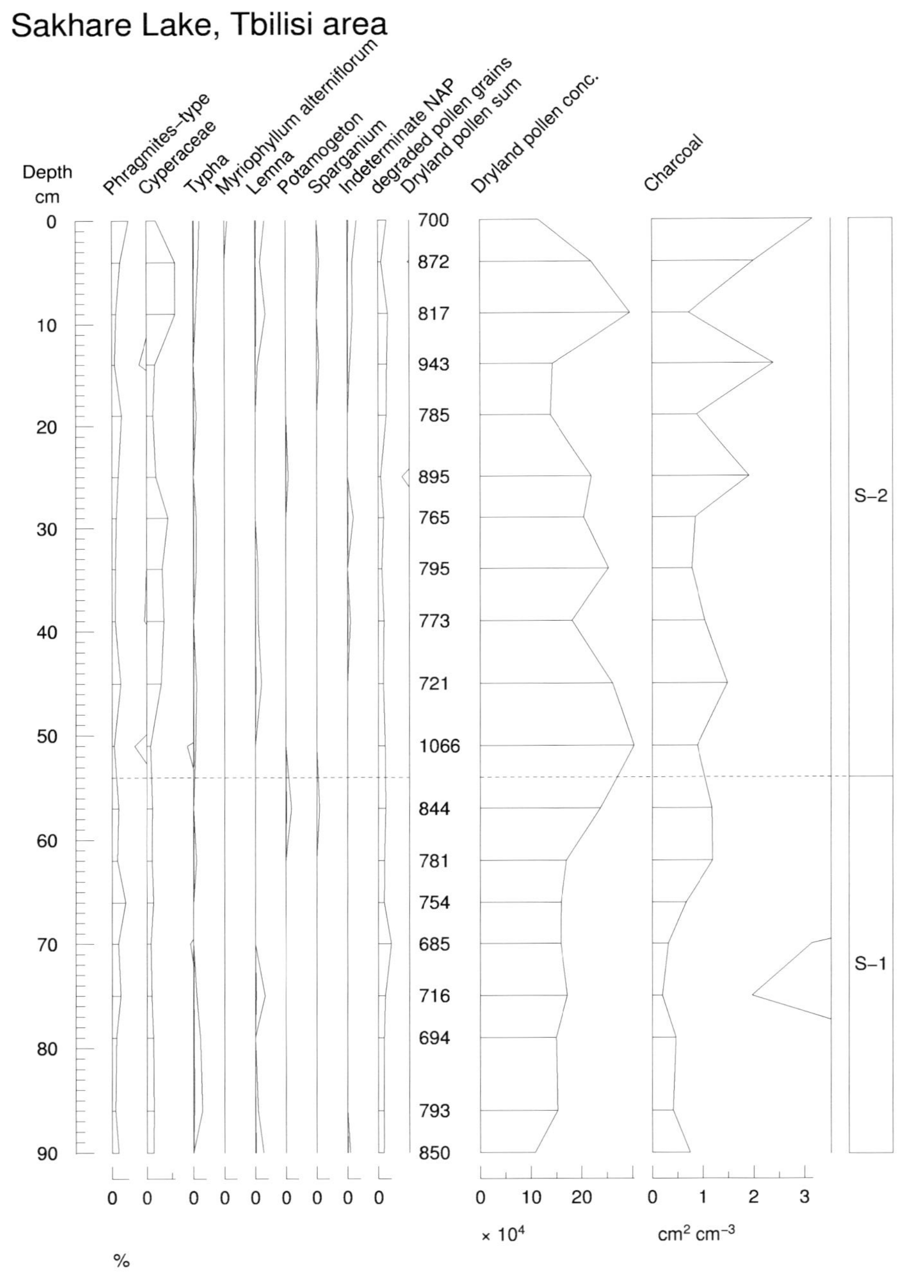

Sakhare Lake, Tbilisi area
Depth cm
Phragmites–type
Cyperaceae
Typha
Myriophyllum alterniflorum
Lemna
Potamogeton
Sparganium
Indeterminate NAP
degraded pollen grains
Dryland pollen sum
Dryland pollen conc.
Charcoal
700
872
817
943
785
895
765
795
773
721
1066
844
781
754
685
716
694
793
850
S–2
S–1
× 10^4
cm² cm⁻³
%

Lake Jvari, Tbilisi area

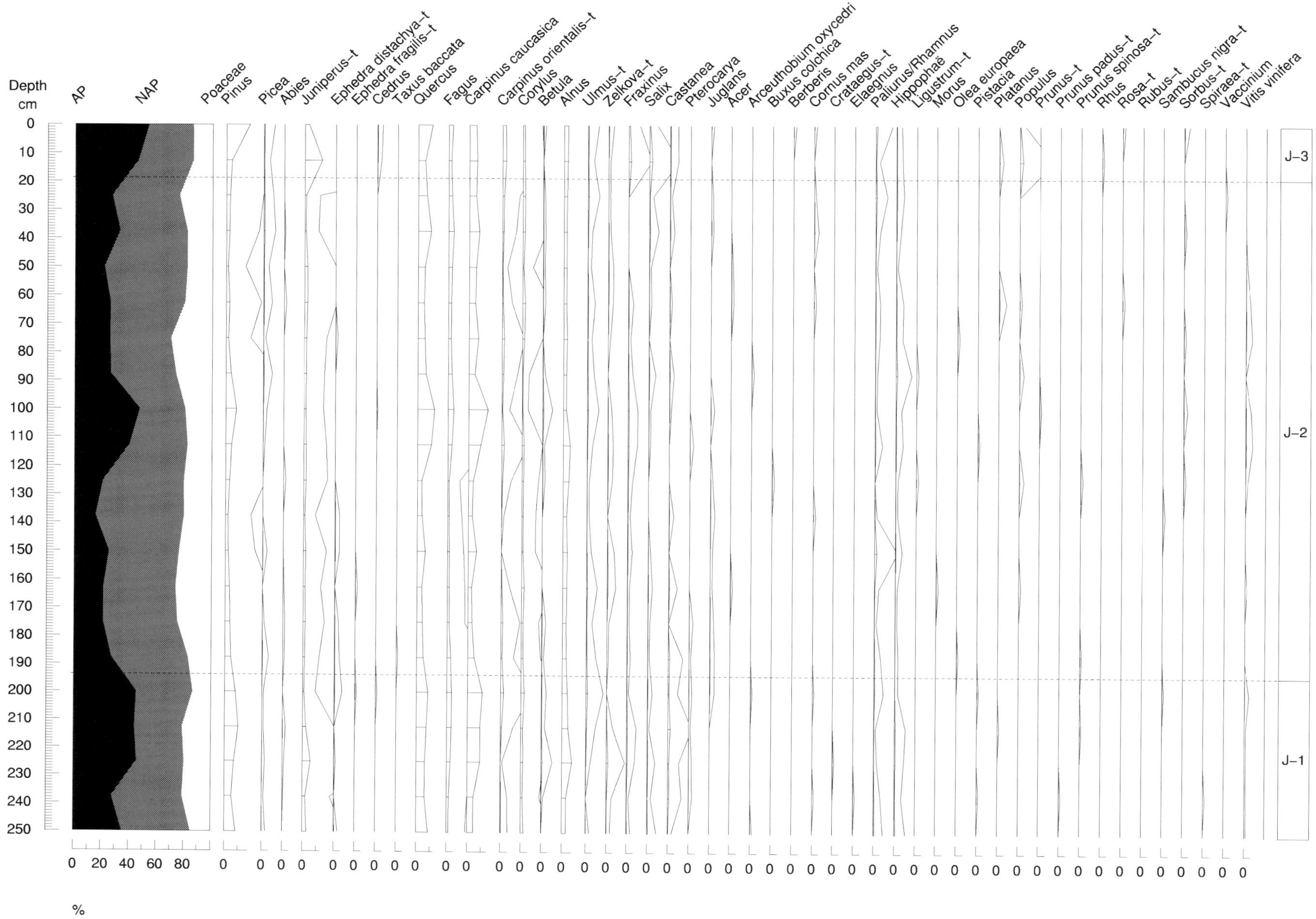

Lake Jvari, Tbilisi area

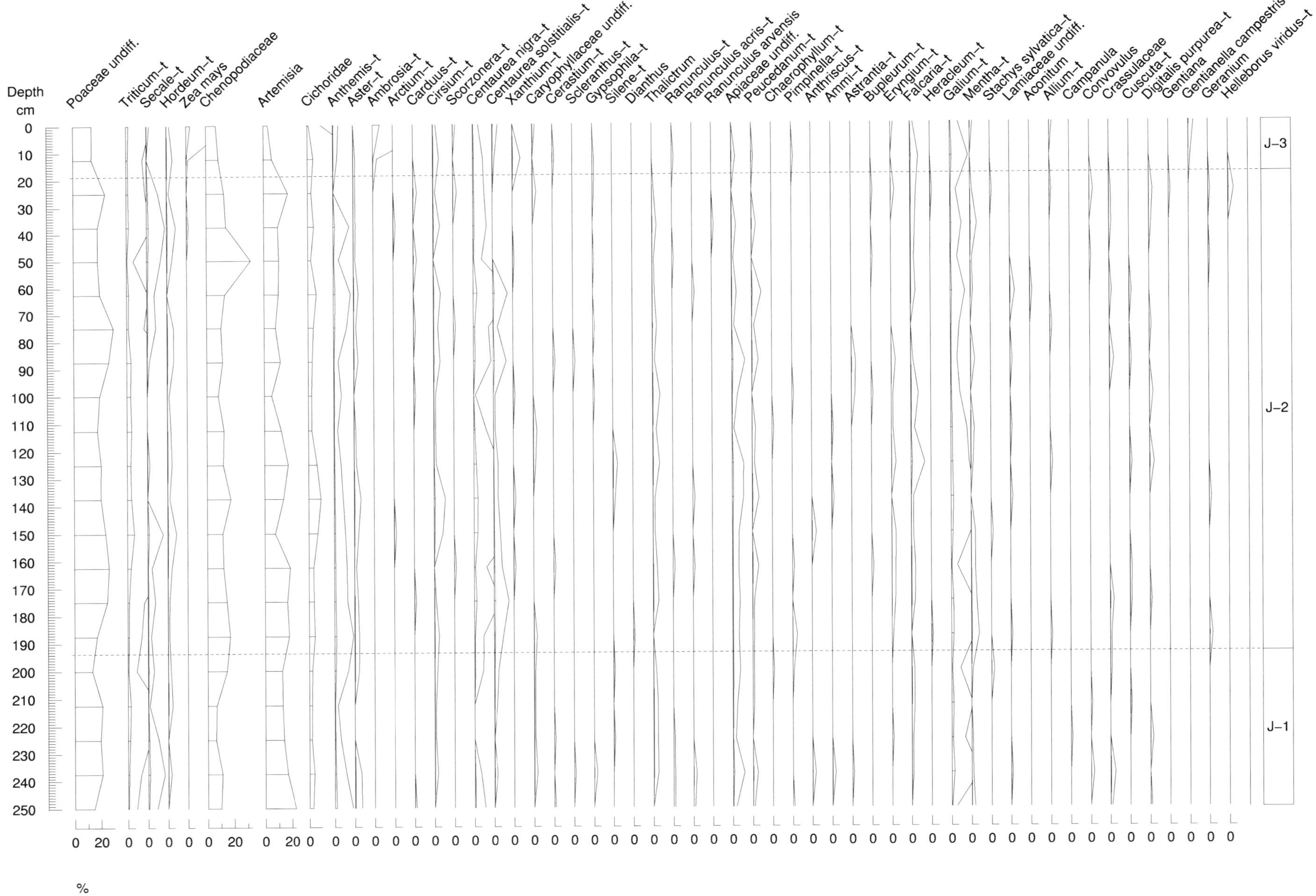

Lake Jvari, Tbilisi area

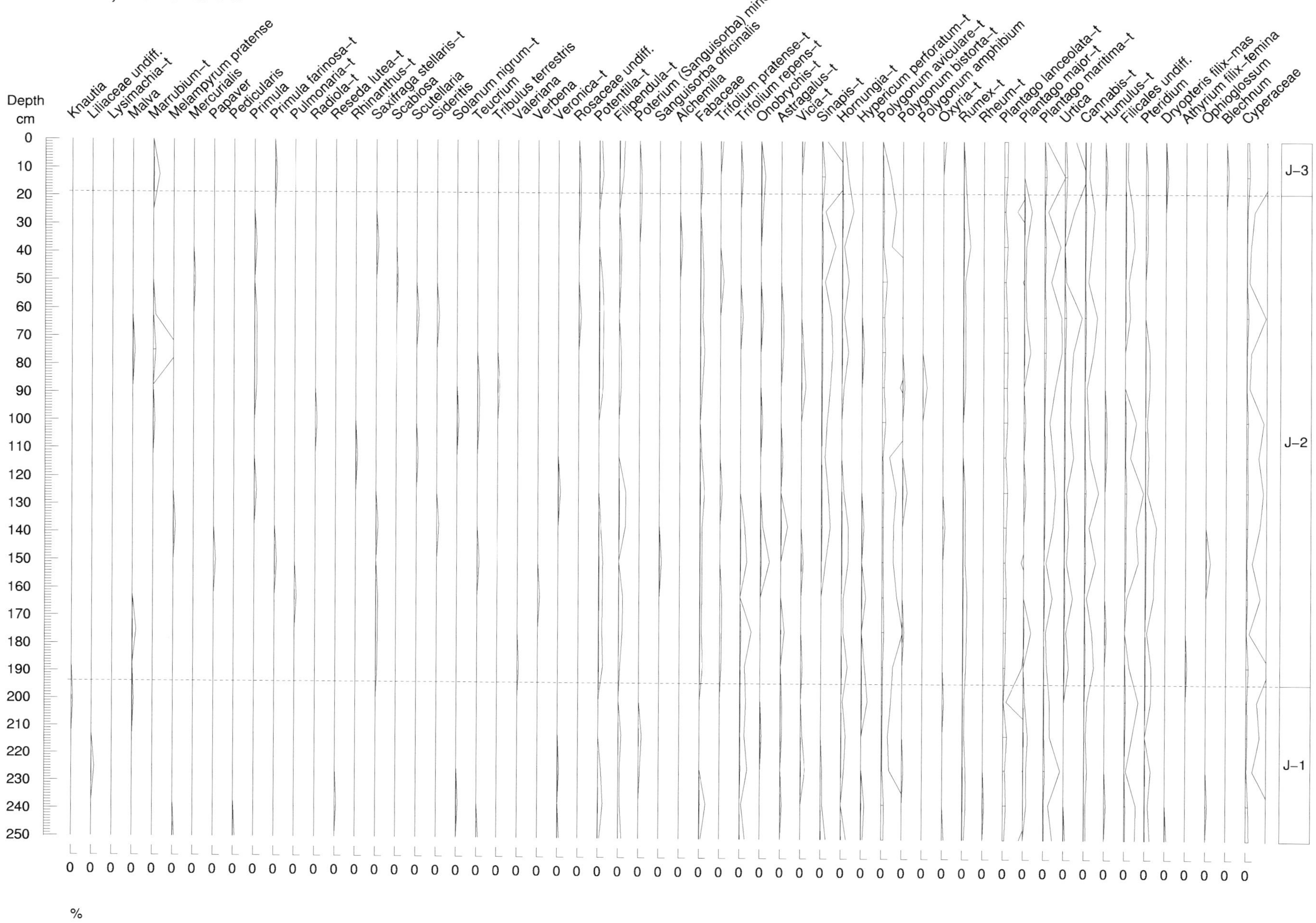

Lake Jvari, Tbilisi area

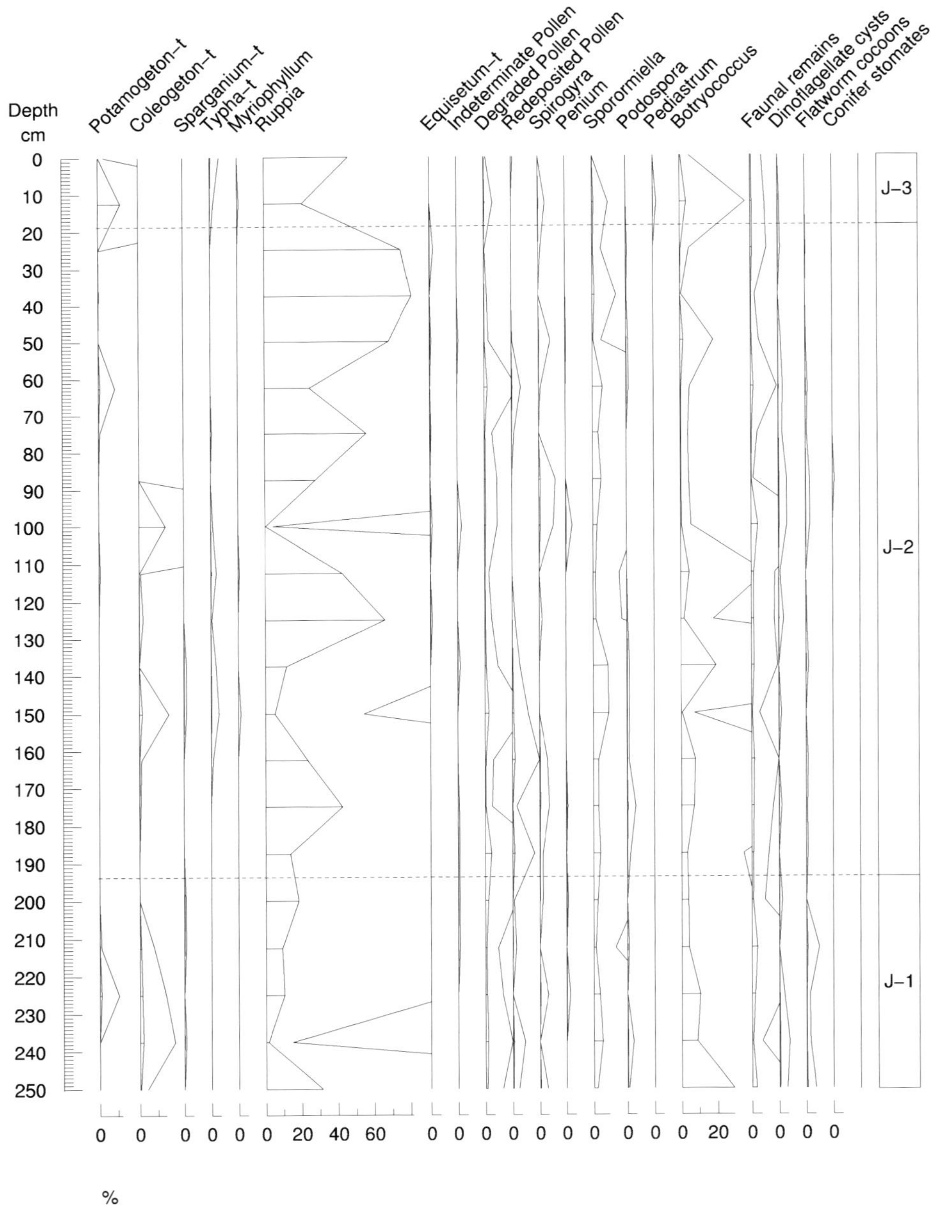

Tsavkisi Wetland, Tbilisi area

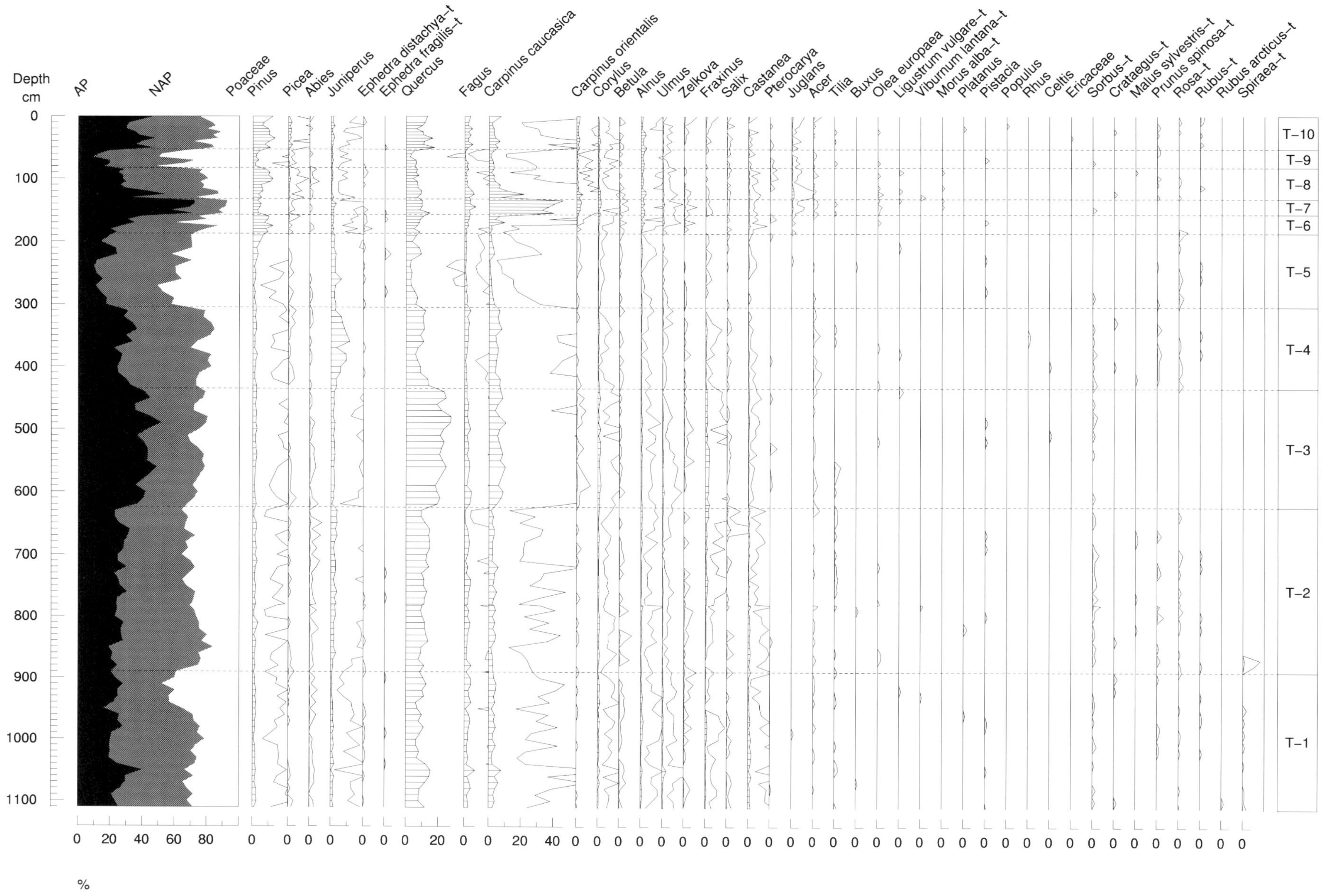

Tsavkisi Wetland, Tbilisi area

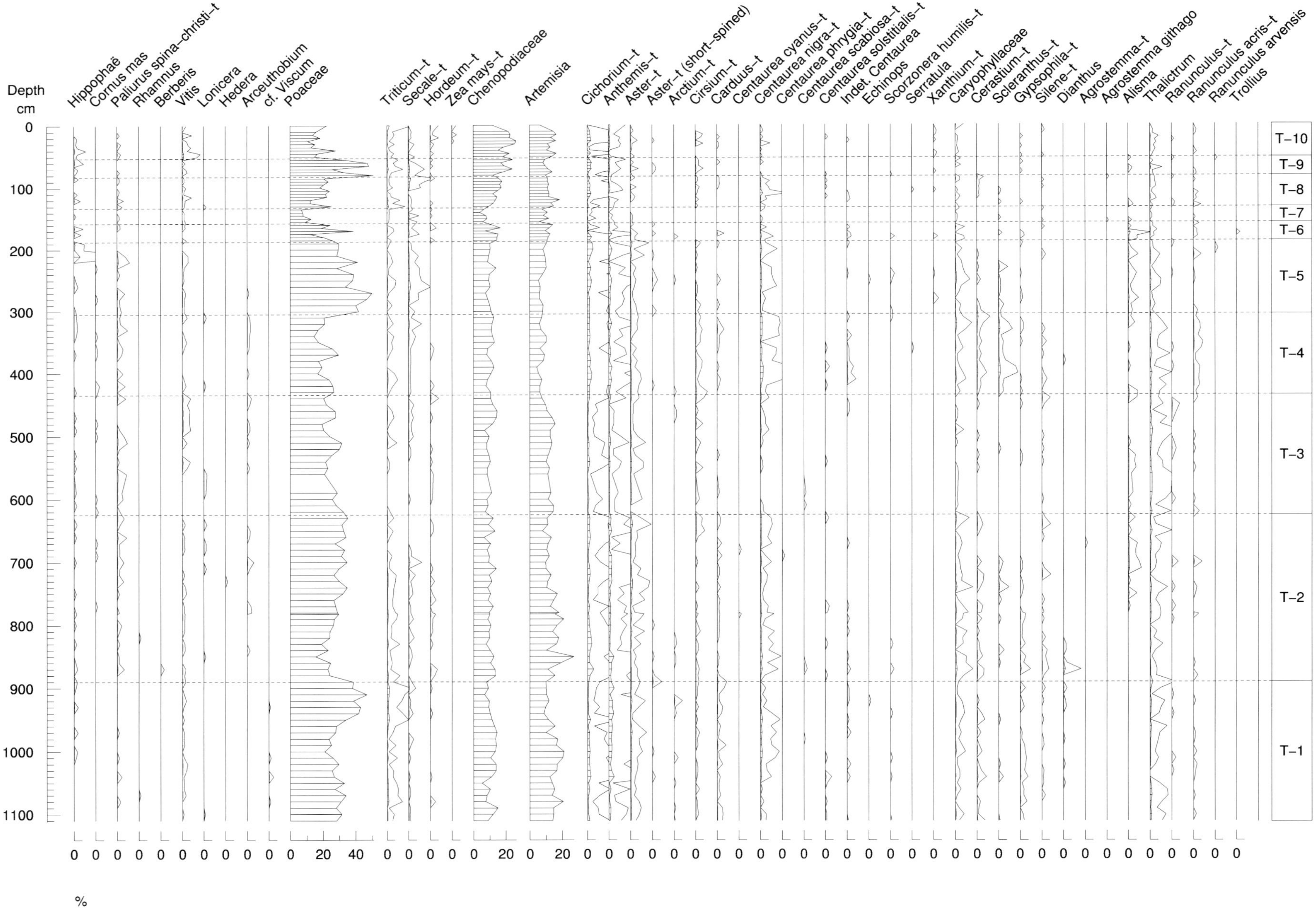

Tsavkisi Wetland, Tbilisi area

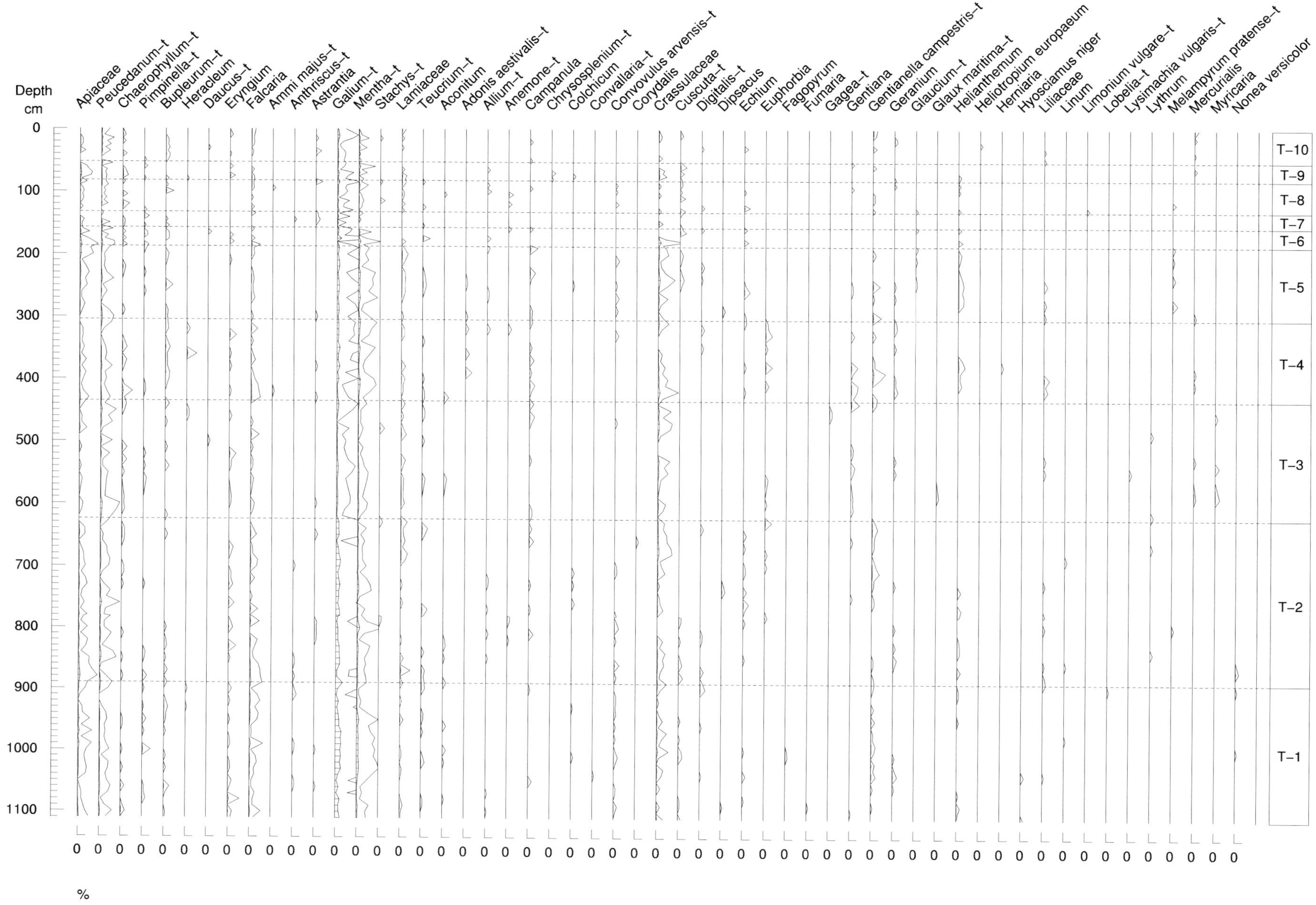

Tsavkisi Wetland, Tbilisi area

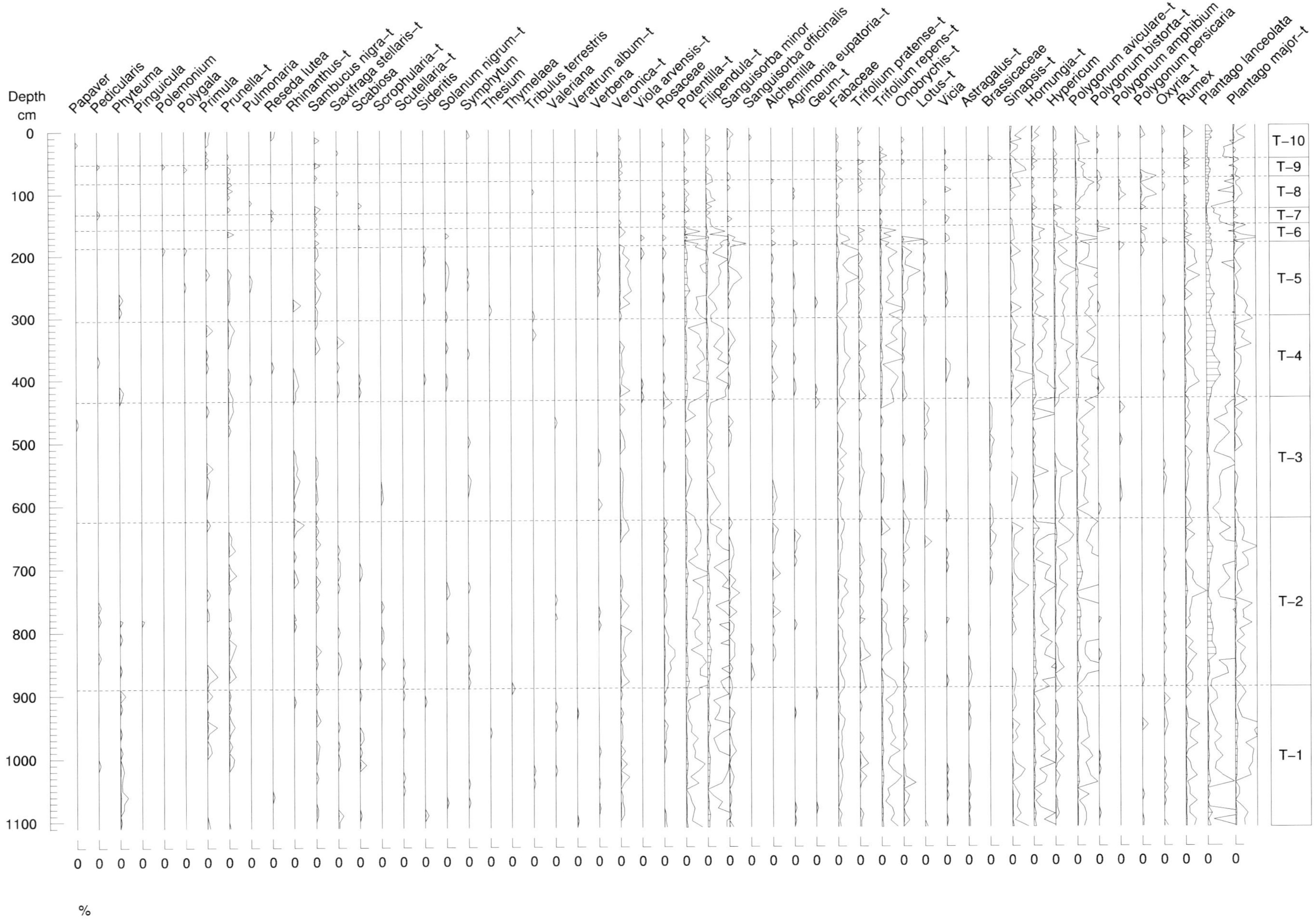

Tsavkisi Wetland, Tbilisi area

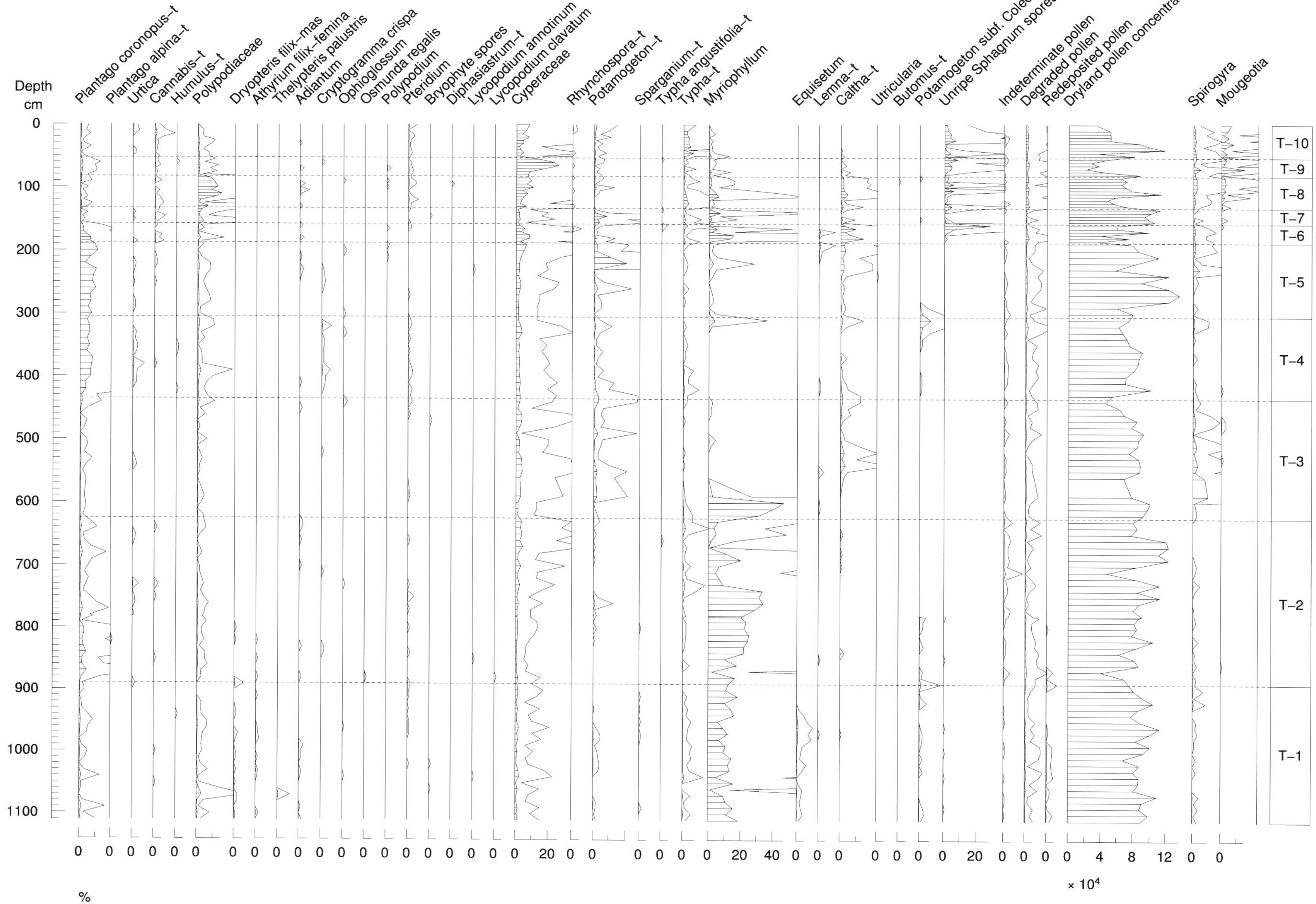

Tsavkisi Wetland, Tbilisi area

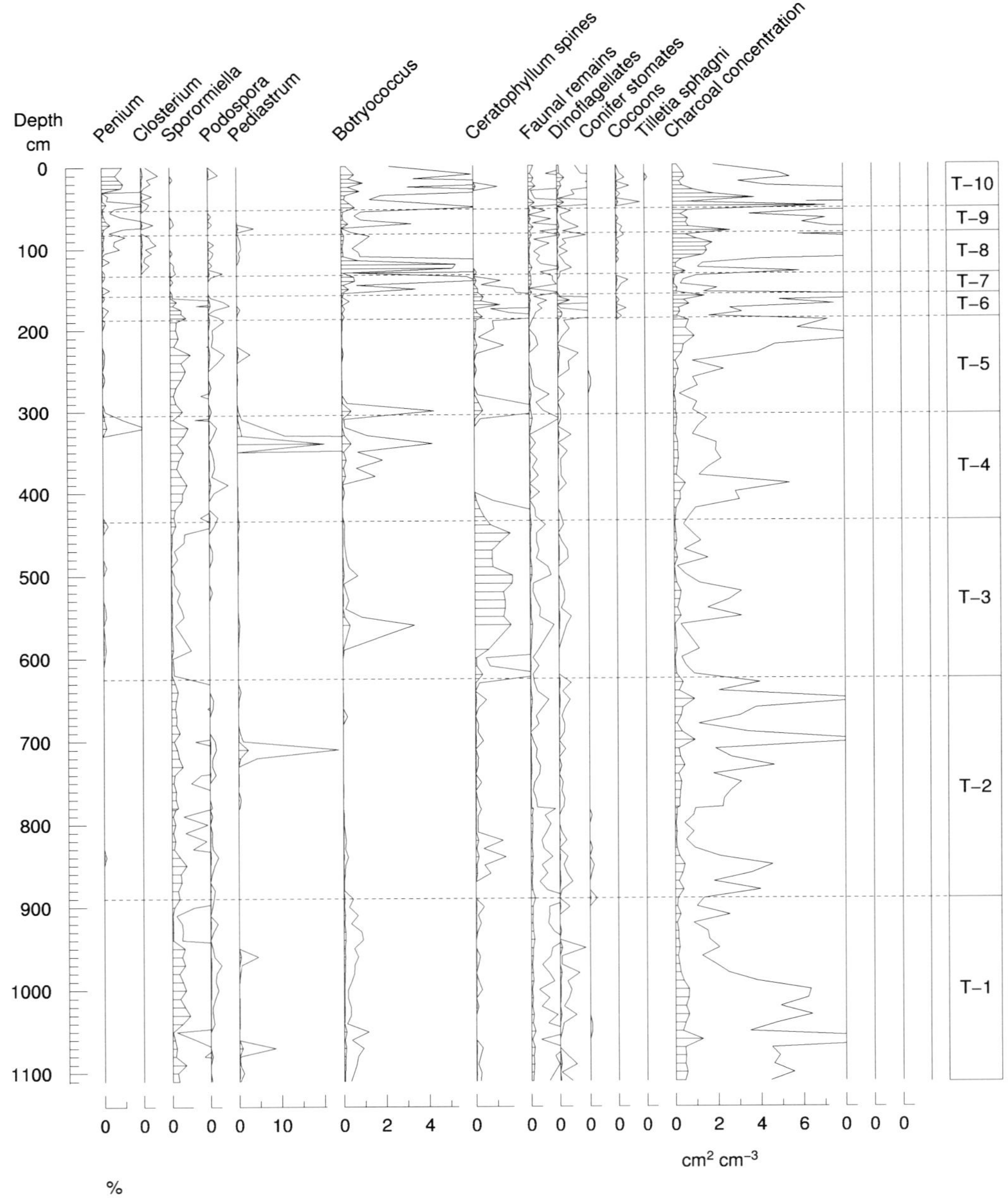

Lake Imera, Tsalka Plateau

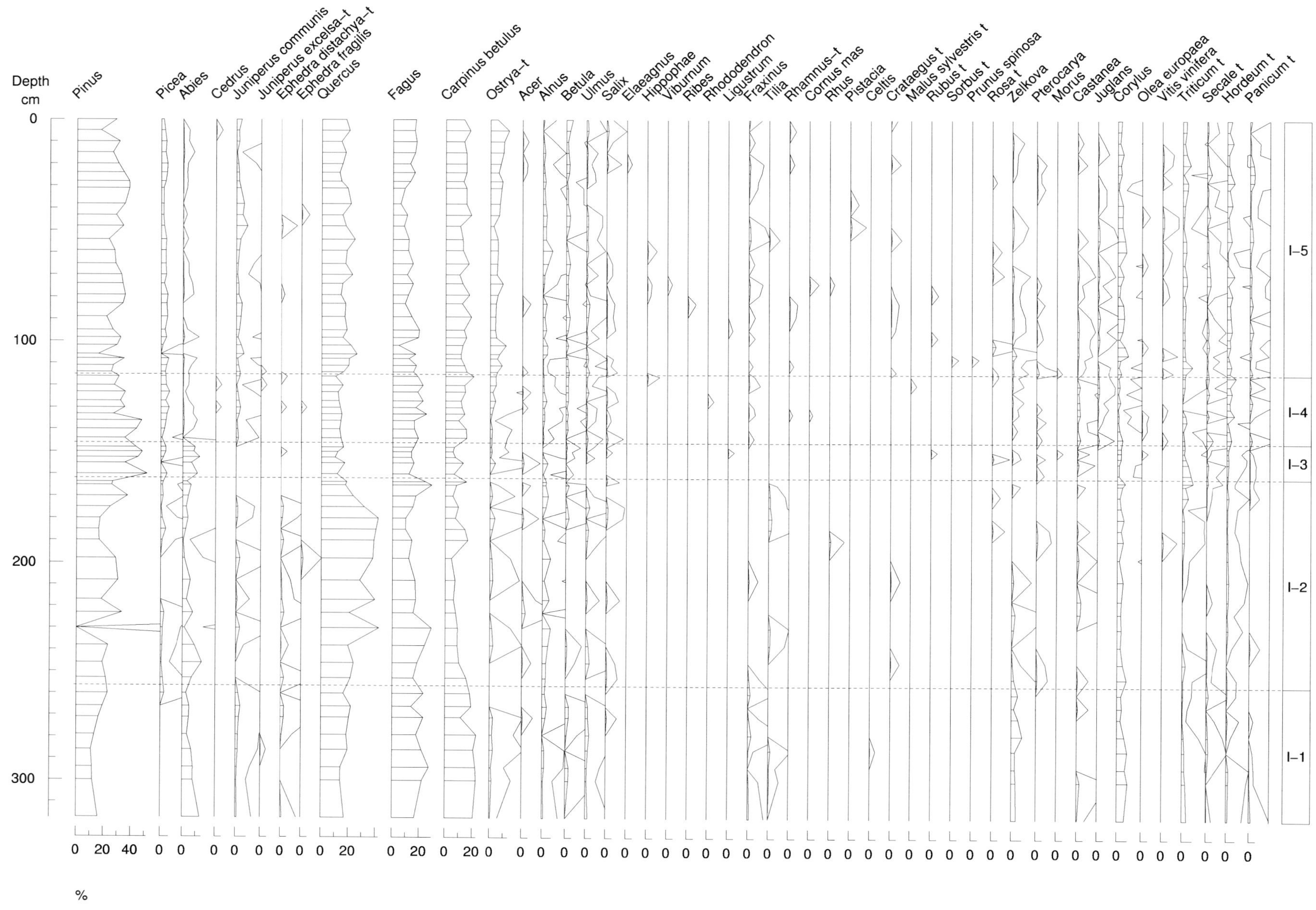

Lake Imera, Tsalka Plateau

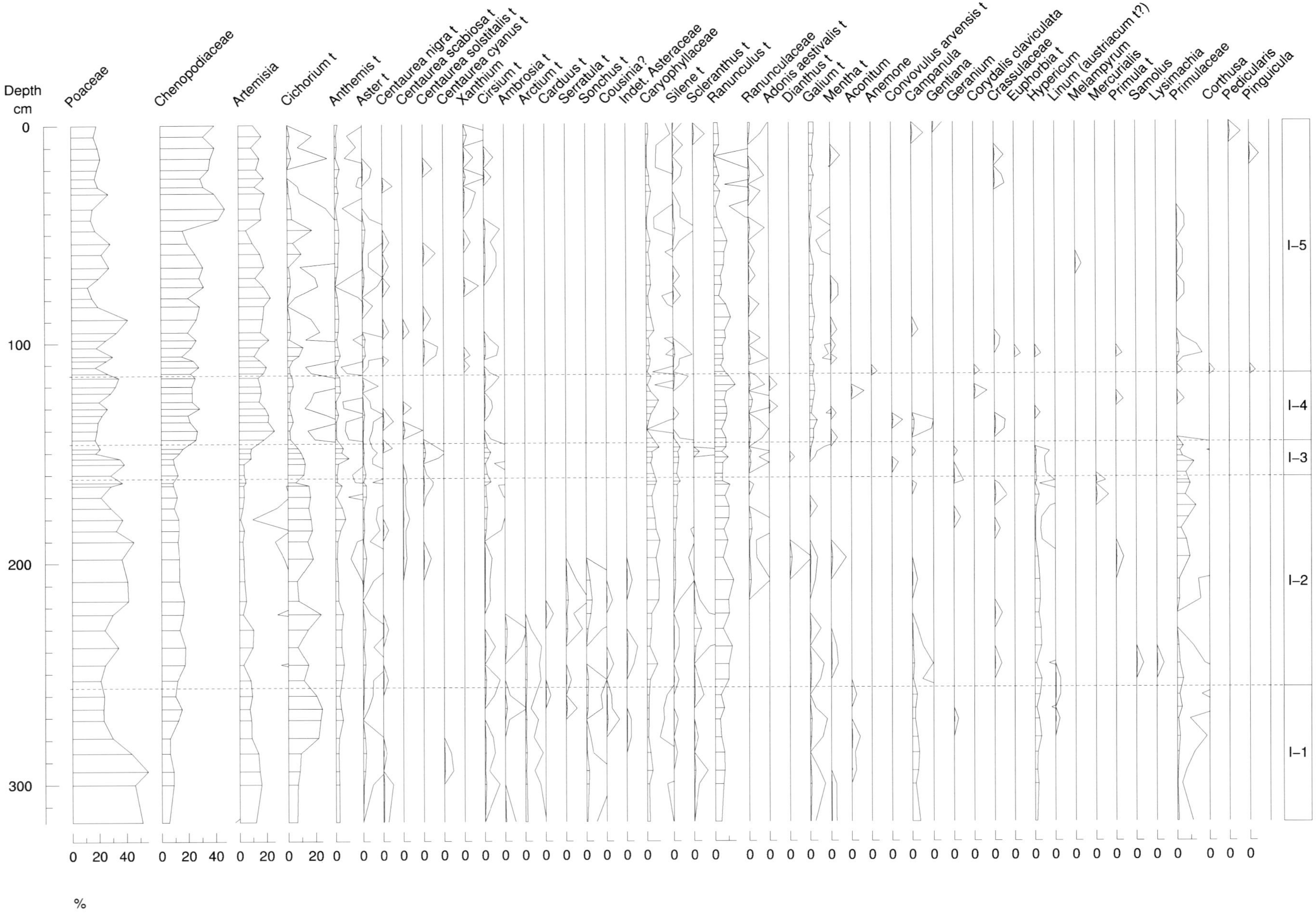

Lake Imera, Tsalka Plateau

Lake Imera, Tsalka Plateau

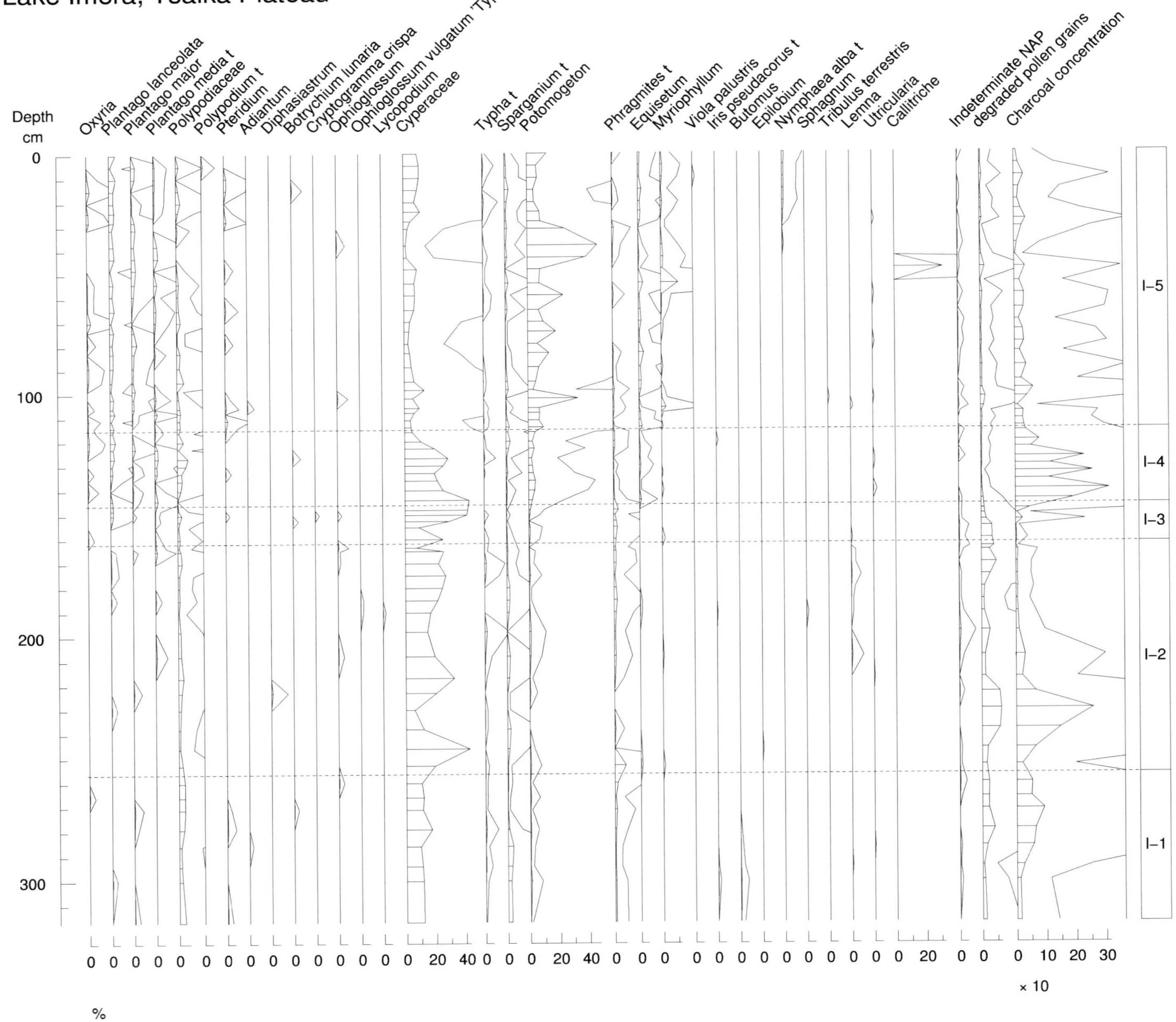

Lake Bareti, Tsalka Plateau

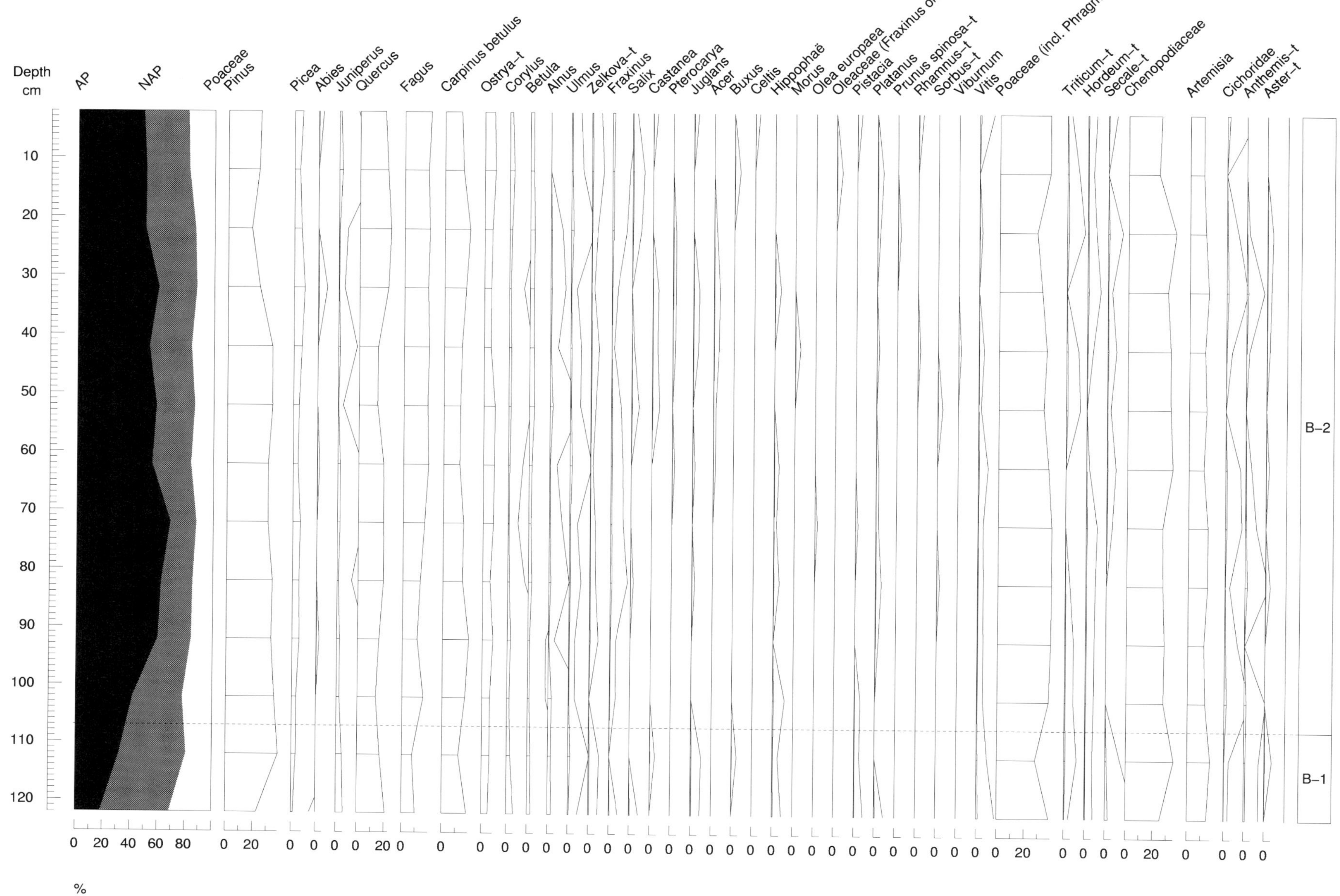

Lake Bareti, Tsalka Plateau

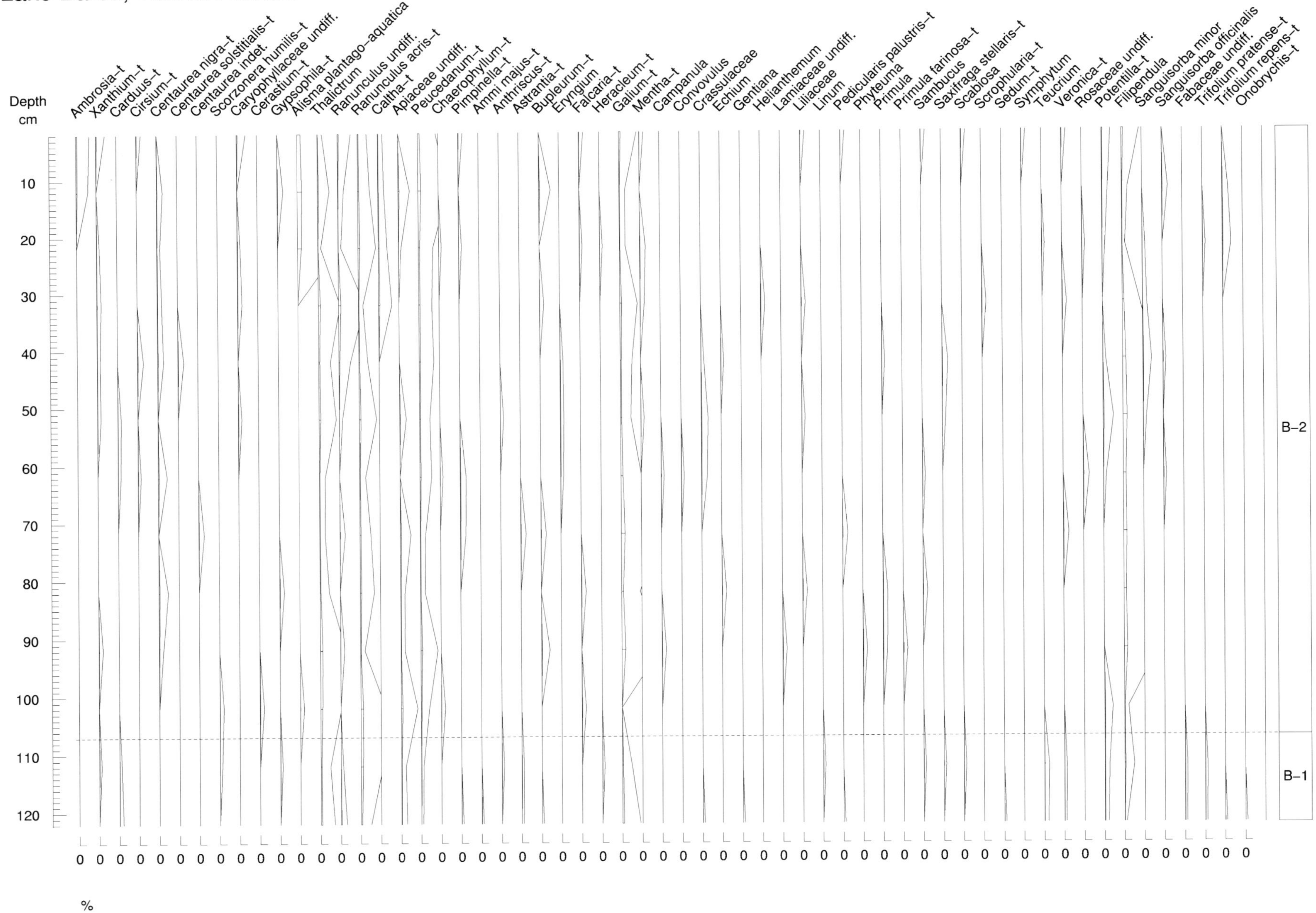

Lake Bareti, Tsalka Plateau

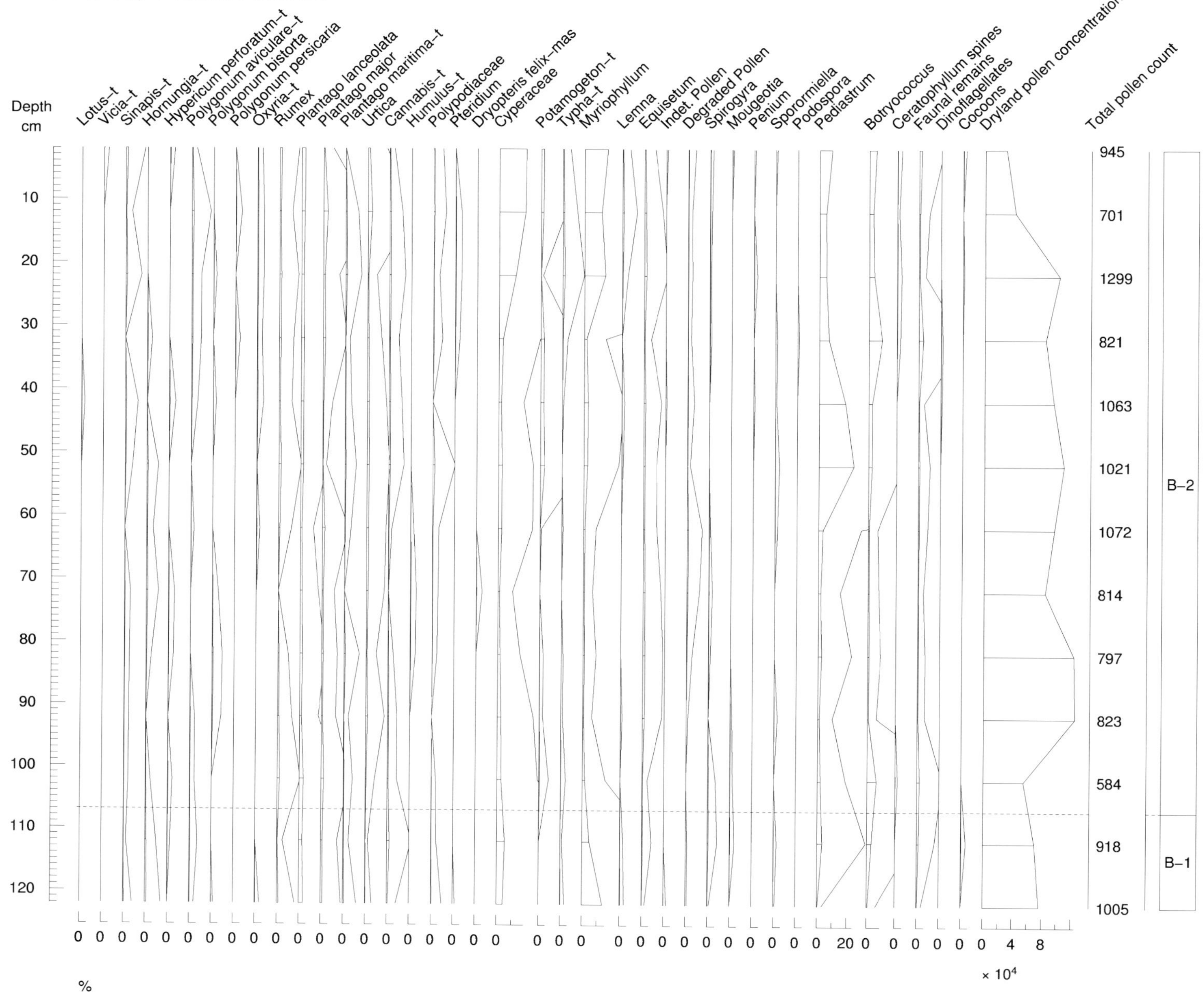

Aligol, Tsalka Plateau

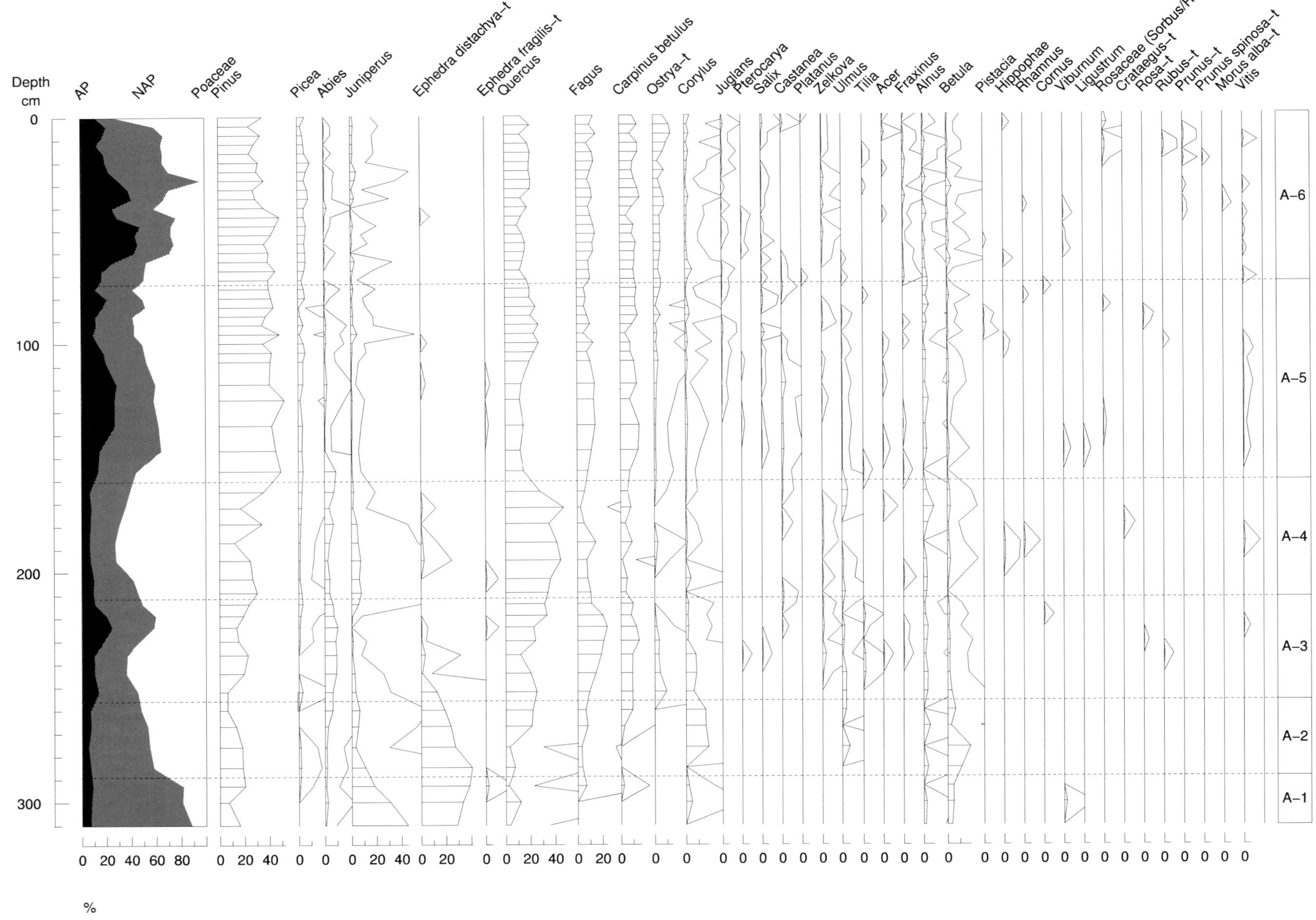

Aligol, Tsalka Plateau

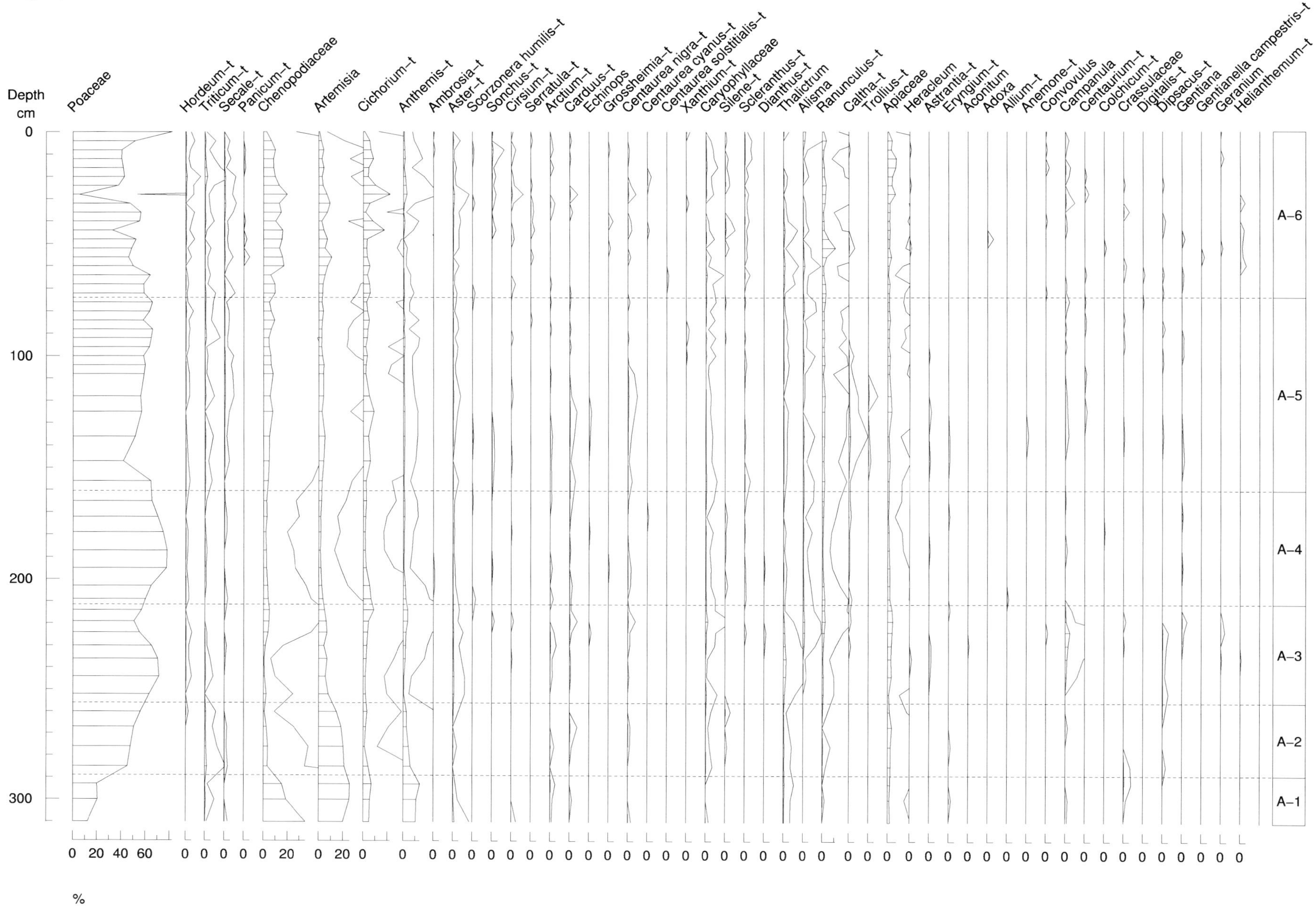

Aligol, Tsalka Plateau

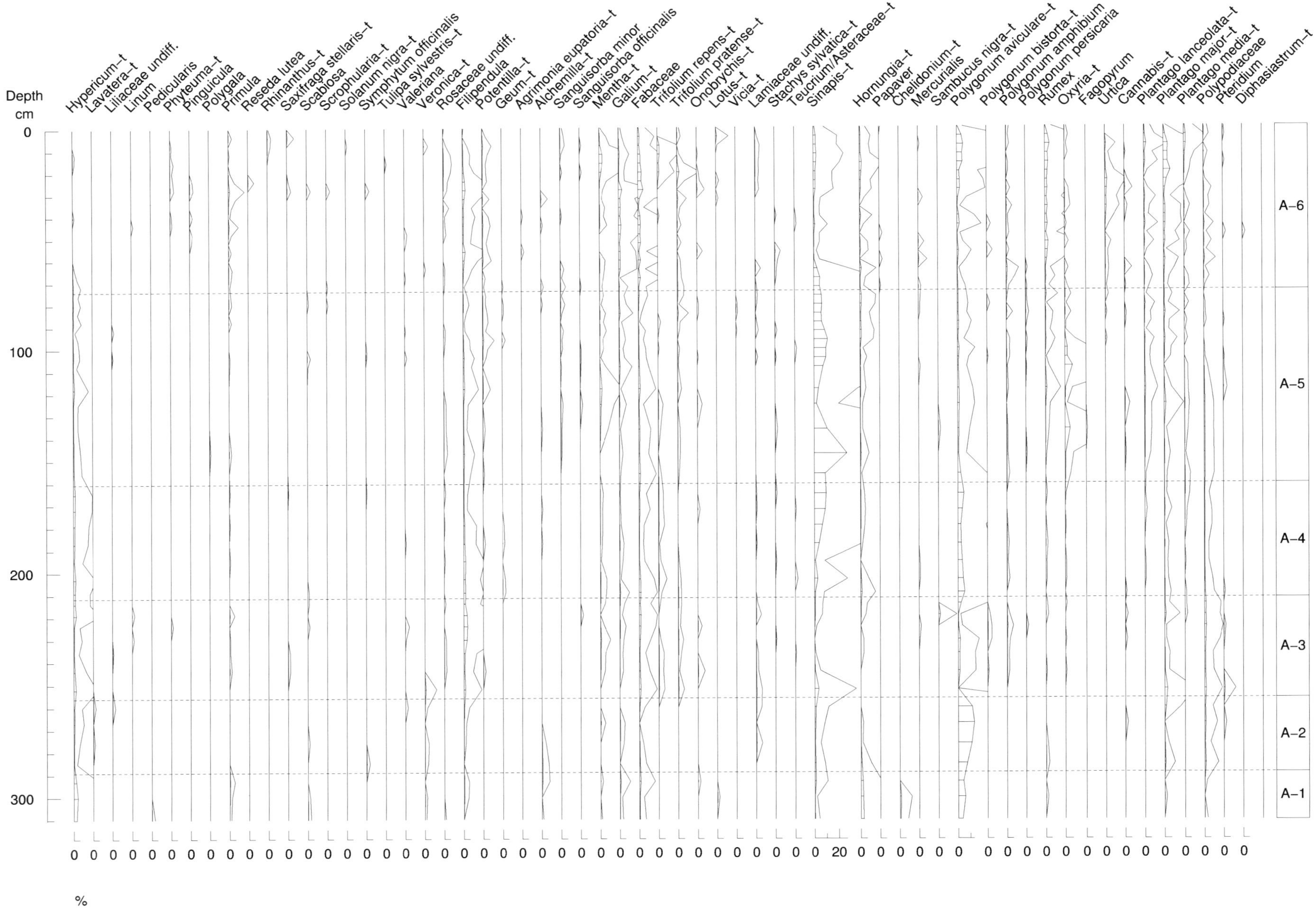

Aligol, Tsalka Plateau

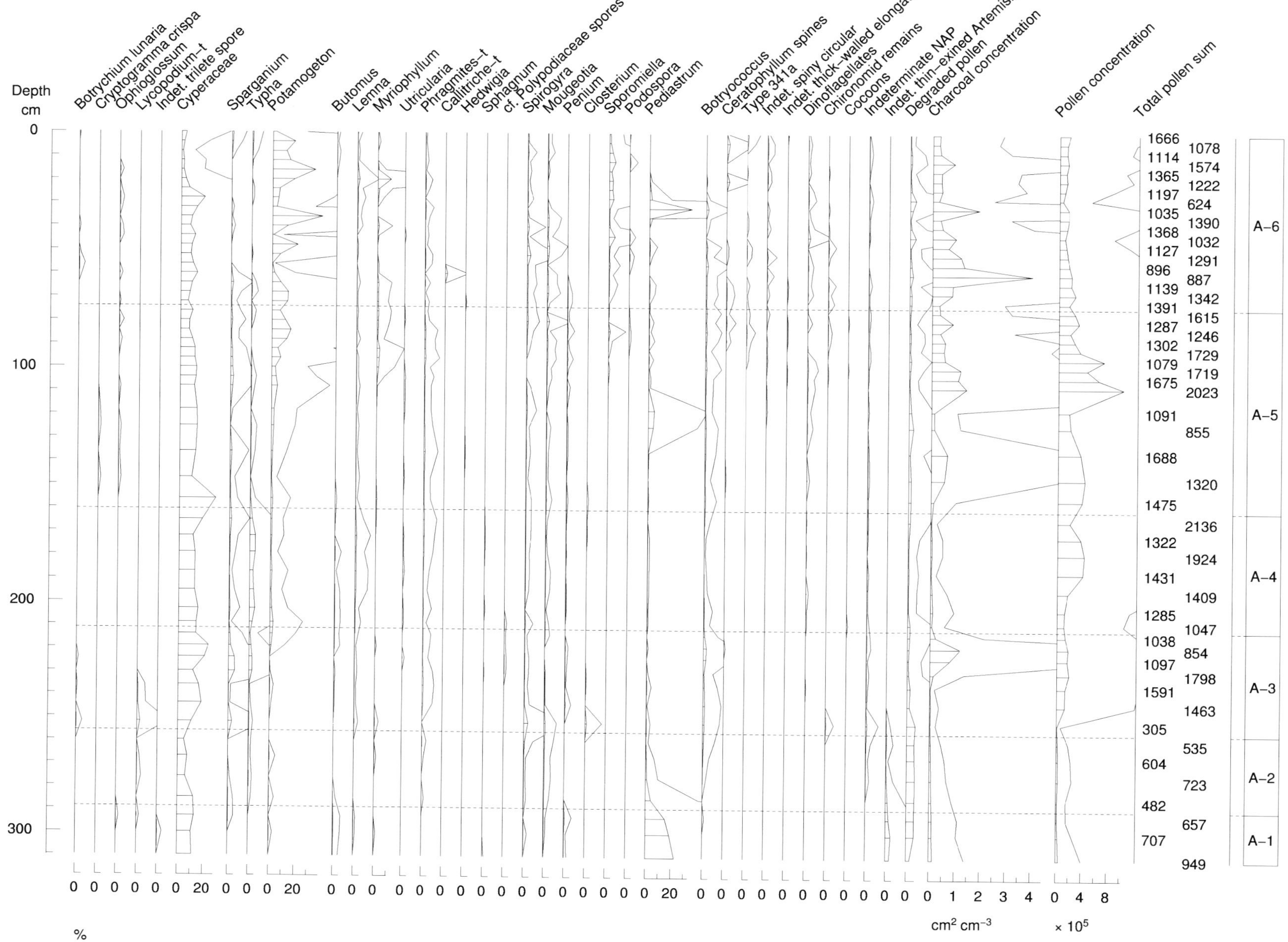

ANCIENT NEAR EASTERN STUDIES

(FORMERLY ABR-NAHRAIN)

Supplement Series

Series Editors: Claudia & Antonio Sagona

1. **Mayer L.A.**, *Bibliography of the Samaritans*, 1964, 50 p. 30 EURO
2. **Bunnens G.**, *Tell Ahmar, 1988 Season*, 1990, 151 p. 38 EURO
3. **Muraoka T.**, *Studies in Qumran Aramaic*, 1992, 167 p. 45 EURO
4. **Muraoka T.**, *Studies in Ancient Hebrew Semantics*, 1995, VIII — 107 p. 35 EURO
5. **Bunnens G.**, *Cultural Interaction in the Ancient Near East. Papers Read at a Symposium Held at the University of Melbourne, Department of Classics and Archaeology (29-30 September 1994)*, 1996, VIII — 155 p. 40 EURO
6. **Muraoka T.**, *Semantics of Ancient Hebrew*, 1998, 151 p. 35 EURO
7. **Bunnens G.**, *Essays on Syria in the Iron Age*, 2000, X — 557 p. 105 EURO
8. **Adamthwaite M.R.**, *Late Hittite Emar. The Chronology, Synchronisms and Socio-Political Aspects of a Late Bronze Age Fortress Town*, 2001, XXIV — 294 p. 70 EURO
9. **Sagona C.**, *The Archaeology of Punic Malta*, 2002, XIV — 1165 p. 160 EURO
10. **Sagona C.**, *Punic Antiquities of Malta and Other Ancient Artefacts Held in Ecclesiastic and Private Collections*, 2003, XXII — 374 p. 105 EURO
11. **Hopkins L.**, *Archaeology at the North-East Anatolian Frontier, VI. An Ethno-archaeological Study of Sos Höyük and Yigittasi Village*, 2003, XXVII — 184 p. 90 EURO
12. **Sagona A.**, *A View from the Highlands: Archaeological Studies in Honour of Charles Burney*, 2004, XX — 743 p. 120 EURO
13. **McConchie M.**, *Archaeology at the North-East Anatolian Frontier, V. Iron Technology and Iron-making Communities of the First Millennium BC*, 2004, XXIV — 393 p. 90 EURO
14. **Sagona A., Sagona C.**, *Archaeology at the North-East Anatolian Frontier, I. An Historical Geography and a Field Survey of the Bayburt Province*, 2004, XXIV — 600 p. 110 EURO
15. **Çilingiroğlu A., Derin Z., Abay E., Sağlamtimur H., Kayan İ.**, *Ulucak Höyük. Excavations Conducted between 1995 and 2002*, 2004, XII — 162 p. 60 EURO
16. **Payne M.**, *Urartian Measures of Volume*, 2005, XX — 387 p. 95 EURO
17. **Stoop-van Paridon P.W.T.**, *The Song of Songs. A Philological Analysis of the the Hebrew Book* **שיר השירים**, 2005, XVI — 540 p. 90 EURO

18. **Sagona C., Vella Gregory I., Bugeja A.**, *Punic Antiquities of Malta and Other Ancient Artefacts Held in Private Collections*, 2, 2006, XXII — 276 p. 77 EURO
19. **Sagona A., Abramishvili M.**, *Archaeology in Southern Caucasus: Perspectives from Georgia*, 2008, VIII — 477 p. 94 EURO
20. **Çilingiroğlu A., Sagona A.**, *Anatolian Iron Ages 6: The Proceedings of the Sixth Anatolian Iron Ages Colloquium Held at Eskişehir, 16-20 August 2004*, 2007, X — 348 p. 85 EURO
21. **Azize J., Weeks N.**, *Gilgameš and the World of Assyria. Proceedings of the Conference Held at the Mandelbaum House, The University of Sidney, 21-23 July 2004*, 2007, VIII — 240 p. 84 EURO
22. **Wightman G.**, *Sacred Spaces. Religious Architecture in the Ancient World*, 2007, XX — 1156 p. 120 EURO
23. **Aitken J.K.**, *The Semantics of Blessing and Cursing in Ancient Hebrew*, 2007, XVI — 306 p. 85 EURO
24. **Steiner M., van der Steen E.J.**, *Sacred and Sweet. Studies on the Material Culture of Tell Deir ʿAlla and Tell Abu Sarbut*, 2008, VIII — 286 p. 85 EURO
25. **Yagodin V.N., Betts A.V.G., Blau S.**, *Ancient Nomads of the Aralo-Caspian Region. The Duana Archaeological Complex*, 2007, VIII — 138 p. 65 EURO
26. **McGeough K.**, *Exchange Relationships at Ugarit*, 2007, XVIII — 438 p. 95 EURO
27. **Rubinson K.S., Sagona A.**, *Ceramics in Transitions. Chalcolithic Through Iron Age in the Highlands of the Southern Caucasus and Anatolia*, 2008, VI — 364 p. 90 EURO
28. **Sagona C.**, *Beyond the Homeland: Markers in Phoenician Chronology*, 2008, VIII — 655 p. 98 EURO
29. **Bienkowski P.**, *Studies on Iron Age Moab and Neighbouring Areas in Honour of Michèle Daviau*, 2009, XIV — 273 p. 87 EURO
30. **Kuty R.J.**, *Studies in the Syntax of Targums Jonathan to Samuel*, 2010, XIV — 285 p. 85 EURO
31. **Redford S., Ergin N.**, *Perceptions of the Past in the Turkish Republic: Classical and Byzantine Periods*, 2010, X — 236 p. 79 EURO
32. **McGeough K., Smith, M.S.**, *Ugaritic Economic Tablets. Text, Translation and Notes*, 2011, XXVI — 625 p. Forthcoming
33. **Hartley J.E.**, *The Semantics of Ancient Hebrew Colour Lexemes*, 2010, XVI — 247 p. 78 EURO

PRINTED ON PERMANENT PAPER • IMPRIME SUR PAPIER PERMANENT • GEDRUKT OP DUURZAAM PAPIER - ISO 9706
N.V. PEETERS S.A., WAROTSTRAAT 50, B-3020 HERENT